Chapter 0 Review

0.1 Real Numbers

1. (a) Natural Numbers: 2, 5

(b) Integers: $-6, 2, 5$

(c) Rational numbers: $-6, \dfrac{1}{2}, -1.333\ldots, 2, 5$

(d) Irrational numbers: π

(e) Real numbers: $-6, \dfrac{1}{2}, -1.333\ldots, \pi, 2, 5$

3. (a) Natural Numbers: 1

(b) Integers: 0, 1

(c) Rational numbers: $0, 1, \dfrac{1}{2}, \dfrac{1}{3}, \dfrac{1}{4}$

(d) Irrational numbers: none

(e) Real numbers: $0, 1, \dfrac{1}{2}, \dfrac{1}{3}, \dfrac{1}{4}$

5. (a) Natural Numbers: none
(b) Integers: none
(c) Rational numbers: none

(d) Irrational numbers: $\sqrt{2}, \pi, \sqrt{2}+1, \pi+\dfrac{1}{2}$

(e) Real numbers: $\sqrt{2}, \pi, \sqrt{2}+1, \pi+\dfrac{1}{2}$

7. Number: 18.9526
Rounded: 18.953
Truncated: 18.952

9. Number: 28.65319
Rounded: 28.653
Truncated: 28.653

11. Number: 0.06291
Rounded: 0.063
Truncated: 0.062

13. Number: 9.9985
Rounded: 9.999
Truncated: 9.998

15. Number: $\dfrac{3}{7} = 0.428571\ldots$

Rounded: 0.429
Truncated: 0.428

17. Number: $\dfrac{521}{15} = 34.73333\ldots$

Rounded: 34.733
Truncated: 34.733

19. $3 + 2 = 5$

21. $x + 2 = 3 \cdot 4$

23. $3y = 1 + 2$

25. $x - 2 = 6$

27. $\dfrac{x}{2} = 6$

29. $9 - 4 + 2 = 5 + 2 = 7$

31. $-6 + 4 \cdot 3 = -6 + 12 = 6$

33. $4 + 5 - 8 = 9 - 8 = 1$

35. $4 + \dfrac{1}{3} = \dfrac{12}{3} + \dfrac{1}{3} = \dfrac{13}{3}$

37.
$$6 - [3 \cdot 5 + 2 \cdot (3 - 2)] = 6 - [15 + 2 \cdot 1]$$
$$= 6 - [15 + 2]$$
$$= 6 - 17$$
$$= -11$$

39.
$$2 \cdot (3 - 5) + 8 \cdot 2 - 1 = 2 \cdot (-2) + 16 - 1$$
$$= -4 + 16 - 1$$
$$= 12 - 1 = 11$$

41.
$$10 - [6 - 2 \cdot 2 + (8 - 3)] \cdot 2$$
$$= 10 - [6 - 4 + 5] \cdot 2$$
$$= 10 - [7] \cdot 2$$
$$= 10 - 14$$
$$= -4$$

43. $(5 - 3)\,\dfrac{1}{2} = (2)\,\dfrac{1}{2} = 1$

45. $\dfrac{4 + 8}{5 - 3} = \dfrac{12}{2} = 6$

47. $\dfrac{3}{5} \cdot \dfrac{10}{21} = \dfrac{30}{105} = \dfrac{2}{7}$

49. $\dfrac{6}{25} \cdot \dfrac{10}{27} = \dfrac{2 \cdot 3 \cdot 2 \cdot 5}{5 \cdot 5 \cdot 3 \cdot 9} = \dfrac{4}{45}$

51. $\dfrac{3}{4} + \dfrac{2}{5} = \dfrac{3 \cdot 5}{4 \cdot 5} + \dfrac{2 \cdot 4}{4 \cdot 5} = \dfrac{15 + 8}{20} = \dfrac{23}{20}$

53. $\dfrac{5}{6} + \dfrac{9}{5} = \dfrac{5 \cdot 5}{6 \cdot 5} + \dfrac{9 \cdot 6}{5 \cdot 6} = \dfrac{25 + 54}{30} = \dfrac{79}{30}$

55. $\dfrac{5}{18} + \dfrac{1}{12} = \dfrac{5 \cdot 2}{18 \cdot 2} + \dfrac{1 \cdot 3}{12 \cdot 3} = \dfrac{10 + 3}{36} = \dfrac{13}{36}$

57.
$$\dfrac{1}{30} - \dfrac{7}{18} = \dfrac{1 \cdot 3}{30 \cdot 3} - \dfrac{7 \cdot 5}{18 \cdot 5}$$
$$= \dfrac{3 - 35}{90} = \dfrac{-32}{90} = -\dfrac{16}{45}$$

59. $\dfrac{3}{20} - \dfrac{2}{15} = \dfrac{3 \cdot 3}{20 \cdot 3} - \dfrac{2 \cdot 4}{15 \cdot 4} = \dfrac{9 - 8}{60} = \dfrac{1}{60}$

61. $\dfrac{\tfrac{5}{18}}{\tfrac{11}{27}} = \dfrac{5}{18} \cdot \dfrac{27}{11} = \dfrac{5 \cdot 9 \cdot 3}{9 \cdot 2 \cdot 11} = \dfrac{15}{22}$

63. $6(x + 4) = 6x + 24$

65. $x(x - 4) = x^2 - 4x$

67.
$$(x + 2)(x + 4) = (x + 2)x + (x + 2)4$$
$$= x^2 + 2x + 4x + 8$$
$$= x^2 + 6x + 8$$

69.
$$(x - 2)(x + 1) = (x - 2)x + (x - 2)1$$
$$= x^2 - 2x + x - 2$$
$$= x^2 - x - 2$$

71. $(x-8)(x-2) = (x-8)x + (x-8)(-2)$
$$= x^2 - 8x - 2x + 16$$
$$= x^2 - 10x + 16$$

73. $(x+2)(x-2) = (x+2)x + (x+2)(-2)$
$$= x^2 + 2x - 2x - 4$$
$$= x^2 - 4$$

75. Answers will vary.

77. Answers will vary.

79. Subtraction is not commutative. Examples will vary.

81. Division is not commutative. Examples will vary.

83. Explanations will vary.

85. There are no real numbers that are both rational and irrational.
There are no real numbers that are neither rational nor irrational.
Explanations will vary.

87. $0.9999 \ldots = 1$
To show that $0.9999 \ldots = 1$, we let $n = 0.9999 \ldots$, then $10n = 9.9999 \ldots$

$$10n = 9.9999 \ldots \quad (1)$$
$$\underline{n = 0.9999 \ldots \quad (2)}$$
$$9n = 9.0000 \ldots \qquad \text{Subtract (2) from (1).}$$
$$n = 1 \qquad \text{Divide both sides by 9.}$$

0.2   Algebra Review

1.

3. $\dfrac{1}{2} > 0$

5. $-1 > -2$

7. $\pi > 3.14$

9. $\dfrac{1}{2} = 0.5$

11. $\dfrac{2}{3} < 0.67$

13. $x > 0$

15. $x < 2$

17. $x \leq 1$

19. $x \geq -2$

21. $x > -1$

23. $d(C, D) = |C - D| = |0 - 1| = |-1| = 1$

25. $d(D, E) = |D - E| = |1 - 3| = |-2| = 2$

27. $d(A, E) = |A - E| = |-3 - 3| = |-6| = 6$

29. If $x = -2$ and $y = 3$, then $x + 2y = (-2) + 2(3) = -2 + 6 = 4$

31. If $x = -2$ and $y = 3$, then $5xy + 2 = 5(-2)(3) + 2 = -30 + 2 = -28$

33. If $x = -2$ and $y = 3$, then $\dfrac{2x}{x - y} = \dfrac{2(-2)}{(-2) - 3} = \dfrac{-4}{-5} = \dfrac{4}{5}$

35. If $x = -2$ and $y = 3$, then $\dfrac{3x + 2y}{2 + y} = \dfrac{3(-2) + 2(3)}{2 + 3} = \dfrac{-6 + 6}{5} = 0$

37. If $x = 3$ and $y = -2$, then $|x + y| = |3 + (-2)| = |1| = 1$

39. If $x = 3$ and $y = -2$, then $|x| + |y| = |3| + |-2| = 3 + 2 = 5$

41. If $x = 3$ and $y = -2$, then $\dfrac{|x|}{x} = \dfrac{|3|}{3} = \dfrac{3}{3} = 1$

43. If $x = 3$ and $y = -2$, then $|4x - 5y| = |4(3) - 5(-2)| = |12 - (-10)| = |22| = 22$

45. If $x = 3$ and $y = -2$,
then $\left| |4x| - |5y| \right| = \left| |4(3)| - |5(-2)| \right| = \left| |12| - |-10| \right| = |12 - 10| = |2| = 2$

47. We must exclude values of x that would cause the denominator to equal zero.
$\qquad x \neq 0 \qquad$ **(c)**

49. We must exclude values of x that would cause the denominator to equal zero.
$\qquad x^2 - 9 \neq 0$
$\qquad (x - 3)(x + 3) \neq 0$
$\qquad x \neq 3; \ x \neq -3 \qquad$ **(a)**

51. We must exclude values of x that would cause the denominator to equal zero, but $x^2 + 1$ can never equal zero, so no values are excluded.

53. We must exclude values of x that would cause the denominator to equal zero.
$\qquad x^3 - x \neq 0$
$\qquad x(x^2 - 1) \neq 0$
$\qquad x(x - 1)(x + 1) \neq 0$
$\qquad x \neq 0; \ x \neq 1; \ x \neq -1 \qquad$ **(b), (c),** and **(d)**

55. The domain of the variable x is $\{x \mid x \neq 5\}$.

57. The domain of the variable x is $\{x \mid x \neq -4\}$.

59. $C = \dfrac{5}{9}(F - 32)$ If $F = 32°$, then $C = \dfrac{5}{9}(32 - 32) = 0°.$

61. $C = \dfrac{5}{9}(F - 32)$ If $F = 77°$, then $C = \dfrac{5}{9}(77 - 32) = \dfrac{5}{9}(45) = 25°.$

63. $(-4)^2 = 16$

65. $4^{-2} = \dfrac{1}{4^2} = \dfrac{1}{16}$

67. $3^{-6} \cdot 3^4 = 3^{-6+4} = 3^{-2} = \dfrac{1}{3^2} = \dfrac{1}{9}$

69. $(3^{-2})^{-1} = 3^{(-2)\cdot(-1)} = 3^2 = 9$

71. $\sqrt{25} = 5$

73. $\sqrt{(-4)^2} = \sqrt{16} = 4$

75. $\left(8x^3\right)^2 = 8^2 \cdot x^{3\cdot 2} = 64x^6$

77. $\left(x^2 y^{-1}\right)^2 = x^{2\cdot 2} \cdot y^{(-1)\cdot 2} = x^4 y^{-2} = \dfrac{x^4}{y^2}$

79. $\dfrac{x^2 y^3}{xy^4} = \dfrac{x^2}{x} \cdot \dfrac{y^3}{y^4} = x^{2-1} \cdot y^{3-4} = x \cdot y^{-1} = \dfrac{x}{y}$

81. $\dfrac{(-2)^3 x^4 (yz)^2}{3^2 xy^3 z} = \dfrac{-8x^4 y^2 z^2}{9xy^3 z} = -\dfrac{8}{9} \cdot \dfrac{x^4}{x} \cdot \dfrac{y^2}{y^3} \cdot \dfrac{z^2}{z} = -\dfrac{8}{9} \cdot x^{4-1} \cdot y^{2-3} \cdot z^{2-1}$

$$= -\dfrac{8}{9} \cdot x^3 \cdot y^{-1} \cdot z = -\dfrac{8}{9} \cdot \dfrac{x^3 z}{y} = -\dfrac{8x^3 z}{9y}$$

83. $\left(\dfrac{3x^{-1}}{4y^{-1}}\right)^{-2} = \left(\dfrac{3y}{4x}\right)^{-2} = \left(\dfrac{4x}{3y}\right)^2 = \dfrac{4^2 \cdot x^2}{3^2 \cdot y^2} = \dfrac{16x^2}{9y^2}$

85. If $x = 2$ and $y = -1$, then $2xy^{-1} = 2 \cdot 2 \cdot (-1)^{-1} = -4.$

87. If $x = 2$ and $y = -1$, then $x^2 + y^2 = 2^2 + (-1)^2 = 4 + 1 = 5.$

89. If $x = 2$ and $y = -1$, then $(xy)^2 = (2 \cdot (-1))^2 = (-2)^2 = 4.$

91. If $x = 2$ and $y = -1$, then $\sqrt{x^2} = \sqrt{2^2} = \sqrt{4} = 2.$

93. If $x = 2$ and $y = -1$, then $\sqrt{x^2 + y^2} = \sqrt{2^2 + (-1)^2} = \sqrt{4+1} = \sqrt{5}.$

95. If $x = 2$ and $y = -1$, then $x^y = 2^{-1} = \dfrac{1}{2}.$

97.   If $x = 2$, then $2x^3 - 3x^2 + 5x - 4 = 2(2)^3 - 3(2)^2 + 5(2) - 4 = 10$.

If $x = 1$, then $2x^3 - 3x^2 + 5x - 4 = 2(1)^3 - 3(1)^2 + 5(1) - 4 = 0$.

99. $\dfrac{(666)^4}{(222)^4} = \left(\dfrac{666}{222}\right)^4 = 3^4 = 81$

101. $(8.2)^6 = 304,006.6714 = 304,006.671$

103. $(6.1)^{-3} = 0.0044057 = 0.004$

105. $(-2.8)^6 = 481.890304 = 481.890$

107. $(-8.11)^{-4} = 0.00023116 = 0.000$

109. $454.2 = 4.542 \times 10^2$

111. $0.013 = 1.3 \times 10^{-2}$

113. $32,155 = 3.2155 \times 10^4$

115. $0.000423 = 4.23 \times 10^{-4}$

117. $6.15 \times 10^4 = 61,500$

119. $1.214 \times 10^{-3} = 0.001214$

121. $1.1 \times 10^8 = 110,000,000$

123. $8.1 \times 10^{-2} = 0.081$

125. $A = lw$;   domain: $A > 0, l > 0, w > 0$

127. $C = \pi d$;   domain: $C > 0, d > 0$

129. $A = \dfrac{\sqrt{3}}{4} \cdot x^2$;   domain: $A > 0, x > 0$

131. $V = \dfrac{4}{3}\pi r^3$;   domain: $V > 0, r > 0$

133. $V = x^3$;   domain: $V > 0, x > 0$

135. $C = 4000 + 2x$
    (a) If $x = 1000$ watches are produced, it will cost
       $C = 4000 + 2(1000) = \$6000.00$

    (b) If $x = 2000$ watches are produced, it will cost
       $C = 4000 + 2(2000) = \$8000.00$

137.   (a) If actual voltage is $x = 113$ then
       $|113 - 115| = |-2| = 2$
    Since $2 < 5$, an actual voltage of 113 is acceptable.

    (b) If actual voltage is $x = 109$ then
       $|109 - 115| = |-6| = 6$

Since $6 > 5$, an actual voltage of 109 is not acceptable.

139.   (a) If the radius is $x = 2.999$, then
$$|x - 3| = |2.999 - 3| = |-0.001| = 0.001$$
Since $0.001 < 0.010$ the ball bearing is not acceptable.
(b) If the radius is $x = 2.89$, then
$$|x - 3| = |2.89 - 3| = |-0.11| = 0.11$$
Since $0.11 > 0.01$, the ball bearing is not acceptable.

141. $\dfrac{1}{3} \neq 0.333$; $\dfrac{1}{3} > 0.333$

$\dfrac{1}{3} = 0.333\ldots$; $0.333\ldots - 0.333 = 0.000333\ldots$

143.   The answer is no. Student answers should justify and explain why not.

145.   Answers will vary.

0.3 Polynomials and Rational Expressions

1. $\left(10x^5 - 8x^2\right) + \left(3x^3 - 2x^2 + 6\right) = 10x^5 + 3x^3 - 10x^2 + 6$

3. $(x + a)^2 - x^2 = \left[x^2 + 2ax + a^2\right] - x^2$
$$= 2ax + a^2$$

5. $(x + 8)(2x + 1) = 2x^2 + x + 16x + 8 = 2x^2 + 17x + 8$

7. $\left(x^2 + x - 1\right)\left(x^2 - x + 1\right) = x^2\left(x^2 - x + 1\right) + x\left(x^2 - x + 1\right) - 1\left(x^2 - x + 1\right)$
$$= x^4 - x^3 + x^2 + x^3 - x^2 + x - x^2 + x - 1$$
$$= x^4 - x^2 + 2x - 1$$

9. $(x + 1)^3 - (x - 1)^3 = \left(x^3 + 3x^2 + 3x + 1\right) - \left(x^3 - 3x^2 + 3x - 1\right)$
$$= x^3 + 3x^2 + 3x + 1 - x^3 + 3x^2 - 3x + 1$$
$$= 6x^2 + 2$$

11.   This is the difference of 2 squares.
$$x^2 - 36 = (x - 6)(x + 6)$$

13.   This is the difference of 2 squares.
$$1 - 4x^2 = (1 - 2x)(1 + 2x)$$

15. $x^2 + 7x + 10 = (x + 5)(x + 2)$ The product of 5 and 2 is 10; the sum of 5 and 2 is 7.

17. This polynomial cannot be factored; it is prime.

19. This polynomial cannot be factored; it is prime.

21. $15 + 2x - x^2 = (5 - x)(3 + x)$ The product $5 \cdot 3 = 15$; the sum $5 \cdot 1 + 3 \cdot (-1) = 2$.

23. $3x^2 - 12x - 36 = 3(x^2 - 4x - 12)$ Factor out the common factor of 3.

$= 3(x - 6)(x + 2)$ The product $-6 \cdot 2 = -12$; the sum $-6 + 2 = -4$.

25. $y^4 + 11y^3 + 30y^2 = y^2(y^2 + 11y + 30)$ Factor out the common factor of y^2.

$= y^2(y + 5)(y + 6)$ The product $5 \cdot 6 = 30$; the sum $5 + 6 = 11$.

27. This polynomial is a perfect square.

$4x^2 + 12x + 9 = (2x + 3)(2x + 3) = (2x + 3)^2$

29. $3x^2 + 4x + 1 = (3x + 1)(x + 1)$ The products $3 \cdot 1 = 3$ and $1 \cdot 1 = 1$; the sum $3 \cdot 1 + 1 \cdot 1 = 4$.

31. $x^4 - 81 = (x^2 - 9)(x^2 + 9)$ Treat $x^4 - 81$ as the difference of 2 squares, and factor.

$= (x - 3)(x + 3)(x^2 + 9)$ $x^2 - 9$ is the difference of 2 squares.

33. Let $u = x^3$. Then

$$x^6 - 2x^3 + 1 = u^2 - 2u + 1 = (u - 1)(u - 1) = (u - 1)^2$$

Now substitute back for u. That is, replace u with x^3.

$$x^6 - 2x^3 + 1 = (x^3 - 1)^2$$

35. $x^7 - x^5 = x^5(x^2 - 1)$ Factor out the common factor of x^5.

$= x^5(x - 1)(x + 1)$ $x^2 - 1$ is the difference of 2 squares.

37. $5 + 16x - 16x^2 = (1 + 4x)(5 - 4x)$ The products $4 \cdot (-4) = -16$ and $1 \cdot 5 = 5$;
the sum $1 \cdot (-4) + 4 \cdot 5 = 16$.

39. $4y^2 - 16y + 15 = (2y - 5)(2y - 3)$ The products $2 \cdot 2 = 4$ and $-5 \cdot (-3) = 15$;
the sum $2 \cdot (-3) + (-5) \cdot 2 = -16$.

41. Let $u = x^2$, then

$$1 - 8x^2 - 9x^4 = 1 - 8u - 9u^2 = (1 - 9u)(1 + u)$$

Now substitute back for u. That is, replace u with x^2.

$$1 - 8x^2 - 9x^4 = (1 - 9x^2)(1 + x^2)$$

$1 - 9x^2$ is the difference of 2 squares.

$$1 - 8x^2 - 9x^4 = \left(1 - 9x^2\right)\left(1 + x^2\right) = \left(1 - 3x\right)\left(1 + 3x\right)\left(1 + x^2\right)$$

43. $x(x+3) - 6(x+3) = (x+3)(x-6)$ Factor out the common factor $x + 3$.

45. $(x+2)^2 - 2(x+2) = (x+2)\big[(x+2) - 5\big]$ Factor out the common factor $x + 2$.
$ = (x+2)(x-3)$ Simplify.

47. $6x(2-x)^4 - 9x^2(2-x)^3 = 3x(2-x)^3\big[2(2-x) - 3x\big]$ Factor out the common factors.
$ = 3x(2-x)^3(4 - 2x - 3x)$ Simplify.
$ = 3x(2-x)^3(4 - 5x)$ Simplify.
$ = 3x(x-2)^3(5x - 4)$ Multiply $(x-2)^3$ by $(-1)^3$ and
$ $ $(5x-4)$ by -1. $[(-1)^3 \cdot (-1) = 1]$.

49. $x^3 + 2x^2 - x - 2 = \left(x^3 + 2x^2\right) - (x+2)$ Group the polynomial into the difference of 2 binomials.
$ = x^2(x+2) - 1(x+2)$ Factor the common factors from each binomial.
$ = \left(x^2 - 1\right)(x+2)$ Factor the common factor $x + 2$.
$ = (x-1)(x+1)(x+2)$ Factor the difference of 2 squares.

51. $x^4 - x^3 + x - 1 = \left(x^4 - x^3\right) + (x-1)$ Group the polynomial into the sum of 2 binomials.
$ = x^3(x-1) + (x-1)$ Factor the common factors from each binomial.
$ = (x-1)\left(x^3 + 1\right)$ Factor the common factor $x - 1$.
$ = (x-1)(x+1)\left(x^2 - x + 1\right)$ Factor the sum of 2 cubes.

53. $\dfrac{3x-6}{5x} \cdot \dfrac{x^2 - x - 6}{x^2 - 4} = \dfrac{3\cancel{(x-2)}}{5x} \cdot \dfrac{(x-3)\cancel{(x+2)}}{\cancel{(x-2)}\cancel{(x+2)}}$ Factor.
$\phantom{\dfrac{3x-6}{5x} \cdot \dfrac{x^2 - x - 6}{x^2 - 4}} = \dfrac{3(x-3)}{5x}$ Simplify.

55. $\dfrac{4x^2 - 1}{x^2 - 16} \cdot \dfrac{x^2 - 4x}{2x+1} = \dfrac{(2x-1)\cancel{(2x+1)}}{\cancel{(x-4)}(x+4)} \cdot \dfrac{x\cancel{(x-4)}}{\cancel{2x+1}} = \dfrac{(2x-1)x}{x+4}$

57. $\dfrac{x}{x^2 - 7x + 6} - \dfrac{x}{x^2 - 2x - 24} = \dfrac{x}{(x-6)(x-1)} - \dfrac{x}{(x-6)(x+4)}$
$\phantom{\dfrac{x}{x^2 - 7x + 6} - \dfrac{x}{x^2 - 2x - 24}} = \dfrac{x(x+4)}{(x-6)(x-1)(x+4)} - \dfrac{x(x-1)}{(x-6)(x+4)(x-1)}$

$$= \frac{x^2 + 4x}{(x-6)(x-1)(x+4)} - \frac{x^2 - x}{(x-6)(x+4)(x-1)}$$

$$= \frac{x^2 + 4x - x^2 + x}{(x-6)(x-1)(x+4)}$$

$$= \frac{5x}{(x-6)(x-1)(x+4)}$$

59. $\dfrac{4}{x^2 - 4} - \dfrac{2}{x^2 + x - 6} = \dfrac{4}{(x-2)(x+2)} - \dfrac{2}{(x+3)(x-2)}$

$$= \frac{4(x+3)}{(x-2)(x+2)(x+3)} - \frac{2(x+2)}{(x+3)(x-2)(x+2)}$$

$$= \frac{4(x+3) - 2(x+2)}{(x-2)(x+2)(x+3)}$$

$$= \frac{4x + 12 - 2x - 4}{(x-2)(x+2)(x+3)}$$

$$= \frac{2x + 8}{(x-2)(x+2)(x+3)} = \frac{2(x+4)}{(x-2)(x+2)(x+3)}$$

61. $\dfrac{1}{x} - \dfrac{2}{x^2 + x} + \dfrac{3}{x^3 - x^2} = \dfrac{1}{x} - \dfrac{2}{x(x+1)} + \dfrac{3}{x^2(x-1)}$

$$= \frac{x(x+1)(x-1)}{x^2(x+1)(x-1)} - \frac{2x(x-1)}{x^2(x+1)(x-1)} + \frac{3(x+1)}{x^2(x-1)(x+1)}$$

$$= \frac{x(x+1)(x-1) - 2x(x-1) + 3(x+1)}{x^2(x+1)(x-1)}$$

$$= \frac{x^3 - x - 2x^2 + 2x + 3x + 3}{x^2(x+1)(x-1)}$$

$$= \frac{x^3 - 2x^2 + 4x + 3}{x^2(x+1)(x-1)}$$

63. $\dfrac{1}{h}\left(\dfrac{1}{x+h} - \dfrac{1}{x}\right) = \dfrac{1}{h}\left(\dfrac{x}{(x+h)x} - \dfrac{1(x+h)}{x(x+h)}\right)$

$$= \frac{1}{h}\left(\frac{x - x - h}{(x+h)x}\right)$$

$$= \frac{1}{\cancel{h}}\left(\frac{-\cancel{h}}{(x+h)x}\right)$$

$$= -\frac{1}{x(x+h)}$$

65.
$$2(3x+4)^2+(2x+3)\cdot 2(3x+4)\cdot 3 = (3x+4)\big[2(3x+4)+6(2x+3)\big]$$
$$= (3x+4)(6x+8+12x+18)$$
$$= (3x+4)(18x+26)$$
$$= 2(3x+4)(9x+13)$$

67.
$$2x(2x+5)+x^2\cdot 2 = 2x(2x+5+x)$$
$$= 2x(3x+5)$$

69.
$$2(x+3)(x-2)^3+(x+3)^2\cdot 3(x-2)^2 = (x+3)(x-2)^2\big[2(x-2)+3(x+3)\big]$$
$$= (x+3)(x-2)^2(2x-4+3x+9)$$
$$= (x+3)(x-2)^2(5x+5)$$
$$= 5(x+3)(x-2)^2(x+1)$$

71.
$$(4x-3)^2+x\cdot 2(4x-3)\cdot 4 = (4x-3)\big[(4x-3)+8x\big]$$
$$= (4x-3)(12x-3)$$
$$= 3(4x-3)(4x-1)$$

73.
$$2(3x-5)\cdot 3(2x+1)^3+(3x-5)^2\cdot 3(2x+1)^2\cdot 2 = 6(3x-5)(2x+1)^2\big[(2x+1)+(3x-5)\big]$$
$$= 6(3x-5)(2x+1)^2(5x-4)$$

75.
$$\frac{(2x+3)\cdot 3-(3x-5)\cdot 2}{(3x-5)^2} = \frac{6x+9-6x+10}{(3x-5)^2} = \frac{19}{(3x-5)^2}$$

77.
$$\frac{x\cdot 2x-(x^2+1)\cdot 1}{(x^2+1)^2} = \frac{2x^2-x^2-1}{(x^2+1)^2} = \frac{x^2-1}{(x^2+1)^2} = \frac{(x-1)(x+1)}{(x^2+1)^2}$$

79.
$$\frac{(3x+1)\cdot 2x-x^2\cdot 3}{(3x+1)^2} = \frac{6x^2+2x-3x^2}{(3x+1)^2} = \frac{3x^2+2x}{(3x+1)^2}$$

81.
$$\frac{(x^2+1)\cdot 3-(3x+4)\cdot 2x}{(x^2+1)^2} = \frac{3x^2+3-6x^2-8x}{(x^2+1)^2} = \frac{-3x^2-8x+3}{(x^2+1)^2}$$

0.4 Solving Equations

1. $3x = 21$
$x = 7$

3. $5x + 15 = 0$
$5x = -15$
$x = -3$

5. $2x - 3 = 5$
$2x = 8$
$x = 4$

7. $\dfrac{1}{3}x = \dfrac{5}{12}$

$x = \dfrac{5}{\cancel{12}\,4} \cdot \dfrac{\cancel{3}}{1}$

$x = \dfrac{5}{4}$

9. $6 - x = 2x + 9$
$-3x = 3$
$x = -1$

11. $2(3 + 2x) = 3(x - 4)$
$6 + 4x = 3x - 12$
$x = -18$

13. $8x - (2x + 1) = 3x - 10$
$6x - 1 = 3x - 10$
$3x = -9$
$x = -3$

15. $\dfrac{1}{2}x - 4 = \dfrac{3}{4}x$

$4 \cdot \left(\dfrac{1}{2}x - 4\right) = 4 \cdot \left(\dfrac{3}{4}x\right)$

$2x - 16 = 3x$
$x = -16$

17. $0.9t = 0.4 + 0.1t$
$0.8t = 0.4$
$t = \dfrac{0.4}{0.8} = 0.5$

19. $\dfrac{2}{y} + \dfrac{4}{y} = 3$

$\dfrac{6}{y} = 3$

$3y = 6$
$y = 2$

21. $(x + 7)(x - 1) = (x + 1)^2$
$x^2 - x + 7x - 7 = x^2 + 2x + 1$
$x^2 + 6x - 7 = x^2 + 2x + 1$
$6x - 7 = 2x + 1$
$4x = 8$
$x = 2$

23. $z(z^2 + 1) = 3 + z^3$
$z^3 + z = 3 + z^3$
$z = 3$

25. $x^2 = 9x$
$x^2 - 9x = 0$
$x(x - 9) = 0$
$x = 0$ or $x - 9 = 0$
$x = 9$

The solution set is $\{0, 9\}$.

27. $t^3 - 9t^2 = 0$
$t^2(t - 9) = 0$
$t^2 = 0$ or $t - 9 = 0$
$t = 0$ or $t = 9$

The solution set is $\{0, 9\}$.

29.
$$\frac{3}{2x-3} = \frac{2}{x+5}$$
$$3(x+5) = 2(2x-3)$$
$$3x+15 = 4x-6$$
$$x = 21$$

31.
$$(x+2)(3x) = (x+2)(6)$$
$$3x^2 + 6x = 6x + 12$$
$$3x^2 - 12 = 0$$
$$x^2 - 4 = 0$$
$$x^2 = 4$$
$$x = -2 \quad \text{or} \quad x = 2$$

The solution set is $\{-2, 2\}$.

33.
$$\frac{2}{x-2} = \frac{3}{x+5} + \frac{10}{(x+5)(x-2)}$$

L.C.D.: $(x+5)(x-2)$

$$\frac{2(x+5)}{(x-2)(x+5)} = \frac{3(x-2)}{(x+5)(x-2)} + \frac{10}{(x+5)(x-2)}$$

Write with the common denominator.

$$2(x+5) = 3(x-2) + 10$$

Solve the equation formed by the numerators.

$$2x + 10 = 3x - 6 + 10$$
$$x = 6$$

Check the answer for extraneous solutions.

The solution set is $\{6\}$.

35.
$$|2x| = 6$$
Either
$$2x = 6 \quad \text{or} \quad 2x = -6$$
$$x = 3 \quad \text{or} \quad x = -3$$
The solution set is $\{-3, \ 3\}$.

37.
$$|2x+3| = 5$$
Either
$$2x+3 = 5 \quad \text{or} \quad 2x+3 = -5$$
$$2x = 2 \quad \text{or} \qquad 2x = -8$$
$$x = 1 \quad \text{or} \qquad x = -4$$
The solution set is $\{-4, 1\}$.

39.
$$|1-4t| = 5$$
Either
$$1-4t = 5 \quad \text{or} \quad 1-4t = -5$$
$$-4t = 4 \qquad \text{or} \qquad -4t = -6$$
$$t = -1 \quad \text{or} \qquad t = \frac{3}{2}$$
The solution set is $\left\{-1, \frac{3}{2}\right\}$.

41.
$$|-2x| = 8$$
Either
$$-2x = 8 \quad \text{or} \quad -2x = -8$$
$$x = -4 \quad \text{or} \quad x = 4$$

The solution set is $\{-4, 4\}$.

43.
$$|-2|x = 4$$
$$2x = 4$$
$$x = 2$$

45.
$$|x-2| = -\frac{1}{2}$$

This equation has no solution. Absolute values are always nonnegative.

47. $\left| x^2 - 4 \right| = 0$

$x^2 - 4 = 0$

$x^2 = 4$

$x = \pm 2$

The solution set is $\{-2, 2\}$.

49. $\left| x^2 - 2x \right| = 3$

Either

$\begin{array}{ll} x^2 - 2x = 3 & \text{or} \\ x^2 - 2x - 3 = 0 & \\ (x-3)(x+1) = 0 & \\ x - 3 = 0 \quad \text{or} \quad x + 1 = 0 & \\ x = 3 \quad \text{or} \quad x = -1 & \end{array}$

$\begin{array}{l} x^2 - 2x = -3 \\ x^2 - 2x + 3 = 0 \qquad a = 1, b = -2, c = 3 \\ \text{The discriminant, } b^2 - 4ac = 4 - 12 = -8 \text{ is} \\ \text{negative; the equation has no real solutions.} \end{array}$

The solution set is $\{-1, 3\}$.

51. $\left| x^2 + x - 1 \right| = 1$

Either

$\begin{array}{ll} x^2 + x - 1 = 1 & \text{or} \\ x^2 + x - 2 = 0 & \\ (x-1)(x+2) = 0 & \\ x - 1 = 0 \quad \text{or} \quad x + 2 = 0 \quad \text{or} \\ x = 1 \quad \text{or} \quad x = -2 \quad \text{or} \end{array}$

$\begin{array}{l} x^2 + x - 1 = -1 \\ x^2 + x = 0 \\ x(x+1) = 0 \\ x = 0 \quad \text{or} \quad x + 1 = 0 \\ x = 0 \quad \text{or} \quad x = -1 \end{array}$

The solution set is $\{-2, -1, 0, 1\}$.

53. $x^2 = 4x$

$x^2 - 4x = 0$

$x(x - 4) = 0$

$x = 0 \quad \text{or} \quad x - 4 = 0$

$x = 4$

The solution set is $\{0, 4\}$.

55. $z^2 + 4z - 12 = 0$

$(z - 2)(z + 6) = 0$

$z - 2 = 0 \quad \text{or} \quad z + 6 = 0$

$z = 2 \quad \text{or} \quad z = -6$

The solution set is $\{-6, 2\}$.

57. $2x^2 - 5x - 3 = 0$

$(2x + 1)(x - 3) = 0$

$2x + 1 = 0 \quad \text{or} \quad x - 3 = 0$

$x = -\dfrac{1}{2} \quad \text{or} \quad x = 3$

The solution set is $\left\{ -\dfrac{1}{2}, 3 \right\}$.

59. $x(x - 7) + 12 = 0$

$x^2 - 7x + 12 = 0$

$(x - 3)(x - 4) = 0$

$x - 3 = 0 \quad \text{or} \quad x - 4 = 0$

$x = 3 \quad \text{or} \quad x = 4$

The solution set is $\{3, 4\}$.

61.
$$4x^2 + 9 = 12x$$
$$4x - 12x + 9 = 0$$
$$(2x - 3)(2x - 3) = 0$$
$$2x - 3 = 0$$
$$x = \frac{3}{2}$$

The solution set is $\left\{\dfrac{3}{2}\right\}$.

63.
$$6x - 5 = \frac{6}{x}$$
$$6x^2 - 5x = 6$$
$$6x^2 - 5x - 6 = 0$$
$$(3x + 2)(2x - 3) = 0$$
$$3x + 2 = 0 \quad \text{or} \quad 2x - 3 = 0$$
$$x = -\frac{2}{3} \quad \text{or} \quad x = \frac{3}{2}$$

The solution set is $\left\{-\dfrac{2}{3}, \dfrac{3}{2}\right\}$.

65.
$$\frac{4(x-2)}{x-3} + \frac{3}{x} = \frac{-3}{x(x-3)}$$

$$\frac{4x(x-2)}{x(x-3)} + \frac{3(x-3)}{x(x-3)} = \frac{-3}{x(x-3)}$$

$$4x(x-2) + 3(x-3) = -3$$

$$4x^2 - 8x + 3x - 9 = -3$$

$$4x^2 - 5x - 6 = 0$$
$$(4x + 3)(x - 2) = 0$$
$$4x + 3 = 0 \quad \text{or} \quad x - 2 = 0$$

$$x = -\frac{3}{4} \quad \text{or} \quad x = 2$$

The solution set is $\left\{-\dfrac{3}{4}, 2\right\}$.

The lowest common denominator is $x(x - 3)$.

Write the equation with the common denominator.

Consider the equation formed by the numerator.

Simplify.

Put the quadratic equation in standard form.
Factor.
Use the Zero-Product Property.

Solve; be sure to check for extraneous solutions.

67.
$$x^2 = 25$$
$$x = \pm\sqrt{25}$$
$$x = \pm 5$$
The solution set is $\{-5, 5\}$.

69.
$$(x-1)^2 = 4$$
$$x - 1 = \pm\sqrt{4}$$
$$x - 1 = \pm 2$$
$$x = 2 + 1 \quad \text{or} \quad x = -2 + 1$$
$$x = 3 \qquad \text{or} \quad x = -1$$
The solution set is $\{-1, 3\}$.

71.
$$(2x+3)^2 = 9$$
$$2x + 3 = \pm\sqrt{9}$$
$$2x + 3 = \pm 3$$
$$2x = -3 + 3 \quad \text{or} \quad 2x = -3 - 3$$
$$2x = 0 \qquad \text{or} \quad 2x = -6$$
$$x = 0 \qquad \text{or} \quad x = -3$$
The solution set is $\{-3, 0\}$.

73.
$$x^2 + 8x$$

Add $\left(\dfrac{8}{2}\right)^2 = 16$.

Result $x^2 + 8x + 16$

75.

$$x^2 + \frac{1}{2}x$$

Add $\left(\frac{1}{4}\right)^2 = \frac{1}{16}$

Result $x^2 + \frac{1}{2}x + \frac{1}{16}$

77.

$$x^2 - \frac{2}{3}x$$

Add $\left(\frac{1}{3}\right)^2 = \frac{1}{9}$

Result $x^2 - \frac{2}{3}x + \frac{1}{9}$

79.

$$x^2 + 4x = 21$$

$$x^2 + 4x + 4 = 21 + 4 \qquad \text{Add 4 to both sides.}$$

$$(x+2)^2 = 25 \qquad \text{Factor.}$$

$$x + 2 = \pm 5 \qquad \text{Use the Square Root Method.}$$

$$x = -2 \pm 5$$

The solution set is $\{-7, 3\}$.

81.

$$x^2 - \frac{1}{2}x - \frac{3}{16} = 0$$

$$x^2 - \frac{1}{2}x = \frac{3}{16}$$

$$x^2 - \frac{1}{2}x + \frac{1}{16} = \frac{3}{16} + \frac{1}{16} \qquad \text{Add } \frac{1}{16} \text{ to both sides.}$$

$$\left(x - \frac{1}{4}\right)^2 = \frac{4}{16}$$

$$x - \frac{1}{4} = \pm\sqrt{\frac{4}{16}}$$

$$x = \frac{1}{4} \pm \frac{2}{4}$$

The solution set is $\left\{-\frac{1}{4}, \frac{3}{4}\right\}$.

83.

$$3x^2 + x - \frac{1}{2} = 0$$

$$x^2 + \frac{1}{3}x - \frac{1}{6} = 0$$

$$x^2 + \frac{1}{3}x = \frac{1}{6}$$

$$x^2 + \frac{1}{3}x + \frac{1}{36} = \frac{1}{6} + \frac{1}{36}$$

$$\left(x + \frac{1}{6}\right)^2 = \frac{7}{36}$$

$$x + \frac{1}{6} = \pm\sqrt{\frac{7}{36}}$$

$$x = -\frac{1}{6} \pm \frac{\sqrt{7}}{6}$$

The solution set is $\left\{\dfrac{-1-\sqrt{7}}{6}, \dfrac{-1+\sqrt{7}}{6}\right\}$.

85. $x^2 - 4x + 2 = 0 \qquad a = 1, b = -4, \text{ and } c = 2$

The discriminant $b^2 - 4ac = (-4)^2 - 4(1)(2) = 16 - 8 = 8$ is positive, so there are 2 real solutions to the equation.

$$x = \frac{-b \pm \sqrt{b^2 - 4ac}}{2a} = \frac{4 \pm \sqrt{8}}{2} = \frac{4 \pm 2\sqrt{2}}{2} = 2 \pm \sqrt{2}$$

The solution set is $\left\{2 - \sqrt{2}, 2 + \sqrt{2}\right\}$.

87. $x^2 - 5x - 1 = 0 \qquad a = 1, b = -5, \text{ and } c = -1$

The discriminant $b^2 - 4ac = (-5)^2 - 4(1)(-1) = 25 + 4 = 29$ is positive, so there are 2 real solutions to the equation.

$$x = \frac{-b \pm \sqrt{b^2 - 4ac}}{2a} = \frac{5 \pm \sqrt{29}}{2}$$

The solution set is $\left\{\dfrac{5 - \sqrt{29}}{2}, \dfrac{5 + \sqrt{29}}{2}\right\}$.

89. $2x^2 - 5x + 3 = 0 \qquad a = 2, b = -5, \text{ and } c = 3$

The discriminant $b^2 - 4ac = (-5)^2 - 4(2)(3) = 25 - 24 = 1$ is positive, so there are 2 real solutions to the equation.

$$x = \frac{-b \pm \sqrt{b^2 - 4ac}}{2a} = \frac{5 \pm \sqrt{1}}{4} = \frac{5 \pm 1}{4}$$

The solution set is $\left\{1, \dfrac{3}{2}\right\}$.

91. $4y^2 - y + 2 = 0 \qquad a = 4, b = -1, \text{ and } c = 2$

The discriminant $b^2 - 4ac = (-1)^2 - 4(4)(2) = 1 - 32 = -31$ is negative, so the equation has no real solution.

93. $4x^2 = 1 - 2x \quad$ First we rewrite the equation in standard form.

$4x^2 + 2x - 1 = 0 \qquad a = 4, b = 2, \text{ and } c = -1$

The discriminant $b^2 - 4ac = (2)^2 - 4(4)(-1) = 4 + 16 = 20$ is positive, so there are 2 real solutions to the equation.

$$x = \frac{-b \pm \sqrt{b^2 - 4ac}}{2a} = \frac{-2 \pm \sqrt{20}}{8} = \frac{-2 \pm 2\sqrt{5}}{8} = \frac{-1 \pm \sqrt{5}}{4}$$

The solution set is $\left\{\dfrac{-1-\sqrt{5}}{4}, \dfrac{-1+\sqrt{5}}{4}\right\}$.

95. $x^2 + \sqrt{3}x - 3 = 0$ $a = 1, b = \sqrt{3}$, and $c = -3$

The discriminant $b^2 - 4ac = \left(\sqrt{3}\right)^2 - 4(1)(-3) = 3 + 12 = 15$ is positive, so there are 2 real solutions to the equation.

$$x = \frac{-b \pm \sqrt{b^2 - 4ac}}{2a} = \frac{-\sqrt{3} \pm \sqrt{15}}{2}$$

The solution set is $\left\{\dfrac{-\sqrt{3}-\sqrt{15}}{2}, \dfrac{-\sqrt{3}+\sqrt{15}}{2}\right\}$.

97. $x^2 - 5x + 7 = 0$ $a = 1, b = -5$, and $c = 7$

The discriminant $b^2 - 4ac = (-5)^2 - 4(1)(7) = 25 - 28 = -3$ is negative, so the equation has no real solution.

99. $9x^2 - 30x + 25 = 0$ $a = 9, b = -30$, and $c = 25$

The discriminant $b^2 - 4ac = (-30)^2 - 4(9)(25) = 900 - 900 = 0$, so the equation has a repeated solution, a root of multiplicity 2.

101. $3x^2 + 5x - 8 = 0$ $a = 3, b = 5$, and $c = -8$

The discriminant $b^2 - 4ac = (5)^2 - 4(3)(-8) = 25 + 96 = 121$ is positive, so there are 2 real solutions to the equation.

103. $ax - b = c$

$$ax = b + c$$
$$x = \frac{b+c}{a}$$

105. $\dfrac{x}{a} + \dfrac{x}{b} = c$

$$\frac{bx}{ab} + \frac{ax}{ab} = \frac{abc}{ab}$$
$$bx + ax = abc$$
$$(b + a)x = abc$$
$$x = \frac{abc}{a+b}$$

107. $\dfrac{1}{x-a} + \dfrac{1}{x+a} = \dfrac{2}{x-1}$

$$\frac{(x+a)(x-1)}{(x-a)(x+a)(x-1)} + \frac{(x-a)(x-1)}{(x-a)(x+a)(x-1)} = \frac{2(x-a)(x+a)}{(x-1)(x-a)(x+a)}$$
$$(x+a)(x-1)+(x-a)(x-1) = 2(x-a)(x+a)$$
$$x^2 + ax - x - a + x^2 - x - ax + a = 2x^2 + 2ax - 2ax - 2a^2$$
$$2x^2 - 2x = 2x^2 - 2a^2$$
$$-2x = -2a^2$$
$$x = a^2$$

109.
$$\frac{1}{R} = \frac{1}{R_1} + \frac{1}{R_2}$$
$$\frac{R_1 R_2}{RR_1 R_2} = \frac{RR_2}{RR_1 R_2} + \frac{RR_1}{RR_1 R_2}$$
$$R_1 R_2 = RR_2 + RR_1$$
$$R_1 R_2 = R(R_2 + R_1)$$
$$R = \frac{R_1 R_2}{R_1 + R_2}$$

111.
$$F = \frac{mv^2}{R}$$
$$FR = mv^2$$
$$R = \frac{mv^2}{F}$$

113.
$$S = \frac{a}{1-r}$$
$$S(1-r) = a$$
$$S - Sr = a$$
$$Sr = S - a$$
$$r = \frac{S-a}{S}$$

115. The roots of the quadratic function $ax^2 + bx + c = 0$ are
$$x_1 = \frac{-b - \sqrt{b^2 - 4ac}}{2a} \quad \text{and} \quad x_2 = \frac{-b + \sqrt{b^2 - 4ac}}{2a}$$
The sum
$$x_1 + x_2 = \frac{-b - \sqrt{b^2 - 4ac}}{2a} + \frac{-b + \sqrt{b^2 - 4ac}}{2a}$$
$$= \frac{-b - \sqrt{b^2 - 4ac} + (-b) + \sqrt{b^2 - 4ac}}{2a}$$
$$= \frac{-2b}{2a} = -\frac{b}{a}$$

117. If $kx^2 + x + k = 0$ has a repeated real solution, then its discriminant is zero.

$$a = k,\ b = 1,\ \text{and } c = k$$

discriminant: $\quad b^2 - 4ac = 1^2 - 4(k)(k) = 0$

$$1 - 4k^2 = 0$$
$$4k^2 = 1$$
$$2k = \pm 1$$

So the equation has one repeated root if $k = \dfrac{1}{2}$ or $k = -\dfrac{1}{2}$.

119. The real solutions of the equation $ax^2 + bx + c = 0$ are

$$x_1 = \frac{-b - \sqrt{b^2 - 4ac}}{2a} \quad \text{and} \quad x_2 = \frac{-b + \sqrt{b^2 - 4ac}}{2a}$$

$$= -\frac{b + \sqrt{b^2 - 4ac}}{2a} \qquad\qquad = -\frac{b - \sqrt{b^2 - 4ac}}{2a}$$

The real solutions of the equation $ax^2 - bx + c = 0$ are

$$x_3 = \frac{b - \sqrt{b^2 - 4ac}}{2a} \quad \text{and} \quad x_4 = \frac{b + \sqrt{b^2 - 4ac}}{2a}$$

So $x_1 = -x_4$ and $x_2 = -x_3$.

121. (a) $x^2 = 9$ and $x = 3$ are not equivalent. The solution set of $x^2 = 9$ is $\{-3, 3\}$, but the solution set of $x = 3$ is $\{3\}$.

(b) $x = \sqrt{9}$ and $x = 3$ are equivalent since they both have the same solution set, $\{3\}$.

(c) $(x - 1)(x - 2) = (x - 1)^2$ and $x - 2 = x - 1$ are not equivalent. The solution of the first equation is $\{1\}$, but the second equation has no solution.

123. – 127. Answers will vary.

0.5 Intervals; Solving Inequalities

1. The graph represents $[0, 2]$ or $0 \le x \le 2$.

3. The graph represents $(-1, 2)$ or $-1 < x < 2$.

5. The graph represents $[0, 3)$ or $0 \le x < 3$.

7. $[0, 4]$

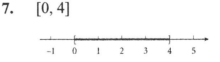

9. $[4, 6)$

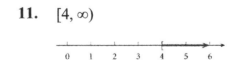

11. $[4, \infty)$

13. $(-\infty, -4)$

15. $2 \le x \le 5$

17. $-3 < x < -2$

19. $4 \le x < \infty$

21. $-\infty < x < -3$

23. $3 < 5$

(a) $3 + 3 < 5 + 3$
$\quad\quad 6 < 8$

(b) $3 - 5 < 5 - 5$
$\quad\quad -2 < 0$

(c) $(3)(3) < (3)(5)$
$\quad\quad 9 < 15$

(d) $(-2)(3) > (-2)(5)$
$\quad\quad -6 > -10$

25. $4 > -3$

(a) $4 + 3 > -3 + 3$
$\quad\quad 7 > 0$

(b) $4 - 5 > -3 - 5$
$\quad\quad -1 > -8$

(c) $(3)(4) > (3)(-3)$
$\quad\quad 12 > -9$

(d) $(-2)(4) < (-2)(-3)$
$\quad\quad -8 < 6$

27. $2x + 1 < 2$

(a) $(2x + 1) + 3 < 2 + 3$
$\quad\quad 2x + 4 < 5$

(b) $(2x + 1) - 5 < 2 - 5$
$\quad\quad 2x - 4 < -3$

(c) $(3)(2x + 1) < (3)(2)$
$\quad\quad 6x + 3 < 6$

(d) $(-2)(2x + 1) > (-2)(2)$
$\quad\quad -4x - 2 > -4$

29. $<$ $x < 5$
$\quad\quad x - 5 < 5 - 5$
$\quad\quad\quad x < 0$

31. $>$ $x > -4$
$\quad\quad x + 4 > -4 + 4$
$\quad\quad x + 4 > 0$

33. $\ge$ $x \ge -4$
$\quad\quad 3x \ge (3)(-4)$
$\quad\quad 3x \ge -12$

35. $<$ $x > 6$
$\quad\quad -2x < (-2)(6)$
$\quad\quad -2x < -12$

37. $\le$ $x \ge 5$
$\quad\quad -4x \le (-4)(5)$
$\quad\quad -4x \le -20$

39. $>$ $2x > 6$
$\quad\quad \dfrac{2x}{2} > \dfrac{6}{2}$
$\quad\quad\quad x > 3$

41. $\ge$

$$-\frac{1}{2}x \le 3$$

$$(-2) \cdot \left(-\frac{1}{2}x\right) \ge (-2)(3)$$

$$x \ge -6$$

43.

$$x + 1 < 5$$
$$x + 1 - 1 < 5 - 1$$
$$x < 4$$

The solution set is $\{x \mid x < 4\}$ or the interval $(-\infty, 4)$.

45.

$$1 - 2x \le 3$$
$$1 - 2x - 1 \le 3 - 1$$
$$-2x \le 2$$
$$\frac{-2x}{-2} \ge \frac{2}{-2}$$
$$x \ge -1$$

The solution set is $\{x \mid x \ge -1\}$ or the interval $[-1, \infty)$.

47.

$$3x - 7 > 2$$
$$3x - 7 + 7 > 2 + 7$$
$$3x > 9$$
$$\frac{3x}{3} > \frac{9}{3}$$
$$x > 3$$

The solution set is $\{x \mid x > 3\}$ or the interval $(3, \infty)$.

49.

$$3x - 1 \ge 3 + x$$
$$3x - 1 + 1 \ge 3 + x + 1$$
$$3x \ge 4 + x$$
$$3x - x \ge 4 + x - x$$
$$2x \ge 4$$
$$x \ge 2$$

The solution set is $\{x \mid x \ge 2\}$ or the interval $[2, \infty)$.

51.

$$-2(x + 3) < 8$$
$$\frac{-2(x+3)}{-2} > \frac{8}{-2}$$
$$x + 3 > -4$$
$$x + 3 - 3 > -4 - 3$$
$$x > -7$$

The solution set is $\{x \mid x > -7\}$ or the interval $(-7, \infty)$.

53.

$$4 - 3(1 - x) \le 3$$
$$4 - 3 + 3x \le 3$$
$$1 + 3x \le 3$$
$$1 + 3x - 1 \le 3 - 1$$
$$3x \le 2$$
$$x \le \frac{2}{3}$$

The solution set is $\left\{x \mid x \le \frac{2}{3}\right\}$ or the interval $\left(-\infty, \frac{2}{3}\right]$.

55.

$$\frac{1}{2}(x - 4) > x + 8$$
$$2 \cdot \left[\frac{1}{2}(x - 4)\right] > 2 \cdot (x + 8)$$
$$x - 4 > 2x + 16$$
$$x - 4 - x > 2x + 16 - x$$
$$-4 > x + 16$$
$$-4 - 16 > x + 16 - 16$$
$$-20 > x \quad \text{or} \quad x < -20$$

The solution set is $\{x \mid x < -20\}$ or the interval $(-\infty, -20)$.

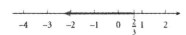   

57.

$$\frac{x}{2} \geq 1 - \frac{x}{4}$$

$$4 \cdot \left(\frac{x}{2}\right) \geq 4 \cdot \left(1 - \frac{x}{4}\right)$$

$$2x \geq 4 - x$$

$$2x + x \geq 4 - x + x$$

$$3x \geq 4$$

$$x \geq \frac{4}{3}$$

The solution set is $\left\{x \mid x \geq \frac{4}{3}\right\}$ or the

interval $\left[\frac{4}{3}, \infty\right)$.

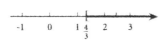

59. $0 \leq 2x - 6 \leq 4$ is equal to the two inequalities

$0 \leq 2x - 6$	and	$2x - 6 \leq 4$
$0 + 6 \leq 2x - 6 + 6$		$2x - 6 + 6 \leq 4 + 6$
$6 \leq 2x$		$2x \leq 10$
$3 \leq x$		$x \leq 5$

The solution set consists of all x for which $x \geq 3$ and $x \leq 5$ which is written either as $\{x \mid 3 \leq x \leq 5\}$ or as the interval $[3, 5]$.

61. $-5 \leq 4 - 3x \leq 2$ is equal to the two inequalities

$-5 \leq 4 - 3x$	and	$4 - 3x \leq 2$
$-9 \leq -3x$		$-3x \leq -2$
$3 \geq x$		$x \geq \dfrac{2}{3}$

The solution set consists of all x for which $x \geq \dfrac{2}{3}$ and $x \leq 3$ which is written either as

$\left\{x \mid \dfrac{2}{3} \leq x \leq 3\right\}$ or as the interval $\left[\dfrac{2}{3}, 3\right]$.

63. $-3 < \dfrac{2x-1}{4} < 0$ is equal to the two inequalities

$$-3 < \dfrac{2x-1}{4} \qquad\qquad\text{and}\qquad\qquad \dfrac{2x-1}{4} < 0$$

$$-12 < 2x - 1 \qquad\qquad\qquad\qquad\qquad\quad 2x - 1 < 0$$

$$-11 < 2x \qquad\qquad\qquad\qquad\qquad\qquad 2x < 1$$

$$-\dfrac{11}{2} < x \qquad\qquad\qquad\qquad\qquad\qquad x < \dfrac{1}{2}$$

The solution set consists of all x for which $x \geq -\dfrac{11}{2}$ and $x \leq \dfrac{1}{2}$ which is written either as

$\left\{x \left|\; -\dfrac{11}{2} < x < \dfrac{1}{2}\right.\right\}$ or as the interval $\left(-\dfrac{11}{2},\ \dfrac{1}{2}\right)$.

65. $1 < 1 - \dfrac{1}{2}x < 4$ is equal to the two inequalities

$$1 < 1 - \dfrac{1}{2}x \qquad\qquad\text{and}\qquad\qquad 1 - \dfrac{1}{2}x < 4$$

$$0 < -\dfrac{1}{2}x \qquad\qquad\qquad\qquad\qquad -\dfrac{1}{2}x < 3$$

$$0 > x \qquad\qquad\qquad\qquad\qquad\qquad\quad x > -6$$

The solution set consists of all x for which $x > -6$ and $x < 0$ which is written either as $\{x \,|\, -6 < x < 0\}$ or as the interval $(-6, 0)$.

67. $(4x + 2)^{-1} = \dfrac{1}{4x + 2} < 0$ is satisfied if $4x + 2 < 0$. That is, when $4x < -2$ or $x < -\dfrac{1}{2}$.

The solution set is written either as $\left\{x \left|\; x < -\dfrac{1}{2}\right.\right\}$ or as the interval $\left(-\infty,\ -\dfrac{1}{2}\right)$.

69. $0 < \dfrac{2}{x} < \dfrac{3}{5}$ is equal to the two inequalities

$$0 < \dfrac{2}{x} \qquad\qquad\qquad\text{and}\qquad\qquad \dfrac{2}{x} < \dfrac{3}{5}$$

$$0 < \dfrac{2}{x} \text{ when } x > 0. \qquad\qquad\qquad\qquad 10 < 3x$$

$$\qquad\qquad\qquad\qquad\qquad\qquad\qquad\qquad \dfrac{10}{3} < x$$

The solution set consists of all x for which $x > 0$ and $x > \dfrac{10}{3}$ which is written either as

$\left\{ x \mid x > \dfrac{10}{3} \right\}$ or as the interval $\left(\dfrac{10}{3}, \ \infty \right)$.

71. First we solve the equation $(x-3)(x+1)=0$ and use the solutions to separate the real number line.

$$(x-3)(x+1)=0$$
$$x-3=0 \quad \text{or} \quad x+1=0$$
$$x=3 \quad \text{or} \quad x=-1$$

We separate the number line into the intervals
$$(-\infty, -1) \qquad (-1, 3) \qquad (3, \infty)$$
In each interval we select a number and evaluate the expression $(x-3)(x+1)$ at that value. We choose $-5, 0, 5$.
For $x = -5$: $\quad (-5-3)(-5+1) = 32$, a positive number.
For $x = 0$: $\quad (0-3)(0+1) = -3$, a negative number.
For $x = 5$: $\quad (5-3)(5+1) = 12$, a positive number.

Since $(x-3)(x+1) < 0$ for $-1 < x < 3$, we write the solution set either as $\{x \mid -1 < x < 3\}$ or as the interval $(-1, 3)$.

73. First we solve the equation $-x^2 + 9 = 0$ and use the solutions to separate the real number line.

$$-x^2 + 9 = 0$$
$$(-x+3)(x+3) = 0$$
$$-x+3=0 \quad \text{or} \quad x+3=0$$
$$x=3 \quad \text{or} \quad x=-3$$

We separate the number line into the intervals
$$(-\infty, -3) \qquad (-3, 3) \qquad (3, \infty)$$
We select a number in each interval and evaluate the expression $(-x+3)(x+3)$ at that value. We choose $-5, 0, 5$.
For $x = -5$: $\quad (-(-5)+3)(-5+3) = -16$, a negative number.
For $x = 0$: $\quad (0+3)(0+3) = 9$, a positive number.
For $x = 5$: $\quad (-5+3)(5+3) = -16$, a negative number.

Since the expression $-x^2 + 9 > 0$ for $-3 < x < 3$, we write the solution set either as $\{x \mid -3 < x < 3\}$ or as the interval $(-3, 3)$.

75. First we solve the equation $x^2 + x = 12$ and use the solutions to separate the real number line.

$$x^2 + x = 12$$

$$x^2 + x - 12 = 0$$
$$(x - 3)(x + 4) = 0$$

$$x - 3 = 0 \quad \text{or} \quad x + 4 = 0$$
$$x = 3 \quad \text{or} \quad x = -4$$

We separate the number line into the intervals
$$(-\infty, -4) \qquad (-4, 3) \qquad (3, \infty)$$
We select a number in each interval and evaluate the expression $x^2 + x - 12$ at that value. We choose $-5, 0, 5$.

For $x = -5$: $\quad (-5)^2 + (-5) - 12 = 8$, a positive number.

For $x = 0$: $\quad 0^2 + 0 - 12 = -12$, a negative number.

For $x = 5$: $\quad 5^2 + 5 - 12 = 18$, a positive number.

The expression $x^2 + x - 12 > 0$ for $x < -4$ or $x > 3$. We write the solution set either as $\{x \mid x < -4 \text{ or } x > 3\}$ or as all x in the interval $(-\infty, -4)$ or $(3, \infty)$.

77. First we solve the equation $x(x - 7) = -12$ and use the solutions to separate the real number line.

$$x(x - 7) = -12$$
$$x^2 - 7x = -12$$
$$x^2 - 7x + 12 = 0$$
$$(x - 3)(x - 4) = 0$$
$$x - 3 = 0 \quad \text{or} \quad x - 4 = 0$$
$$x = 3 \quad \text{or} \quad x = 4$$

We separate the number line into the intervals
$$(-\infty, 3) \qquad (3, 4) \qquad (4, \infty)$$
We select a number in each interval and evaluate the expression $x^2 - 7x + 12$ at that value. We choose $0, 3.5,$ and 5.

For $x = 0$: $\quad 0^2 - 7(0) + 12 = 12$, a positive number.

For $x = 3.5$: $\quad (3.5)^2 - 7(3.5) + 12 = -0.25$, a negative number.

For $x = 5$: $\quad 5^2 - 7(5) + 12 = 2$, a positive number.

The expression $x^2 - 7x + 12 > 0$ for $x < 3$ or $x > 4$. We write the solution set either as $\{x \mid x < 3 \text{ or } x > 4\}$ or as all x in the interval $(-\infty, 3)$ or $(4, \infty)$.

79. First we solve the equation $4x^2 + 9 = 6x$ and use the solutions to separate the real number line.

$$4x^2 + 9 = 6x$$
$$4x^2 - 6x + 9 = 0$$

This equation has no real solutions. Its discriminant, $b^2 - 4ac = 36 - 144 = -108$, is negative. The value of $4x^2 - 6x + 9$ either is always positive or always negative. To see which is true, we test $x = 0$. Since $4(0)^2 - 6(0) + 9 = 9$ is positive, we conclude that expression is always positive, and the inequality

$$4x^2 + 9 < 6x \text{ has no solution.}$$

81. First we solve the equation $(x-1)(x^2 + x + 1) = 0$ and use the solutions to separate the real number line.

$$(x-1)(x^2 + x + 1) = 0$$
$$x - 1 = 0 \quad \text{or} \quad x^2 + x + 1 = 0$$
$$x = 1$$

$x = 1$ is the only solution, since the equation $x^2 + x + 1 = 0$ has a negative discriminant. We use $x = 1$ to separate the number line into two parts:

$$-\infty < x < 1 \qquad \text{and} \qquad 1 < x < \infty$$

In each part select a test number and evaluate the expression $(x-1)(x^2 + x + 1)$.

For $x = 0$: $(0-1)(0^2 + 0 + 1) = -1$, a negative number.

For $x = 2$: $(2-1)(2^2 + 2 + 1) = 7$, a positive number.

The expression $(x-1)(x^2 + x + 1) > 0$ for $x > 1$. The solution set is $\{x \mid x > 1\}$, or for all x in the interval $(1, \infty)$.

83. First we solve the equation $(x-1)(x-2)(x-3) = 0$ and use the solutions to separate the real number line.

$$(x-1)(x-2)(x-3) = 0$$
$$x - 1 = 0 \quad \text{or} \quad x - 2 = 0 \quad \text{or} \quad x - 3 = 0$$
$$x = 1 \quad \text{or} \quad x = 2 \quad \text{or} \quad x = 3$$

We separate the number line into the following 4 parts, choose a test number in each part, and evaluate the expression $(x-1)(x-2)(x-3)$ at each test number.

Parts: $\quad -\infty < x < 1 \qquad 1 < x < 2 \qquad 2 < x < 3 \qquad 3 < x < \infty$

For $x = 0$: $(0-1)(0-2)(0-3) = -6$, a negative number.

For $x = 1.5$ $(1.5-1)(1.5-2)(1.5-3) = 0.375$, a positive number.

For $x = 2.5$: $(2.5-1)(2.5-2)(2.5-3) = -0.375$, a negative number.

For $x = 5$: $(5-1)(5-2)(5-3) = 24$, a positive number.

The expression $(x-1)(x-2)(x-3) < 0$ for $x < 1$ or for $2 < x < 3$. The solution set is $\{x \mid x < 1 \text{ or } 2 < x < 3\}$, or for all x in the interval $(-\infty, 1)$ or $(2, 3)$.

85. First we solve the equation $-x^3 + 2x^2 + 8x = 0$ and use the solutions to separate the real number line.

$$-x^3 + 2x^2 + 8x = 0$$
$$-x(x^2 - 2x - 8x) = 0$$
$$-x(x+2)(x-4) = 0$$
$$-x = 0 \quad \text{or} \quad x+2 = 0 \quad \text{or} \quad x-4 = 0$$
$$x = 0 \quad \text{or} \quad x = -2 \quad \text{or} \quad x = 4$$

We separate the number line into the following 4 parts, choose a test number in each part, and evaluate the expression $-x^3 + 2x^2 + 8x$ at each test number.

Parts: $-\infty < x < -2$ $-2 < x < 0$ $0 < x < 4$ $4 < x < \infty$

For $x = -3$: $-(-3)^3 + 2(-3)^2 + 8(-3) = 21$, a positive number.

For $x = -1$: $-(-1)^3 + 2(-1)^2 + 8(-1) = -5$, a negative number.

For $x = 1$: $-(1)^3 + 2(1)^2 + 8(1) = 9$, a positive number.

For $x = 10$: $-(10)^3 + 2(10)^2 + 8(10) = -720$, a negative number.

The expression $-x^3 + 2x^2 + 8x < 0$ for $-2 < x < 0$ or for $x > 4$. The solution set is $\{x \mid -2 < x < 0 \text{ or } x > 4\}$, or for all x in the interval $(-2, 0)$ or $(4, \infty)$.

87. First we solve the equation $x^3 = x$ and use the solutions to separate the real number line.

$$x^3 = x$$
$$x^3 - x = 0$$
$$x(x^2 - 1) = 0$$
$$x(x-1)(x+1) = 0$$
$$x = 0 \quad \text{or} \quad x-1 = 0 \quad \text{or} \quad x+1 = 0$$
$$x = 0 \quad \text{or} \quad x = 1 \quad \text{or} \quad x = -1$$

We separate the number line into the following 4 parts, choose a test number in each part, and evaluate the expression $x^3 - x$ at each test number.

Parts: $-\infty < x < -1$ $-1 < x < 0$ $0 < x < 1$ $1 < x < \infty$

For $x = -2$: $(-2)^3 - (-2) = -6$, a negative number.

For $x = -0.5$: $(-0.5)^3 - (-0.5) = 0.375$, a positive number.

For $x = 0.5$: $(0.5)^3 - (0.5) = -0.375$, a negative number.

For $x = 2$: $(2)^3 - (2) = 6$, a positive number.

The expression $x^3 - x > 0$ for $-1 < x < 0$ or for $x > 1$. The solution set is $\{x \mid -1 < x < 0 \text{ or } x > 1\}$, or for all x in the interval $(-1, 0)$ or $(1, \infty)$.

89. First we solve the equation $x^3 = x^2$ and use the solutions to separate the real number line.

$$x^3 = x^2$$
$$x^3 - x^2 = 0$$
$$x^2(x - 1) = 0$$
$$x^2 = 0 \quad \text{or} \quad x - 1 = 0$$
$$x = 0 \quad \text{or} \quad x = 1$$

We separate the number line into the following 3 parts, choose a test number in each part, and evaluate the expression $x^3 - x^2$ at each test number.

Parts: $-\infty < x < 0$ $0 < x < 1$ $1 < x < \infty$

For $x = -1$: $(-1)^3 - (-1)^2 = -2$, a negative number.

For $x = 0.5$: $0.5^3 - 0.5^2 = -0.125$, a negative number.

For $x = 2$: $2^3 - 2^2 = 4$, a positive number.

The expression $x^3 - x^2 > 0$ for $x > 1$. The solution set is $\{x \mid x > 1\}$, or for all x in the interval $(1, \infty)$.

91. $\dfrac{x+1}{1-x}$ is not defined when $1 - x = 0$ or when $x = 1$.

$\dfrac{x+1}{1-x} = 0$ when $x + 1 = 0$ or when $x = -1$.

We use these two numbers to separate the number line into three parts. We then choose a test number in each part and evaluate the expression $\dfrac{x+1}{1-x}$ at the test number.

Parts: $-\infty < x < -1$ $-1 < x < 1$ $1 < x < \infty$

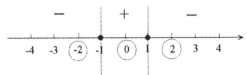

For $x = -2$: $\dfrac{(-2)+1}{1-(-2)} = -\dfrac{1}{3}$, which is a negative number.

For $x = 0$: $\dfrac{0+1}{1-0} = 1$, which is a positive number.

For $x = 2$: $\dfrac{2+1}{1-2} = -3$, which is a negative number.

The expression $\dfrac{x+1}{1-x} < 0$ for $x < -1$ or for $x > 1$. The solution set is $\{x \mid x < -1 \text{ or } x > 1\}$, or for all x in the interval $(-\infty, -1)$ or $(1, \infty)$.

93. $\dfrac{(x-1)(x+1)}{x}$ is not defined for $x = 0$. $\dfrac{(x-1)(x+1)}{x} = 0$ for $x = 1$ or $x = -1$.

We use these three numbers to separate the number line into four parts.

Parts: $-\infty < x < -1$ $-1 < x < 0$ $0 < x < 1$ $1 < x < \infty$

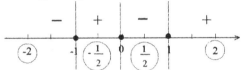

For $x = -2$: $\dfrac{[(-2)-1][(-2)+1]}{-2} = -1.5$, which is a negative number.

For $x = -\dfrac{1}{2}$: $\dfrac{[(-0.5)-1][(-0.5)+1]}{-0.5} = 1.5$, which is a positive number.

For $x = \dfrac{1}{2}$: $\dfrac{(0.5-1)(0.5+1)}{0.5} = -1.5$, which is a negative number.

For $x = 2$: $\dfrac{(2-1)(2+1)}{2} = 1.5$, which is a positive number.

The expression $\dfrac{(x-1)(x+1)}{x} < 0$ for $x < -1$ or for $0 < x < 1$. The solution set is $\{x \mid x < -1 \text{ or } 0 < x < 1\}$, or for all x in the interval $(-\infty, -1)$ or $(0, 1)$.

95. $\dfrac{x-2}{x^2-1} = \dfrac{x-2}{(x-1)(x+1)}$ is not defined for $x=1$ or $x=-1$. $\dfrac{x-2}{x^2-1}=0$ for $x=2$.

We use these three numbers to separate the real number line into four parts.

Parts: $-\infty < x < -1$ $\qquad$ $-1 < x < 1$ $\qquad$ $1 < x < 2$ $\qquad$ $2 < x < \infty$

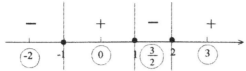

For $x=-2$: $\dfrac{(-2)-2}{(-2)^2-1} = -\dfrac{4}{3}$, which is a negative number.

For $x=0$: $\dfrac{0-2}{0^2-1} = 2$, which is a positive number.

For $x=\dfrac{3}{2}=1.5$: $\dfrac{1.5-2}{1.5^2-1} = -0.4$, which is a negative number.

For $x=3$: $\dfrac{3-2}{3^2-1} = 0.125$, which is a positive number.

The expression $\dfrac{x-2}{x^2-1} \geq 0$ for $-1 < x < 1$ or for $2 \leq x < \infty$. The solution set is $\{x \mid -1 < x < 1 \text{ or } x \geq 2\}$, or for all x in the interval $(-1, 1)$ or $[2, \infty)$.

97. First we rewrite $\dfrac{x+4}{x-2} \leq 1$ so it has a 0 on the right.

$$\dfrac{x+4}{x-2} - 1 \leq 0$$

or $\qquad$ $\dfrac{x+4}{x-2} - \dfrac{x-2}{x-2} = \dfrac{6}{x-2} \leq 0$

The expression $\dfrac{6}{x-2}$ is not defined for $x=2$; it is never zero.

We use $x=2$ to separate the number line into two parts,

$$-\infty < x < 2 \qquad \text{and} \qquad 2 < x < \infty$$

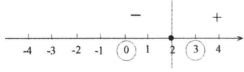

For $x=0$: $\dfrac{0+4}{0-2} - 1 = -3$, which is a negative number.

For $x=3$: $\dfrac{3+4}{3-2} - 1 = 6$, which is a positive number.

The expression $\dfrac{x+4}{x-2} - 1 \le 0$ for $x < 2$. The solution set is $\{x \mid x < 2\}$, or for all x in the interval $(-\infty, 2)$.

99. First we rewrite $\dfrac{2x+5}{x+1} > \dfrac{x+1}{x-1}$ so it has a 0 on the right.

$$\frac{2x+5}{x+1} - \frac{x+1}{x-1} > 0$$

Then we write the expression with a single denominator.

$$\frac{(2x+5)(x-1)}{(x+1)(x-1)} - \frac{(x+1)(x+1)}{(x-1)(x+1)} = \frac{(2x+5)(x-1)-(x+1)(x+1)}{(x+1)(x-1)}$$

$$= \frac{2x^2+3x-5-(x^2+2x+1)}{(x+1)(x-1)} = \frac{x^2+x-6}{(x+1)(x-1)} = \frac{(x+3)(x-2)}{(x+1)(x-1)} > 0$$

The expression $\dfrac{(x+3)(x-2)}{(x+1)(x-1)}$ is not defined for $x = 1$ or $x = -1$. $\dfrac{(x+3)(x-2)}{(x+1)(x-1)} = 0$

for $x = 2$ or $x = -3$. We use these numbers to separate the real number line into 5 parts

$$-\infty < x < -3 \qquad -3 < x < -1 \qquad -1 < x < 1 \qquad 1 < x < 2 \qquad 2 < x < \infty$$

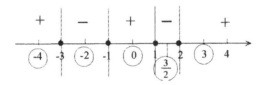

For $x = -4$: $\dfrac{[(-4)+3][(-4)-2]}{[(-4)+1][(-4)-1]} = \dfrac{2}{5}$

For $x = -2$: $\dfrac{[(-2)+3][(-2)-2]}{[(-2)+1][(-2)-1]} = -\dfrac{4}{3}$

For $x = 0$: $\dfrac{(0+3)(0-2)}{(0+1)(0-1)} = 6$

For $x = \dfrac{3}{2} = 1.5$: $\dfrac{(1.5+3)(1.5-2)}{(1.5+1)(1.5-1)} = -1.8$

For $x = 3$: $\dfrac{(3+3)(3-2)}{(3+1)(3-1)} = 0.75$

The expression $\dfrac{2x+5}{x+1} - \dfrac{x+1}{x-1} > 0$ for $x < -3$ or $-1 < x < 1$ or $2 < x < \infty$. The solution set is $\{x \mid x < -3 \text{ or } -1 < x < 1 \text{ or } x > 2\}$ or for all x in the interval $(-\infty, -3)$ or $(-1, 1)$ or $(2, \infty)$.

101. Let x represent the score on the last test. To get a B you need $80 < $ average < 90.

The average is $\dfrac{68 + 82 + 87 + 89 + x}{5} = \dfrac{326 + x}{5}$.

We will solve the inequality

$$80 \le \frac{326 + x}{5} < 90$$

which is equivalent to the two inequalities

$$80 \le \frac{326 + x}{5} \qquad \text{and} \qquad \frac{326 + x}{5} < 90$$

Solving each inequality we find

$$80 \le \frac{326 + x}{5} \qquad\qquad \frac{326 + x}{5} < 90$$
$$400 \le 326 + x \qquad\qquad 326 + x < 450$$
$$74 \le x \qquad\qquad\qquad x < 124$$

The solution set to the combined inequality is $\{x \mid 74 \le x < 124\}$, but since 100 is usually the highest score possible, you need to score between a 74 and 100 to get a B.

103. If we let x represent the selling price of the property, then we can write an equation relating the commission C to the selling price.

$$C = 45,000 + 0.25(x - 900,000)$$
$$= 45,000 + 0.25x - 225,000$$
$$= 0.25x - 180,000$$

We are told that $900,000 \le x \le 1,100,000$, so the commission varies between

$$0.25(900,000) - 180,000 \le C \le 0.25(1,100,000) - 180,000$$
$$225,000 - 180,000 \le C \le 275,000 - 180,000$$
$$45,000 \ \le C \le 95,000$$

The commission on the sale varies between \$45,000 and \$95,000 inclusive.

If the apartment complex sells for \$900,000, then the \$45,000 commission represents

$$\frac{45,000}{900,000} = 0.05 = 5\% \text{ of the sale.}$$

If the apartment complex sells for \$1,100,000, then the \$95,000 commission represents

$$\frac{95,000}{1,100,000} = 0.086 = 8.6\% \text{ of the sale.}$$

The sale's commission varies between 5% and 8.6% of the sale.

105. If we let x represent the weekly wages, then an equation relating the withholding W to x is

$$W = 74.35 + 0.25(x - 592)$$
$$= 74.35 + 0.25x - 148$$
$$= 0.25x - 73.65$$

If wages are between \$600 and \$800 inclusive, then $600 \le x \le 800$, and

$600 \le x$	and	$x \le 800$
$0.25(600) - 73.65 \le 0.25x - 73.65$		$0.25x - 73.65 \le 0.25(800) - 73.65$
$150 - 73.65 \le W$		$W \le 200 - 73.65$
$76.35 \le W$		$W \le 126.35$

The tax withholdings are between \$76.35 and \$126.35 inclusive.

107. If x represents the monthly electric usage, then an equation relating the monthly cost C to x is

$$C = 0.08275x + 7.58$$

If the monthly bills ranged between \$63.47 and \$214.53, then

$63.47 \le 0.08275x + 7.58$	and	$0.08275x + 7.58 \le 214.53$
$55.89 \le 0.08275x$		$0.08275x \le 206.95$
$675.41 \le x$		$x \le 2500.91$

The monthly electricity usage ranged between 675.41 kilowatt-hours and 2500.91 kilowatt-hours.

109. The price of a car is determined by the dealer's cost x plus the markup. We are told the price is \$8800, and that the markup ranges between 12% of the dealer's cost and 18% of the dealer's cost.

If the markup is 12%, we have

$$8800 = x + 0.12x$$
$$8800 = 1.12x$$
$$x = 7857.14$$

If the markup is 18%, we have

$$8800 = x + 0.18x$$
$$8800 = 1.18x$$
$$x = 7457.63$$

So the dealer's cost varies between \$7457.63 and \$7857.14 inclusive.

111. – 113. Answers will vary.

0.6 *n*th Roots; Rational Exponents

1. $\sqrt[3]{27} = \sqrt[3]{3^3} = 3$

3. $\sqrt[3]{-8} = \sqrt[3]{(-2)^3} = -2$

5. $\sqrt{8} = \sqrt{4} \cdot \sqrt{2} = 2\sqrt{2}$

7. $\sqrt[3]{-8x^4} = \sqrt[3]{(-2x)^3 x}$
$$= \sqrt[3]{(-2x)^3} \cdot \sqrt[3]{x}$$
$$= -2x \cdot \sqrt[3]{x}$$

9.
$$\sqrt[4]{x^{12}y^8} = \sqrt[4]{(x^3)^4 \cdot (y^2)^4}$$
$$= \sqrt[4]{(x^3)^4} \cdot \sqrt[4]{(y^2)^4}$$
$$= x^3 y^2$$

11.
$$\sqrt[4]{\frac{x^9 y^7}{xy^3}} = \sqrt[4]{x^{9-1} y^{7-3}}$$
$$= \sqrt[4]{x^8 y^4}$$
$$= \sqrt[4]{(x^2)^4} \cdot \sqrt[4]{y^4}$$
$$= x^2 y$$

13.
$$\sqrt{36x} = \sqrt{6^2 \cdot x} = 6\sqrt{x}$$

15.
$$\sqrt{3x^2}\sqrt{12x} = \sqrt{36x^3}$$
$$= \sqrt{6^2 \cdot x^2 \cdot x}$$
$$= \sqrt{(6x)^2} \cdot \sqrt{x}$$
$$= 6x \cdot \sqrt{x}$$

17.
$$\left(\sqrt{5}\sqrt[3]{9}\right)^2 = \sqrt{5^2} \cdot \sqrt[3]{81}$$
$$= 5 \cdot \sqrt[3]{3^3 \cdot 3}$$
$$= 5 \cdot 3\sqrt[3]{3}$$
$$= 15\sqrt[3]{3}$$

19.
$$\left(3\sqrt{6}\right)\left(2\sqrt{2}\right) = (3 \cdot 2)\left(\sqrt{6} \cdot \sqrt{2}\right)$$
$$= 6\sqrt{12}$$
$$= 6\sqrt{4 \cdot 3}$$
$$= 6 \cdot 2\sqrt{3}$$
$$= 12\sqrt{3}$$

21.
$$\left(\sqrt{3}+3\right)\left(\sqrt{3}-1\right) = \left(\sqrt{3}\right)^2 - \sqrt{3} + 3\sqrt{3} - 3$$
$$= 3 + 2\sqrt{3} - 3$$
$$= 2\sqrt{3}$$

23.
$$\left(\sqrt{x}-1\right)^2 = \left(\sqrt{x}-1\right)\left(\sqrt{x}-1\right) = \left(\sqrt{x}\right)^2 - \sqrt{x} - \sqrt{x} + 1$$
$$= x - 2\sqrt{x} + 1$$

25.
$$3\sqrt{2} - 4\sqrt{8} = 3\sqrt{2} - 4\sqrt{4 \cdot 2}$$
$$= 3\sqrt{2} - 4 \cdot 2\sqrt{2}$$
$$= 3\sqrt{2} - 8\sqrt{2}$$
$$= -5\sqrt{2}$$

27.
$$\sqrt[3]{16x^4} - \sqrt[3]{2x} = \sqrt[3]{(2x)^4} - \sqrt[3]{2x}$$
$$= \sqrt[3]{(2x)^3 \cdot 2x} - \sqrt[3]{2x}$$
$$= \sqrt[3]{(2x)^3} \cdot \sqrt[3]{2x} - \sqrt[3]{2x}$$
$$= 2x\sqrt[3]{2x} - \sqrt[3]{2x}$$
$$= (2x-1)\sqrt[3]{2x}$$

29. $\dfrac{1}{\sqrt{2}} \cdot \dfrac{\sqrt{2}}{\sqrt{2}} = \dfrac{\sqrt{2}}{\sqrt{2^2}} = \dfrac{\sqrt{2}}{2}$

31. $\dfrac{-\sqrt{3}}{\sqrt{5}} \cdot \dfrac{\sqrt{5}}{\sqrt{5}} = \dfrac{-\sqrt{15}}{\sqrt{5^2}} = -\dfrac{\sqrt{15}}{5}$

33. $\dfrac{\sqrt{3}}{5-\sqrt{2}} \cdot \dfrac{5+\sqrt{2}}{5+\sqrt{2}} = \dfrac{5\sqrt{3}+\sqrt{6}}{25-2} = \dfrac{5\sqrt{3}+\sqrt{6}}{23}$

35.
$$\dfrac{2-\sqrt{5}}{2+3\sqrt{5}} \cdot \dfrac{2-3\sqrt{5}}{2-3\sqrt{5}} = \dfrac{4-6\sqrt{5}-2\sqrt{5}+3\left(\sqrt{5}\right)^2}{4-9\left(\sqrt{5}\right)^2}$$
$$= \dfrac{4-8\sqrt{5}+15}{4-45}$$
$$= -\dfrac{19-8\sqrt{5}}{41} = \dfrac{8\sqrt{5}-19}{41}$$

37. $\dfrac{5}{\sqrt[3]{2}} \cdot \dfrac{\sqrt[3]{2^2}}{\sqrt[3]{2^2}} = \dfrac{5\sqrt[3]{4}}{\sqrt[3]{2^3}} = \dfrac{5\sqrt[3]{4}}{2}$

39.
$$\dfrac{\sqrt{x+h}-\sqrt{x}}{\sqrt{x+h}+\sqrt{x}} \cdot \dfrac{\sqrt{x+h}-\sqrt{x}}{\sqrt{x+h}-\sqrt{x}} = \dfrac{\left(\sqrt{x+h}-\sqrt{x}\right)^2}{\left(\sqrt{x+h}\right)^2-\left(\sqrt{x}\right)^2} = \dfrac{\left(\sqrt{x+h}-\sqrt{x}\right)^2}{x+h-x}$$
$$= \dfrac{\left(\sqrt{x+h}-\sqrt{x}\right)^2}{h} = \dfrac{(x+h)-2\sqrt{(x+h)x}+x}{h}$$
$$= \dfrac{2x+h-2\sqrt{\left(x^2+xh\right)}}{h}$$

41.
$$\sqrt[3]{2t-1} = 2$$
$$\left(\sqrt[3]{2t-1}\right)^3 = (2)^3$$
$$2t-1 = 8$$
$$2t = 9$$
$$t = \dfrac{9}{2}$$

43.
$$\sqrt{15-2x} = x$$
$$\left(\sqrt{15-2x}\right)^2 = x^2$$
$$15-2x = x^2$$
$$x^2+2x-15 = 0$$
$$(x+5)(x-3) = 0$$

Check: $x = -5$
$$\sqrt{15-2(-5)} = \sqrt{15+10} = \sqrt{25} = 5 \neq x$$
So $x = -5$ is not a solution.

$x = 3$
$$\sqrt{15-2(3)} = \sqrt{15-6} = \sqrt{9} = 3 = x$$
So $x = 3$ is a solution.

$$x + 5 = 0 \quad \text{or} \quad x - 3 = 0$$
$$x = -5 \qquad\qquad x = 3$$

The solution set is $\{3\}$.

45.
$$8^{2/3} = \left(\sqrt[3]{8}\right)^2 = 2^2 = 4$$

47.
$$(-27)^{1/3} = \sqrt[3]{-27} = -3$$

49.
$$16^{3/2} = \left(\sqrt{16}\right)^3 = 4^3 = 64$$

51.
$$9^{-3/2} = \frac{1}{9^{3/2}} = \frac{1}{\left(\sqrt{9}\right)^3} = \frac{1}{3^3} = \frac{1}{27}$$

53.
$$\left(\frac{9}{8}\right)^{3/2} = \left(\sqrt{\frac{9}{8}}\right)^3 = \left(\frac{\sqrt{9}}{\sqrt{8}}\right)^3 = \frac{3^3}{\left(2\sqrt{2}\right)^3} = \frac{27}{8 \cdot \left(\sqrt{2}\right)^2 \cdot \sqrt{2}}$$
$$= \frac{27}{16\sqrt{2}} \cdot \frac{\sqrt{2}}{\sqrt{2}} = \frac{27\sqrt{2}}{16 \cdot 2} = \frac{27\sqrt{2}}{32}$$

55.
$$\left(\frac{8}{9}\right)^{-3/2} = \left(\frac{9}{8}\right)^{3/2} = \frac{27\sqrt{2}}{32}$$
(See problem 53.)

57.
$$x^{3/4} x^{1/3} x^{-1/2} = x^{9/12} x^{4/12} x^{-6/12}$$
$$= x^{9/12 + 4/12 - 6/12}$$
$$= x^{7/12}$$

59.
$$\left(x^3 y^6\right)^{1/3} = \left(x^3\right)^{1/3} \left(y^6\right)^{1/3} = x y^2$$

61.
$$\left(x^2 y\right)^{1/3} \left(x y^2\right)^{2/3} = x^{2/3} y^{1/3} x^{2/3} y^{4/3} = \left(x^{2/3} x^{2/3}\right)\left(y^{1/3} y^{4/3}\right) = x^{4/3} y^{5/3}$$

63.
$$\left(16 x^2 y^{-1/3}\right)^{3/4} = 16^{3/4} x^{6/4} y^{-1/4} = 2^3 x^{3/2} \cdot \frac{1}{y^{1/4}} = \frac{8 x^{3/2}}{y^{1/4}}$$

65.
$$\frac{x}{\left(1+x\right)^{1/2}} + 2\left(1+x\right)^{1/2} = \frac{x}{\left(1+x\right)^{1/2}} + \frac{2\left(1+x\right)^{1/2}\left(1+x\right)^{1/2}}{\left(1+x\right)^{1/2}}$$
$$= \frac{x + 2\left(1+x\right)}{\left(1+x\right)^{1/2}}$$
$$= \frac{3x + 2}{\left(1+x\right)^{1/2}}$$

67.
$$2x\left(x^2+1\right)^{1/2} + x^2 \cdot \frac{1}{\cancel{2}}\left(x^2+1\right)^{-1/2} \cdot \cancel{2}x = 2x\left(x^2+1\right)^{1/2} + x^2 \cdot \frac{1}{\left(x^2+1\right)^{1/2}} \cdot x$$
$$= 2x\left(x^2+1\right)^{1/2} + \frac{x^3}{\left(x^2+1\right)^{1/2}}$$

$$= \frac{2x\left(x^2+1\right)^{1/2} \cdot \left(x^2+1\right)^{1/2}}{\left(x^2+1\right)^{1/2}} + \frac{x^3}{\left(x^2+1\right)^{1/2}}$$

$$= \frac{2x\left(x^2+1\right)}{\left(x^2+1\right)^{1/2}} + \frac{x^3}{\left(x^2+1\right)^{1/2}}$$

$$= \frac{2x^3+2x+x^3}{\left(x^2+1\right)^{1/2}} = \frac{3x^3+2x}{\left(x^2+1\right)^{1/2}}$$

69.

$$\sqrt{4x+3} \cdot \frac{1}{2\sqrt{x-5}} + \sqrt{x-5} \cdot \frac{1}{5\sqrt{4x+3}} = \frac{\sqrt{4x+3}}{2\sqrt{x-5}} + \frac{\sqrt{x-5}}{5\sqrt{4x+3}}$$

$$= \frac{\left(\sqrt{4x+3}\right)\left(5\sqrt{4x+3}\right)}{\left(2\sqrt{x-5}\right)\left(5\sqrt{4x+3}\right)} + \frac{\left(2\sqrt{x-5}\right)\left(\sqrt{x-5}\right)}{\left(2\sqrt{x-5}\right)\left(5\sqrt{4x+3}\right)}$$

$$= \frac{\left(\sqrt{4x+3}\right)\left(5\sqrt{4x+3}\right) + \left(2\sqrt{x-5}\right)\left(\sqrt{x-5}\right)}{\left(2\sqrt{x-5}\right)\left(5\sqrt{4x+3}\right)}$$

$$= \frac{5(4x+3) + 2(x-5)}{10\sqrt{(x-5)(4x+3)}}$$

$$= \frac{20x+15+2x-10}{10\sqrt{(x-5)(4x+3)}}$$

$$= \frac{22x+5}{10\sqrt{(x-5)(4x+3)}} = \frac{22x+5}{10\sqrt{4x^2-17x-15}}$$

71.

$$\frac{\sqrt{1+x} - x \cdot \frac{1}{2\sqrt{1+x}}}{1+x} = \frac{\frac{2\left(\sqrt{1+x}\right)\left(\sqrt{1+x}\right) - x}{2\sqrt{1+x}}}{1+x} = \frac{\frac{2(1+x) - x}{2\sqrt{1+x}}}{1+x}$$

$$= \frac{2+x}{2(1+x)\sqrt{1+x}} = \frac{2+x}{2(1+x)^{3/2}}$$

73.

$$\frac{\left(x+4\right)^{1/2} - 2x\left(x+4\right)^{-1/2}}{x+4} = \frac{\left(x+4\right)^{1/2} - \frac{2x}{\left(x+4\right)^{1/2}}}{x+4} = \frac{1}{x+4}\left[\left(x+4\right)^{1/2} - \frac{2x}{\left(x+4\right)^{1/2}}\right]$$

$$= \frac{1}{x+4}\left[\frac{\left(x+4\right)^{1/2} \cdot \left(x+4\right)^{1/2}}{\left(x+4\right)^{1/2}} - \frac{2x}{\left(x+4\right)^{1/2}}\right]$$

$$= \frac{1}{x+4}\left[\frac{x+4}{\left(x+4\right)^{1/2}} - \frac{2x}{\left(x+4\right)^{1/2}}\right]$$

$$= \frac{1}{x+4}\left[\frac{x+4-2x}{(x+4)^{1/2}}\right] = \frac{4-x}{(x+4)^{3/2}}$$

75.

$$\frac{\dfrac{x^2}{(x^2-1)^{1/2}} - (x^2-1)^{1/2}}{x^2} = \frac{1}{x^2}\left[\frac{x^2}{(x^2-1)^{1/2}} - (x^2-1)^{1/2}\right]$$

$$= \frac{1}{x^2}\left[\frac{x^2}{(x^2-1)^{1/2}} - \frac{(x^2-1)^{1/2}\cdot(x^2-1)^{1/2}}{(x^2-1)^{1/2}}\right]$$

$$= \frac{1}{x^2}\left[\frac{x^2}{(x^2-1)^{1/2}} - \frac{x^2-1}{(x^2-1)^{1/2}}\right]$$

$$= \frac{1}{x^2}\left[\frac{x^2-x^2+1}{(x^2-1)^{1/2}}\right] = \frac{1}{x^2(x^2-1)^{1/2}}$$

77.

$$\frac{\dfrac{1+x^2}{2\sqrt{x}} - 2x\sqrt{x}}{(1+x^2)^2} = \frac{1}{(1+x^2)^2}\cdot\left[\frac{1+x^2}{2\sqrt{x}} - 2x\sqrt{x}\right] = \frac{1}{(1+x^2)^2}\cdot\left[\frac{1+x^2}{2\sqrt{x}} - \frac{2x\sqrt{x}\cdot 2\sqrt{x}}{2\sqrt{x}}\right]$$

$$= \frac{1}{(1+x^2)^2}\cdot\left[\frac{1+x^2}{2\sqrt{x}} - \frac{4x\cdot x}{2\sqrt{x}}\right]$$

$$= \frac{1}{(1+x^2)^2}\cdot\left[\frac{1+x^2-4x^2}{2\sqrt{x}}\right]$$

$$= \frac{1}{(1+x^2)^2}\cdot\left[\frac{1-3x^2}{2\sqrt{x}}\right] = \frac{1-3x^2}{2\sqrt{x}(1+x^2)^2}$$

79.

$$(x+1)^{3/2} + x\cdot\frac{3}{2}(x+1)^{1/2} = (x+1)^{1/2}\left[x+1+\frac{3}{2}x\right]$$

$$= (x+1)^{1/2}\left(\frac{5}{2}x+1\right) = \frac{1}{2}(x+1)^{1/2}(5x+2)$$

81.

$$6x^{1/2}(x^2+x) - 8x^{3/2} - 8x^{1/2} = 2x^{1/2}\left[3(x^2+x)-4x-4\right]$$

$$= 2x^{1/2}(3x^2-x-4)$$

$$= 2x^{1/2}(3x-4)(x+1)$$

83.

$$3\left(x^2+4\right)^{4/3}+x\cdot 4\left(x^2+4\right)^{1/3}\cdot 2x=3\left(x^2+4\right)^{4/3}+8x^2\left(x^2+4\right)^{1/3}$$

$$=\left(x^2+4\right)^{1/3}\left[3\left(x^2+4\right)+8x^2\right]$$

$$=\left(x^2+4\right)^{1/3}\left[3x^2+12+8x^2\right]$$

$$=\left(x^2+4\right)^{1/3}\left(11x^2+12\right)$$

85.

$$4\left(3x+5\right)^{1/3}\left(2x+3\right)^{3/2}+3\left(3x+5\right)^{4/3}\left(2x+3\right)^{1/2}$$

$$=\left(3x+5\right)^{1/3}\left(2x+3\right)^{1/2}\left[4\left(2x+3\right)+3\left(3x+5\right)\right]$$

$$=\left(3x+5\right)^{1/3}\left(2x+3\right)^{1/2}\left[8x+12+9x+15\right]$$

$$=\left(3x+5\right)^{1/3}\left(2x+3\right)^{1/2}\left(17x+27\right)$$

87.

$$3x^{-1/2}+\frac{3}{2}x^{1/2}=\frac{3}{x^{1/2}}+\frac{3}{2}x^{1/2}$$

$$=\frac{3}{x^{1/2}}+\frac{3}{2}\frac{x^{1/2}\cdot x^{1/2}}{x^{1/2}}$$

$$=\frac{2\cdot 3}{2x^{1/2}}+\frac{3x}{2x^{1/2}}$$

$$=\frac{3\left(2+x\right)}{2x^{1/2}}$$

89.

$$x\left(\frac{1}{2}\right)\left(8-x^2\right)^{-1/2}\left(-2x\right)+\left(8-x^2\right)^{1/2}=-x^2\left(8-x^2\right)^{-1/2}+\left(8-x^2\right)^{1/2}$$

$$=\frac{-x^2}{\left(8-x^2\right)^{1/2}}+\left(8-x^2\right)^{1/2}$$

$$=\frac{-x^2}{\left(8-x^2\right)^{1/2}}+\frac{\left(8-x^2\right)^{1/2}\left(8-x^2\right)^{1/2}}{\left(8-x^2\right)^{1/2}}$$

$$=\frac{-x^2}{\left(8-x^2\right)^{1/2}}+\frac{8-x^2}{\left(8-x^2\right)^{1/2}}$$

$$=\frac{-x^2+8-x^2}{\left(8-x^2\right)^{1/2}}$$

$$=\frac{-2x^2+8}{\left(8-x^2\right)^{1/2}}=\frac{2\left(2-x\right)\left(2+x\right)}{\left(8-x^2\right)^{1/2}}$$

0.7 Geometry Review

1. $a = 5, \ b = 12$
    $c^2 = a^2 + b^2$
    $c^2 = 5^2 + 12^2 = 25 + 144 = 169$
    $c = \sqrt{169} = 13$

3. $a = 10, \ b = 24$
    $c^2 = a^2 + b^2$
    $c^2 = 10^2 + 24^2 = 100 + 576 = 676$
    $c = \sqrt{676} = 26$

5. $a = 7, \ b = 24$
    $c^2 = a^2 + b^2$
    $c^2 = 7^2 + 24^2 = 49 + 576 = 625$
    $c = \sqrt{625} = 25$

7. Square the sides of the triangle. $3^2 = 9$ $4^2 = 16$ $5^2 = 25$
    Since $9 + 16 = 25$, the triangle is a right triangle. The hypotenuse is 5 (the longest side).

9. Square the sides of the triangle. $4^2 = 16$ $5^2 = 25$ $6^2 = 36$
    The sum $16 + 25 = 41 \neq 36$, so the triangle is not a right triangle.

11. Square the sides of the triangle. $7^2 = 49$ $24^2 = 576$ $25^2 = 625$
    Since $49 + 576 = 625$, the triangle is a right triangle. The hypotenuse is 25 (the longest side).

13. Square the sides of the triangle. $6^2 = 36$ $4^2 = 16$ $3^2 = 9$
    The sum $16 + 9 = 25 \neq 36$, the triangle is not a right triangle.

15. The area A of a rectangle is $A = lw = 4 \cdot 2 = 8$ square inches.

17. The area A of a triangle is $A = \dfrac{1}{2} bh = \dfrac{1}{2} \cdot 2 \cdot 4 = 4$ square inches.

19. The area A of a circle is $A = \pi r^2 = \pi \cdot 5^2 = 25\pi$ square meters.
    The circumference of a circle is $C = 2\pi r = 2\pi \cdot 5 = 10\pi$ meters.

21. The volume V of a rectangular box is $V = lwh = 8 \cdot 4 \cdot 7 = 224$ cubic feet.

    The surface area SA of a rectangular box is
    $$SA = 2lw + 2lh + 2wh = 2 \cdot 8 \cdot 4 + 2 \cdot 8 \cdot 7 + 2 \cdot 4 \cdot 7$$
    $$= 64 + 112 + 56 = 232 \text{ square feet.}$$

23. The volume V of a sphere is $V = \dfrac{4}{3}\pi r^3 = \dfrac{4}{3}\pi \cdot 4^3 = \dfrac{4^4}{3}\pi = \dfrac{256}{3}\pi$ centimeters cubed.

    The surface area S of a sphere is $4\pi r^2 = 4\pi 4^2 = 4^3\pi = 64\pi$ centimeters squared.

25. The volume V of a right circular cylinder is $V = \pi r^2 h = \pi \cdot 9^2 \cdot 8 = 648\pi$ cubic inches.

The surface area S of a right circular cylinder is $S = 2\pi r^2 + 2\pi rh$.
$$S = 2\pi \cdot 9^2 + 2\pi \cdot 9 \cdot 8 = 162\pi + 144\pi = 306\pi \text{ square inches}$$

27.   The shaded region is a circle of radius $r = 1$.

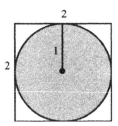

$$A = \pi r^2$$
$$A = \pi \cdot 1^2$$
$$A = \pi \text{ square units}$$

29.   The shaded region is the area of the circle. First we use the Pythagorean theorem to find its diameter.

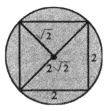

$$c^2 = a^2 + b^2$$
$$c^2 = 2^2 + 2^2 = 4 + 4 = 8$$
$$c = \sqrt{8} = 2\sqrt{2}$$

The radius is half the diameter, so $r = \sqrt{2}$, and
$$A = \pi r^2$$
$$A = \pi \cdot \left(\sqrt{2}\right)^2$$
$$A = 2\pi \text{ square units}$$

31.   In 1 revolution the wheel travels its circumference C. So in 4 revolutions it travels a distance $D = 4C$.
$$D = 4C = 4\left(\pi d\right)$$
$$= 4\left(\pi \cdot 16\right) = 64\pi \approx 201 \text{ inches}$$

33.   The area of the border is the difference between the area A of the outer square which has a side $S = 10 = 6 + 2 + 2$ feet and the area a of the inner square which has a side $s = 6$ feet.

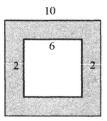

$$A - a = S^2 - s^2$$
$$= 10^2 - 6^2$$
$$= 100 - 36$$
$$= 64 \text{ square feet}$$

35.   The area of the window is the sum of the area of half the circle and the area of the rectangle.
$$\text{Area} = \frac{1}{2}\left(\pi r^2\right) + lw$$
$$\text{Area} = \frac{1}{2} \cdot \pi \cdot 2^2 + 6 \cdot 4$$
$$= 2\pi + 24 \approx 30.28 \text{ square feet.}$$

The amount of wood frame is measured by the perimeter of the window. The perimeter is the sum of the half the circumference of the circle and the 3 outer sides of the rectangle.

$$\text{Perimeter} = \frac{1}{2}(\pi d) + 2l + w$$

$$= \frac{1}{2} \cdot \pi \cdot 4 + 2 \cdot 6 + 4$$

$$= 2\pi + 16 \approx 22.28 \text{ feet}$$

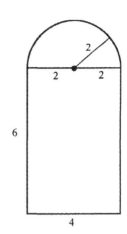

37. Since 1 mile = 5280 feet, 20 feet = $\dfrac{20}{5280}$ mile.

We use the Pythagorean theorem to find the distance d that we can see.

$$d^2 + (3960)^2 = \left(3960 + \frac{20}{5280}\right)^2$$

$$d^2 = \left(3960 + \frac{20}{5280}\right)^2 - (3960)^2 \approx 30.0000$$

$$d \approx 5.48 \text{ miles}$$

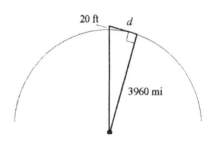

39. Since 1 mile = 5280 feet, 150 feet = $\dfrac{150}{5280}$ mile.

We use the Pythagorean theorem to find the distance d that we can see.
From the deck:

$$d^2 + (3960)^2 = \left(3960 + \frac{100}{5280}\right)^2$$

$$d^2 = \left(3960 + \frac{100}{5280}\right)^2 - (3960)^2 \approx 150.0004$$

$$d \approx 12.25 \text{ miles}$$

From the bridge:

$$d^2 + (3960)^2 = \left(3960 + \frac{150}{5280}\right)^2$$

$$d^2 = \left(3960 + \frac{150}{5280}\right)^2 - (3960)^2 \approx 225.0008$$

$$d \approx 15.00 \text{ miles}$$

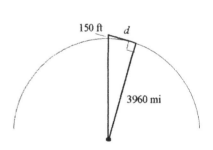

41.   Answers will vary.

0.8 Rectangular Coordinates

1. (a) $A = (-3, 2)$ 2nd quadrant

    (b) $B = (6, 0)$ x-axis

    (c) $C = (-2, -2)$ 3rd quadrant

    (d) $D = (6, 5)$ 1st quadrant

    (e) $E = (0, -3)$ y-axis

    (f) $F = (6, -3)$ 4th quadrant

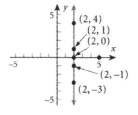

3.   The set of points of the form, $(2, y)$, where y is a real number, is a vertical line passing through 2 on the x-axis.

   The equation of the line is $x = 2$.

5.   When $P_1 = (0, 0)$ and $P_2 = (2, 1)$ the distance is

$$d(P_1, P_2) = \sqrt{(1-0)^2 + (2-0)^2} = \sqrt{1+4} = \sqrt{5} \approx 2.24$$

7.   When $P_2 = (-2, 2)$ and $P_1 = (1, 1)$ the distance is

$$d(P_1, P_2) = \sqrt{[1-(-2)]^2 + (1-2)^2} = \sqrt{3^2 + (-1)^2} = \sqrt{9+1} = \sqrt{10} \approx 3.16$$

9.   When $P_1 = (3, -4)$ and $P_2 = (5, 4)$, the distance is

$$d(P_1, P_2) = \sqrt{(5-3)^2 + [4-(-4)]^2} = \sqrt{2^2 + 8^2} = \sqrt{4+64} = \sqrt{68} \approx 8.25$$

11.   When $P_1 = (-3, 2)$ and $P_2 = (6, 0)$, the distance is

$$d(P_1, P_2) = \sqrt{[6-(-3)]^2 + (0-2)^2} = \sqrt{9^2 + (-2)^2} = \sqrt{81+4} = \sqrt{85} \approx 9.22$$

13.   When $P_1 = (4, -3)$ and $P_2 = (6, 4)$, the distance is

$$d\left(P_1,\ P_2\right) = \sqrt{\left(6-4\right)^2 + \left[4-\left(-3\right)\right]^2} = \sqrt{2^2 + 7^2} = \sqrt{4+49} = \sqrt{53} \approx 7.28$$

15.   When $P_1 = (-0.2, 0.3)$ and $P_2 = (2.3, 1.1)$, the distance is

$$d\left(P_1,\ P_2\right) = \sqrt{\left[2.3-\left(-0.2\right)\right]^2 + \left(1.1-0.3\right)^2}$$
$$= \sqrt{2.5^2 + 0.8^2} = \sqrt{6.25+0.64} = \sqrt{6.89} \approx 2.62$$

17.   When $P_1 = (a, b)$ and $P_2 = (0, 0)$, the distance is

$$d\left(P_1,\ P_2\right) = \sqrt{\left(0-a\right)^2 + \left(0-b\right)^2} = \sqrt{\left(-a\right)^2 + \left(-b\right)^2} = \sqrt{a^2 + b^2}$$

19.   We first find the length of each side of the triangle.

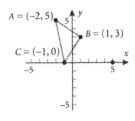

$$d(A, B) = \sqrt{\left[1-\left(-2\right)\right]^2 + \left(3-5\right)^2} = \sqrt{9+4} = \sqrt{13}$$

$$d(B, C) = \sqrt{\left[\left(-1\right)-1\right]^2 + \left(0-3\right)^2} = \sqrt{4+9} = \sqrt{13}$$

$$d(C, A) = \sqrt{\left[\left(-2\right)-\left(-1\right)\right]^2 + \left(5-0\right)^2} = \sqrt{1+25}$$
$$= \sqrt{26}$$

To verify that the triangle is right, we show

$$[d(A, B)]^2 + [d(B, C)]^2 = \left(\sqrt{13}\right)^2 + \left(\sqrt{13}\right)^2$$
$$= 13 + 13 = 26 = \left[\sqrt{26}\right]^2 = [d(C, A)]^2$$

From the converse of the Pythagorean Theorem, triangle ABC is a right triangle.

The area of a triangle Area $= \dfrac{1}{2}\,bh = \dfrac{1}{2}\cdot\sqrt{13}\ \cdot\sqrt{13} = \dfrac{13}{2} = 7.5$ square units.

21.   We first find the length of each side of the triangle.

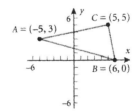

$$d(A, B) = \sqrt{\left[6-\left(-5\right)\right]^2 + \left(0-3\right)^2}$$
$$= \sqrt{11^2 + \left(-3\right)^2} = \sqrt{130}$$

$$d(B, C) = \sqrt{\left(5-6\right)^2 + \left(5-0\right)^2} = \sqrt{\left(-1\right)^2 + 5^2}$$
$$= \sqrt{1+25} = \sqrt{26}$$

$$d(C, A) = \sqrt{\left(-5-5\right)^2 + \left(3-5\right)^2}$$

$$= \sqrt{(-10)^2 + (-2)^2} = \sqrt{100 + 4} = \sqrt{104}$$

To verify that the triangle is right, we show

$$[d(B, C)]^2 + [d(C, A)]^2 = \left(\sqrt{26}\right)^2 + \left(\sqrt{104}\right)^2 = 26 + 104 = 130 = \left(\sqrt{130}\right)^2 = [d(A, B)]^2$$

From the converse of the Pythagorean Theorem, triangle ABC is a right triangle.

The area of the triangle $= \dfrac{1}{2} bh = \dfrac{1}{2}\left(\sqrt{26}\right)\left(\sqrt{104}\right) = \dfrac{1}{2}\sqrt{2704} = \dfrac{1}{2} \cdot 52 = 26$ square units.

23. We first find the length of each side of the triangle.

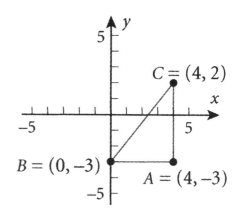

$$d(A, B) = \sqrt{(0-4)^2 + \left[-3 - (-3)\right]^2}$$
$$= \sqrt{(-4)^2 + 0} = \sqrt{16} = 4$$

$$d(B, C) = \sqrt{(4-0)^2 + \left[2 - (-3)\right]^2} = \sqrt{4^2 + 5^2}$$
$$= \sqrt{16 + 25} = \sqrt{41}$$

$$d(C, A) = \sqrt{(4-4)^2 + (-3-2)^2} = \sqrt{0 + (-5)^2}$$
$$= \sqrt{25} = 5$$

To verify that the triangle is right, we show
$$[d(A, B)]^2 + [d(C, A)]^2 = 4^2 + 5^2 = 16 + 25 = 41 = [d(B, C)]^2$$
From the converse of the Pythagorean Theorem, triangle ABC is a right triangle.

The area of the triangle $= \dfrac{1}{2} bh = \dfrac{1}{2}(4)(5) = 10$ square units.

25. We want the points $(2, y)$ for which the distance between $(2, y)$ and $(-2, -1)$ is 5
$$d^2 = \left[2 - (-2)\right]^2 + \left[y - (-1)\right]^2 = 5^2$$
$$4^2 + y^2 + 2y + 1 = 25$$
$$y^2 + 2y - 8 = 0$$
$$(y - 2)(y + 4) = 0$$
$$y - 2 = 0 \quad \text{or} \quad y + 4 = 0$$
$$y = 2 \quad \text{or} \quad y = -4$$
The points that are 5 units from $(-2, -1)$ are $(2, 2)$ and $(2, -4)$.

27. We want the points $(x, 0)$ that are 5 units away from $(4, -3)$.
$$d^2 = (x - 4)^2 + \left[0 - (-3)\right]^2 = 5^2$$
$$x^2 - 8x + 16 + 9 = 25$$
$$x^2 - 8x = 0$$

$$x\,(x-8)=0$$
$$x=0 \quad \text{or} \quad x-8=0$$
$$x=8$$

The points on the x-axis that are a distance of 5 from $(4,-3)$ are $(0, 0)$ and $(8, 0)$.

29. When the points on the y-axis that are 5 units from (4.4) are (0.1) and (0.7), $P_1 = (1, 3)$ and $P_2 = (5, 15)$, then the length of the line segment is

$$d\left(P_1,\ P_2\right) = \sqrt{(5-1)^2 + (15-3)^2} = \sqrt{(4)^2 + (12)^2} = \sqrt{16+144} = \sqrt{160} \approx 12.65$$

31. When $P_1 = (-4, 6)$ and $P_2 = (4, -8)$, then the length of the line segment is

$$d\left(P_1,\ P_2\right) = \sqrt{\left[4-(-4)\right]^2 + (-8-6)^2} = \sqrt{8^2 + (-14)^2} = \sqrt{64+196} = \sqrt{260} \approx 16.12$$

33. Since the baseball "diamond" is a square, the baselines meet at right angles, and the triangle formed by home plate, first base, and second base is a right triangle. The distance from home plate to second base is the hypotenuse of the right triangle.

$$c^2 = a^2 + b^2$$
$$c^2 = 90^2 + 90^2 = 8100 + 8100 = 16,200$$
$$c = \sqrt{16,200} \approx 127.28 \text{ feet}$$

35.

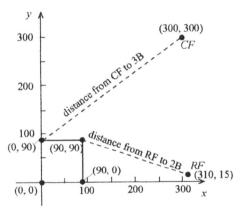

(a) The coordinates of first base are (90 feet, 0 feet), second base are (90 feet, 90 feet), third base are (0 feet, 90 feet).

(b) The distance between the right fielder and second base is the distance between the points (310, 15) and (90, 90).

$$d = \sqrt{(90-310)^2 + (90-15)^2} = \sqrt{220^2 + 75^2} = \sqrt{54,025} \approx 232.42 \text{ feet}$$

(c) The distance between the center fielder and third base is the distance between the points (300, 300) and (0, 90).

$$d = \sqrt{(300-0)^2 + (300-90)^2} = \sqrt{300^2 + 210^2} = \sqrt{134,100} \approx 366.20 \text{ feet}$$

37. After t hours the Intrepid has traveled $30t$ miles to the east, and the truck has traveled $40t$ miles south. Since east and south are 90° apart, the distance between the car and the truck is the hypotenuse of a right triangle. See the diagram.

$$d = \sqrt{(30t)^2 + (40t)^2}$$

$$= \sqrt{900t^2 + 1600t^2}$$

$$= \sqrt{2500t^2}$$

$$= 50t \text{ miles}$$

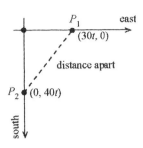

0.9 Lines

1. $y = 2x + 4$

x	0	-2	2	-2	4	-4
y	4	0	8	0	12	-4

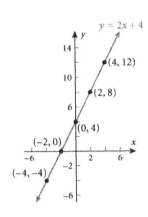

3. $2x - y = 6$

x	0	3	2	-2	4	-4
y	-6	0	-2	-10	2	-14

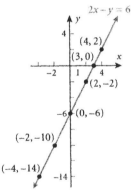

5. (a) The vertical line containing the point $(2, -3)$ is $x = 2$.

　　(b) The horizontal line containing the point $(2, -3)$ is $y = -3$.

7. (a) The vertical line containing the point $(-4, 1)$ is $x = -4$.

　　(b) The horizontal line containing the point $(-4, 1)$ is $y = 1$.

9. $m = \dfrac{y_2 - y_1}{x_2 - x_1} = \dfrac{1 - 0}{2 - 0} = \dfrac{1}{2}$

We interpret the slope to mean that for every 2 unit change in x, y changes 1 unit. That is, for every 2 units x increases, y increases by 1 unit.

11. $m = \dfrac{y_2 - y_1}{x_2 - x_1} = \dfrac{3 - 1}{-1 - 1} = -1$

We interpret the slope to mean that for every 1 unit change in x, y changes by (-1) unit. That is, for every 1 unit increase in x, y decreases by 1 unit.

13. $$m = \frac{y_2 - y_1}{x_2 - x_1} = \frac{3-0}{2-1} = \frac{3}{1} = 3$$

A slope = 3 means that for every 1 unit change in x, y will change 3 units.

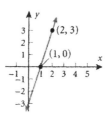

15. $$m = \frac{y_2 - y_1}{x_2 - x_1} = \frac{1-3}{2-(-2)} = \frac{-2}{4} = -\frac{1}{2}$$

A slope $= -\dfrac{1}{2}$ means that for every 2 unit change in x, y will change (-1) unit.

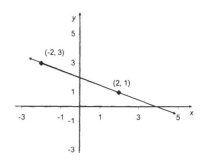

17. $$m = \frac{y_2 - y_1}{x_2 - x_1} = \frac{(-1)-(-1)}{2-(-3)} = \frac{0}{5} = 0$$

A slope of zero indicates that regardless of how x changes, y remains constant.

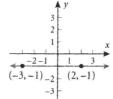

19. $$m = \frac{y_2 - y_1}{x_2 - x_1} = \frac{(-2)-2}{(-1)-(-1)} = \frac{-4}{0}$$

The slope is not defined.

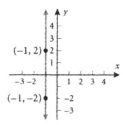

21.

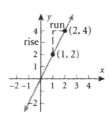

23.

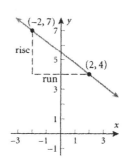

25.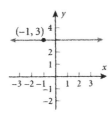

27.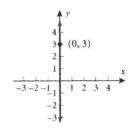

29. Use the points $(0, 0)$ and $(2, 1)$ to compute the slope of the line:

$$m = \frac{y_2 - y_1}{x_2 - x_1} = \frac{1 - 0}{2 - 0} = \frac{1}{2}$$

Since the y-intercept, $(0, 0)$, is given, use the slope-intercept form of the equation of the line:

$$y = mx + b$$
$$y = \frac{1}{2}x + 0$$
$$y = \frac{1}{2}x$$
$$2y = x$$
$$x - 2y = 0$$

31. Use the points $(1, 1)$ and $(-1, 3)$ to compute the slope of the line:

$$m = \frac{y_2 - y_1}{x_2 - x_1} = \frac{3 - 1}{(-1) - 1} = \frac{2}{-2} = -1$$

We now use the point $(1, 1)$ and the slope $m = -1$ to write the point-slope form of the equation of the line:

$$y - y_1 = m(x - x_1)$$
$$y - 1 = (-1)(x - 1)$$
$$y - 1 = -x + 1$$
$$x + y = 2$$

33. Since the slope and a point are given, use the point-slope form of the line:

$$y - y_1 = m(x - x_1)$$
$$y - 1 = 2(x - (-4))$$
$$y - 1 = 2x + 8$$
$$2x - y = -9$$

35. Since the slope and a point are given, use the point-slope form of the line:

$$y - y_1 = m(x - x_1)$$
$$y - (-1) = -\frac{2}{3}(x - 1)$$
$$3y + 3 = -2(x - 1)$$
$$3y + 3 = -2x + 2$$
$$2x + 3y = -1$$

37. Since we are given two points, $(1, 3)$ and $(-1, 2)$, first find the slope.

$$m = \frac{3 - 2}{1 - (-1)} = \frac{1}{2}$$

39. Since we are given the slope, $m = -2$, and the y-intercept, $(0, 3)$, we use the slope-intercept form of the line:

$$y = mx + b$$

Then with the slope and one of the points, (1, 3), we use the point-slope form of the line:

$$y - y_1 = m(x - x_1)$$

$$y - 3 = \frac{1}{2}(x - 1)$$

$$2y - 6 = x - 1$$

$$x - 2y = -5$$

$$y = -2x + 3$$

$$2x + y = 3$$

41. Since we are given the slope $m = 3$ and the x-intercept $(-4, 0)$, we use the point-slope form of the line:

$$y - y_1 = m(x - x_1)$$

$$y - 0 = 3(x - (-4))$$

$$y = 3x + 12$$

$$3x - y = -12$$

43. We are given the slope $m = \frac{4}{5}$ and the point $(0, 0)$, which is the y-intercept. So, we use the slope-intercept form of the line:

$$y = mx + b$$

$$y = \frac{4}{5}x + 0$$

$$5y = 4x$$

$$4x - 5y = 0$$

45. We are given two points, the x-intercept $(2, 0)$ and the y-intercept $(0, -1)$ so we need to find the slope and then to use the slope-intercept form of the line to get the equation.

$$\text{slope} = \frac{0 - (-1)}{2 - 0} = \frac{1}{2}$$

$$y = mx + b$$

$$y = \frac{1}{2}x - 1$$

$$2y = x - 2$$

$$x - 2y = 2$$

47. Since the slope is undefined, the line is vertical. The equation of the vertical line containing the point $(1, 4)$ is:

$$x = 1$$

49. Since the slope = 0, the line is horizontal. The equation of the horizontal line containing the point $(1, 4)$ is:

$$y = 4$$

51. $y = 2x + 3$,
slope: $m = 2$; y-intercept: $(0, 3)$

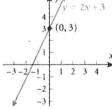

53. To obtain the slope and y-intercept, we transform the equation into its slope-intercept form. To do this we solve for y.

$$\frac{1}{2}y = x - 1$$
$$y = 2x - 2$$

slope: $m = 2$; y-intercept: $(0, -2)$

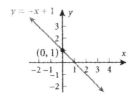

55. To obtain the slope and y-intercept, we transform the equation into its slope-intercept form. To do this we solve for y.

$$2x - 3y = 6$$
$$y = \frac{2}{3}x - 2$$

slope: $m = \frac{2}{3}$; y-intercept: $(0, -2)$

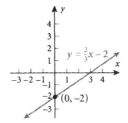

57. To obtain the slope and y-intercept, we transform the equation into its slope-intercept form. To do this we solve for y.

$$x + y = 1$$
$$y = -x + 1$$

slope: $m = -1$; y-intercept: $(0, 1)$

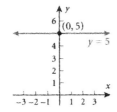

59. $x = -4$

The slope is not defined; there is no y-intercept.

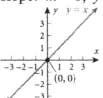

61. $y = 5$

slope: $m = 0$; y-intercept: $(0, 5)$

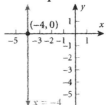

63. To obtain the slope and y-intercept, we transform the equation into its slope-intercept form. To do this we solve for y.

$$y - x = 0$$
$$y = x$$

slope: $m = 1$; y-intercept $= (0, 0)$

65. To obtain the slope and y-intercept, we transform the equation into its slope-intercept form. To do this we solve for y.

$$2y - 3x = 0$$
$$2y = 3x$$
$$y = \frac{3}{2}x$$

slope: $m = \frac{3}{2}$; y-intercept $= (0, 0)$

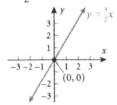

67. The equation of a horizontal line containing the point $(-1, -3)$ is given by $y = -3$.

69. The average cost of operating the car is given as $0.122 per mile. This is the slope of the equation. So the equation will be $C = 0.122x$, where x is the number of miles the car is driven.

71. The fixed cost of electricity for the month is $7.58. In addition, the electricity costs $0.08275 (8.275 cents) for every kilowatt-hour (KWH) used. If x represents the number of KWH of electricity used in a month,
(a) the total monthly is represented by the equation:
$$C = 0.08275x + 7.58, \quad 0 \le x \le 400$$

(b)

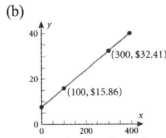

(c) The charge for using 100 KWH of electricity is found by substituting 100 in part (a):
$$C = 0.08275(100) + 7.58$$
$$= 8.275 + 7.58$$
$$= \$15.86$$

(d) The charge for using 300 KWH of electricity is found by substituting 300 in part (a):
$$C = 0.08275(300) + 7.58$$
$$= 24.825 + 7.58$$
$$= \$32.41$$

(e) The slope of the line, $m = 0.08275$, indicates that for every extra KWH used (up to 400 KWH), the electric bill increases by 8.275 cents.

73. Two points are given, $(h_1, w_1) = (67, 139)$ and $(h_2, w_2) = (70, 151)$, and we are told they are linearly related. So we will first compute the slope of the line:

$$m = \frac{w_2 - w_1}{h_2 - h_1} = \frac{151 - 139}{70 - 67} = \frac{12}{3} = 4$$

We use the point $(70, 151)$ and the fact that the slope $m = 4$ to get the point-slope form of the equation of the line:

$$w - 151 = 4(h - 70)$$
$$w - 151 = 4h - 280$$
$$w = 4h - 129$$

75. The delivery cost of the Sunday *Chicago Tribune* is $1,070,000 plus $0.53 for each of the x copies delivered. The total cost of delivering the papers is :

$$C = 0.53x + 1,070,000$$

77. Since we are told the relationship is linear, we will use the two points to get the slope of the line:

$$m = \frac{C_2 - C_1}{F_2 - F_1} = \frac{100 - 0}{212 - 32} = \frac{100}{180} = \frac{5}{9}$$

We use the point $(32, 0)$ and the fact that the slope $m = \frac{5}{9}$ to get the point-slope form of the equation:

$$C - C_1 = m(F - F_1)$$
$$C - 0 = \frac{5}{9}(F - 32)$$
$$C = \frac{5}{9}(F - 32)$$

To find the Celsius measure of 68 °F we substitute 68 for F in the equation and simplify:

$$C = \frac{5}{9}(68 - 32) = 20\,°$$

79. Since the problem states that the *rate* of the loss of water remains constant, we can assume that the relationship is linear.

(a) We are given two points, the amount of water (in billions of gallons) on November 8, 2002 ($t = 8$) and the amount of water on December 8, 2002 ($t = 38$). We use these two points and the fact that the relation is linear to find the slope of the line:

$$m = \frac{y_2 - y_1}{t_2 - t_1} = \frac{52.5 - 52.9}{38 - 8} = \frac{-0.4}{30} = -\frac{1}{75}$$

We use the point $(8, 52.9)$ and the slope $m = -\frac{1}{75}$, to get the point-slope form of the equation of the line:

$$y - y_1 = m(t - t_1)$$

$$y - 52.9 = -\frac{1}{75}(t - 8)$$

$$y = -\frac{1}{75}t + 53.007$$

(b) To find the amount of water in the reservoir on November 20, let $t = 20$ in the equation from part (a).

$$y = -\frac{1}{75}(20) + 53.007$$

$$y = 52.74 \text{ billion gallons of water}$$

(c) The slope tells us that the reservoir loses one billion gallons of water every 75 days.

(d) To find the amount of water in the reservoir on December 31, 2002, let $t = 61$ in the equation from (a).

$$y = -\frac{1}{75}(61) + 53.007$$

$$y = 52.194 \text{ billion gallons of water}$$

(e) To determine when the reservoir will be empty, assume $y = 0$ gallons and solve for t.

$$y = -\frac{1}{75}(t) + 53.007$$

$$0 = -\frac{1}{75}(t) + 53.007$$

$$t = (53.007)(75)$$

$$= 3975.525 \text{ days, or about 10 years, 10 months and 22 days.}$$

(f) Answers will vary.

81. To graph an equation on a graphing utility, first solve the equation for y.

Window: Xmin $= -10$; Xmax $= 10$
Ymin $= -10$; Ymax $= 10$

$$1.2x + 0.8y = 2$$

$$0.8y = -1.2x + 2$$

$$y = -1.5x + 2.5$$

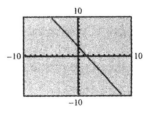

The x-intercept is $(1.67, 0)$.
The y-intercept is $(0, 2.50)$.

83. To graph an equation on a graphing utility, first solve the equation for y.

Window: Xmin $= -10$; Xmax $= 10$
Ymin $= -10$; Ymax $= 10$

$$21x - 15y = 53$$
$$15y = 21x - 53$$
$$y = \frac{21}{15}x - \frac{53}{15}$$
$$y = \frac{7}{5}x - \frac{53}{15}$$

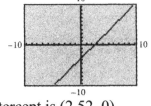

The x-intercept is $(2.52, 0)$.
The y-intercept is $(0, -3.53)$.

85. To graph an equation on a graphing utility, first solve the equation for y.

$$\frac{4}{17}x + \frac{6}{23}y = \frac{2}{3}$$
$$\frac{6}{23}y = -\frac{4}{17}x + \frac{2}{3}$$
$$y = \frac{23}{6}\left(-\frac{4}{17}x + \frac{2}{3}\right)$$
$$y = -\frac{46}{51}x + \frac{23}{9}$$

Window: Xmin $= -10$; Xmax $= 10$
Ymin $= -10$; Ymax $= 10$

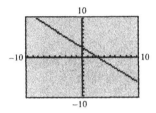

The x-intercept is $(2.83, 0)$.
The y-intercept is $(0, 2.56)$.

87. To graph an equation on a graphing utility, first solve the equation for y.

$$\pi x - \sqrt{3}y = \sqrt{6}$$
$$\sqrt{3}y = \pi x - \sqrt{6}$$
$$y = \frac{\pi}{\sqrt{3}}x - \sqrt{2}$$

Window: Xmin $= -10$; Xmax $= 10$
Ymin $= -10$; Ymax $= 10$

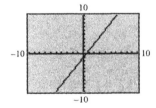

The x-intercept is $(0.78, 0)$.
The y-intercept is $(0, -1.41)$.

89. The graph passes through the points $(0, 0)$ and $(4, 8)$. We use the points to find the slope of the line:

$$m = \frac{y_2 - y_1}{x_2 - x_1} = \frac{8 - 0}{4 - 0} = \frac{8}{4} = 2$$

The y-intercept $(0, 0)$ is given, so we use the y-intercept and the slope $m = 2$, to get the slope-intercept form of the line:

$$y = mx + b$$
$$y = 2x + 0$$
$$y = 2x \quad \text{which is answer (b)}$$

91. The graph passes through the points $(0, 0)$ and $(2, 8)$. We use the points to find the slope of the line:

$$m = \frac{y_2 - y_1}{x_2 - x_1} = \frac{8 - 0}{2 - 0} = \frac{8}{2} = 4$$

The y-intercept $(0, 0)$ is given, so we use the y-intercept and the slope $m = 4$, to get the slope-intercept form of the line:

$$y = mx + b$$
$$y = 4x + 0$$
$$y = 4x \quad \text{which is answer (d)}$$

93. From the graph we can read the two intercepts, $(-2, 0)$ and $(0, 2)$. Use these points to compute the slope of the line:

$$m = \frac{y_2 - y_1}{x_2 - x_1} = \frac{2 - 0}{0 - (-2)} = \frac{2}{2} = 1$$

We use the y-intercept $(0, 2)$ and the slope $m = 1$, to find the slope-intercept form of the equation:

$$y = mx + b$$
$$y = 1x + 2$$
$$y = x + 2$$

The general form of the equation is:

$$x - y = -2$$

95. From the graph we can read the two intercepts, $(3, 0)$ and $(0, 1)$. Use these points to compute the slope of the line:

$$m = \frac{y_2 - y_1}{x_2 - x_1} = \frac{1 - 0}{0 - 3} = \frac{1}{-3} = -\frac{1}{3}$$

We use the y-intercept $(0, 1)$ and the slope $m = -\dfrac{1}{3}$, to find the slope-intercept form of the equation:

$$y = mx + b$$
$$y = -\frac{1}{3}x + 1$$

The general form of the equation is:

$$x + 3y = 3$$

97. From the graph we can see that the line has a positive slope and a y-intercept of the form $(0, b)$ where b is a positive number. Put each of the equations into slope-intercept form and choose those with both positive slope and positive y-intercept.

(b) $y = \dfrac{2}{3}x + 2$ (c) $y = \dfrac{3}{4}x + 3$ (e) $y = x + 1$ (g) $y = 2x + 3$

99. The x-axis is a horizontal line; its slope is zero. The general equation of the x-axis is $y = 0$.

101. Answers vary.

103. Not every line has two distinct intercepts. Every line has at least one intercept. Explanations will vary.

105. If two lines have the same x-intercept and the same y-intercept, and x-intercept is not $(0, 0)$, then the two lines have equal slopes. Lines that have equal slopes and equal y-intercepts have equivalent equations and identical graphs.

107. Two lines can have the same y-intercept but different slopes only if their y-intercept is the point $(0, 0)$.

Chapter 1
Functions and Their Graphs

1.1 Graphs of Equations

1. The point: $(3, 4)$
   (a) The point symmetric with respect to the x-axis: $(3, -4)$.

   (b) The point symmetric with respect to the y-axis: $(-3, 4)$.

   (c) The point symmetric with respect to the origin: $(-3, -4)$.

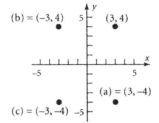

3. The point: $(-2, 1)$
   (a) The point symmetric with respect to the x-axis: $(-2, -1)$.

   (b) The point symmetric with respect to the y-axis: $(2, 1)$.

   (c) The point symmetric with respect to the origin: $(2, -1)$.

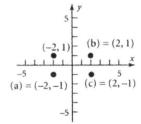

5. The point: $(1, 1)$
   (a) The point symmetric with respect to the x-axis: $(1, -1)$.

   (b) The point symmetric with respect to the y-axis: $(-1, 1)$.

   (c) The point symmetric with respect to the origin: $(-1, -1)$.

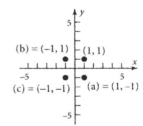

7. The point: $(-3, -4)$
   (a) The point symmetric with respect to the x-axis: $(-3, 4)$.

   (b) The point symmetric with respect to the y-axis: $(3, -4)$.

   (c) The point symmetric with respect to the origin: $(3, 4)$.

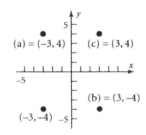

9. The point: $(0, -3)$
    (a) The point symmetric with respect to the x-axis: $(0, 3)$.

    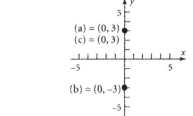

    (b) The point symmetric with respect to the y-axis: $(0, -3)$.

    (c) The point symmetric with respect to the origin: $(0, 3)$.

11. (a) The x-intercepts are $(1, 0)$ and $(-1, 0)$. There is no y-intercept.

    (b) The graph is symmetric with respect to the x-axis, the y-axis, and the origin.

13.
    (a) The x-intercepts are $\left(-\dfrac{\pi}{2}, 0\right)$ and $\left(\dfrac{\pi}{2}, 0\right)$. The y-intercept is $(0, 1)$.

    (b) The graph is symmetric only with respect to the y-axis.

15. (a) The x-intercept is $(0, 0)$. The y-intercept is also $(0, 0)$.

    (b) The graph is symmetric only with respect to the x-axis.

17. (a) The x-intercept is $(1, 0)$; there is no y-intercept.

    (b) The graph has no symmetry with respect to the x-axis, the y-axis, or the origin.

19. (a) The x-intercepts are $(1, 0)$ and $(-1, 0)$. The y-intercept is $(0, -1)$.

    (b) The graph is symmetric only with respect to the y-axis.

21. (a) There is no x-intercept, and there is no y-intercept.

    (b) The graph is symmetric only with respect to the origin.

23. For each point we check to see if the point satisfies the equation $y = x^4 - \sqrt{x}$.
    (a) $(0, 0)$: $0^4 - \sqrt{0} = 0$
    The equation is satisfied so the point $(0, 0)$ is on the graph.

    (b) $(1, 1)$: $1^4 - \sqrt{1} = 0 \neq 1$
    The equation is not satisfied so the point $(1, 1)$ is not on the graph.

    (c) $(-1, 0)$: $y = (-1)^4 - \sqrt{(-1)}$ is not a real number, so the point $(-1, 0)$ is not on the graph.

25. For each point we check to see if the point satisfies the equation $y^2 = x^2 + 9$ or $y^2 - x^2 = 9$.

(a) $(0, 3)$: $3^2 - 0^2 = 9$
The equation is satisfied so the point $(0, 3)$ is on the graph.

(b) $(3, 0)$: $0^2 - 3^2 = -9 \neq 9$
The equation is not satisfied so the point $(0, 3)$ is not on the graph.

(c) $(-3, 0)$: $0^2 - (-3)^2 = -9 \neq 9$
The equation is not satisfied so the point $(0, -3)$ is not on the graph.

27. For each point we check to see if the point satisfies the equation $x^2 + y^2 = 4$.

    (a) $(0, 2)$: $0^2 + 2^2 = 4$
    The equation is satisfied so the point $(0, 2)$ is on the graph.

    (b) $(-2, 2)$: $(-2)^2 + 2^2 = 8 \neq 4$
    The equation is not satisfied so the point $(-2, 2)$ is not on the graph.

    (c) $\left(\sqrt{2}, \sqrt{2}\right)$: $\left(\sqrt{2}\right)^2 + \left(\sqrt{2}\right)^2 = 2 + 2 = 4$

    The equation is satisfied so the point $\left(\sqrt{2}, \sqrt{2}\right)$ is on the graph.

29. To find the x-intercept(s) we let $y = 0$ and solve the equation $x^2 = 0$ or $x = 0$. So the x-intercept is $(0, 0)$.

    To find the y-intercept(s) we let $x = 0$ and get $0 = y$. So the y-intercept is also $(0, 0)$.

    To test the graph of the equation $x^2 = y$ for symmetry with respect to the x-axis, we replace y by $-y$ in the equation $x^2 = y$. Since the resulting equation $x^2 = -y$ is not equivalent to the original equation, the graph is not symmetric with respect to the x-axis.

    To test the graph of the equation $x^2 = y$ for symmetry with respect to the y-axis, we replace x by $-x$ in the equation and simplify.
    $$x^2 = y$$
    $$(-x)^2 = y \text{ or } x^2 = y$$
    Since the resulting equation is equivalent to the original equation, the graph is symmetric with respect to the y-axis.

    To test the graph of the equation $x^2 = y$ for symmetry with respect to the origin, we replace x by $-x$ and y by $-y$ in the equation and simplify.
    $$x^2 = y$$
    $$(-x)^2 = -y \text{ or } x^2 = -y$$
    Since the resulting equation is not equivalent to the original equation, the graph is not symmetric with respect to the origin.

31. To find the x-intercept(s) we let $y = 0$ and solve the equation $0 = 3x$ or $x = 0$. So the x-intercept is $(0, 0)$.

    To find the y-intercept(s) we let $x = 0$ and get $y = 3 \cdot 0 = 0$. So the y-intercept is also $(0, 0)$.

To test the graph of the equation $y = 3x$ for symmetry with respect to the x-axis, we replace y by $-y$ in the equation. Since the resulting equation $-y = 3x$ is not equivalent to the original equation, the graph is not symmetric with respect to the x-axis.

To test the graph of the equation $y = 3x$ for symmetry with respect to the y-axis, we replace x by $-x$ in the equation and simplify.

$$y = 3x$$
$$y = 3 \cdot (-x) = -3x$$

Since the resulting equation is not equivalent to the original equation, the graph is not symmetric with respect to the y-axis.

To test the graph of the equation $y = 3x$ for symmetry with respect to the origin, we replace x by $-x$ and y by $-y$ in the equation and simplify.

$$y = 3x$$
$$-y = 3 \cdot (-x)$$
$$-y = -3x$$
$$y = 3x$$

Since the resulting equation is equivalent to the original equation, the graph is symmetric with respect to the origin.

33. To find the x-intercept(s) we let $y = 0$ and solve the equation

$$x^2 + 0 - 9 = 0$$
$$x^2 = 9$$
$$x = -3 \text{ or } x = 3$$

So the x-intercepts are $(-3, 0)$ and $(3, 0)$.

To find the y-intercept(s) we let $x = 0$ and solve the equation

$$0^2 + y - 9 = 0$$
$$y = 9$$

So the y-intercept is $(0, 9)$.

To test the graph of the equation $x^2 + y - 9 = 0$ for symmetry with respect to the x-axis, we replace y by $-y$ in the equation.

$$x^2 + (-y) - 9 = 0$$
$$x^2 - y - 9 = 0$$

Since the resulting equation is not equivalent to the original equation, the graph is not symmetric with respect to the x-axis.

To test the graph of the equation $x^2 + y - 9 = 0$ for symmetry with respect to the y-axis, we replace x by $-x$ in the equation and simplify.

$$x^2 + y - 9 = 0$$
$$(-x)^2 + y - 9 = 0$$
$$x^2 + y - 9 = 0$$

Since the resulting equation is equivalent to the original equation, the graph is symmetric with respect to the y-axis.

To test the graph of the equation $x^2 + y - 9 = 0$ for symmetry with respect to the origin, we replace x by $-x$ and y by $-y$ in the equation and simplify.

$$x^2 + y - 9 = 0$$
$$(-x)^2 + (-y) - 9 = 0$$
$$x^2 - y - 9 = 0$$

Since the resulting equation is not equivalent to the original equation, the graph is not symmetric with respect to the origin.

35. To find the x-intercept(s) we let $y = 0$ and solve the equation
$$9x^2 + 4 \cdot 0^2 = 36$$
$$9x^2 = 36$$
$$x^2 = 4$$
$$x = -2 \text{ or } x = 2$$
So the x-intercepts are $(-2, 0)$ and $(2, 0)$.

To find the y-intercept(s) we let $x = 0$ and solve the equation
$$9 \cdot 0^2 + 4y^2 = 36$$
$$4y^2 = 36$$
$$y^2 = 9$$
$$y = -3 \text{ or } y = 3$$
So the y-intercepts are $(0, -3)$ and $(0, 3)$.

To test the graph of the equation $9x^2 + 4y^2 = 36$ for symmetry with respect to the x-axis, we replace y by $-y$ in the equation.
$$9x^2 + 4y^2 = 36$$
$$9x^2 + 4(-y)^2 = 36$$
$$9x^2 + 4y^2 = 36$$
Since the resulting equation is equivalent to the original equation, the graph is symmetric with respect to the x-axis.

To test the graph of the equation $9x^2 + 4y^2 = 36$ for symmetry with respect to the y-axis, we replace x by $-x$ in the equation and simplify.
$$9x^2 + 4y^2 = 36$$
$$9(-x)^2 + 4y^2 = 36$$
$$9x^2 + 4y^2 = 36$$
Since the resulting equation is equivalent to the original equation, the graph is symmetric with respect to the y-axis.

To test the graph of the equation $9x^2 + 4y^2 = 36$ for symmetry with respect to the origin, we replace x by $-x$ and y by $-y$ in the equation and simplify.
$$9x^2 + 4y^2 = 36$$
$$9(-x)^2 + 4(-y)^2 = 36$$
$$9x^2 + 4y^2 = 36$$
Since the resulting equation is equivalent to the original equation, the graph is symmetric with respect to the origin.

37. To find the x-intercept(s) we let $y = 0$ and solve the equation
$$0 = x^3 - 27$$
$$x^3 = 27$$
$$x = 3$$
So the x-intercept is $(3, 0)$.

To find the y-intercept(s) we let $x = 0$ and solve the equation
$$y = 0^3 - 27$$
$$y = -27$$
So the y-intercept is $(0, -27)$.

To test the graph of the equation $y = x^3 - 27$ for symmetry with respect to the x-axis, we replace y by $-y$ in the equation.
$$y = x^3 - 27$$
$$-y = x^3 - 27$$
Since the resulting equation is not equivalent to the original equation, the graph is not symmetric with respect to the x-axis.

To test the graph of the equation $y = x^3 - 27$ for symmetry with respect to the y-axis, we replace x by $-x$ in the equation and simplify.
$$y = x^3 - 27$$
$$y = (-x)^3 - 27$$
$$y = -x^3 - 27$$
Since the resulting equation is not equivalent to the original equation, the graph is not symmetric with respect to the y-axis.

To test the graph of the equation $y = x^3 - 27$ for symmetry with respect to the origin, we replace x by $-x$ and y by $-y$ in the equation and simplify.
$$y = x^3 - 27$$
$$(-y) = (-x)^3 - 27$$
$$-y = -x^3 - 27$$
$$y = x^3 + 27$$
Since the resulting equation is not equivalent to the original equation, the graph is not symmetric with respect to the origin.

39. To find the x-intercept(s) we let $y = 0$ and solve the equation
$$0 = x^2 - 3x - 4$$
$$0 = (x - 4)(x + 1)$$
$$x = 4 \text{ or } x = -1$$
So the x-intercepts are $(4, 0)$ and $(-1, 0)$.

To find the y-intercept(s) we let $x = 0$ and solve the equation
$$y = 0^2 - 3 \cdot 0 - 4$$
$$y = -4$$
So the y-intercept is $(0, -4)$.

To test the graph of the equation $y = x^2 - 3x - 4$ for symmetry with respect to the x-axis, we replace y by $-y$ in the equation.
$$y = x^2 - 3x - 4$$
$$-y = x^2 - 3x - 4$$
Since the resulting equation is not equivalent to the original equation, the graph is not symmetric with respect to the x-axis.

To test the graph of the equation $y = x^2 - 3x - 4$ for symmetry with respect to the y-axis, we replace x by $-x$ in the equation and simplify.

$$y = x^2 - 3x - 4$$
$$y = (-x)^2 - 3 \cdot (-x) - 4$$
$$y = x^2 + 3x - 4$$

Since the resulting equation is not equivalent to the original equation, the graph is not symmetric with respect to the y-axis.

To test the graph of the equation $y = x^2 - 3x - 4$ for symmetry with respect to the origin, we replace x by $-x$ and y by $-y$ in the equation and simplify.

$$y = x^2 - 3x - 4$$
$$(-y) = (-x)^2 - 3 \cdot (-x) - 4$$
$$-y = x^2 + 3x - 4$$

Since the resulting equation is not equivalent to the original equation, the graph is not symmetric with respect to the origin.

41. To find the x-intercept(s) we let $y = 0$ and solve the equation

$$0 = \frac{3x}{x^2 + 9}$$

$$0 = 3x \qquad\qquad \text{Since } x^2 + 9 \neq 0, \text{ multiply both sides by } x^2 + 9.$$

$$x = 0$$

So the x-intercept is $(0, 0)$.

To find the y-intercept(s) we let $x = 0$ and solve the equation

$$y = \frac{3 \cdot 0}{0^2 + 9} = 0$$

So the y-intercept is also $(0, 0)$.

To test the graph of the equation $y = \dfrac{3x}{x^2 + 9}$ for symmetry with respect to the x-axis, we replace y by $-y$ in the equation.

$$y = \frac{3x}{x^2 + 9}$$

$$-y = \frac{3x}{x^2 + 9}$$

Since the resulting equation is not equivalent to the original equation, the graph is not symmetric with respect to the x-axis.

To test the graph of the equation $y = \dfrac{3x}{x^2 + 9}$ for symmetry with respect to the y-axis, we replace x by $-x$ in the equation and simplify.

$$y = \frac{3x}{x^2 + 9}$$

$$y = \frac{3(-x)}{(-x)^2 + 9} = -\frac{3x}{x^2 + 9}$$

Since the resulting equation is not equivalent to the original equation, the graph is not symmetric with respect to the y-axis.

To test the graph of the equation $y = \dfrac{3x}{x^2 + 9}$ for symmetry with respect to the origin, we replace x by $-x$ and y by $-y$ in the equation and simplify.

$$y = \frac{3x}{x^2 + 9}$$

$$-y = \frac{3(-x)}{(-x)^2 + 9}$$

$$-y = -\frac{3x}{x^2 + 9}$$

$$y = \frac{3x}{x^2 + 9}$$

Since the resulting equation is equivalent to the original equation, the graph is symmetric with respect to the origin.

43. To find the x-intercept(s) we let $y = 0$ and solve the equation

$$0 = |x|$$
$$0 = x$$

So the x-intercept is $(0, 0)$.

To find the y-intercept(s) we let $x = 0$ and solve the equation

$$y = |0| = 0$$

So the y-intercept is also $(0, 0)$.

To test the graph of the equation $y = |x|$ for symmetry with respect to the x-axis, we replace y by $-y$ in the equation.

$$y = |x|$$
$$-y = |x|$$

Since the resulting equation is not equivalent to the original equation, the graph is not symmetric with respect to the x-axis.

To test the graph of the equation $y = |x|$ for symmetry with respect to the y-axis, we replace x by $-x$ in the equation and simplify.

$$y = |x|$$
$$y = |-x| = |x|$$

Since the resulting equation is equivalent to the original equation, the graph is symmetric with respect to the y-axis.

To test the graph of the equation $y = |x|$ for symmetry with respect to the origin, we replace x by $-x$ and y by $-y$ in the equation and simplify.

$$y = |x|$$
$$-y = |-x| = |x|$$

Since the resulting equation is not equivalent to the original equation, the graph is not symmetric with respect to the origin.

45.

x	$x^3 - 1 = y$	(x, y)
-2	$(-2)^3 - 1 = -9$	$(-2, -9)$
-1	$(-1)^3 - 1 = -2$	$(-1, -2)$
0	$0^3 - 1 = -1$	$(0, -1)$
1	$1^3 - 1 = 0$	$(1, 0)$
2	$2^3 - 1 = 7$	$(2, 7)$

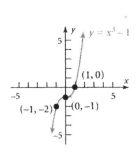

47.

x	$2\sqrt{x} = y$	(x, y)
0	$2\sqrt{0} = 0$	$(0, 0)$
1	$2\sqrt{1} = 2$	$(1, 2)$
4	$2\sqrt{4} = 4$	$(4, 4)$
9	$2\sqrt{9} = 6$	$(9, 6)$

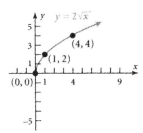

49. If $(a, 2)$ is a point on the graph, then $x = a$ and $y = 2$ must satisfy the equation

$$y = 3x + 5$$
$$2 = 3a + 5$$
$$3a = -3$$
$$a = -1$$

51. If (a, b) is a point on the graph, then $x = a$ and $y = b$ must satisfy the equation

$$2x + 3y = 6$$
$$2a + 3b = 6$$
$$2a = 6 - 3b$$
$$a = \frac{6 - 3b}{2} = 3 - \frac{3}{2}b$$

53. (a)

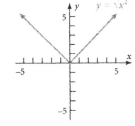

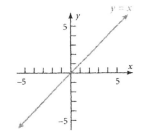

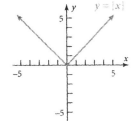

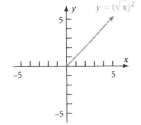

The graphs of $y = \sqrt{x^2}$ and $y = |x|$ are the same.

(b) - (d) Answers will vary.

55. Let (x, y) be a point on the graph of the equation.
(1) Assume the graph is symmetric with respect to both axes. Then because of symmetry with respect to the y-axis, the point $(-x, y)$ is on the graph. Similarly, because of symmetry with respect to the x-axis, the point $(-x, -y)$ is also on the graph. So the graph is symmetric with respect to the origin.

(2) Assume the graph is symmetric with respect to the x-axis and the origin. Then because of symmetry with respect to the x-axis, the point $(x, -y)$ is on the graph. Similarly, because of symmetry with respect to the origin, the point $(-x, y)$ is also on the graph. So the graph of the equation is symmetric with respect to the y-axis.

(3) Assume the graph is symmetric with respect to the y-axis and the origin. Then because of symmetry with respect to the y-axis, the point $(-x, y)$ is on the graph. Similarly, because of symmetry with respect to the origin, the point $(x, -y)$ is also on the graph. So the graph of the equation is symmetric with respect to the x-axis.

1.2 Functions

1. (a) We substitute 0 for x in the equation for f to get
$$f(0) = 3 \cdot 0^2 + 2 \cdot 0 - 4 = -4$$

(b) We substitute 1 for x in the equation for f to get
$$f(1) = 3 \cdot 1^2 + 2 \cdot 1 - 4 = 1$$

(c) We substitute (-1) for x in the equation for f to get
$$f(-1) = 3 \cdot (-1)^2 + 2 \cdot (-1) - 4 = -3$$

(d) We substitute $(-x)$ for x in the equation for f to get
$$f(-x) = 3 \cdot (-x)^2 + 2 \cdot (-x) - 4 = 3x^2 - 2x - 4$$

(e) $-f(x) = -(3x^2 + 2x - 4) = -3x^2 - 2x + 4$

(f) We substitute $(x + 1)$ for x in the equation for f to get
$$f(x+1) = 3(x+1)^2 + 2(x+1) - 4 = (3x^2 + 6x + 3) + (2x + 2) - 4 = 3x^2 + 8x + 1$$

(g) We substitute $2x$ for x in the equation for f to get
$$f(2x) = 3 \cdot (2x)^2 + 2 \cdot 2x - 4 = 12x^2 + 4x - 4$$

(h) We substitute $(x + h)$ for x in the equation for f to get
$$f(x+h) = 3(x+h)^2 + 2(x+h) - 4 = (3x^2 + 6xh + 3h^2) + (2x + 2h) - 4$$

$$= 3x^2 + 2x + 2h + 6xh + 3h^2 - 4$$

3. (a) We substitute 0 for x in the equation for f to get
$$f(0) = \frac{0}{0^2 + 1} = 0$$

(b) We substitute 1 for x in the equation for f to get
$$f(1) = \frac{1}{1^2 + 1} = \frac{1}{2}$$

(c) We substitute -1 for x in the equation for f to get
$$f(-1) = \frac{-1}{(-1)^2 + 1} = -\frac{1}{2}$$

(d) We substitute $-x$ for x in the equation for f to get
$$f(-x) = \frac{-x}{(-x)^2 + 1} = -\frac{x}{x^2 + 1}$$

(e) $-f(x) = -\dfrac{x}{x^2 + 1}$

(f) We substitute $(x + 1)$ for x in the equation for f to get
$$f(x+1) = \frac{x+1}{(x+1)^2 + 1} = \frac{x+1}{x^2 + 2x + 2}$$

(g) We substitute $2x$ for x in the equation for f to get
$$f(2x) = \frac{2x}{(2x)^2 + 1} = \frac{2x}{4x^2 + 1}$$

(h) We substitute $(x + h)$ for x in the equation for f to get
$$f(x+h) = \frac{x+h}{(x+h)^2 + 1} = \frac{x+h}{x^2 + 2xh + h^2 + 1}$$

5. (a) We substitute 0 for x in the equation for f to get
$$f(0) = |0| + 4 = 4$$

(b) We substitute 1 for x in the equation for f to get
$$f(1) = |1| + 4 = 5$$

(c) We substitute -1 for x in the equation for f to get
$$f(-1) = |-1| + 4 = 5$$

(d) We substitute $-x$ for x in the equation for f to get
$$f(-x) = |-x| + 4 = |x| + 4 = f(x)$$

(e) $-f(x) = -\big[|x|+4\big] = -|x|-4$

(f) We substitute $(x + 1)$ for x in the equation for f to get
$$f(x + 1) = |x+1|+4$$

(g) We substitute $2x$ for x in the equation for f to get
$$f(2x) = |2x|+4 = 2|x|+4$$

(h) We substitute $(x + h)$ for x in the equation for f to get
$$f(x+h) = |x+h|+4$$

7. (a) We substitute 0 for x in the equation for f to get
$$f(0) = \frac{2\cdot 0+1}{3\cdot 0-5} = \frac{1}{-5} = -\frac{1}{5}$$

(b) We substitute 1 for x in the equation for f to get
$$f(1) = \frac{2\cdot 1+1}{3\cdot 1-5} = \frac{3}{-2} = -\frac{3}{2}$$

(c) We substitute -1 for x in the equation for f to get
$$f(-1) = \frac{2\cdot(-1)+1}{3\cdot(-1)-5} = \frac{-1}{-8} = \frac{1}{8}$$

(d) We substitute $-x$ for x in the equation for f to get
$$f(-x) = \frac{2(-x)+1}{3(-x)-5} = \frac{-2x+1}{-3x-5} = \frac{2x-1}{3x+5}$$

(e) $-f(x) = -\dfrac{2x+1}{3x-5}$

(f) We substitute $(x + 1)$ for x in the equation for f to get
$$f(x + 1) = \frac{2(x+1)+1}{3(x+1)-5} = \frac{2x+3}{3x-2}$$

(g) We substitute $2x$ for x in the equation for f to get
$$f(2x) = \frac{2(2x)+1}{3(2x)-5} = \frac{4x+1}{6x-5}$$

(h) We substitute $(x + h)$ for x in the equation for f to get
$$f(x+h) = \frac{2(x+h)+1}{3(x+h)-5} = \frac{2x+2h+1}{3x+3h-5}$$

9. If $f(x) = 4x + 3$ then $f(x + h) = 4(x + h) + 3 = 4x + 4h + 3$, and the difference quotient is
$$\frac{f(x+h) - f(x)}{h} = \frac{4x + 4h + 3 - (4x + 3)}{h} = \frac{4h}{h} = 4$$

11. If $f(x) = x^2 - x + 4$ then $f(x + h) = (x + h)^2 - (x + h) + 4 = x^2 + 2xh + h^2 - x - h + 4$, and the difference quotient is
$$\frac{f(x+h) - f(x)}{h} = \frac{x^2 + 2xh + h^2 - x - h + 4 - (x^2 - x + 4)}{h} = \frac{2xh + h^2 - h}{h}$$
$$= \frac{h(2x + h - 1)}{h} = 2x + h - 1$$

13. If $f(x) = x^3$ then $f(x + h) = (x + h)^3 = x^3 + 3x^2h + 3xh^2 + h^3$, and the difference quotient is
$$\frac{f(x+h) - f(x)}{h} = \frac{x^3 + 3x^2h + 3xh^2 + h^3 - x^3}{h} = \frac{3x^2h + 3xh^2 + h^3}{h}$$
$$= \frac{h(3x^2 + 3xh + h^2)}{h} = 3x^2 + 3xh + h^2$$

15. If $f(x) = x^4$ then $f(x + h) = (x + h)^4 = x^4 + 4x^3h + 6x^2h^2 + 4xh^3 + h^4$, and the difference quotient is
$$\frac{f(x+h) - f(x)}{h} = \frac{(x^4 + 4x^3h + 6x^2h^2 + 4xh^3 + h^4) - x^4}{h}$$
$$= \frac{4x^3h + 6x^2h^2 + 4xh^3 + h^4}{h} = \frac{h(4x^3 + 6x^2h + 4xh^2 + h^3)}{h}$$
$$= 4x^3 + 6x^2h + 4xh^2 + h^3$$

17. Since there is only one y-value for each x-value, $y = x^2$ is a function.

19. Since there is only one y-value for each x-value, $y = \dfrac{1}{x}$ is a function.

21. To determine whether the equation $y^2 = 4 - x^2$ is a function, we need to solve the equation for y.
$$y = \pm\sqrt{4 - x^2}$$
For values of x between -2 and 2, two values of y result. This means the equation is not a function.

23. To determine whether the equation $x = y^2$ is a function, we need to solve the equation for y.
$$y = \pm\sqrt{x}$$
For values of $x > 0$, two values of y result. This means the equation is not a function.

25. Since there is only one y-value for each x-value, $y = 2x^2 - 3x + 4$ is a function.

27. To determine whether the equation $2x^2 + 3y^2 = 1$ is a function, we need to solve the equation for y.

$$3y^2 = 1 - 2x^2$$

$$y = \pm\sqrt{\frac{1-2x^2}{3}}$$

For values of x between $-\dfrac{\sqrt{2}}{2}$ and $\dfrac{\sqrt{2}}{2}$, two values of y result. This means the equation is not a function.

29. Since the operations of multiplication and addition can be performed on any real number, the domain of $f(x) = -5x + 4$ is all real numbers. (The domain of a polynomial function is always all real numbers.)

31. The function $f(x) = \dfrac{x}{x^2 + 1}$ is defined provided the denominator is not equal to zero. Since $x^2 + 1$ never equals zero, the domain of f is all real numbers.

33. The function $g(x) = \dfrac{x}{x^2 - 16}$ is defined provided the denominator is not equal to zero. $x^2 - 16 = 0$ when $x = -4$ or when $x = 4$, so the domain of function g is the set $\{x \mid x \neq -4, x \neq 4\}$.

35. The function $F(x) = \dfrac{x-2}{x^3 + x}$ is defined provided the denominator is not equal to zero. $x^3 + x = x(x^2 + 1) = 0$ when $x = 0$, so the domain of function F is the set $\{x \mid x \neq 0\}$.

37. The function $h(x) = \sqrt{3x - 12}$ is defined provided the radicand is nonnegative. $3x - 12 \geq 0$ when $3x \geq 12$ or when $x \geq 4$. So the domain of function h is the set $\{x \mid x \geq 4\}$ or the interval $[4, \infty)$.

39. The function $f(x) = \dfrac{4}{\sqrt{x-9}}$ is defined provided the radicand is positive. Zero must be eliminated to avoid dividing by zero.
$x - 9 > 0$ when $x > 9$. So the domain of function f is the set $\{x \mid x > 9\}$ or the interval $(9, \infty)$.

41. The function $p(x) = \sqrt{\dfrac{2}{x-1}}$ is defined provided the denominator is positive. (Radicals must be nonnegative and division by zero is not defined.)
$x - 1 > 0$ when $x > 1$. So the domain of function p is the set $\{x \mid x > 1\}$ or the interval $(1, \infty)$.

43. If $f(2) = 5$ then $x = 2$, $y = 5$ must satisfy the equation $y = 2x^3 + Ax^2 + 4x - 5$. We substitute for x and y and solve for A.

$$5 = 2 \cdot 2^3 + A \cdot 2^2 + 4 \cdot 2 - 5$$

$$5 = 16 + 4A + 8 - 5$$
$$5 = 4A + 19$$
$$-14 = 4A$$
$$A = -\frac{7}{2}$$

45. If $f(0) = 2$ then $x = 0$, $y = 2$ must satisfy the equation $y = \dfrac{3x+8}{2x-A}$. We substitute for x and y and solve for A.

$$2 = \frac{3 \cdot 0 + 8}{2 \cdot 0 - A} = \frac{8}{-A}$$
$$A = -4$$

The function f is undefined when $2x - A = 0$ because division by zero is undefined. Since $A = -4$, f is undefined when $2x + 4 = 0$ or when $x = -2$.

47. If $f(4) = 0$ then $x = 4$, $y = 0$ must satisfy the equation $y = \dfrac{2x-A}{x-3}$. We substitute for x and y and solve for A.

$$0 = \frac{2 \cdot 4 - A}{4-3} = 8 - A$$
$$A = 8$$

The function f is undefined when $x - 3 = 0$ or when $x = 3$ because division by zero is undefined.

49. $G(x) = 10x$ dollars where \$10 is the hourly wage and x is the number of hours worked. The domain of G is $\{x \mid x \geq 0\}$ since a person cannot work a negative number of hours.

51. Revenue $R = R(x) = xp$.

$$R(x) = x\left(-\frac{1}{5}x + 100\right)$$
$$= -\frac{1}{5}x^2 + 100x \qquad 0 \leq x \leq 500$$

53. Revenue $R = R(x) = xp$. Here the demand equation is expressed in terms of p, so we first solve for p.

$$x = -20p + 100 \qquad\qquad 0 \leq p \leq 5$$
$$x - 100 = -20p$$
$$p = \frac{x-100}{-20} = -\frac{1}{20}x + 5 \qquad 0 \leq x \leq 100$$

We then construct the revenue function R.

$$R(x) = x\left(-\frac{1}{20}x + 5\right) = -\frac{1}{20}x^2 + 5x \qquad 0 \leq x \leq 100$$

55. If 1990 is represented by $t = 0$, then 2010, $(1990 + 20 = 2010)$, will be represented by $t = 20$, $(0 + 20 = 20)$. To project the number (in thousands) of acres of wheat that will be planted in 2010 we evaluate the function A at $t = 20$.

$$A(20) = -119(20^2) + 113(20) + 73,367 = 28,027$$

In 2010, it is projected that 28,027,000 acres of wheat will be planted.

57. 2010 is 16 years after 1994. The expected SAT mathematics score will be

$$A(16) = -0.04(16^3) + 0.43(16^2) + 0.24(16) + 506 = 456$$

59. (a) With no wind, $x = 500$ miles per hour, the cost per passenger is

$$C(500) = 100 + \frac{500}{10} + \frac{36,000}{500} = 222 \text{ dollars.}$$

(b) With a head wind of 50 miles per hour, $x = 500 - 50 = 450$ miles per hour, and the cost per passenger is

$$C(450) = 100 + \frac{450}{10} + \frac{36,000}{450} = 225 \text{ dollars.}$$

(c) With a tail wind of 100 miles per hour, $x = 500 + 100 = 600$ miles per hour and the cost per passenger is

$$C(600) = 100 + \frac{600}{10} + \frac{36,000}{600} = 220 \text{ dollars.}$$

(d) With a head wind of 100 miles per hour, $x = 500 - 100 = 400$ miles per hour, and the cost per passenger is

$$C(400) = 100 + \frac{400}{10} + \frac{36,000}{400} = 230 \text{ dollars.}$$

61. (a) $h(x) = 2x$
$$h(a+b) = 2(a+b)$$
$$= 2a + 2b$$
$$= h(a) + h(b)$$

(b) $g(x) = x^2$
$$g(a+b) = (a+b)^2$$
$$= a^2 + 2ab + b^2$$
$$= g(a) + g(b) + 2ab$$
$$\neq g(a) + g(b)$$

(c) $F(x) = 5x - 2$
$$F(a+b) = 5(a+b) - 2$$
$$= 5a + 5b - 2$$
$$= 5a - 2 + 2 + 5b - 2$$
$$= F(a) + F(b) + 2$$
$$\neq F(a) + F(b)$$

(d) $G(x) = \frac{1}{x}$
$$G(a+b) = \frac{1}{a+b}$$
$$\neq G(a) + G(B) = \frac{1}{a} + \frac{1}{b} = \frac{b+a}{ab}$$

Only (a) has the property, $f(a + b) = f(a) + f(b)$.

63. Research sources and explanations may vary.

1.3 Graphs of Functions; Properties of Functions

1. The graph fails the vertical line test. It is not the graph of a function.

3. This is the graph of a function.
    (a) The domain is the set $\{x \mid -\pi \le x \le \pi\}$ or the interval $[-\pi, \pi]$; the range is the set $\{y \mid -1 \le y \le 1\}$ or the interval $[-1, 1]$.

    (b) The x-intercepts are $\left(-\dfrac{\pi}{2}, 0\right)$ and $\left(\dfrac{\pi}{2}, 0\right)$; the y-intercept is the point $(0, 1)$.

    (c) The graph is symmetric with respect to the y-axis.

5. The graph fails the vertical line test. It is not the graph of a function.

7. This is the graph of a function.
    (a) The domain is the set $\{x \mid x > 0\}$ or the interval $(0, \infty)$; the range is all real numbers.

    (b) There is no y-intercept; the x-intercept is the point $(1, 0)$.

    (c) The graph has no symmetries with respect to the x-axis, the y-axis, or the origin.

9. This is the graph of a function.
    (a) The domain is all real numbers; the range is the set $\{y \mid y \le 2\}$ or the interval $(-\infty, 2]$.

    (b) The x-intercepts are the points $(-3, 0)$ and $(3, 0)$; the y-intercept is the point $(0, 2)$.

    (c) The graph is symmetric with respect to the y-axis.

11. This is the graph of a function.
    (a) The domain is the set of all real numbers; the range is the set $\{y \mid y \ge -3\}$ or the interval $[-3, \infty)$.

    (b) The x-intercepts are the points $(1, 0)$ and $(3, 0)$; the y-intercept is the point $(0, 9)$.

    (c) The graph has no symmetries with respect to the x-axis, the y-axis, or the origin.

13. (a) $f(0) = 3; f(-6) = -3$

    (b) $f(6) = 0; f(11) = 1$

    (c) $f(3)$ is above the x-axis so it is positive.

    (d) $f(-4) = -1$ is below the x-axis so it is negative.

    (e) $f(x) = 0$ whenever the graph crosses or touches the x-axis. So $f(x) = 0$ when $x = -3$, when $x = 6$, and when $x = 10$.

(f) $f(x) > 0$ whenever the graph of f is above the x-axis. $f(x) > 0$ on the intervals $[-3, 6]$ and $[10, 11]$.

(g) The domain of f is the set $\{x \mid -6 \leq x \leq 11\}$ or the interval $[-6, 11]$.

(h) The range of f is the set $\{y \mid -3 \leq y \leq 4\}$ or the interval $[-3, 4]$.

(i) The x-intercepts are the points $(-3, 0)$, $(6, 0)$, and $(10, 0)$.

(j) The y-intercept is the point $(0, 3)$.

(k) Draw the horizontal line $y = \dfrac{1}{2}$ on the same axes as the graph. Count the number of times the two graphs intersect. They intersect 3 times.

(l) Draw the vertical line $x = 5$ on the same axes as the graph of f. Count the number of times the two graphs intersect. They intersect once.

(m) $f(x) = 3$ whenever the y-value equals 3. This occurs when $x = 0$ and when $x = 4$.

(n) $f(x) = -2$ whenever the y-value equals -2. This occurs when $x = -5$ and when $x = 8$.

15. (a) The point $(-1, 2)$ is on the graph of f if $x = -1$, $y = 2$ satisfies the equation.
$$f(x) = 2x^2 - x - 1$$
$$2(-1)^2 - (-1) - 1 = 2$$
So the point $(-1, 2)$ is on the graph of f.

(b) If $x = -2$, $f(-2) = 2(-2)^2 - (-2) - 1 = 9$. The point $(-2, 9)$ is on the graph of f.

(c) If $f(x) = -1$, then
$$2x^2 - x - 1 = -1$$
$$2x^2 - x = 0$$
$$x(2x - 1) = 0$$
$$x = 0 \quad \text{or} \quad 2x - 1 = 0$$
$$x = 0 \quad \text{or} \quad x = \frac{1}{2}$$
The points $(0, -1)$ and $\left(\dfrac{1}{2}, -1\right)$ are on the graph of f.

(d) The domain of f is all real numbers.

(e) To find the x-intercepts, we let $f(x) = y = 0$ and solve for x.
$$2x^2 - x - 1 = 0$$
$$(2x + 1)(x - 1) = 0$$

$$x = -\frac{1}{2} \quad \text{or} \quad x = 1$$

The x-intercepts are $\left(-\frac{1}{2}, 0\right)$ and $(1, 0)$.

(f) To find the y-intercept, we let $x = 0$ and solve for y.
$$f(0) = 2(0^2) - 0 - 1 = -1$$
The y-intercept is $(0, -1)$.

17. (a) The point $(3, 14)$ is on the graph of f if $x = 3, y = 14$ satisfies the equation.
$$f(x) = \frac{x+2}{x-6}$$
$$\frac{3+2}{3-6} = \frac{5}{-3} = -\frac{5}{3}$$

The point $\left(3, -\frac{5}{3}\right)$ is on the graph of f, but the point $(3, 14)$ is not.

(b) If $x = 4, f(4) = \frac{4+2}{4-6} = \frac{6}{-2} = -3$. The point $(4, -3)$ is on the graph of f.

(c) If $f(x) = 2$, then
$$\frac{x+2}{x-6} = 2$$
$$x + 2 = 2(x-6)$$
$$x + 2 = 2x - 12$$
$$x = 14$$
The point $(14, 2)$ is on the graph of f.

(d) The function f is a rational function. Rational functions are not defined at values of x that would cause the denominator to equal zero. So the domain of f is the set $\{x \mid x \neq 6\}$.

(e) To find the x-intercepts, we let $f(x) = y = 0$ and solve for x.
$$\frac{x+2}{x-6} = 0$$
$$x + 2 = 0$$
$$x = -2$$
The x-intercept is $(-2, 0)$.

(f) To find the y-intercept, we let $x = 0$ and solve for y.
$$f(0) = \frac{0+2}{0-6} = -\frac{1}{3}$$
The y-intercept is $\left(0, -\frac{1}{3}\right)$.

19.   (a) The point $(-1, 1)$ is on the graph of f if $x = -1, y = 1$ satisfies the equation.

$$f(x) = \frac{2x^2}{x^4 + 1}$$

$$\frac{2(-1)^2}{(-1)^4 + 1} = \frac{2}{2} = 1$$

The point $(-1, 1)$ is on the graph of f.

(b) If $x = 2, f(2) = \dfrac{2 \cdot 2^2}{2^4 + 1} = \dfrac{8}{17}$. The point $\left(2, \dfrac{8}{17}\right)$ is on the graph of f.

(c) If $f(x) = 1$, then

$$\frac{2x^2}{x^4 + 1} = 1$$

$$2x^2 = x^4 + 1$$

$$x^4 - 2x^2 + 1 = 0$$

$$(x^2 - 1)(x^2 - 1) = 0$$

$$(x - 1)(x + 1) = 0$$

$$x - 1 = 0 \quad \text{or} \quad x + 1 = 0$$

$$x = 1 \quad \text{or} \quad x = -1$$

The points $(1, 1)$ and $(-1, 1)$ are on the graph of f.

(d) The function f is a rational function. Rational functions are not defined at values of x that would cause the denominator to equal zero, but this denominator can never equal zero. So the domain of f is the set of all real numbers.

(e) To find the x-intercepts, we let $f(x) = y = 0$ and solve for x. The x-intercept is $(0, 0)$.

(f) To find the y-intercept, we let $x = 0$ and solve for y.

$$f(0) = \frac{2 \cdot 0^2}{0^4 + 1} = 0$$

The y-intercept is also $(0, 0)$.

21.   Yes, the function is increasing when $-8 < x < -2$.

23.   No, the function is decreasing on the interval $(2, 5)$ and then it increases on the interval $(5, 10)$.

25.   The function f is increasing on the intervals $(-8, -2)$; $(0, 2)$; and $(5, \infty)$ or for $-8 < x < -2$; $0 < x < 2$; and $x > 5$.

27.   There is a local maximum at $x = 2$. The maximum value is $f(2) = 10$.

29.   The function f has local maxima at $x = -2$ and at $x = 2$. The local maxima are $f(-2) = 6$ and $f(2) = 10$.

31.   (a) The x-intercepts of the graph are $(-2, 0)$ and $(2, 0)$. The y-intercept is $(0, 3)$.

(b) The domain of f is the set $\{x \mid -4 \le x \le 4\}$ or the interval $[-4, 4]$.
The range of f is the set $\{y \mid 0 \le y \le 3\}$ or the interval $[0, 3]$.

(c) The function is increasing on the intervals $(-2, 0)$ and $(2, 4)$ or when $-2 < x < 0$
and $2 < x < 4$. The function is decreasing on the intervals $(-4, -2)$ and $(0, 2)$ or when
$-4 < x < -2$ and $0 < x < 2$.

(d) The graph of the function is symmetric with respect to the y-axis, so the function is
even.

33. (a) There is no x-intercept of the graph. The y-intercept is $(0, 1)$.

(b) The domain of f is the set of all real numbers.
The range of f is the set $\{y \mid y > 0\}$ or the interval $(0, \infty)$.
(c) The function is always increasing.

(d) The graph of the function is not symmetric with respect to the y-axis or the origin, so
the function is neither even nor odd.

35. (a) The x-intercepts of the graph are $(-\pi, 0)$, $(0, 0)$, and $(\pi, 0)$. The y-intercept is $(0, 0)$.

(b) The domain of f is the set $\{x \mid -\pi \le x \le \pi\}$ or the interval $[-\pi, \pi]$. The range of f is
the set $\{y \mid -1 \le y \le 1\}$ or the interval $[-1, 1]$.

(c) The function is increasing on the interval $\left(-\dfrac{\pi}{2}, \dfrac{\pi}{2}\right)$ or for $-\dfrac{\pi}{2} < x < \dfrac{\pi}{2}$. The

function is decreasing on the intervals $\left(-\pi, -\dfrac{\pi}{2}\right)$ and $\left(\dfrac{\pi}{2}, \pi\right)$ or for $-\pi < x < -\dfrac{\pi}{2}$ and

$\dfrac{\pi}{2} < x < \pi$.

(d) The graph of the function is symmetric with respect to the origin, so the function is
odd.

37. (a) The x-intercepts of the graph are $\left(\dfrac{1}{2}, 0\right)$ and $\left(\dfrac{5}{2}, 0\right)$. The y-intercept is $\left(0, \dfrac{1}{2}\right)$.

(b) The domain of f is the set $\{x \mid -3 \le x \le 3\}$ or the interval $[-3, 3]$. The range of f is
the set $\{y \mid -1 \le y \le 2\}$ or the interval $[-1, 2]$.

(c) The function is increasing on the interval $(2, 3)$ or $2 < x < 3$; it is decreasing on the
interval $(-1, 1)$ or $-1 < x < 1$; and it is constant on the intervals $(-3, -1)$ and $(1, 2)$ or
when $-3 < x < -1$ and $1 < x < 2$.

(d) The graph of the function is not symmetric with respect to the y-axis or the origin, so the function is neither even nor odd.

39. (a) f has a local maximum at 0 since for all x close to 0, $x \ne 0$, $f(x) < f(0)$. The local maximum is $f(0) = 3$.

(b) f has local minima at -2 and at 2. The local minima are $f(-2) = 0$ and $f(2) = 0$.

41. (a) f has a local maximum at $\dfrac{\pi}{2}$ since for all x close to $\dfrac{\pi}{2}$, $x \ne \dfrac{\pi}{2}$, $f(x) < f\left(\dfrac{\pi}{2}\right)$. The local maximum is $f\left(\dfrac{\pi}{2}\right) = 1$.

(b) f has local minimum at $-\dfrac{\pi}{2}$ since for all x close to $-\dfrac{\pi}{2}$, $x \ne -\dfrac{\pi}{2}$, $f(x) > f\left(-\dfrac{\pi}{2}\right)$. The local minimum is $f\left(-\dfrac{\pi}{2}\right) = -1$.

43. (a) The average rate of change from 0 to 2 is
$$\frac{\Delta y}{\Delta x} = \frac{f(2) - f(0)}{2 - 0} = \frac{\left[-2 \cdot 2^2 + 4\right] - \left[-2 \cdot 0^2 + 4\right]}{2} = \frac{-4 - 4}{2} = -4$$

(b) The average rate of change from 1 to 3 is
$$\frac{\Delta y}{\Delta x} = \frac{f(3) - f(1)}{3 - 1} = \frac{\left[-2 \cdot 3^2 + 4\right] - \left[-2 \cdot 1^2 + 4\right]}{2} = \frac{-16}{2} = -8$$

(c) The average rate of change from 1 to 4 is
$$\frac{\Delta y}{\Delta x} = \frac{f(4) - f(1)}{4 - 1} = \frac{\left[-2 \cdot 4^2 + 4\right] - \left[-2 \cdot 1^2 + 4\right]}{3} = \frac{-30}{3} = -10$$

45. (a) $\dfrac{f(x) - f(1)}{x - 1} = \dfrac{5x - 5(1)}{x - 1} = \dfrac{5(x - 1)}{x - 1} = 5$

(b) Using $x = 2$ in part (a), we get 5. This is the slope of the secant line containing the points $(1, f(1))$ and $(2, f(2))$.

(c) $m_{\text{sec}} = 5$. Using the point-slope form of the line and the point $(1, f(1)) = (1, 5)$, we get
$$\begin{aligned} y - y_1 &= m_{\text{sec}}(x - x_1) \\ y - 5 &= 5(x - 1) \\ y &= 5x - 5 + 5 \\ y &= 5x \end{aligned}$$

47. (a) $\dfrac{f(x) - f(1)}{x - 1} = \dfrac{[1 - 3x] - [1 - 3]}{x - 1} = \dfrac{3 - 3x}{x - 1} = \dfrac{3(1 - x)}{x - 1} = -3$

(b) Using $x = 2$ in part (a), we get -3. This is the slope of the secant line containing the points $(1, f(1))$ and $(2, f(2))$.

(c) $m_{\text{sec}} = -3$. Using the point-slope form of the line and the point $(1, f(1)) = (1, -2)$, we get

$$y - y_1 = m_{\text{sec}}(x - x_1)$$
$$y - (-2) = -3(x - 1)$$
$$y = -3x + 3 - 2$$
$$y = -3x + 1$$

49. (a) $\dfrac{f(x) - f(1)}{x - 1} = \dfrac{\left[x^2 - 2x\right] - \left[1^2 - 2(1)\right]}{x - 1} = \dfrac{\left[x^2 - 2x\right] - [-1]}{x - 1} = \dfrac{x^2 - 2x + 1}{x - 1}$

$$= \dfrac{(x-1)^2}{x-1} = x - 1$$

(b) Using $x = 2$ in part (a), we get 1. This is the slope of the secant line containing the points $(1, f(1))$ and $(2, f(2))$.

(c) $m_{\text{sec}} = 1$. Using the point-slope form of the line and the point $(1, f(1)) = (1, -1)$, we get

$$y - y_1 = m_{\text{sec}}(x - x_1)$$
$$y - (-1) = 1(x - 1)$$
$$y = x - 1 - 1$$
$$y = x - 2$$

51. (a) $\dfrac{f(x) - f(1)}{x - 1} = \dfrac{\left[x^3 - x\right] - \left[1^3 - 1\right]}{x - 1} = \dfrac{\left[x^3 - x\right] - [0]}{x - 1} = \dfrac{x(x^2 - 1)}{x - 1}$

$$= \dfrac{x(x-1)(x+1)}{x-1} = x^2 + x$$

(b) Using $x = 2$ in part (a), we get $2^2 + 2 = 6$. This is the slope of the secant line containing the points $(1, f(1))$ and $(2, f(2))$.

(c) $m_{\text{sec}} = 6$. Using the point-slope form of the line and the point $(1, f(1)) = (1, 0)$, we get

$$y - y_1 = m_{\text{sec}}(x - x_1)$$
$$y - 0 = 6(x - 1)$$
$$y = 6x - 6$$

53. (a) $\dfrac{f(x) - f(1)}{x - 1} = \dfrac{\left[\dfrac{2}{x+1}\right] - \left[\dfrac{2}{1+1}\right]}{x - 1} = \dfrac{\left[\dfrac{2}{x+1}\right] - [1]}{x - 1} = \dfrac{2 - (1)(x+1)}{(x-1)(x+1)} = \dfrac{2 - x - 1}{(x-1)(x+1)}$

$$= \dfrac{-x+1}{(x-1)(x+1)} = -\dfrac{1}{x+1}$$

(b) Using $x = 2$ in part (a), we get $-\dfrac{1}{2+1} = -\dfrac{1}{3}$. This is the slope of the secant line containing the points $(1, f(1))$ and $(2, f(2))$.

(c) $m_{\text{sec}} = -\dfrac{1}{3}$. Using the point-slope form of the line and the point $(1, f(1)) = (1, 1)$, we get

$$y - y_1 = m_{\text{sec}}(x - x_1)$$
$$y - 1 = -\frac{1}{3}(x - 1)$$
$$y = -\frac{1}{3}x + \frac{1}{3} + 1$$
$$y = -\frac{1}{3}x + \frac{4}{3}$$

55. (a) $\dfrac{f(x) - f(1)}{x - 1} = \dfrac{\sqrt{x} - \sqrt{1}}{x - 1} = \dfrac{\sqrt{x} - 1}{(\sqrt{x} - 1)(\sqrt{x} + 1)} = \dfrac{1}{\sqrt{x} + 1}$

(b) Using $x = 2$ in part (a), we get $\dfrac{1}{\sqrt{2} + 1}$. This is the slope of the secant line containing the points $(1, f(1))$ and $(2, f(2))$.

(c) $m_{\text{sec}} = \dfrac{1}{\sqrt{2} + 1}$

Using the point-slope form of the line and the point $(1, f(1)) = (1, 1)$, we get
$$y - y_1 = m_{\text{sec}}(x - x_1)$$
$$y - 1 = \frac{1}{\sqrt{2} + 1}(x - 1)$$
$$y \approx 0.414x - 0.414 + 1$$
$$y \approx 0.414x + 0.586$$

57. To determine algebraically whether a function is even, odd, or neither we replace x with $-x$ and simplify.
$$f(-x) = 4(-x)^3 = 4 \cdot (-x^3) = -4x^3 = -f(x)$$
Since $f(-x) = -f(x)$, the function is odd.

59. To determine algebraically whether a function is even, odd, or neither we replace x with $-x$ and simplify.
$$g(-x) = -3(-x)^2 - 5 = -3x^2 - 5 = g(x)$$
Since $g(-x) = g(x)$, the function is even.

61. To determine algebraically whether a function is even, odd, or neither we replace x with $-x$ and simplify.

$$F(-x) = \sqrt[3]{-x} = -\sqrt[3]{x} = -F(x)$$

Since $F(-x) = -F(x)$, the function is odd.

63. To determine algebraically whether a function is even, odd, or neither we replace x with $-x$ and simplify.

$$f(-x) = -x + |-x| = -x + |x|$$

Since $f(-x)$ equals neither $f(x)$ nor $-f(x)$, the function is neither even nor odd.

65. To determine algebraically whether a function is even, odd, or neither we replace x with $-x$ and simplify.

$$g(-x) = \frac{1}{(-x)^2} = \frac{1}{x^2} = g(x)$$

Since $g(-x) = g(x)$, the function is even.

67. To determine algebraically whether a function is even, odd, or neither we replace x with $-x$ and simplify.

$$h(-x) = \frac{-(-x)^3}{3(-x)^2 - 9} = \frac{x^3}{3x^2 - 9} = -\left(\frac{-x^3}{3x^2 - 9}\right) = -h(x)$$

Since $h(-x) = -h(x)$, the function is odd.

69. Using 2nd, calculate, we find there is a local maximum at -1. The local maximum is $f(-1) = 4$.
There is a local minimum at 1. The local minimum is $f(1) = 0$.

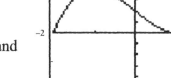

The function is increasing on the intervals $(-2, -1)$ and $(1, 2)$.

The function is decreasing on the interval $(-1, 1)$.

71. Using 2nd, CALCULATE, we find there is a local maximum at -0.77. The local maximum is $f(-0,77) = 0.19$.
There is a local minimum at 0.77. The local minimum is $f(0.77) = -0.19$.

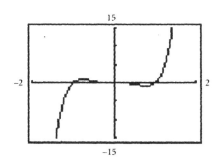

The function is increasing on the intervals $(-2, -0.77)$ and $(0.77, 2)$.

The function is decreasing on the interval $(-0.77, 0.77)$.

73. Using 2nd, calculate, we find there is a local maximum at 1.77. The local maximum is $f(1.77) = -1.91$.
There is a local minimum at -3.77. The local minimum is $f(-3.77) = -18.89$.

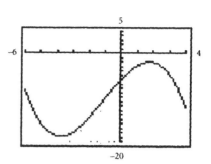

The function is increasing on the interval $(-3.77, 1.77)$.

The function is decreasing on the intervals $(-6, -3.77)$ and $(1.77, 4)$.

75. Using 2nd, calculate, we find there is a local maximum at 0. The local maximum is $f(0) = 3$.
There is a local minimum at -1.87. The local minimum is $f(-1.87) = 0.95$.
There is another local minimum at 0.97. The local minimum is $f(0.97) = 2.65$.

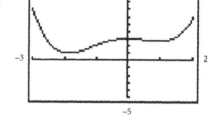

The function is increasing on the intervals $(-1.87, 0)$ and $(0.97, 2)$.

The function is decreasing on the intervals $(-3, -1.87)$ and $(0, 0.97)$.

77. The average rate of change of f from 0 to x is

$$\frac{\Delta y}{\Delta x} = \frac{f(x) - f(0)}{x - 0} = \frac{x^2 - 0}{x} = x \qquad x \neq 0$$

(a) The average rate of change from 0 to 1 is 1.
(b) The average rate of change from 0 to 0.5 is 0.5.
(c) The average rate of change from 0 to 0.1 is 0.1.
(d) The average rate of change from 0 to 0.01 is 0.01.
(e) The average rate of change from 0 to 0.001 is 0.001.

(g) and (h) Answers will vary.

(f)

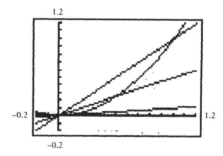

79.

(a) $h(100) = \dfrac{-32(100)^2}{130^2} + 100 = 81.07$ feet high

(b) $h(300) = \dfrac{-32(300)^2}{130^2} + 300 = 129.59$ feet high

(c) $h(500) = \dfrac{-32(500)^2}{130^2} + 500 = 26.63$ feet high

(d) To determine how far the ball was hit, we find $h(x) = 0$.

$$\frac{-32x^2}{130^2} + x = 0$$
$$-32x^2 + 130^2 x = 0$$
$$x(-32x + 130^2) = 0$$
$$x = 0 \quad \text{or} \quad -32x + 130^2 = 0$$
$$x = 0 \quad \text{or} \quad x = 528.125$$

The ball traveled 528.125 feet.

(e)

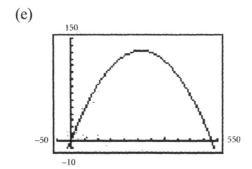

(f)

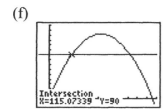 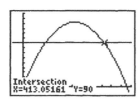

The golf ball is 90 feet high after it has traveled about 115 feet and again when it has traveled about 413 feet.

(g)

X	Y₁
0	0
25	23.817
50	45.266
75	64.349
100	81.065
125	95.414
150	107.4

X=0

X	Y₁	Y₂
200	124.26	90
225	129.14	90
250	131.66	90
275	131.8	90
300	129.59	90
325	125	90
350	118.05	90

X=275

(h)   The ball travels about 275 feet when it reaches its maximum height of about 131.8 feet.

(i) The ball actually travels only 264 feet before reaching its maximum height of 132.03 feet.

X	Y₁	Y₂
260	132	90
261	132.01	90
262	132.02	90
263	132.03	90
264	132.03	90
265	132.03	90
266	132.02	90

X=264

81. (a) Volume of a box is the product of its length, width, and height. The box has height x inches and length and width equal to $24 - 2x$ inches.
$$V = V(x) = x(24 - 2x)^2 \text{ cubic inches}$$

(b) If $x = 3$ inches, then $V = 3(24 - 2 \cdot 3)^2 = 972$ cubic inches.

(c) If $x = 10$ inches, then $V = 10(24 - 2 \cdot 10)^2 = 10(4)^2 = 160$ cubic inches.

(d)
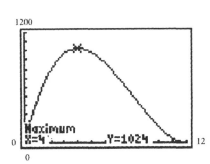

83. (a)
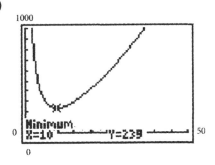

(b) The average cost is minimized when 10 riding mowers are produced.

(c) The average cost of producing each of the 10 mowers is $239.

85. Best matches are are (a) graph (II), (b) graph (V), (c) graph (IV), (d) graph (III), (e) graph (I). Reasons may vary.

87. Graphs will vary.

89. Explanations and graphs will vary.

91. Descriptions will vary.

93. Functions will vary.

1.4 Library of Functions; Piecewise-defined Functions

1. (C) Graphs of square functions are parabolas.

3. (E) Graphs of square root functions are defined for $x \geq 0$ and are increasing.

5. (B) Graphs of linear functions are straight lines.

7. (F) Graph of a reciprocal function is not defined at zero.

9.

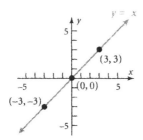

11.

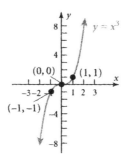

13.

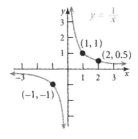

15.

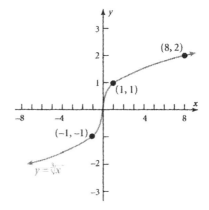

17. (a) When $x = -2$, the equation for f is $f(x) = x^2$, so $f(-2) = (-2)^2 = 4$

(b) When $x = 0$, the equation for f is $f(x) = 2$, so $f(0) = 2$

(c) When $x = 2$, the equation for f is $f(x) = 2x + 1$, so $f(2) = 2(2) + 1 = 5$

19. (a) $f(1.2) = \text{int}(2 \cdot 1.2) = \text{int}(2.4) = 2$

(b) $f(1.6) = \text{int}(2 \cdot 1.6) = \text{int}(3.2) = 3$

(c) $f(-1.8) = \text{int}(2 \cdot (-1.8)) = \text{int}(-3.6) = -4$

21. (a) The domain of f is all real numbers or the interval $(-\infty, \infty)$.

(b) To find the y-intercept we let $x = 0$ and solve. When $x = 0$, the equation for f is $f(0) = 1$. So the y-intercept is $(0, 1)$.

To find the x-intercept, we let $y = 0$ and solve for x. $f(x)$ never equals zero, so there is no x-intercept.

(c)

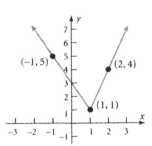

(d) The range of f is the set $\{y \mid y \neq 0\}$.

23. (a) The domain of f is all real numbers or the interval $(-\infty, \infty)$.

(b) To find the y-intercept we let $x = 0$ and solve. When $x = 0$, the equation for f is $f(x) = -2x + 3$ and $f(0) = 3$. So the y-intercept is $(0, 3)$.

To find the x-intercept, we let $y = 0$ and solve. If $x < 1$, then $y = -2x + 3 = 0$ or
$$x = \frac{3}{2} = 1.5.$$
Since $x = 1.5$ is not less than 1, we ignore this solution. If $x \geq 1$, then $y = 3x - 2 = 0$ or $x = \frac{2}{3}$.

Since $x = \frac{2}{3} < 1$, we ignore this solution and conclude $f(x)$ never equals zero and there is no x-intercept.

(c)

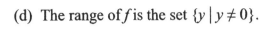

(d) The range of f is the set $\{y \mid y \geq 1\}$ or $[1, \infty)$.

25. (a) The domain of f is the set $\{x \mid x \geq -2\}$ or the interval $[-2, \infty)$.

(b) To find the y-intercept we let $x = 0$ and solve. When $x = 0$, the equation for f is $f(x) = x + 3$, and $f(0) = 3$. So the y-intercept is $(0, 3)$.

To find the x-intercept(s), we let $y = 0$ and solve.
If $-2 \leq x < 1$, $y = f(x) = x + 3 = 0$ or $x = -3$. Since $x = -3$ is not in the interval $[-2, 1)$, we ignore this solution.
If $x > 1$, $y = -x + 2 = 0$ or $x = 2$. So there is an x-intercept at $(2, 0)$.

(c)

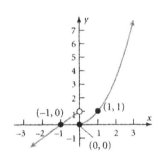

(d) The range of *f* is the set $\{y \mid y = 5; y < 4\}$.

27.   (a) The domain of *f* is all real numbers.

(b) To find the *y*-intercept we let $x = 0$ and solve. When $x = 0$, the equation for *f*
is $f(x) = x^2$, and $f(0) = 0$. So the *y*-intercept is (0, 0).

To find the *x*-intercept(s), we let $y = 0$ and solve.
If $x < 0$, $y = f(x) = 1 + x = 0$ or $x = -1$. So there is an *x*-intercept at (−1, 0).
If $x > 0$, $y = f(x) = x^2 = 0$ or $x = 0$. So there is an *x*-intercept at (0, 0).

(c)

(d) The range of *f* is the set of all real numbers.

29.   (a) The domain of *f* is the set $\{x \mid x \geq -2\}$ or the interval $[-2, \infty)$.

(b) To find the *y*-intercept we let $x = 0$ and solve. When $x = 0$, the equation for *f*
is $f(x) = 1$, and $f(0) = 1$. So the *y*-intercept is (0, 1).

To find the *x*-intercept(s), we let $y = 0$ and solve. If $-2 \leq x < 0$, $y = f(x) = |x| = 0$ or $x = 0$.
Since $x = 0$ is not in the interval $[-2, 0)$, we ignore this solution.
If $x > 0$, $y = x^3 = 0$ or $x = 0$. Since $x = 0$ is not in the interval $(0, \infty)$, we also ignore this
solution, and conclude there is no *x*-intercept.

(c)

(d) The range of *f* is the set $\{y \mid y > 0\}$ or the interval $(0, \infty)$.

31. (a) The domain of f is the set of all real numbers.

(b) To find the y-intercept we let $x = 0$ and solve. When $x = 0$, the equation for f is $f(x) = 2$ int (x), and $f(0) = 0$. So the y-intercept is $(0, 0)$.

To find the x-intercept(s), we let $y = 0$ and solve. $y = f(x) = 2$ int $(x) = 0$ for all x in the set $\{x \mid 0 \le x < 1\}$ or the interval $[0, 1)$. So there are x-intercepts at all points in the interval.

(c)   (d) The range of f is the set of even integers.

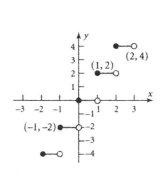

33. This graph is made of two line segments. We find each.
For $-1 \le x \le 0$, we have two points $(-1, 1)$ and $(0, 0)$. The slope of the line between the points is

$$m = \frac{y_2 - y_1}{x_2 - x_1} = \frac{1 - 0}{-1 - 0} = -1$$

Using the point-slope form of an equation of a line and the point $(0, 0)$, we get

$$y - y_1 = m(x - x_1)$$
$$y - 0 = -1(x - 0)$$
$$y = -x$$

For $0 < x \le 2$, we have two points $(0, 0)$ and $(2, 1)$. The slope of the line between the points is

$$m = \frac{y_2 - y_1}{x_2 - x_1} = \frac{1 - 0}{2 - 0} = \frac{1}{2}$$

Using the point-slope form of an equation of a line and the point $(0, 0)$, we get

$$y - y_1 = m(x - x_1)$$
$$y - 0 = \frac{1}{2}(x - 0)$$
$$y = \frac{1}{2}x$$

The piecewise function is

$$f(x) = \begin{cases} -x & -1 \le x \le 0 \\ \dfrac{1}{2}x & 0 < x \le 2 \end{cases}$$

35. This graph is made of two line segments. We find each.
For $x \leq 0$, we have two points $(-1, 1)$ and $(0, 0)$. The slope of the line between the points is

$$m = \frac{y_2 - y_1}{x_2 - x_1} = \frac{1 - 0}{-1 - 0} = -1$$

Using the point-slope form of an equation of a line and the point $(0, 0)$, we get

$$y - y_1 = m(x - x_1)$$
$$y - 0 = -1(x - 0)$$
$$y = -x$$

For $0 < x \leq 2$, we have two points $(0, 2)$ and $(2, 0)$. The slope of the line between the points is

$$m = \frac{y_2 - y_1}{x_2 - x_1} = \frac{2 - 0}{0 - 2} = -1$$

Using the point-slope form of an equation of a line and the point $(0, 2)$, we get

$$y - y_1 = m(x - x_1)$$
$$y - 2 = -1(x - 0)$$
$$y = -x + 2$$

The piecewise function is

$$f(x) = \begin{cases} -x & x \leq 0 \\ -x + 2 & 0 < x \leq 2 \end{cases}$$

37. (a) If $x = 200$ minutes are used, the equation for C is $C(x) = 39.99$. So $C(200) = \$39.99$.

(b) If $x = 365$ minutes are used, the equation for C is $C(x) = 0.25x - 47.51$. So $C(365) = 0.25(365) - 47.51 = \$43.74$.

(c) If $x = 351$ minutes are used, the equation for C is $C(x) = 0.25x - 47.51$. So $C(351) = 0.25(351) - 47.51 = \$40.24$.

39. (a) If 50 therms are used, the charge C will be
$$C(50) = 9.45 + 0.36375(50) + 0.6338(50) = \$59.33$$

(b) If 500 therms are used the charge will be
$$C(500) = 9.45 + 0.36375(50) + 0.11445(500 - 50) + 0.6338(500) = \$396.04$$

(c) When $0 \leq x \leq 50$ the equation that relates the monthly charge for using x therms of gas is
$$C(x) = 9.45 + 0.36375x + 0.6338x = 0.99755x + 9.45$$
When $x > 50$ the equation that relates the monthly charge for using x therms of gas is
$$C(x) = 9.45 + 0.36375(50) + 0.11445(x - 50) + 0.6338x$$
$$= 21.915 + 0.74825\,x$$
The function that relates the monthly charge C for x therms of gas is
$$C(x) = \begin{cases} 9.45 + 0.99755x & 0 \leq x \leq 50 \\ 21.915 + 0.74825x & x > 50 \end{cases}$$

(d)

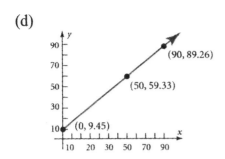

41. (a) When $v = 1$ meter per second and $t = 10°$, the equation representing the wind chill is $W = t = 10°$.

(b) When $v = 5$ m/sec. and $t = 10°$, the equation representing the wind chill is

$$W = 33 - \frac{(10.45 + 10\sqrt{v} - v)(33 - t)}{22.04} = 33 - \frac{(10.45 + 10\sqrt{5} - 5)(33 - 10)}{22.04}$$

$$= 33 - \frac{(5.45 + 10\sqrt{5})(23)}{22.04} = 3.98°$$

(c) When $v = 15$ m/sec. and $t = 10°$, the equation representing the wind chill is

$$W = 33 - \frac{(10.45 + 10\sqrt{v} - v)(33 - t)}{22.04} = 33 - \frac{(10.45 + 10\sqrt{15} - 15)(33 - 10)}{22.04}$$

$$= 33 - \frac{(10\sqrt{15} - 4.55)(23)}{22.04} = -2.67°$$

(d) When $v = 25$ m/sec. and $t = 10°$, the equation representing the wind chill is
$W = 33 - 1.5958(33 - t) = 33 - 1.5958(33 - 10) = 33 - 1.5958(23) = -3.70°$

(e) Answers may vary.

(f) Answers may vary.

43.
$$y = f(x) = \begin{cases} 0.10x & 0 < x < 7000 \\ 700 + 0.15(x - 7000) & 7000 \le x < 28,400 \\ 3910 + 0.25(x - 28,400) & 28,400 \le x < 68,800 \\ 14,010 + 0.28(x - 68,800) & 68,800 \le x < 143,500 \\ 34,926 + 0.33(x - 143,500) & 143,500 \le x < 311,950 \\ 90,514 + 0.35(x - 311,950) & x \ge 311,950 \end{cases}$$

45.

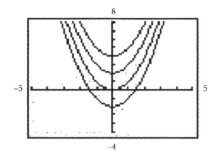

Answers may vary.

47.

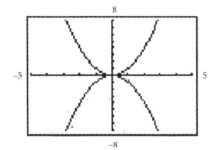

Answers may vary.

49.

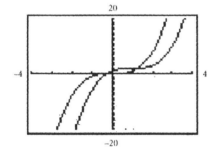

Answers may vary.

51.

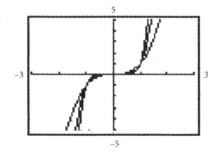

Answers may vary.

1.5 Graphing Techniques: Shifts and Reflections

1. (B) This is the graph of a square function reflected around the x-axis and shifted up two units.

3. (H) This is the graph of an absolute value function shifted to the left two units and then reflected over the x-axis.

5. (A) This is the graph of a square function shifted up two units.

7. (F) This is the graph of a square function shifted to the left two units and then reflected over the x-axis.

9. $y = (x - 4)^3$ 11. $y = x^3 + 4$

13. $y = (-x)^3 = -x^3$

15. (1) Shift up 2 units.   Add 2. $y = \sqrt{x} + 2$
     (2) Reflect about the x-axis.   Multiply y by -1. $y = -\left(\sqrt{x} + 2\right) = -\sqrt{x} - 2$
     (3) Reflect about the y-axis.   Replace x by $-x$. $y = -\sqrt{-x} - 2$

17. (1) Reflect about the x-axis.   Multiply y by -1. $y = -\sqrt{x}$
     (2) Shift up 2 units.   Add 2. $y = -\sqrt{x} + 2$
     (3) Shift left 3 units.   Replace x by $x + 3$. $y = -\sqrt{x+3} + 2$

19. Ans: (c) $-f(x)$ changes the sign of y. So if $(3, 0)$ is a point on f, $(3, 0)$ is a point on $-f(x)$.

21.

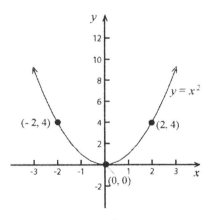

(a) $y = x^2$

$\rightarrow$

Subtract 1; vertical
shift down 1 unit.

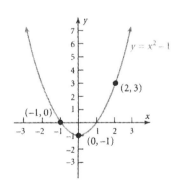

(b) $y = x^2 - 1$

23.

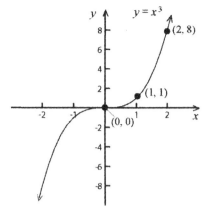

(a) $y = x^3$ $\rightarrow$ Add 1; vertical   (b) $y = x^3 + 1$

shift up 1 unit.

25.

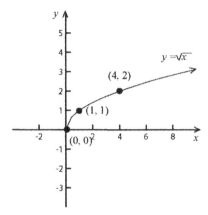

(a) $y = \sqrt{x}$ $\rightarrow$ Replace x by $x - 2$;   (b) $y = \sqrt{x - 2}$

horizontal shift right 2 units.

27.

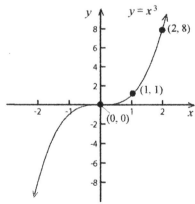

(a) $y = x^3$ $\rightarrow$ Replace x by $x - 1$;   (b) $y = (x - 1)^3$

horizontal shift right

1 unit.

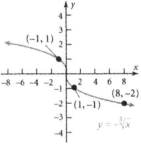

Add 2: vertical shift
up 2 units.
$\longrightarrow$

(c) $y = (x-1)^3 + 2$

29.

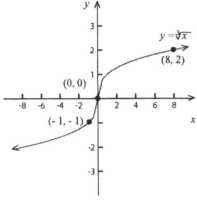

(a) $y = \sqrt[3]{x}$

$\longrightarrow$
Multiply $f(x)$ by -1;
reflect about the x-axis.

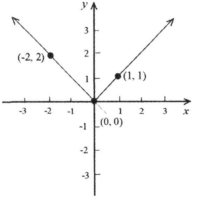

(b) $y = -\sqrt[3]{x}$

31.

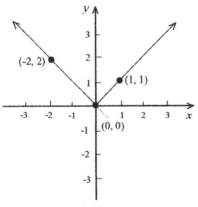

(a) $y = |x|$

$\longrightarrow$
Replace x by $-x$;
reflect about y-axis.

(b) $y = |-x|$

33.

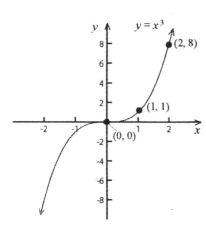

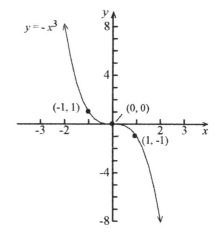

(a) $y = x^3$

Multiply $f(x)$ by $-x$;
reflect about y-axis. $\longrightarrow$

(b) $y = -x^3$

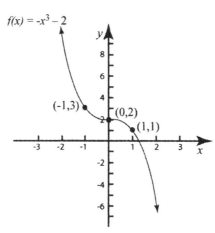

Add 2; vertical
shift 2 units. $\longrightarrow$

(c) $y = -x^3 + 2$

35.

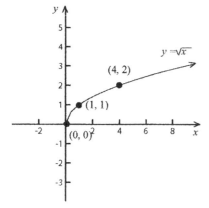

(a) $y = \sqrt{x}$

Replace x by $x - 2$;
horizontal shift right 2 units. $\longrightarrow$

(b) $y = \sqrt{x - 2}$

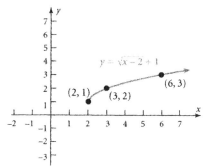

Add 1; vertical shift
up 1 unit.

(c) $y = \sqrt{x-2} + 1$

37.

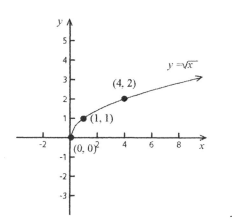

(a) $y = \sqrt{x}$

Replace x by $-x$;
reflect about y-axis.

(b) $y = \sqrt{-x}$

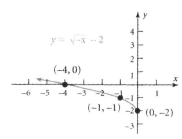

Subtract 2; vertical
down 2 units.

(c) $y = \sqrt{-x} - 2$

39.

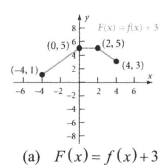

(a) $F(x) = f(x) + 3$

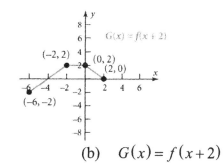

(b) $G(x) = f(x+2)$

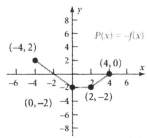

(c) $P(x) = -f(x)$

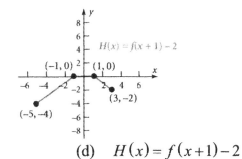

(d) $H(x) = f(x+1) - 2$

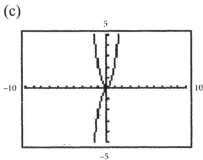

(e) $g(x) = f(-x)$

41. (a)   (b)

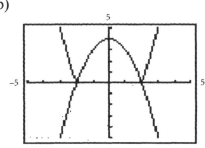

(c)   (d)   Answers may vary.

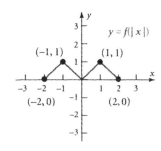

43.

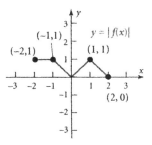

(a) $y = |f(x)|$ (b) $y = f(|x|)$

Chapter 1 Review

TRUE-FALSE ITEMS

1. False   **3.** False

5. True

FILL IN THE BLANKS

1. independent; dependent   **3.** 5; −3

5. (−5, 0), (−2, 0), (2, 0)

REVIEW EXERCISES

1.

x	$x^2 + 4$	(x, y)
−2	$(-2)^2 + 4 = 8$	$(-2, 8)$
−1	$(-1)^2 + 4 = 5$	$(-1, 5)$
0	$0^2 + 4 = 4$	$(0, 4)$
1	$1^2 + 4 = 5$	$(1, 5)$
2	$2^2 + 4 = 8$	$(2, 8)$

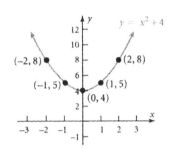

3. To find the x-intercept(s) we let $y = 0$ and solve the equation
$$2x = 3 \cdot 0^2$$
$$x = 0$$
So the x-intercept is (0, 0).

To find the y-intercept(s) we let $x = 0$ and solve the equation
$$2 \cdot 0 = 3y^2$$
$$y = 0$$
So the y-intercept is (0, 0).

To test the graph of the equation $2x = 3y^2$ for symmetry with respect to the x-axis, we replace y by $-y$ in the equation.
$$2x = 3y^2$$
$$2x = 3(-y)^2$$
$$2x = 3y^2$$
Since the resulting equation is equivalent to the original equation, the graph is symmetric with respect to the x-axis.

To test the graph of the equation $2x = 3y^2$ for symmetry with respect to the y-axis, we replace x by $-x$ in the equation and simplify.

$$2x = 3y^2$$
$$2(-x) = 3y^2$$
$$-2x = 3y^2$$

Since the resulting equation is not equivalent to the original equation, the graph is not symmetric with respect to the y-axis.

To test the graph of the equation $2x = 3y^2$ for symmetry with respect to the origin, we replace x by $-x$ and y by $-y$ in the equation and simplify.
$$2(-x) = 3(-y)^2$$
$$-2x = 3y^2$$
Since the resulting equation is not equivalent to the original equation, the graph is not symmetric with respect to the origin.

5. To find the x-intercept(s) we let $y = 0$ and solve the equation
$$x^2 + 4 \cdot 0^2 = 16$$
$$x^2 = 16$$
$$x = -4 \text{ or } x = 4$$
So the x-intercepts are $(-4, 0)$ and $(4, 0)$.

To find the y-intercept(s) we let $x = 0$ and solve the equation
$$0^2 + 4y^2 = 16$$
$$4y^2 = 16$$
$$y^2 = 4$$
$$y = -2 \text{ or } y = 2$$
So the y-intercepts are $(0, -2)$ and $(0, 2)$.

To test the graph of the equation $x^2 + 4y^2 = 16$ for symmetry with respect to the x-axis, we replace y by $-y$ in the equation.
$$x^2 + 4y^2 = 16$$
$$x^2 + 4(-y)^2 = 16$$
$$x^2 + 4y^2 = 16$$

Since the resulting equation is equivalent to the original equation, the graph is symmetric with respect to the x-axis.

To test the graph of the equation $x^2 + 4y^2 = 16$ for symmetry with respect to the y-axis, we replace x by $-x$ in the equation and simplify.
$$x^2 + 4y^2 = 16$$
$$(-x)^2 + 4y^2 = 16$$
$$x^2 + 4y^2 = 16$$
Since the resulting equation is equivalent to the original equation, the graph is symmetric with respect to the y-axis.

To test the graph of the equation $x^2 + 4y^2 = 16$ for symmetry with respect to the origin, we replace x by $-x$ and y by $-y$ in the equation and simplify.
$$x^2 + 4y^2 = 16$$
$$(-x)^2 + 4(-y)^2 = 16$$
$$x^2 + 4y^2 = 16$$

Since the resulting equation is equivalent to the original equation, the graph is symmetric with respect to the origin.

7. To find the x-intercept(s) we let $y = 0$ and solve the equation
$$x^4 + 2x^2 + 1 = 0$$
$$(x^2 + 1)(x^2 + 1) = 0$$
$$x^2 + 1 = 0$$
has no real solution. So there is no x-intercept.

To find the y-intercept(s) we let $x = 0$ and solve the equation
$$0^4 + 2 \cdot 0^2 + 1 = y$$
$$y = 1$$
So the y-intercept is $(0, 1)$.

To test the graph of the equation $x^4 + 2x^2 + 1 = y$ for symmetry with respect to the x-axis, we replace y by $-y$ in the equation.
$$x^4 + 2x^2 + 1 = -y$$
Since the resulting equation is not equivalent to the original equation, the graph is not symmetric with respect to the x-axis.

To test the graph of the equation $x^4 + 2x^2 + 1 = y$ for symmetry with respect to the y-axis, we replace x by $-x$ in the equation and simplify.
$$(-x)^4 + 2(-x)^2 + 1 = y$$
$$x^4 + 2x^2 + 1 = y$$
Since the resulting equation is equivalent to the original equation, the graph is symmetric with respect to the y-axis.

To test the graph of the equation $x^4 + 2x^2 + 1 = y$ for symmetry with respect to the origin, we replace x by $-x$ and y by $-y$ in the equation and simplify.
$$(-x)^4 + 2(-x)^2 + 1 = -y$$
$$x^4 + 2x^2 + 1 = -y$$
Since the resulting equation is not equivalent to the original equation, the graph is not symmetric with respect to the origin.

9. To find the x-intercept(s) we let $y = 0$ and solve the equation
$$x^2 + x + 0^2 + 2 \cdot 0 = 0$$
$$x^2 + x = x(x+1) = 0$$
$$x = 0 \quad \text{or} \quad x + 1 = 0$$
$$x = -1$$
So the x-intercepts are $(-1, 0)$ and $(0, 0)$.

To find the y-intercept(s) we let $x = 0$ and solve the equation
$$0^2 + 0 + y^2 + 2y = 0$$
$$y^2 + 2y = y(y+2) = 0$$

$$y = 0 \quad \text{or} \quad y + 2 = 0$$
$$y = -2$$

So the y-intercepts are $(0, -2)$ and $(0, 0)$.

To test the graph of the equation $x^2 + x + y^2 + 2y = 0$ for symmetry with respect to the x-axis, we replace y by $-y$ in the equation.
$$x^2 + x + (-y)^2 + 2(-y) = 0$$
$$x^2 + x + y^2 - 2y = 0$$

Since the resulting equation is not equivalent to the original equation, the graph is not symmetric with respect to the x-axis.

To test the graph of the equation $x^2 + x + y^2 + 2y = 0$ for symmetry with respect to the y-axis, we replace x by $-x$ in the equation and simplify.
$$(-x)^2 + (-x) + y^2 + 2y = 0$$
$$x^2 - x + y^2 + 2y = 0$$

Since the resulting equation is not equivalent to the original equation, the graph is not symmetric with respect to the y-axis.

To test the graph of the equation $x^2 + x + y^2 + 2y = 0$ for symmetry with respect to the origin, we replace x by $-x$ and y by $-y$ in the equation and simplify.
$$(-x)^2 + (-x) + (-y)^2 + 2(-y) = 0$$
$$x^2 - x + y^2 - 2y = 0$$

Since the resulting equation is not equivalent to the original equation, the graph is not symmetric with respect to the origin.

11. (a) We substitute 2 for x in the equation for f to get
$$f(2) = \frac{3(2)}{2^2 - 1} = \frac{6}{3} = 2$$

(b) We substitute -2 for x in the equation for f to get
$$f(-2) = \frac{3(-2)}{(-2)^2 - 1} = \frac{-6}{3} = -2$$

(c) We substitute $(-x)$ for x in the equation for f to get
$$f(-x) = \frac{3(-x)}{(-x)^2 - 1} = \frac{-3x}{x^2 - 1}$$

(d) $-f(x) = -\dfrac{3x}{x^2 - 1}$

(e) We substitute $(x - 2)$ for x in the equation for f to get
$$f(x-2) = \frac{3(x-2)}{(x-2)^2 - 1} = \frac{3x - 6}{x^2 - 4x + 4 - 1} = \frac{3x - 6}{x^2 - 4x + 3}$$

(f) We substitute $2x$ for x in the equation for f to get
$$f(2x) = \frac{3(2x)}{(2x)^2 - 1} = \frac{6x}{4x^2 - 1}$$

13. (a) We substitute 2 for x in the equation for f to get
$$f(2) = \sqrt{2^2 - 4} = 0$$

(b) We substitute -2 for x in the equation for f to get
$$f(-2) = \sqrt{(-2)^2 - 4} = 0$$

(c) We substitute $(-x)$ for x in the equation for f to get
$$f(-x) = \sqrt{(-x)^2 - 4} = \sqrt{x^2 - 4} = f(x)$$

(d) $-f(x) = -\sqrt{x^2 - 4}$

(e) We substitute $(x - 2)$ for x in the equation for f to get
$$f(x-2) = \sqrt{(x-2)^2 - 4} = \sqrt{x^2 - 4x + 4 - 4} = \sqrt{x^2 - 4x}$$

(f) We substitute $2x$ for x in the equation for f to get
$$f(2x) = \sqrt{(2x)^2 - 4} = \sqrt{4x^2 - 4} = 2\sqrt{x^2 - 1}$$

15. (a) We substitute 2 for x in the equation for f to get
$$f(2) = \frac{2^2 - 4}{2^2} = 0$$

(b) We substitute -2 for x in the equation for f to get
$$f(-2) = \frac{(-2)^2 - 4}{(-2)^2} = 0$$

(c) We substitute $(-x)$ for x in the equation for f to get
$$f(-x) = \frac{(-x)^2 - 4}{(-x)^2} = \frac{x^2 - 4}{x^2} = f(x)$$

(d) $-f(x) = -\dfrac{x^2 - 4}{x^2}$

(e) We substitute $(x - 2)$ for x in the equation for f to get
$$f(x-2) = \frac{(x-2)^2 - 4}{(x-2)^2} = \frac{x^2 - 4x + 4 - 4}{x^2 - 4x + 4} = \frac{x^2 - 4x}{x^2 - 4x + 4}$$

(f) We substitute $2x$ for x in the equation for f to get

$$f(2x) = \frac{(2x)^2 - 4}{(2x)^2} = \frac{4x^2 - 4}{4x^2} = \frac{x^2 - 1}{x^2}$$

17. The denominator of the f cannot equal 0, so $x^2 - 9 \neq 0$ or $x \neq 3$ and $x \neq -3$. The domain is the set $\{x \mid x \neq 3 \text{ and } x \neq -3\}$.

19. The radicand must be nonnegative, so $2 - x \geq 0$ or $x \leq 2$. The domain is the set $\{x \mid x \leq 2\}$ or the interval $(-\infty, -2]$.

21. The radicand must be nonnegative and the denominator cannot equal 0. The domain is the set $\{x \mid x > 0\}$ or the interval $(0, \infty)$.

23. The denominator of the function f cannot equal 0, so $x^2 + 2x - 3 \neq 0$.
$$x^2 + 2x - 3 = 0$$
$$(x - 1)(x + 3) = 0$$
$$x - 1 = 0 \text{ or } x + 3 = 0$$
$$x = 1 \text{ or } x = -3$$
The domain is the set $\{x \mid x \neq 1 \text{ and } x \neq -3\}$.

25.
$$\frac{f(x+h) - f(x)}{h} = \frac{-2(x+h)^2 + (x+h) + 1 - [-2x^2 + x + 1]}{h}$$
$$= \frac{-2x^2 - 4xh - 2h^2 + x + h + 1 + 2x^2 - x - 1}{h}$$
$$= \frac{-4xh - 2h^2 + h}{h}$$
$$= \frac{\cancel{h}(-4x - 2h + 1)}{\cancel{h}}$$
$$= -4x - 2h + 1$$

27. (a) The domain of f is the set $\{x \mid -4 \leq x \leq 3\}$ or the interval $[-4, 3]$. The range of f is the set $\{y \mid -3 \leq y \leq 3\}$ or the interval $[-3, 3]$.

   (b) The x-intercept is $(0, 0)$, and the y-intercept is $(0, 0)$.

   (c) $f(-2) = -1$

   (d) $f(x) = -3$ when $x = -4$.

   (e) $f(x) > 0$ on the interval $(0, 3]$.

29. (a) The domain of f is the set of all real numbers or the interval $(-\infty, \infty)$. The range of f is the set $\{y \mid y \leq 1\}$ or the interval $(-\infty, 1]$.

   (b) f is increasing on the intervals $(-\infty, -1)$ and $(3, 4)$; f is decreasing on the intervals $(-1, 3)$ and $(4, \infty)$.

(c) The local maxima are 1 at $f(-1) = 1$ and 0 at $f(4) = 0$. There is a local minimum of -3 at $f(3) = -3$.

(d) The graph is not symmetric with respect to the x-axis, y-axis, or the origin.

(e) Since the graph of the function has no symmetry, the function is neither even nor odd.

(f) The x-intercepts are $(-2, 0)$, $(0, 0)$ and $(4, 0)$; the y-intercept is $(0, 0)$.

31. $f(-x) = (-x)^3 - 4(-x) = -x^3 + 4x = -f(x)$ Since $f(-x) = -f(x)$, the function is odd.

33. $h(-x) = \dfrac{1}{(-x)^4} + \dfrac{1}{(-x)^2} + 1 = \dfrac{1}{x^4} + \dfrac{1}{x^2} + 1 = h(x)$ Since $h(-x) = h(x)$, the function is even.

35. $G(-x) = 1 - (-x) + (-x)^3 = 1 + x - x^3$ Since $G(-x)$ is not equal to $G(x)$ or to $-G(x)$, the function is neither even nor odd.

37. $f(-x) = \dfrac{(-x)}{1 + (-x)^2} = \dfrac{-x}{1+x^2} = -f(x)$

Since $f(-x) = -f(x)$, the function is odd.

39. There is a local maximum of 4.043 at $x = -0.913$, and a local minimum of -2.043 at $x = 0.913$.

The function is increasing on the intervals $(-3, -0.913)$ and $(0.913, 3)$. The function is decreasing on the interval $(-0.913, 0.913)$.

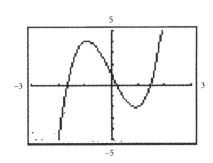

41. There is a local maximum of 1.532 at $x = 0.414$, and local minima of 0.543 at $x = -0.336$ and of -3.565 at $x = 1.798$.

The function is increasing on the intervals $(-0.336, 0.414)$ and $(1.798, 3)$. The function is decreasing on the intervals $(-2, -0.336)$ and $(0.414, 1.798)$.

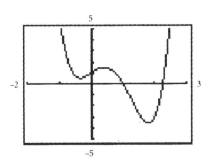

43.

(a) $\dfrac{\Delta y}{\Delta x} = \dfrac{f(2) - f(1)}{2 - 1} = \dfrac{[8 \cdot 2^2 - 2] - [8 \cdot 1^2 - 1]}{1} = 30 - 7 = 23$

(b) $\dfrac{\Delta y}{\Delta x} = \dfrac{f(1)-f(0)}{1-0} = \dfrac{[8 \cdot 1^2 - 1]-[8 \cdot 0^2 - 0]}{1} = 7 - 0 = 7$

(c) $\dfrac{\Delta y}{\Delta x} = \dfrac{f(4)-f(2)}{4-2} = \dfrac{[8 \cdot 4^2 - 4]-[8 \cdot 2^2 - 2]}{2} = \dfrac{124-30}{2} = 47$

45. $\dfrac{\Delta y}{\Delta x} = \dfrac{f(x)-f(2)}{x-2} = \dfrac{[2-5x]-[2-5 \cdot 2]}{x-2} = \dfrac{2-5x-2+10}{x-2}$

$$= \dfrac{-5x+10}{x-2} = \dfrac{-5(x-2)}{x-2} = -5$$

47. $\dfrac{\Delta y}{\Delta x} = \dfrac{f(x)-f(2)}{x-2} = \dfrac{[3x-4x^2]-[3 \cdot 2 - 4 \cdot 2^2]}{x-2} = \dfrac{3x-4x^2-6+16}{x-2}$

$$= \dfrac{-4x^2+3x+10}{x-2} = \dfrac{-(4x+5)(x-2)}{x-2} = -4x-5$$

49. (b), (c), (d), and (e) are graphs of functions because they pass the vertical line test.

51.

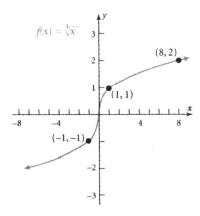

53. (a) The domain of f is the set $\{x \mid x > -2\}$ or the interval $(-2, \infty)$.

(b) To find the y-intercept we let $x = 0$ and solve. When $x = 0$, the equation for f is $f(x) = 3x$, and $f(0) = 0$. So the y-intercept is $(0, 0)$.

To find the x-intercept(s), we let $y = 0$ and solve.
If $-2 \le x < 1$, $y = f(x) = 3x = 0$ or $x = 0$. So there is an x-intercept at $(0, 0)$
If $x > 1$, $y = x + 1 = 0$ or $x = -1$. Since $x = -1$ is not in the interval $(1, \infty)$, we ignore this solution.

(c)

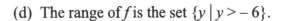

(d) The range of f is the set $\{y \mid y > -6\}$.

55. (a) The domain of f is the set $\{x \mid x \geq -4\}$ or the interval $[-4, \infty)$.

(b) To find the y-intercept we let $x = 0$ and solve. When $x = 0$, the equation for f is $f(x) = 1$. So the y-intercept is $(0, 1)$.

To find the x-intercept(s), we let $y = 0$ and solve.
If $-4 \leq x < 0$, $y = f(x) = x = 0$. Since $x = 0$ is not in the interval $[-4, 0)$, we ignore this solution.
If $x > 0$, $y = 3x = 0$ or $x = 0$. Since $x = 0$ is not in the interval $(0, \infty)$, we ignore this solution and we conclude that there is no x-intercept on this graph.

(c)

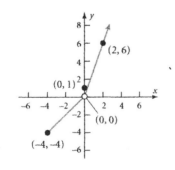

(d) The range of f is the set $\{y \mid y \geq -4$, but $y \neq 0\}$ or the interval $[-4, 0)$ and $(0, \infty)$.

57.

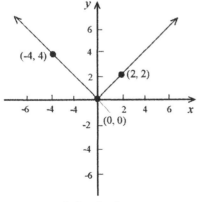

(a) $F(y) = |x|$

Subtract 4; vertical shift down 4 units.

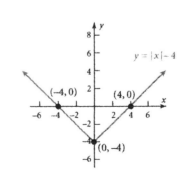

(b) $F(x) = |x| - 4$

The x-intercepts of F are $(-4, 0)$ and $(4, 0)$; the y-intercept is $(0, -4)$. The domain is all real numbers and the range is $\{y \mid y \geq -4\}$.

59.

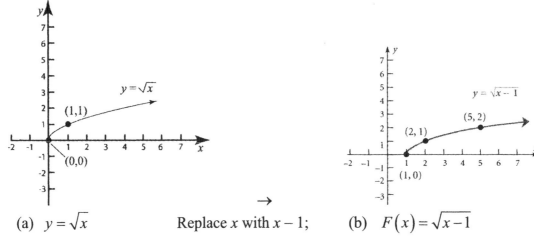

(a) $y = \sqrt{x}$ Replace x with $x - 1$;   (b) $F(x) = \sqrt{x-1}$
horizontal shift right 1 unit.

The x-intercept of F is $(1, 0)$; there is no y-intercept. The domain of F is the interval $[1, \infty)$ and the range is $[0, \infty)$.

61.

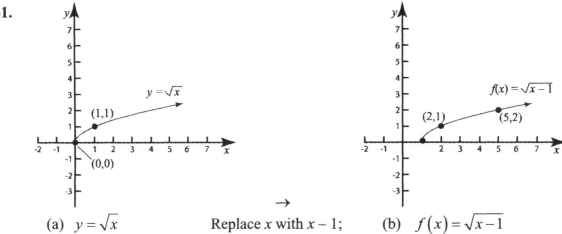

(a) $y = \sqrt{x}$ Replace x with $x - 1$;   (b) $f(x) = \sqrt{x-1}$
horizontal shift right 1 unit.

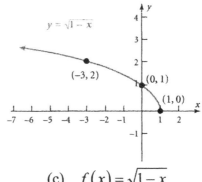

Replace $x - 1$ by $-(x - 1)$;   (c) $f(x) = \sqrt{1-x}$
reflect about the y-axis.

The x-intercept of f is $(1, 0)$; the y-intercept is $(0, 1)$. The domain of f is the set $\{x \mid x \le 1\}$ or the interval $[-\infty, 1)$. The range is the set $\{y \mid y \ge 0\}$ or the interval $[0, \infty)$.

63.

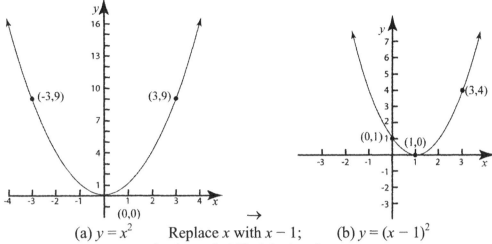

(a) $y = x^2$ Replace x with $x - 1$;   (b) $y = (x - 1)^2$
horizontal shift right 1 unit.

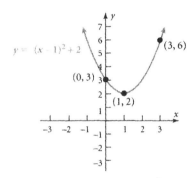

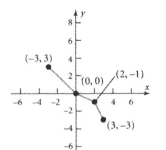

Add -2; vertical   (c) $f(x) = (x-1)^2 + 2$

shift up 2 units.

The y-intercept is $(0, 3)$; there is no x-intercept. The domain of f is all real numbers. The range is the set $\{y \mid y \geq 2\}$ or the interval $[2, \infty)$.

65.   (a) $y = f(-x)$ (b) $y = -f(x)$

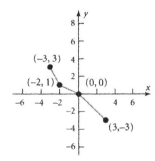

(c) $y = f(x + 2)$ (d) $y = f(x) + 2$

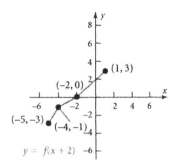

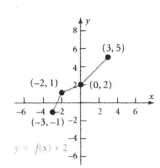

67.   Since f is linear, we first find the slope of the function using the 2 given points $(4, -5)$ and $(0, 3)$.

$$m = \frac{y_2 - y_1}{x_2 - x_1} = \frac{-5 - 3}{4 - 0} = \frac{-8}{4} = -2$$

Then we get the point-slope form of a line by using m and the point $(0, 3)$.

$$y - y_1 = m(x - x_1)$$
$$y - 3 = -2(x - 0)$$
$$y = -2x + 3$$

So the linear function is $f(x) = -2x + 3$.

69. $f(1) = 4$ means the point $(1, 4)$ satisfies the equation for f. So

$$4 = \frac{A(1) + 5}{6(1) - 2} = \frac{A + 5}{4}$$
$$16 = A + 5$$
$$A = 11$$

71.   Since the height is twice the radius, we can write $h = 2r$. Then the volume of the cylinder can be expressed as $V = \pi r^2 (2r) = 2\pi r^3$.

73.   (a) $R = R(x) = xp$

$$R(x) = x\left(-\frac{1}{6}x + 100\right)$$
$$= -\frac{1}{6}x^2 + 100x \qquad 0 \le x \le 600$$

(b) If 200 units are sold, $x = 200$ and the revenue is

$$R(200) = -\frac{1}{6}(200)^2 + 100(200) = \$13{,}333.33$$

75.   (a) To find the revenue function $R = R(x)$, we first solve the demand equation for p.

$$x = -5p + 100 \qquad (1)$$
$$5p = 100 - x$$
$$p = 20 - \frac{1}{5}x \qquad 0 \le x \le 100$$

We find the domain by using equation (1) and solving

$$\text{when } p = 0, \quad x = -5(0) + 100 = 100$$
$$\text{when } p = 20, \quad x = -5(20) + 100 = 0$$

So the revenue R can be expressed as

$$R(x) = x\left(20 - \frac{1}{5}x\right)$$

$$R(x) = 20x - \frac{1}{5}x^2 \qquad 0 \le x \le 100$$

(b) If 15 units are sold, $x = 15$, and the revenue is

$$R(15) = 20(15) - \frac{1}{5}(15)^2 = \$255.00$$

77. (a) The cost of making the drum is the sum of the costs of making the top and bottom and the side.

The amount of material used in the top and bottom is the area of the two circles,
$$A_{\text{top}} + A_{\text{bottom}} = \pi r^2 + \pi r^2 = 2\pi r^2 \text{ square centimeters.}$$
At \$0.06 per square centimeter, the cost of the top and the bottom of the drum is
$$C = 0.06(2\pi r^2) = 0.12\pi r^2 \text{ dollars.}$$

The amount of material used in the side of the drum is the area of the rectangle of material measured by the circumference of the top and the height of the drum.
$$A_{\text{side}} = 2\pi r h \text{ square centimeters.}$$
To express the area of the side as a function of r, we use the fact that we are told the volume of the drum is 500 cubic centimeters.
$$V = \pi r^2 h = 500$$
$$h = \frac{500}{\pi r^2}$$
So $A_{\text{side}} = 2\pi r h = 2\pi r\left(\dfrac{500}{\pi r^2}\right) = \dfrac{1000}{r}$ square centimeters.

At \$0.04 per square centimeter, the cost of making the side of the drum is
$$C = 0.04\left(\frac{1000}{r}\right) = \frac{40}{r} \text{ dollars.}$$

The total cost of making the drum is $C = C(r) = 0.12\pi r^2 + \dfrac{40}{r}$ dollars.

(b) If the radius is 4 cm, the cost of making the drum is
$$C(4) = 0.12\pi(4)^2 + \frac{40}{4} = \$16.03$$

(c) If the radius is 8 cm, the cost of making the drum is
$$C(8) = 0.12\pi(8)^2 + \frac{40}{8} = \$29.13$$

(d)

Making the can with a radius of 3.758 centimeters minimizes the cost of making the drum. The minimum cost is $15.97.

Chapter 1 Project

1. Since Avis has unlimited mileage, the cost of driving an Avis car x miles is
$$A = A(x) = 64.99$$
A is a constant function.

3.

If you drive more than 226 miles, Avis becomes the better choice.

5. If driving fewer than 130 miles, SaveALot Car Rental costs the least.
   If driving between 130 and 227 miles Enterprise is the most economical.
   If driving more than 227 miles Avis is the best buy.

7. If driving fewer than 53 miles, USave Car Rental is the cheapest.
   If driving between 53 and 130 miles, SaveALot Car Rental costs the least.
   If driving between 130 and 227 miles Enterprise is the most economical.
   If driving more than 227 miles Avis is the best buy.

Mathematical Questions from Professional Exams

1. (d) On the open interval $(-1, 2)$, The minimum value of f is 0, and the maximum value of f approaches 4. So the range of f is $0 \leq y < 4$.

3. (c) The domain of f is nonnegative. That is,
$$x^3 - x \geq 0$$
$$x\left(x^2 - 1\right) \geq 0$$
$$x(x-1)(x+1) \geq 0$$
Solving the related equation for 0 and testing a point in each interval gives
$$x(x-1)(x+1) = 0$$
$$x = 0 \quad \text{or} \quad x = 1 \quad \text{or} \quad x = -1$$

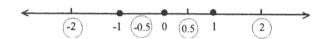

When $x = -2$, $(-2)^3 - (-2) = -6 < 0$, so the interval $(-\infty, -1)$ is not part of the domain.
When $x = -0.5$, $(-0.5)^3 - (-0.5) = 0.375 \geq 0$, so the interval $[-1, 0]$ is part of the domain.
When $x = 0.5$, $(0.5)^3 - (0.5) = -0.375 < 0$, so the interval $(0, 1)$ is not part of the domain.
When $x = 2$, $(2)^3 - (2) = 6 \geq 0$, so the interval $[1, \infty)$ is part of the domain.

Chapter 2
Classes of Functions

2.1 Quadratic Functions

1. (C) f is a quadratic function whose graph opens up, and whose vertex is
$$\left(-\frac{b}{2a}, f\left(-\frac{b}{2a}\right)\right) = (0, -1).$$
$$-\frac{b}{2a} = -\frac{0}{2} = 0; \qquad f(0) = 0^2 - 1 = -1$$

3. (F) f is a quadratic function whose graph opens up, and whose vertex is
$$\left(-\frac{b}{2a}, f\left(-\frac{b}{2a}\right)\right) = (1, 0).$$
$$-\frac{b}{2a} = \frac{2}{2} = 1; \qquad f(1) = 1^2 - 2 \cdot 1 + 1 = 0$$

5. (G) f is a quadratic function whose graph opens up, and whose vertex is
$$\left(-\frac{b}{2a}, f\left(-\frac{b}{2a}\right)\right) = (1, 1).$$
$$-\frac{b}{2a} = \frac{2}{2} = 1; \qquad f(1) = 1^2 - 2 \cdot 1 + 2 = 1$$

7. (H) f is a quadratic function whose graph opens up, and whose vertex is
$$\left(-\frac{b}{2a}, f\left(-\frac{b}{2a}\right)\right) = (1, -1).$$
$$-\frac{b}{2a} = \frac{2}{2} = 1; \qquad f(1) = 1^2 - 2 \cdot 1 = -1$$

9. $a = 1$, $b = 2$, $c = 0$. Since $a > 0$, the parabola opens up.

The x-coordinate of the vertex is $-\frac{b}{2a} = -\frac{2}{2(1)} = -1$.

The y-coordinate of the vertex is $f(-1) = (-1)^2 + 2(-1) = -1$.

So the vertex is $(-1, -1)$ and the axis of symmetry is the line $x = -1$.

Since $f(0) = c = 0$, the y-intercept is $(0, 0)$.

The x-intercepts are found by solving $f(x) = 0$.
$$x^2 + 2x = 0$$
$$x(x + 2) = 0$$
$$x = 0 \text{ or } x = -2$$

The x-intercepts are $(0, 0)$ and $(-2, 0)$.

The domain is the set of all real numbers or the interval $(-\infty, \infty)$; the range is the set $\{y \mid$

$y \geq -1$} or the interval $[-1, \infty)$.

The function is increasing to the right of the axis or on the interval $(-1, \infty)$, and it is decreasing to the left of the axis or on the interval $(-\infty, -1)$.

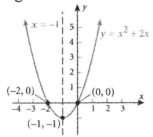

11. $a = -1, b = -6, c = 0$. Since $a < 0$, the parabola opens down.

The x-coordinate of the vertex is $-\dfrac{b}{2a} = -\dfrac{-6}{2(-1)} = -3$.

The y-coordinate of the vertex is $f(-3) = -(-3)^2 - 6(-3) = 9$.

So the vertex is $(-3, 9)$ and the axis of symmetry is the line $x = -3$.

Since $f(0) = c = 0$, the y-intercept is $(0, 0)$.

The x-intercepts are found by solving $f(x) = 0$.

$$-x^2 - 6x = 0$$
$$-x(x+6) = 0$$
$$x = 0 \text{ or } x = -6$$

The x-intercepts are $(0, 0)$ and $(-6, 0)$.

The domain is the set of all real numbers or the interval $(-\infty, \infty)$; the range is the set {$y \mid y \leq 9$} or the interval $(-\infty, 9]$.

The function is increasing to the left of the axis or on the interval $(-\infty, -3)$, and it is decreasing to the right of the axis or on the interval $(-3, \infty)$.

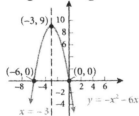

13. $a = 2, b = -8, c = 0$. Since $a > 0$, the parabola opens up.

The x-coordinate of the vertex is $-\dfrac{b}{2a} = -\dfrac{-8}{2(2)} = 2$.

The y-coordinate of the vertex is $f(2) = 2(2)^2 - 8(2) = -8$.

So the vertex is $(2, -8)$ and the axis of symmetry is the line $x = 2$.

Since $f(0) = c = 0$, the y-intercept is $(0, 0)$.

The x-intercepts are found by solving $f(x) = 0$.

$$2x^2 - 8x = 0$$
$$2x(x-4) = 0$$
$$x = 0 \text{ or } x = 4$$

The x-intercepts are $(0, 0)$ and $(4, 0)$.

The domain is the set of all real numbers or the interval $(-\infty, \infty)$; the range is the set {$y \mid y \geq -8$} or the interval $[-8, \infty)$.

The function is increasing to the right of the axis or on the interval $(2, \infty)$, and it is decreasing to the left of the axis or on the interval $(-\infty, 2)$.

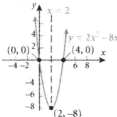

15. $a = 1$, $b = 2$, $c = -8$. Since $a > 0$, the parabola opens up.

The x-coordinate of the vertex is $-\dfrac{b}{2a} = -\dfrac{2}{2(1)} = -1$.

The y-coordinate of the vertex is $f(-1) = (-1)^2 + 2(-1) - 8 = -9$.

So the vertex is $(-1, -9)$ and the axis of symmetry is the line $x = -1$.

Since $f(0) = c = -8$, the y-intercept is $(0, -8)$.

The x-intercepts are found by solving $f(x) = 0$.

$$x^2 + 2x - 8 = 0$$
$$(x - 2)(x + 4) = 0$$
$$x = 2 \ \text{ or } \ x = -4$$

The x-intercepts are $(2, 0)$ and $(-4, 0)$.

The domain is the set of all real numbers or the interval $(-\infty, \infty)$; the range is the set $\{y \mid y \geq -9\}$ or the interval $[-9, \infty)$.

The function is increasing to the right of the axis or on the interval $(-1, \infty)$, and it is decreasing to the left of the axis or on the interval $(-\infty, -1)$.

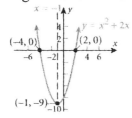

17. $a = 1$, $b = 2$, $c = 1$. Since $a > 0$, the parabola opens up.

The x-coordinate of the vertex is $-\dfrac{b}{2a} = -\dfrac{2}{2(1)} = -1$.

The y-coordinate of the vertex is $f(-1) = (-1)^2 + 2(-1) + 1 = 0$.

So the vertex is $(-1, 0)$ and the axis of symmetry is the line $x = -1$.

Since $f(0) = c = 1$, the y-intercept is $(0, 1)$.

The x-intercepts are found by solving $f(x) = 0$.

$$x^2 + 2x + 1 = 0$$
$$(x + 1)(x + 1) = 0$$
$$x = -1$$

The x-intercept is $(-1, 0)$.

The vertex and the x-intercept are the same, so we use symmetry and the y-intercept to obtain a third point $(-2, 1)$ on the graph.

The domain is the set of all real numbers or the interval $(-\infty, \infty)$; the range is the set

$\{y \mid y \geq 0\}$ or the interval $[0, \infty)$.
The function is increasing to the right of the axis or on the interval $(-1, \infty)$, and it is decreasing to the left of the axis or on the interval $(-\infty, -1)$.

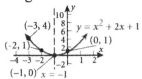

19. $a = 2$, $b = -1$, $c = 2$. Since $a > 0$, the parabola opens up.

The x-coordinate of the vertex is $-\dfrac{b}{2a} = -\dfrac{-1}{2(2)} = \dfrac{1}{4}$.

The y-coordinate of the vertex is $f\left(\dfrac{1}{4}\right) = 2\left(\dfrac{1}{4}\right)^2 - \left(\dfrac{1}{4}\right) + 2 = \dfrac{15}{8}$.

So the vertex is $\left(\dfrac{1}{4}, \dfrac{15}{8}\right) = (0.25, 1.875)$ and the axis of symmetry is the line $x = 0.25$.

Since $f(0) = c = 2$, the y-intercept is $(0, 2)$.

The x-intercepts are found by solving $f(x) = 0$. Since the discriminant $b^2 - 4ac = (-1)^2 - 4(2)(2) = -15$ is negative, the equation $f(x) = 0$ has no real solution, and therefore, the parabola has no x-intercept.

To graph the function, we choose a point and use symmetry. If we choose $x = 1$, $f(1) = 2(1)^2 - (1) + 2 = 3$. Using symmetry, we obtain the point $(-0.5, 3)$.

The domain is the set of all real numbers or the interval $(-\infty, \infty)$; the range is the set $\{y \mid y \geq 1.875\}$ or the interval $[1.875, \infty)$.
The function is increasing to the right of the axis or on the interval $(0.25, \infty)$, and it is decreasing to the left of the axis or on the interval $(-\infty, 0.25)$.

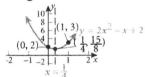

21. $a = -2$, $b = 2$, $c = -3$. Since $a < 0$, the parabola opens down.

The x-coordinate of the vertex is $-\dfrac{b}{2a} = -\dfrac{2}{2(-2)} = \dfrac{1}{2} = 0.5$.

The y-coordinate of the vertex is $f\left(\dfrac{1}{2}\right) = -2\left(\dfrac{1}{2}\right)^2 + 2\left(\dfrac{1}{2}\right) - 3 = -\dfrac{5}{2} = -2.5$.

So the vertex is $\left(\dfrac{1}{2}, -\dfrac{5}{2}\right) = (0.5, -2.5)$, and the axis of symmetry is the line $x = 0.5$.

Since $f(0) = c = -3$, the y-intercept is $(0, -3)$.

The x-intercepts are found by solving $f(x) = 0$. Since the discriminant

$$b^2 - 4ac = 2^2 - 4(-2)(-3) = -20$$

is less than zero, the equation $f(x) = 0$ has no real solution, and therefore, the parabola has no x-intercept.

To graph the function, we choose an additional point and use symmetry. If we choose $x = 2$, $f(2) = -2(2)^2 + 2(2) - 3 = -7$. Using symmetry, we obtain the point $(-1, -7)$.

The domain is the set of all real numbers or the interval $(-\infty, \infty)$; the range is the set $\{y \,|\, y \le -2.5\}$ or the interval $(-\infty, -2.5]$.

The function is increasing to the left of the axis or on the interval $(-\infty, 0.5)$, and it is decreasing to the right of the axis or on the interval $(0.5, \infty)$.

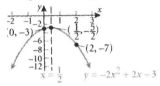

23. $a = 3$, $b = 6$, $c = 2$. Since $a > 0$, the parabola opens up.

The x-coordinate of the vertex is $-\dfrac{b}{2a} = -\dfrac{6}{2(3)} = -1$.

The y-coordinate of the vertex is $f(-1) = 3(-1)^2 + 6(-1) + 2 = -1$.
So the vertex is $(-1, -1)$ and the axis of symmetry is the line $x = -1$.

Since $f(0) = c = 2$, the y-intercept is $(0, 2)$.

The x-intercepts are found by solving $f(x) = 0$. Using the quadratic formula, we obtain

$$x = \frac{-6 \pm \sqrt{6^2 - 4(3)(2)}}{2(3)} = \frac{-6 \pm \sqrt{36 - 24}}{6} = \frac{-6 \pm \sqrt{12}}{6} = \frac{-3 \pm \sqrt{3}}{3}$$

$$x \approx -1.58 \text{ or } x \approx -0.42$$

The x-intercepts are approximately $(-1.58, 0)$ and $(-0.42, 0)$.

The domain is the set of all real numbers or the interval $(-\infty, \infty)$; the range is the set $\{y \,|\, y \ge -1\}$ or the interval $[-1, \infty)$.

The function is increasing to the right of the vertex or on the interval $(-1, \infty)$, and it is decreasing to the left of the vertex or on the interval $(-\infty, -1)$.

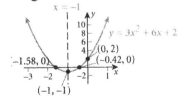

25. $a = -4$, $b = -6$, $c = 2$. Since $a < 0$, the parabola opens down.

The x-coordinate of the vertex is $-\dfrac{b}{2a} = -\dfrac{-6}{2(-4)} = -\dfrac{3}{4} = -0.75$.

The y-coordinate of the vertex is $f\left(-\dfrac{3}{4}\right) = -4\left(-\dfrac{3}{4}\right)^2 - 6\left(-\dfrac{3}{4}\right) + 2 = \dfrac{17}{4}$.

So the vertex is $\left(-\dfrac{3}{4}, \dfrac{17}{4}\right) = (-0.75, 4.25)$, and the axis of symmetry is the line $x = -0.75$.

Since $f(0) = c = 2$, the y-intercept is $(0, 2)$.

The x-intercepts are found by solving $f(x) = 0$. Using the quadratic formula, we obtain

$$x = \frac{6 \pm \sqrt{(-6)^2 - 4(-4)(2)}}{2(-4)} = \frac{6 \pm \sqrt{36 + 32}}{-8} = \frac{6 \pm \sqrt{68}}{-8} = \frac{3 \pm \sqrt{17}}{-4}$$

$$x \approx -1.78 \quad \text{or} \quad x \approx 0.28$$

The x-intercepts are approximately $(-1.78, 0)$ and $(0.28, 0)$.

The domain is the set of all real numbers or the interval $(-\infty, \infty)$; the range is the set $\{y \mid y \le 4.25\}$ or the interval $(-\infty, 4.25]$.

The function is increasing to the left of the axis or on the interval $(-\infty, -0.75)$, and it is decreasing to the right of the axis or on the interval $(-0.75, \infty)$.

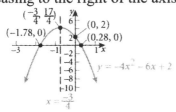

27. $a = 2$, $b = 12$, $c = 0$. Since $a > 0$, the parabola opens up, and the function has a minimum value. The minimum value occurs at

$$x = -\frac{b}{2a} = -\frac{12}{2(2)} = -3$$

The minimum value is $f(-3) = 2(-3)^2 + 12(-3) = -18$.

29. $a = 2$, $b = 12$, $c = -3$. Since $a > 0$, the parabola opens up, and the function has a minimum value. The minimum value occurs at

$$x = -\frac{b}{2a} = -\frac{12}{2(2)} = -3$$

The minimum value is $f(-3) = 2(-3)^2 + 12(-3) - 3 = -21$.

31. $a = -1$, $b = 10$, $c = -4$. Since $a < 0$, the parabola opens down, and the function has a maximum value. The maximum value occurs at

$$x = -\frac{b}{2a} = -\frac{10}{2(-1)} = 5$$

The maximum value is $f(5) = -(5)^2 + 10(5) - 4 = 21$.

33. $a = -3$, $b = 12$, $c = 1$. Since $a < 0$, the parabola opens down and the function has a maximum value. The maximum value occurs at

$$x = -\frac{b}{2a} = -\frac{12}{2(-3)} = 2$$

The maximum value is $f(2) = -3(2)^2 + 12(2) + 1 = 13$.

35. If $r_1 = -3$, and $r_2 = 1$, and $f(x) = a(x - r_1)(x - r_2)$

(a) Then if $a = 1$, $f(x) = 1(x - (-3))(x - 1) = (x + 3)(x - 1)$.
If $a = 2$, $f(x) = 2(x - (-3))(x - 1) = 2(x + 3)(x - 1)$.
If $a = -2$, $f(x) = -2(x - (-3))(x - 1) = -2(x + 3)(x - 1)$.
If $a = 5$, $f(x) = 5(x - (-3))(x - 1) = 5(x + 3)(x - 1)$.

(b) The x-intercepts are found by solving $f(x) = 0$.
$$a(x - r_1)(x - r_2) = 0$$
$$(x - r_1)(x - r_2) = 0$$
So the value of a, $a \neq 0$, has no affect on the x-intercept.

The y-intercept is found by letting $x = 0$ and simplifying.
$$f(x) = a(x - r_1)(x - r_2)$$
$$f(0) = a(0 - r_1)(0 - r_2)$$
$$f(0) = ar_1r_2$$
So the y-intercept is $(0, ar_1r_2)$ is the product of the a and the x-intercepts.

(c) The axis of symmetry is the line $x = -\dfrac{b}{2a}$. To determine b, we multiply out the factors of f.

$$\begin{aligned} f(x) &= a(x - r_1)(x - r_2) \\ &= a(x^2 - r_2 x - r_1 x + r_1 r_2) \\ &= a[x^2 - (r_1 + r_2) x + r_1 r_2] \\ &= ax^2 - a(r_1 + r_2) x + ar_1 r_2 \end{aligned}$$

We find $b = -a(r_1 + r_2)$. The line of symmetry is
$$x = -\frac{b}{2a} = -\frac{-a(r_1 + r_2)}{2a} = \frac{r_1 + r_2}{2}$$
which does not involve a. So the value of a does not affect the axis of symmetry.

(d) The vertex is the point $\left(\dfrac{r_1 + r_2}{2}, f\left(\dfrac{r_1 + r_2}{2} \right) \right)$. If we evaluate f at $\dfrac{r_1 + r_2}{2}$, we get

$$f\left(\frac{r_1 + r_2}{2} \right) = a\left(\frac{r_1 + r_2}{2} - r_1 \right)\left(\frac{r_1 + r_2}{2} - r_2 \right)$$
$$= a\left[\left(\frac{r_1 + r_2}{2} - r_1 \right)\left(\frac{r_1 + r_2}{2} - r_2 \right) \right]$$

We see the y-value of the vertex is changed by a factor of a.

(e) The x-coordinate of the vertex is $x = \dfrac{r_1 + r_2}{2}$. The midpoint of the x-intercepts is

$$\text{midpoint} = \left(\frac{x_1 + x_2}{2}, \frac{y_1 + y_2}{2} \right)$$

$$= \left(\frac{r_1 + r_2}{2}, \frac{0 + 0}{2} \right) = \left(\frac{r_1 + r_2}{2}, 0 \right)$$

The x-coordinate of the vertex and the midpoint of the x-intercepts is the same.

37. Since R is a quadratic function with $a = -4 < 0$, the vertex will give the maximum revenue.

The unit price to be charged should be $p = -\dfrac{b}{2a} = -\dfrac{4000}{2(-4)} = 500$ dollars.

If the dryers cost $500, the revenue R will be maximized. The maximum revenue will be
$$R(500) = -4(500)^2 + 4000(500) = -1,000,000 + 2,000,000 = \$1,000,000$$

39. (a) $R(x) = xp$
$$R(x) = x\left(-\frac{1}{6}x + 100 \right) = -\frac{1}{6}x^2 + 100x \qquad 0 \le x \le 600$$

(b) If 200 units are sold, $x = 200$, and the revenue R is
$$R(200) = -\frac{1}{6}(200)^2 + 100(200) = \frac{-40,000 + 120,000}{6} = \frac{80,000}{6} = \$13,333.33$$

(c) Since R is a quadratic function with $a = -\dfrac{1}{6} < 0$, the vertex will give the maximum

revenue. Revenue is maximized when $x = -\dfrac{b}{2a} = -\dfrac{100}{2\left(-\dfrac{1}{6}\right)} = 300$ units are sold. The

maximum revenue is $R(300) = -\dfrac{1}{6}(300)^2 + 100(300) = -15,000 + 30,000 = \$15,000$

(d) The company should charge $p = -\dfrac{1}{6}(300) + 100 = -50 + 100 = \$50.00$ per unit to

maximize revenue.

41. (a) We first solve the demand equation for x.
$$x = -5p + 100 \qquad\qquad 0 \le p \le 20,$$
$$5p = 100 - x$$
$$p = \frac{100 - x}{5} = 20 - \frac{1}{5}x$$

Since when $p = 0$, $x = 100$ and when $p = 20$, $x = 0$, the domain of the function p is $0 \le x \le 100$.

The revenue function R is
$$R(x) = x\left(20 - \frac{1}{5}x \right) = 20x - \frac{1}{5}x^2 \qquad 0 \le x \le 100$$

(b) If 15 units are sold, $x = 15$, and the revenue R is
$$R(15) = 20(15) - \frac{1}{5}(15)^2 = 300 - 45 = \$255$$

(c) Since R is a quadratic function with $a = -\frac{1}{5} < 0$, the vertex will give the maximum revenue. Revenue is maximized when $x = -\frac{b}{2a} = -\frac{20}{2\left(-\frac{1}{5}\right)} = 50$ units are sold. The maximum revenue is $R(50) = 20(50) - \frac{1}{5}(50)^2 = 1000 - 500 = \$500$

(d) The company should charge $p = 20 - \frac{1}{5}(50) = 20 - 10 = \$10$ per unit to maximize revenue.

43. If the rectangle is shown at right and we are told the width is x and the perimeter, or the distance around the edge of the rectangle, is 400 yards, then we can find the length of the rectangle.
$$P = 2l + 2w$$
$$400 = 2l + 2x$$
$$l = \frac{400 - 2x}{2} = 200 - x \text{ yards}$$

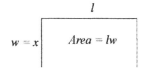

(a) $A(x) = lw = (200 - x)x = 200x - x^2$ square yards.

(b) A is a quadratic function with $a = -1$, $b = 200$, and $c = 0$. Since $a < 0$, the vertex gives the maximum area. The area is maximum when $x = -\frac{b}{2a} = -\frac{200}{2(-1)} = 100$ yards.

(c) The maximum area is $A(100) = 200(100) - 100^2 = 10{,}000$ square yards.

45. From the figure we see that the width of the plot measures x meters and the length of the plot measures $4000 - 2x$ meters. The area of the plot is
$$A(x) = lw = (4000 - 2x)x = 4000x - 2x^2 \text{ meters squared.}$$

A is a quadratic function with $a = -2$, $b = 4000$, and $c = 0$. Since $a < 0$, the vertex gives the maximum area. The area is maximum when $x = -\frac{b}{2a} = -\frac{4000}{2(-2)} = 1000$ meters. The maximum area is $A(1000) = 4000(1000) - 2(1000)^2 = 2{,}000{,}000$ meters squared.

47.

(a) h is a quadratic function with $a = -\frac{32}{50^2}$, $b = 1$, and $c = 200$. Since $a < 0$, the vertex gives the maximum height.

The height is maximum when $x = -\dfrac{b}{2a} = -\dfrac{1}{2\left(\dfrac{-32}{50^2}\right)} = \dfrac{2500}{64} = 39.06$ feet from the cliff.

(b) The maximum height of the projectile is
$$h(39.06) = \dfrac{-32(39.06)^2}{50^2} + 39.06 + 200 = 219.53 \text{ feet above the water.}$$

(c) The projectile will hit the water when $h(x) = 0$.

$$-\dfrac{32}{50^2}x^2 + x + 200 = 0$$
$$x = 170.02 \text{ feet from the cliff.}$$

(d)

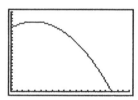

(e)

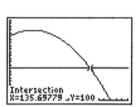

When the projectile is 100 feet above the water, it is 135.70 feet from the base of the cliff.

49. If we denote the depth of the rain gutter by x, then the area A of the cross-section is given by
$$A = lw$$
$$A(x) = (12 - 2x)x$$
$$= 12x - 2x^2$$
The function A is quadratic, with $a = -2 < 0$, so the vertex is the maximum point. The cross-sectional area is maximum when
$$x = -\dfrac{b}{2a} = -\dfrac{12}{2(-2)} = 3 \text{ inches.}$$

51. If x denotes the width of the rectangle (and the diameter of the circle), then the perimeter of the track is
$$P = 2l + 2\left(\dfrac{1}{2}\pi x\right) = 2l + \pi x = 400$$

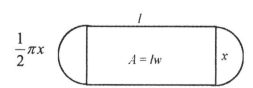

We can express l in terms of x by solving the equation for l.

$$l = \frac{400 - \pi x}{2}$$

The area of the rectangle is

$$A = \left(\frac{400 - \pi x}{2}\right)x = 200x - \frac{\pi}{2}x^2$$

A is a maximum when

$$x = -\frac{b}{2a} = -\frac{200}{2 \cdot \left(-\frac{\pi}{2}\right)} = \frac{200}{\pi} \approx 63.66 \text{ meters, and}$$

$$l = \frac{1}{2}\left[400 - \pi\left(\frac{200}{\pi}\right)\right] = \frac{1}{2}(200) = 100 \text{ meters.}$$

53. (a) Since $H(x)$ is a quadratic function, the income level for which there are the most hunters is given by x-value of the vertex. For this function $a = -1.01$, $b = 114.3$, and $c = 451$.

$$x = -\frac{b}{2a} = -\frac{114.3}{2(-1.01)} = 56.584$$

The most hunters have an income level of approximately \$56,584.
There are about

$$H(56.584) = -1.01(56.584)^2 + 114.3(56.584) + 451 = 3684.785$$

About 3685 hunters have an annual income of \$56,584.

(b)

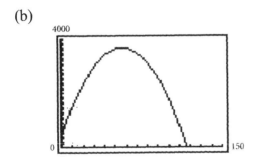

The number of hunters earning between \$20,000 and \$40,000 is increasing.

55. (a) $M(23) = 0.76(23)^2 - 107.00(23) + 3854.18 = 1795.2$

Approximately 1795 males who are 23 years old are murdered.

(b) Using a graphing utility, and finding the intersection of $M(x)$ and the line $y = 1456$, we find that 1456 males age 28 are murdered.

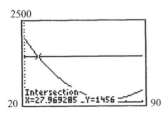

0

(c)

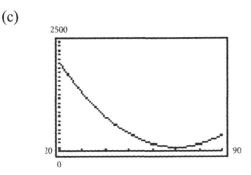

(d) The number of male murder victims decreases with age until age 70, and then it begins to increase.

57. The reaction rate is modeled by the quadratic function V.
$$V = V(x) = akx - kx^2$$
where $a = -k$, $b = ak$, and $c = 0$. The reaction rate is maximum at the vertex of V.
$$x = -\frac{b}{2a} = -\frac{ak}{2(-k)} = \frac{a}{2}$$
The reaction rate is maximum when half the initial amount of the compound is present.

59. $y = f(x) = -5x^2 + 8$

$y_0 = f(-1) = -5(-1)^2 + 8 = 3$

$y_1 = f(0) = -5(0)^2 + 8 = 8$

$y_2 = f(1) = -5(1)^2 + 8 = 3$

$$\text{Area} = \frac{h}{3}(y_0 + 4y_1 + y_2)$$

$$= \frac{1}{3}(3 + 4(8) + 3)$$

$$= \frac{38}{3} \text{ square units}$$

61. $y = f(x) = x^2 + 3x + 5$

$y_0 = f(-4) = (-4)^2 + 3(-4) + 5 = 9$

$y_1 = f(0) = (0)^2 + 3(0) + 5 = 5$

$y_2 = f(4) = (4)^2 + 3(4) + 5 = 33$

$$\text{Area} = \frac{h}{3}(y_0 + 4y_1 + y_2)$$

$$= \frac{4}{3}(9 + 4(5) + 33)$$

$$= \frac{248}{3} \text{ square units}$$

63. The rectangle is drawn on the right. The width of the rectangle is x units, and the length is $y = 10 - x$ units.

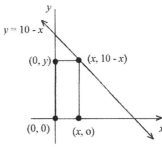

The area A of the rectangle is
$$A = lw$$
$$A(x) = (10 - x)x$$
$$= 10x - x^2$$
A is a quadratic function with $a = -1$, $b = 10$, and $c = 0$. Since $a < 0$, the maximum area occurs at the vertex of A. That is, when the width of the rectangle is
$$x = -\frac{b}{2a} = -\frac{10}{2(-1)} = 5$$

The largest area enclosed by the rectangle is
$$A = 10(5) - 5^2 = 25 \text{ square units.}$$

65. Functions will vary. All answers should have $a < 0$ and be perfect squares.

67.

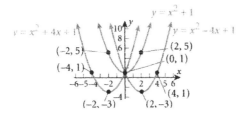

Descriptions may vary.

69. Answers will vary.

2.2 Power Functions; Polynomial Functions; Rational Functions

1. Answers will vary.

3. origin

5.

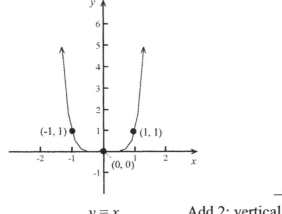

$$y = x \qquad \text{Add 2; vertical} \qquad f(x) = y + 2 = x^6 + 2$$
$$\text{shift up 2 units.}$$

7.

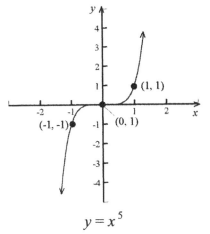

$$y = x^5$$

$\longrightarrow$
Replace y by $-y$;

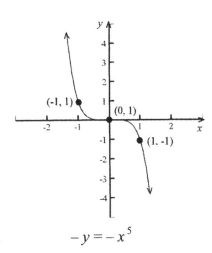

$$-y = -x^5$$

$\longrightarrow$
Add 2; vertical
shift up 2 units.

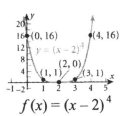

$$f(x) = -y + 2 = -x^5 + 2$$

9.

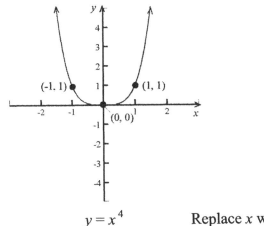

$$y = x^4$$

$\longrightarrow$
Replace x with $x - 2$;
horizontal shift to the
right 2 units.

$$f(x) = (x - 2)^4$$

11. f is a polynomial. Its degree is 3.

13. g is a polynomial. Its degree is 2.

15. f is not a polynomial. The exponent of x is -1.

17. g is not a polynomial. One of the exponents is not an integer.

19. F is a polynomial. Its degree is 4.

21. G is a polynomial. Its degree is 4.

23. $y = 3x^4$

25. $y = -2x^5$

27. Multiply the function out. $5(x+1)^2 (x-2) = 5(x^2 + 2x + 1)(x-2)$
$$= (5x^2 + 10x + 5)(x-2)$$
$$= 5x^3 - 10x^2 + 10x^2 - 20x + 5x - 10$$
$$= 5x^3 - 15x - 10$$
So the power function is $y = 5x^3$.

29. R is a rational function. The domain is all real numbers except those for which the denominator is 0.
$$x - 3 \neq 0 \text{ or } x \neq 3$$
The domain of R is the set $\{x \mid x \neq 3\}$.

31. H is a rational function. The domain is all real numbers except those for which the denominator is 0.
$$(x-2)(x+4) \neq 0$$
$$x - 2 \neq 0 \quad \text{or} \quad x + 4 \neq 0$$
$$x \neq 2 \quad \text{or} \quad x \neq -4$$

The domain of H is the set $\{x \mid x \neq 2 \text{ and } x \neq -4\}$.

33. F is a rational function. The domain is all real numbers except those for which the denominator is 0.
$$2x^2 - 5x - 3 \neq 0$$
$$(2x+1)(x-3) \neq 0$$
$$2x + 1 \neq 0 \quad \text{or} \quad x - 3 \neq 0$$
$$x \neq -\frac{1}{2} \quad \text{or} \quad x \neq 3$$
The domain of F is the set $\left\{x \mid x \neq -\frac{1}{2} \text{ and } x \neq 3\right\}$.

35. R is a rational function. The domain is all real numbers except those for which the denominator is 0.
$$x^3 - 8 \neq 0$$
$$(x-2)(x^2 + 2x + 4) \neq 0 \qquad$$
$$x - 2 \neq 0 \qquad\qquad\qquad x^2 + 2x + 4 = 0 \text{ has a negative discriminant,}$$
$$\qquad\qquad\qquad\qquad\qquad \text{and so has no real solutions.}$$
$$x \neq 2$$

The domain of R is the set $\{x \mid x \neq 2\}$.

37. H is a rational function. The domain is all real numbers except those for which the denominator is 0.
$$x^2 + 4 \neq 0$$

$x^2 + 4 = 0$ has a negative discriminant, and so has no real solutions. The denominator never equals 0.

The domain of H is the set of all real numbers or the interval $(-\infty, \infty)$.

39. R is a rational function. The domain is all real numbers except those for which the denominator is 0.

$$4(x^2 - 9) \neq 0$$
$$4(x - 3)(x + 3) \neq 0$$
$$x - 3 \neq 0 \quad \text{or} \quad x + 3 \neq 0$$
$$x \neq 3 \quad \text{or} \quad x \neq -3$$

The domain of R is the set $\{x \mid x \neq 3 \text{ and } x \neq -3\}$.

41. (a) The year 2000 is 70 years since the year 1930. $(2000 - 1930 = 70)$
The percentage of union membership in 2000 is $u(70)$.
$$u(70) = 11.93 + 1.9(70) - 0.052(70^2) + 0.00037(70^3)$$
$$= 17.04\%$$
(b) $u(75) = 11.93 + 1.9(75) - 0.052(75^2) + 0.00037(75^3) = 18.02\%$
Answers will vary.

2.3 Exponential Functions

1. (a) $3^{2.2} = 11.2116$ (b) $3^{2.23} = 11.5873$
(c) $3^{2.236} = 11.6639$ (d) $3^{\sqrt{5}} = 11.6648$

3. (a) $2^{3.14} = 8.8152$ (b) $2^{3.141} = 8.8214$
(c) $2^{3.1415} = 8.8244$ (d) $2^{\pi} = 8.8250$

5. (a) $3.1^{2.7} = 21.2166$ (b) $3.14^{2.71} = 22.2167$
(c) $3.141^{2.718} = 22.4404$ (d) $\pi^e = 22.4592$

7. $e^{1.2} = 3.3201$

9. $e^{-0.85} = 0.4274$

11.

x	$f(x)$	$\dfrac{f(x)}{f(x-1)}$
-1	3	
0	6	$\dfrac{6}{3}=2$
1	12	$\dfrac{12}{6}=2$
2	18	$\dfrac{18}{12}=\dfrac{3}{2}$
3	30	$\dfrac{30}{18}=\dfrac{5}{3}$

The ratio of consecutive outputs is not constant for unit increases in inputs. So the function f is not exponential.

13.

x	$H(x)$	$\dfrac{H(x)}{H(x-1)}$
-1	$\dfrac{1}{4}$	
0	1	$\dfrac{1}{\frac{1}{4}}=4$
1	4	$\dfrac{4}{1}=4$
2	16	$\dfrac{16}{4}=4$
3	64	$\dfrac{64}{16}=4$

The ratio of consecutive outputs is constant for unit increases in inputs. So the function H is exponential. The base $a=4$.

15.

x	$f(x)$	$\dfrac{f(x)}{f(x-1)}$
-1	$\dfrac{3}{2}$	
0	3	$\dfrac{3}{\frac{3}{2}}=2$
1	6	$\dfrac{6}{3}=2$
2	12	$\dfrac{12}{6}=2$
3	24	$\dfrac{24}{12}=2$

The ratio of consecutive outputs is constant for unit increases in inputs. So the function f is exponential. The base $a=2$.

17.

x	$H(x)$	$\dfrac{H(x)}{H(x-1)}$
-1	2	
0	4	$\dfrac{4}{2} = 2$
1	6	$\dfrac{6}{4} = \dfrac{3}{2}$
2	8	$\dfrac{8}{6} = \dfrac{4}{3}$
3	10	$\dfrac{10}{8} = \dfrac{5}{4}$

The ratio of consecutive outputs is not constant for unit increases in inputs. So the function H is not exponential.

19.   B.   This is the graph of $y = 3^x$ reflected about the y-axis. It is the graph of $y = 3^{-x}$.

21.   D.   This is the graph of $y = 3^x$ reflected over both the x- and the y-axes. It is the graph of $y = -3^{-x}$.

23.   A.   This is the graph of $y = 3^x$.

25.   E.   This is the graph of $y = 3^x$ vertically shifted down one unit. It is the graph of $y = 3^x - 1$.

27.

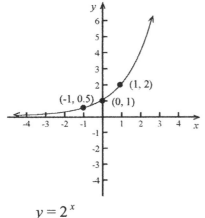

$y = 2^x$

Add 1; vertical shift up 1 unit.

$\rightarrow$

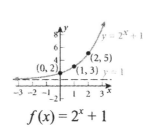

$f(x) = 2^x + 1$

The domain of f is all real numbers or the interval $(-\infty, \infty)$; the range is the set $\{y \mid y > 1\}$ or the interval $(1, \infty)$. The horizontal asymptote is the line $y = 1$.

29.

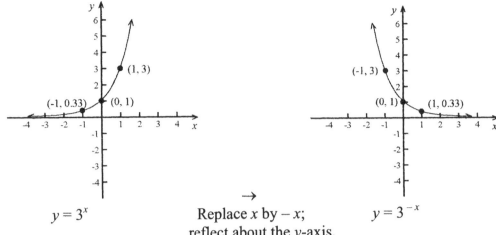

$y = 3^x$ Replace x by $-x$; $y = 3^{-x}$
reflect about the y-axis.

$\longrightarrow$
Subtract 2; vertical $f(x) = 3^{-x} - 2$
shift down 2 units.

The domain of f is all real numbers or the interval $(-\infty, \infty)$; the range is the set $\{y \mid y > -2\}$ or the interval $(-2, \infty)$. The horizontal asymptote is the line $y = -2$.

31.

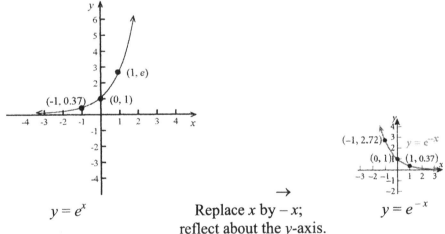

$y = e^x$ Replace x by $-x$; $y = e^{-x}$
reflect about the y-axis.

The domain of f is all real numbers or the interval $(-\infty, \infty)$; the range is the set $\{y \mid y > 0\}$ or the interval $(0, \infty)$. The horizontal asymptote is the line $y = 0$.

33.

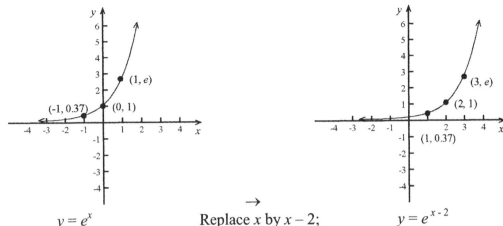

$$y = e^x$$

→

Replace x by $x - 2$;
horizontal shift 2 units to the right.

$$y = e^{x-2}$$

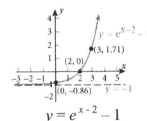

→

Subtract 1; vertical
shift 1 unit down.

$$y = e^{x-2} - 1$$

The domain of f is all real numbers or the interval $(-\infty, \infty)$; the range is the set $\{y \mid y > -1\}$ or the interval $(-1, \infty)$. The horizontal asymptote is the line $y = -1$.

35. $2^{2x+1} = 4 = 2^2$

Since the bases are the same, we obtain
$$2x + 1 = 2$$
$$2x = 1$$
$$x = \frac{1}{2}$$

$$\left\{\frac{1}{2}\right\}$$

37. $3^{x^3} = 9^x = \left(3^2\right)^x = 3^{2x}$

Since the bases are the same, we obtain
$$x^3 = 2x$$
$$x^3 - 2x = 0$$
$$x(x^2 - 2) = 0$$
$$x = 0 \quad \text{or} \quad x^2 - 2 = 0$$
$$x = \pm\sqrt{2}$$

$$\left\{-\sqrt{2}, 0, \sqrt{2}\right\}$$

39.

$$8^{x^2 - 2x} = \frac{1}{2}$$

$$\left(2^3\right)^{x^2 - 2x} = 2^{-1}$$

$$2^{3\left(x^2 - 2x\right)} = 2^{-1}$$

Since the bases are the same, we obtain
$$3(x^2 - 2x) = -1$$
$$3x^2 - 6x + 1 = 0, \text{ where } a = 3, b = -6, c = 1$$

41.

$$2^x \cdot 8^{-x} = 4^x$$

$$2^x \cdot \left(2^3\right)^{-x} = \left(2^2\right)^x$$

$$2^x \cdot 2^{-3x} = 2^{2x}$$

$$2^{x - 3x} = 2^{2x}$$

Since the bases are the same, we obtain
$$x - 3x = 2x$$
$$-4x = 0$$
$$x = 0$$

$$\{0\}$$

Using the quadratic formula we find

$$x = \frac{6 \pm \sqrt{(-6)^2 - 4(3)(1)}}{2(3)} = \frac{6 \pm \sqrt{36 - 12}}{6}$$

$$= \frac{6 \pm \sqrt{24}}{6} = \frac{6 \pm 2\sqrt{6}}{6} = \frac{3 \pm \sqrt{6}}{3}$$

$$\left\{ \frac{3 - \sqrt{6}}{3}, \frac{3 + \sqrt{6}}{3} \right\}$$

43.
$$\left(\frac{1}{5} \right)^{2-x} = 25$$
$$5^{-(2-x)} = 5^2$$

Since the bases are the same, we obtain
$$-(2 - x) = 2$$
$$x = 4$$

{4}

45.
$$4^x = 8$$
$$2^{2x} = 2^3$$

Since the bases are the same, we obtain
$$2x = 3$$

$$x = \frac{3}{2}$$

$$\left\{ \frac{3}{2} \right\}$$

47.
$$e^{x^2} = \left(e^{3x} \right) \cdot \frac{1}{e^2}$$
$$e^{x^2} = e^{3x-2}$$

Since the bases are the same, we obtain
$$x^2 = 3x - 2$$
$$x^2 - 3x + 2 = 0$$
$$(x - 2)(x - 1) = 0$$
$$x - 2 = 0 \quad \text{or} \quad x - 1 = 0$$
$$x = 2 \quad \text{or} \quad x = 1$$

{2, 1}

49.
$$4^{-2x} = (4^x)^{-2}$$
So if $4^x = 7$, $4^{-2x} = 7^{-2} = \frac{1}{7^2} = \frac{1}{49}$

51. If $3^{-x} = 2$, then $3^x = 2^{-1}$, and
$$3^{2x} = (2^{-1})^2$$
$$= 2^{-2} = \frac{1}{4}$$

53. The graph is increasing and $\left(-1, \frac{1}{3} \right)$, (0, 1), and (1, 3) are points on the graph. So the function $f(x) = 3^x$.

55. The graph is decreasing, negative, and $\left(-1, -\dfrac{1}{6}\right)$, $(0, -1)$, and $(1, -6)$ are points on the graph. So the function $f(x) = -6^x$.

57. (a) The percent of light passing through 10 panes of glass is
$$p(10) = 100e^{-0.03(10)} = 74.08\%$$

(b) The percent of light passing through 25 panes of glass is
$$p(25) = 100e^{-0.03(25)} = 47.24\%$$

59. (a) After 30 days there will be
$$w(30) = 50e^{-0.004(30)} = 44.35 \text{ watts of power.}$$

(b) After 365 days there will be
$$w(365) = 50e^{-0.004(365)} = 11.61 \text{ watts of power.}$$

61. After 1 hour, there will be
$$D(1) = 5e^{-0.4(1)} = 3.35 \text{ milligrams of drug will be in the patient's bloodstream.}$$

After 6 hours, there will be
$$D(6) = 5e^{-0.4(6)} = 0.45 \text{ milligrams of drug will be in the patient's bloodstream.}$$

63. (a) The probability that a car will arrive within 10 minutes of 12:00 PM is
$$F(10) = 1 - e^{-0.1(10)} = 0.632$$

(b) The probability that a car will arrive within 40 minutes of 12:00 PM is
$$F(40) = 1 - e^{-0.1(40)} = 0.982$$

(c) As t becomes unbounded in the positive direction, $e^{-0.1t} = \dfrac{1}{e^{-0.1t}}$ approaches 0. So F approaches 1.

(d)

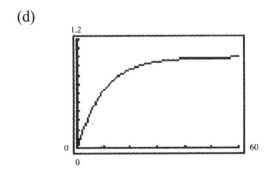

(e)

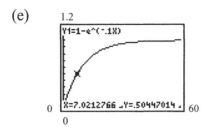

It takes 6.9 minutes for the probability to reach 50%.

65. (a) If 15 cars arrive, then $x = 15$ and
$$P(x) = P(15) = \frac{20^{15}e^{-20}}{15!} = 0.0516$$

The probability that 15 cars arrive between 5:00 PM and 6:00 PM is 0.0516.

(b) The probability that 20 cars arrive between 5:00 PM and 6:00 PM is given by
$$P(x) = P(20) = \frac{20^{20}e^{-20}}{20!} = 0.0888$$

67. (a) If the Civic is 3 years old $x = 3$, and its cost is
$$p(x) = p(3) = 16,630(0.90)^3 = \$12,123.27$$

(b) If the Civic is 9 years old $x = 9$, and its cost is
$$p(x) = p(9) = 16,630(0.90)^9 = \$6442.80$$

69. (a) If $E = 120$ volts, $R = 10$ ohms, and $L = 5$ henrys, then the amperage is given by the function
$$I(t) = \frac{120}{10}\left[1 - e^{-(10/5)t}\right] = 12\left[1 - e^{-2t}\right]$$

After 0.3 second $t = 0.3$ and $I(t) = I(0.3) = 12\left[1 - e^{-2(0.3)}\right] = 5.414$ amps.

After 0.5 second $t = 0.5$ and $I(t) = I(0.5) = 12\left[1 - e^{-2(0.5)}\right] = 7.585$ amps.

After 1 second $t = 1$ and $I(t) = I(1) = 12\left[1 - e^{-2}\right] = 10.375$ amps.

(b)

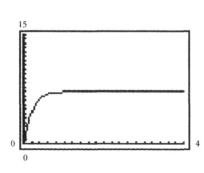

(c) The maximum current of $I_1(t)$ approaches 12 amps as t becomes unbounded in the positive direction.

(d) If $E = 120$ volts, $R = 5$ ohms, and $L = 10$ henrys, then the amperage is given by the function

$$I(t) = \frac{120}{5}\left[1 - e^{-(5/10)t}\right] = 24\left[1 - e^{-(1/2)t}\right]$$

After 0.3 second $t = 0.3$ and $I(t) = I(0.3) = 24\left[1 - e^{-(1/2)(0.3)}\right] = 3.343$ amps.

After 0.5 second $t = 0.5$ and $I(t) = I(0.5) = 24\left[1 - e^{-(1/2)(0.5)}\right] = 5.309$ amps.

After 1 second $t = 1$ and $I(t) = I(1) = 24\left[1 - e^{-1/2}\right] = 9.443$ amps.

(e)

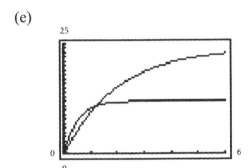

(f) The maximum current of $I_2(t)$ approaches 24 amps as t becomes unbounded in the positive direction.

71. Using a calculator, we get

$n = 4,$ $2 + \dfrac{1}{2!} + \dfrac{1}{3!} + \dfrac{1}{4!} = 2.70833333$

$n = 6,$ $2 + \dfrac{1}{2!} + \dfrac{1}{3!} + \dfrac{1}{4!} + \dfrac{1}{5!} + \dfrac{1}{6!} = 2.71805556$

$n = 8,$ $2 + \dfrac{1}{2!} + \dfrac{1}{3!} + \dfrac{1}{4!} + \dfrac{1}{5!} + \dfrac{1}{6!} + \dfrac{1}{7!} + \dfrac{1}{8!} = 2.71827877$

$n = 10$ $2 + \dfrac{1}{2!} + \dfrac{1}{3!} + \dfrac{1}{4!} + \dfrac{1}{5!} + \dfrac{1}{6!} + \dfrac{1}{7!} + \dfrac{1}{8!} + \dfrac{1}{9!} + \dfrac{1}{10!} = 2.718281801$

and $e = 2.718281828$

Comparisons might differ.

73. $\dfrac{f(x+h)-f(x)}{h} = \dfrac{a^{(x+h)}-a^x}{h}$

$\qquad\qquad = \dfrac{a^x a^h - a^x}{h}$ Use the exponential property $a^r \cdot a^s = a^{r+s}$.

$\qquad\qquad = \dfrac{a^x\left(a^h-1\right)}{h} = a^x\left(\dfrac{a^h-1}{h}\right)$ Factor.

75. If $f(x)=a^x$, then $f(-x)=a^{-x}$

$\qquad\qquad = \dfrac{1}{a^x}$ Use the exponential property $a^{-r}=\dfrac{1}{a^r}$.

$\qquad\qquad = \dfrac{1}{f(x)}$ Substitute.

77. (a) If F $= 50°$ and D $= 41°$, then

$$R = 10^{\frac{4221}{T+459.4}-\frac{4221}{D+459.4}+2} = 10^{\frac{4221}{50+459.4}-\frac{4221}{41+459.4}+2}$$

$$= 10^{\frac{4221}{509.4}-\frac{4221}{500.4}+2}$$

$$= 70.95$$

The relative humidity is 70.95%.

(b) If F $= 68°$ and D $= 59°$, then

$$R = 10^{\frac{4221}{T+459.4}-\frac{4221}{D+459.4}+2} = 10^{\frac{4221}{68+459.4}-\frac{4221}{59+459.4}+2}$$

$$= 10^{\frac{4221}{527.4}-\frac{4221}{518.4}+2}$$

$$= 72.62$$

The relative humidity is 72.62%.

(c) If F $=$ D, then $\dfrac{4221}{T+459.4} = \dfrac{4221}{D+459.4}$. So $\dfrac{4221}{T+459.4} - \dfrac{4221}{D+459.4} = 0$, and

$$R = 10^{0+2} = 100$$

When the temperature and the dew point are equal, the relative humidity is 100%.

79. (a) To show $f(x) = \sinh x$ is an odd function, we evaluate $f(-x)$ and simplify.

$$f(-x) = \frac{1}{2}\left(e^{-x}-e^{-(-x)}\right)$$

$$= \frac{1}{2}\left(e^{-x}-e^x\right)$$ Simplify.

$$= \frac{1}{2}\left(-e^x + e^{-x}\right)$$ Rearrange the terms.

$$= -\frac{1}{2}\left(e^x - e^{-x}\right)$$ Factor out -1.

$$= -f(x)$$

So f is an odd function.

(b)
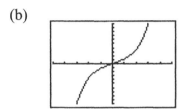

81. It takes 59 minutes. Explanations will vary.

83. There is no power function that increases more rapidly than an exponential function whose base is greater than 1. Explanations will vary.

85. Answers will vary.

2.4 Logarithmic Functions

1. We use the fact that $y = \log_a x$ and $x = a^y$, $a > 0$, $a \neq 1$ are equivalent. If $9 = 3^2$, then $\log_3 9 = 2$.

3. We use the fact that $y = \log_a x$ and $x = a^y$, $a > 0$, $a \neq 1$ are equivalent. If $a^2 = 1.6$, then $\log_a 1.6 = 2$.

5. We use the fact that $y = \log_a x$ and $x = a^y$, $a > 0$, $a \neq 1$ are equivalent. If $1.1^2 = M$, then $\log_{1.1} M = 2$.

7. We use the fact that $y = \log_a x$ and $x = a^y$, $a > 0$, $a \neq 1$ are equivalent. If $2^x = 7.2$, then $\log_2 7.2 = x$.

9. We use the fact that $y = \log_a x$ and $x = a^y$, $a > 0$, $a \neq 1$ are equivalent. If $x^{\sqrt{2}} = \pi$, then $\log_x \pi = \sqrt{2}$.

11. We use the fact that $y = \log_a x$ and $x = a^y$, $a > 0$, $a \neq 1$ are equivalent. If $e^x = 8$, then $\log_e 8 = x$ or $\ln 8 = x$.

13. We use the fact that $x = a^y$, $a > 0$, $a \neq 1$ and $y = \log_a x$ are equivalent.
If $\log_2 8 = 3$, then $2^3 = 8$.

15. We use the fact that $x = a^y$, $a > 0$, $a \neq 1$ and $y = \log_a x$ are equivalent.
If $\log_a 3 = 6$, then $a^6 = 3$.

17. We use the fact that $x = a^y$, $a > 0$, $a \neq 1$ and $y = \log_a x$ are equivalent.
If $\log_3 2 = x$, then $3^x = 2$.

19. We use the fact that $x = a^y$, $a > 0$, $a \neq 1$ and $y = \log_a x$ are equivalent.
If $\log_2 M = 1.3$, then $2^{1.3} = M$.

21. We use the fact that $x = a^y$, $a > 0$, $a \neq 1$ and $y = \log_a x$ are equivalent.
If $\log_{\sqrt{2}} \pi = x$, then $\left(\sqrt{2}\right)^x = \pi$.

23. We use the fact that $x = a^y$, $a > 0$, $a \neq 1$ and $y = \log_a x$ are equivalent.
If $\ln 4 = x$, then $e^x = 4$.

25. To find the exact value of the logarithm, we change the expression to its equivalent exponential expression and simplify.

$$y = \log_2 1$$

$2^y = 1$ Write in exponential form.
$2^y = 2^0$ $1 = 2^0$
$y = 0$ Equate exponents.

Therefore, $\log_2 1 = 0$

27. To find the exact value of the logarithm, we change the expression to its equivalent exponential expression and simplify.

$$y = \log_5 25$$

$5^y = 25$ Write in exponential form.
$5^y = 5^2$ $25 = 5^2$
$y = 2$ Equate exponents.

Therefore, $\log_5 25 = 2$.

29. To find the exact value of the logarithm, we change the expression to its equivalent exponential expression and simplify.

$$y = \log_{1/2} 16$$

$\left(\dfrac{1}{2}\right)^y = 16$ Write in exponential form.

$\left(\dfrac{1}{2}\right)^y = \left(\dfrac{1}{2}\right)^{-4}$ $16 = \left(\dfrac{1}{2}\right)^{-4}$

$y = -4$ Equate exponents.

Therefore, $\log_{1/2} 16 = -4$.

31. To find the exact value of the logarithm, we change the expression to its equivalent exponential expression and simplify.

$$y = \log_{10} \sqrt{10}$$

$10^y = \sqrt{10}$ Write in exponential form.

$10^y = 10^{1/2}$ $\sqrt{10} = 10^{1/2}$

$y = \dfrac{1}{2}$ Equate exponents.

Therefore, $\log_{10} \sqrt{10} = \dfrac{1}{2}$.

33. To find the exact value of the logarithm, we change the expression to its equivalent exponential expression and simplify.

$$y = \log_{\sqrt{2}} 4$$

$\left(\sqrt{2}\right)^y = 4$ Write in exponential form.

$\left(\sqrt{2}\right)^y = \left(\sqrt{2}\right)^4$ $4 = 2^2 = \left[\left(\sqrt{2}\right)^2\right]^2 = \left(\sqrt{2}\right)^4$

$y = 4$ Equate exponents.

Therefore, $\log_{\sqrt{2}} 4 = 4$.

35. To find the exact value of the logarithm, we change the expression to its equivalent exponential expression and simplify.

$$y = \ln \sqrt{e}$$

$e^y = \sqrt{e}$ Write in exponential form.

$e^y = e^{1/2}$ $\sqrt{e} = e^{1/2}$

$y = \dfrac{1}{2}$ Equate exponents.

Therefore, $\ln \sqrt{e} = \dfrac{1}{2}$.

37. The domain of a logarithmic function is limited to all positive real numbers, so for $f(x) = \ln(x - 3)$, $x - 3 > 0$. The domain is all $x > 3$, or using interval notation $(3, \infty)$.

39. The domain of a logarithmic function is limited to all positive real numbers, so for $F(x) = \log_2 x^2$, $x^2 > 0$. x^2 is positive except when $x = 0$, meaning the domain is the set $\{x \mid x \neq 0\}$.

41. The domain of $f(x) = 3 - 2\log_4 \dfrac{x}{2}$ is restricted by $\log_4 \dfrac{x}{2}$ which is defined only when $\dfrac{x}{2}$ is positive. So the domain of f is all $x > 0$, or using interval notation $(0, \infty)$.

43. There are two restrictions on the domain of $f(x) = \sqrt{\ln x}$.

First, $y = \sqrt{x}$ is defined only when x is nonnegative, that is when $x \geq 0$.

Second, $y = \ln x$ is defined only when x is positive, that is when $x > 0$. However, on the interval $(0, 1)$, $\ln x < 0$, so $f(x) = \sqrt{\ln x}$ is not defined.

So the domain of f is all $x \geq 1$, or using interval notation $[1, \infty)$.

45. $\ln \dfrac{5}{3} = 0.511$

47. $\dfrac{\ln \dfrac{10}{3}}{0.04} = 30.099$

49. If the graph of f contains the point $(2, 2)$, then $x = 2$ and $y = 2$ must satisfy the equation
$$y = f(x) = \log_a x$$
$$2 = \log_a 2$$
$$a^2 = 2$$
$$a = \sqrt{2}$$

51. The graph of $y = \log_3 x$ has the properties:

The domain is $(0, \infty)$. The range is $(-\infty, \infty)$.
The x-intercept is $(1, 0)$.
The y-axis is a vertical asymptote.
The graph is increasing since $a = 3 > 1$.

The graph contains the points $(3, 1)$ and $\left(\dfrac{1}{3}, -1 \right)$.

The graph is continuous.

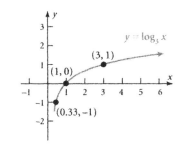

53. The graph of $y = \log_{1/5} x$ has the properties:

The domain is $(0, \infty)$. The range is $(-\infty, \infty)$.
The x-intercept is $(1, 0)$.
The y-axis is a vertical asymptote.

The graph is decreasing since $a = \dfrac{1}{5} < 1$.

The graph contains the points $(5, -1)$ and $\left(\dfrac{1}{5}, 1 \right)$.

The graph is continuous.

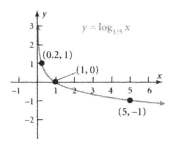

55. B. This graph is reflected over the y-axis.

57. D. This graph was reflected about both the x- and y-axes.

59. A. This graph has not been shifted or reflected.

61. E. This graph was shifted vertically down 1 unit.

63.

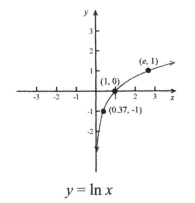

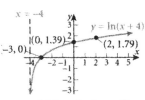

$$y = \ln x$$ Replace x by $x + 4$; $f(x) = \ln(x + 4)$
                       horizontal shift 4 units
                       to the left.

The domain of f is all $x > -4$, or in interval notation, $(-4, \infty)$. The range is all real numbers or $(-\infty, \infty)$. The vertical asymptote is the line $x = -4$.

65.

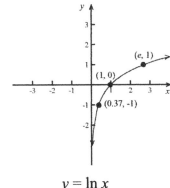

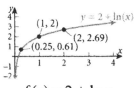

$$y = \ln x$$ Add 2; vertical $f(x) = 2 + \ln x$
                       shift up 2 units.

The domain of f is all $x > 0$, or in interval notation, $(0, \infty)$. The range is all real numbers or $(-\infty, \infty)$. The vertical asymptote is the y-axis, that is the line $x = 0$.

67.

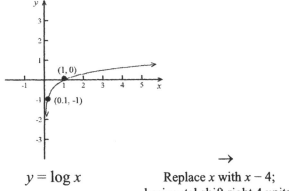

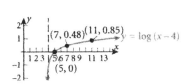

$$y = \log x$$ Replace x with $x - 4$; $f(x) = \log(x - 4)$
                        horizontal shift right 4 units.

The domain of f is all $x > 4$, or in interval notation, $(4, \infty)$. The range is all real numbers or $(-\infty, \infty)$. The vertical asymptote is the line $x = 4$.

69.

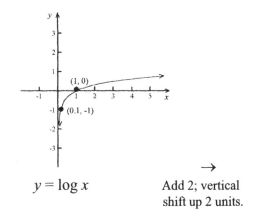

$$y = \log x \qquad \xrightarrow{\begin{array}{c}\text{Add 2; vertical}\\ \text{shift up 2 units.}\end{array}} \qquad f(x) = \log x + 2$$

The domain of f is all $x > 0$, or in interval notation, $(0, \infty)$. The range is all real numbers or $(-\infty, \infty)$. The vertical asymptote is the y-axis, that is the line $x = 0$.

71. $\log_3 x = 2$

$\qquad x = 3^2 = 9$ 　　　　Change to exponential form and simplify.

73. $\log_2 (2x + 1) = 3$

$\qquad 2x + 1 = 2^3$ 　　　Change to exponential form.

$\qquad 2x + 1 = 8$ 　　　　Solve the linear equation.

$\qquad 2x = 7$

$\qquad x = \dfrac{7}{2}$

75. $\log_x 4 = 2$

$\qquad 4 = x^2$ 　　　　Change to exponential form.

$\qquad x = 2$ 　　　　Solve using the Square Root Method. ($x \neq -2$, the base is positive.)

77. $\ln e^x = 5$

$\qquad e^x = e^5$ 　　　　Change to exponential form.

$\qquad x = 5$ 　　　　Since the bases are equal, the exponents are equal.

79. $\log_4 64 = x$

$\qquad 64 = 4^x$ 　　　　Change to exponential form.

$\qquad 4^3 = 4^x$

$\qquad 3 = x$ 　　　　Since the bases are equal, the exponents are equal.

81. $\log_3 243 = 2x + 1$

$\qquad 3^{2x+1} = 243$ 　　　Change to exponential form.

$\qquad 3^{2x+1} = 3^5$

$\qquad 2x + 1 = 5$ 　　　Since the bases are equal, the exponents are equal.

$\qquad 2x = 4$ 　　　　Solve the linear equation.

$\qquad x = 2$

83.
$$e^{3x} = 10$$
$\ln 10 = 3x$ Change to a logarithmic expression.
$$x = \frac{\ln 10}{3}$$ Exact solution.
≈ 0.768 Approximate solution.

85.
$$e^{2x+5} = 8$$
$\ln 8 = 2x + 5$ Change to a logarithmic expression.
$2x = -5 + \ln 8$
$$x = \frac{-5 + \ln 8}{2}$$ Exact solution.
≈ -1.460 Approximate solution.

87. $\log_3\left(x^2 + 1\right) = 2$
$x^2 + 1 = 3^2$ Change to an exponential expression.
$x^2 + 1 = 9$
$x^2 = 8$
$x = \pm\sqrt{8}$ Use the Square Root Method.
$= \pm 2\sqrt{2}$ Simplify.

89. $\log_2 8^x = -3$
$8^x = 2^{-3}$ Change to an exponential expression.
$(2^3)^x = 2^{-3}$
$2^{3x} = 2^{-3}$
$3x = -3$ Since the bases are equal, the exponents are equal.
$x = -1$

91. (a) When $H^+ = 0.1$, $pH = -\log(0.1) = 1$.
(b) When $H^+ = 0.01$, $pH = -\log(0.01) = 2$.
(c) When $H^+ = 0.001$, $pH = -\log(0.001) = 3$.

(d) As the hydrogen ion concentration decreases, pH increases.

(e) If $pH = 3.5$, then $3.5 = -\log x$
$$\log x = -3.5$$
$$10^{-3.5} = x$$
$$x = 0.000316$$

(f) If $pH = 7.4$, then $7.4 = -\log x$
$$\log x = -7.4$$
$$10^{-7.4} = x$$
$$x = 3.981 \times 10^{-8}$$

93. (a) If $p = 320$ mm,
$$320 = 760 e^{-0.145 h}$$

$$e^{-0.145h} = \frac{320}{760} = \frac{8}{19}$$ Divide both sides by 760.

$$-0.145h = \ln\left(\frac{8}{19}\right)$$ Change to a logarithmic expression.

$$h = -\frac{1000}{145} \ln\left(\frac{8}{19}\right)$$ $0.145 = \frac{145}{1000}$

$$\approx 5.965$$

The aircraft is at an altitude of approximately 5.965 kilometers above sea level.

(b) If $p = 667$ mm,

$$667 = 760e^{-0.145h}$$

$$e^{-0.145h} = \frac{667}{760}$$ Divide both sides by 760.

$$-0.145h = \ln\left(\frac{667}{760}\right)$$ Change to a logarithmic expression.

$$h = -\frac{1000}{145} \ln\left(\frac{667}{760}\right)$$ $0.145 = \frac{145}{1000}$

$$\approx 0.900$$

The mountain is approximately 0.9 kilometer above sea level.

95. (a) We want $F(t) = 0.50$.

$$0.50 = 1 - e^{-0.1t}$$
$$e^{-0.1t} = 0.50$$
$$-0.1t = \ln 0.5$$
$$t = -10 \ln 0.5$$
$$\approx 6.93$$

After approximately 6.9 minutes the probability that a car arrives at the drive-thru reaches 50%.

(b) We want $F(t) = 0.80$.

$$0.80 = 1 - e^{-0.1t}$$
$$e^{-0.1t} = 0.20$$
$$-0.1t = \ln 0.2$$
$$t = -10 \ln 0.2$$
$$\approx 16.09$$

After approximately 16.1 minutes the probability that a car arrives at the drive-thru reaches 80%.

97. We want $D = 2$.

$$2 = 5e^{-0.4h}$$

$$e^{-0.4h} = \frac{2}{5} = 0.4$$

$$-0.4h = \ln 0.4$$

$$h = -\frac{5}{2} \ln 0.4$$

$$\approx 2.29$$

The drug should be administered every 2.3 hours (about 2 hours 17 minutes).

99. We want $I = 0.5$

$$0.5 = \frac{12}{10}\left[1 - e^{-(10/5)t}\right] = 1.2\left(1 - e^{-2t}\right)$$

$$1 - e^{-2t} = \frac{0.5}{1.2} = \frac{5}{12}$$

$$e^{-2t} = \frac{7}{12}$$

$$-2t = \ln\left(\frac{7}{12}\right)$$

$$t = -\frac{1}{2}\ln\left(\frac{7}{12}\right)$$

$$\approx 0.269$$

It takes about 0.27 seconds to obtain a current of 0.5 ampere.

Next we want $I = 1.0$

$$1.0 = 1.2\left(1 - e^{-2t}\right)$$

$$1 - e^{-2t} = \frac{1.0}{1.2} = \frac{5}{6}$$

$$e^{-2t} = \frac{1}{6}$$

$$-2t = \ln\left(\frac{1}{6}\right)$$

$$t = -\frac{1}{2}\ln\left(\frac{1}{6}\right)$$

$$\approx 0.896$$

It takes about 0.9 seconds to obtain a current of 1 ampere.

101. We are interested in the population when $t = 11$, so we want $P(11)$.

$$P(10) = 298{,}710{,}000 + 10{,}000{,}000 \log 11$$
$$= 309{,}123{,}926.9$$

On January 1, 2020 the population (to the nearest thousand) will be 309,124,000.

103. Normal conversation: $x = 10^{-7}$ watt per square meter,

$$L\left(10^{-7}\right) = 10 \log \frac{10^{-7}}{10^{-12}} = 50 \text{ decibels}$$

105. Amplified rock music: $x = 10^{-1}$ watt per square meter,

$$L\left(10^{-1}\right) = 10 \log \frac{10^{-1}}{10^{-12}} = 110 \text{ decibels}$$

107. Mexico City, 1985: $x = 125{,}892$ mm.
$$M(125{,}892) = \log\left(\frac{125{,}892}{10^{-3}}\right) = 8.1 = 8.1$$

109. (a) Since $R = 10$ when $x = 0.06$, we can find k.
$$10 = 3\, e^{0.06k}$$
$$0.06k = \ln\left(\frac{10}{3}\right)$$
$$k = \frac{100}{6}\ln\left(\frac{10}{3}\right) = \frac{50}{3}\ln\left(\frac{10}{3}\right) \approx 20.066$$

(b) If $x = 0.17$, then the risk of having a car accident is
$$R = 3e^{\left(\frac{50}{3}\ln\left(\frac{10}{3}\right)\right)(0.17)} = 90.91\%$$

(c) If the risk is 100%, then $R = 100$.
$$3e^{\left(\frac{50}{3}\ln\left(\frac{10}{3}\right)\right)x} = 100$$
$$e^{\left(\frac{50}{3}\ln\left(\frac{10}{3}\right)\right)x} = \frac{100}{3}$$
$$\frac{50}{3}\ln\left(\frac{10}{3}\right)x = \ln\frac{100}{3}$$
$$x \approx 0.1747$$

(d) $$R = 15 = 3e^{\left(\frac{50}{3}\ln\left(\frac{10}{3}\right)\right)x}$$
$$e^{\left(\frac{50}{3}\ln\left(\frac{10}{3}\right)\right)x} = 5$$

$$\frac{50}{3}\ln\left(\frac{10}{3}\right)x = \ln 5$$
$$x \approx 0.080$$

A driver with a blood alcohol concentration greater than or equal to 0.08 should be arrested and charged with DUI.

(e) Answers will vary.

111. Answers may vary.

2.5 Properties of Logarithms

1. $\log_3 3^{71} = 71$ $\qquad \log_a a^r = r$ $\qquad$ **3.** $\ln e^{-4} = -4$ $\qquad \log_a a^r = r$

5. $2^{\log_2 7} = 7$ $\qquad a^{\log_a M} = M$

7. $\log_8 2 + \log_8 4 = \log_8 (2 \cdot 4) = \log_8 8 = 1$

9.

$\log_6 18 - \log_6 3 = \log_6 \left(\dfrac{18}{3}\right) = \log_6 6 = 1$

11. Use a change of base formula,

$$\log_2 6 \cdot \log_6 4 = \dfrac{\ln 6}{\ln 2} \cdot \dfrac{\ln 4}{\ln 6} = \dfrac{\ln 4}{\ln 2}$$

Next simplify $\ln 4$,

$$\dfrac{\ln 4}{\ln 2} = \dfrac{\ln 2^2}{\ln 2} = \dfrac{2 \ln 2}{\ln 2} = 2$$

So, $\log_2 6 \cdot \log_6 4 = 2$

13.

$$3^{\log_3 5 - \log_3 4} = 3^{\log_3 \left(\frac{5}{4}\right)} = \dfrac{5}{4}$$

15. First we write the exponent as $y = \log_{e^2} 16$, and express it as an exponential.

$$y = \log_{e^2} 16$$
$$e^{2y} = 16 \quad \text{or} \quad e^y = 4. \quad \text{So } y = \ln 4.$$
Then, $e^{\log_{e^2} 16} = e^y = e^{\ln 4} = 4.$

17. $\ln 6 = \ln (2 \cdot 3) = \ln 2 + \ln 3 = a + b$ **19.** $\ln 1.5 = \ln \dfrac{3}{2} = \ln 3 - \ln 2 = b - a$

21. $\ln 8 = \ln 2^3 = 3 \ln 2 = 3a$

23. $\ln \sqrt[5]{6} = \dfrac{1}{5} \ln 6 = \dfrac{1}{5} \ln (2 \cdot 3) = \dfrac{1}{5} (\ln 2 + \ln 3) = \dfrac{1}{5} (a + b)$

25. $\log_5 (25x) = \log_5 (5^2 x) = \log_5 5^2 + \log_5 x = 2 \log_5 5 + \log_5 x$

27. $\log_2 z^3 = 3 \log_2 z$

29. $\ln (ex) = \ln e + \ln x = 1 + \ln x$

31. $\ln (xe^x) = \ln x + \ln e^x = \ln x + x \ln e = \ln x + x$

33. $\log_a \left(u^2 v^3\right) = \log_a u^2 + \log_a v^3 = 2 \log_a u + 3 \log_a v$

35. $\ln \left(x^2 \sqrt{1-x}\right) = \ln x^2 + \ln (1-x)^{1/2} = 2 \ln x + \dfrac{1}{2} \ln (1-x)$

37.
$$\log_2\left(\frac{x^3}{x-3}\right) = \log_2 x^3 - \log_2(x-3) = 3\log_2 x - \log_2(x-3)$$

39.
$$\log\left[\frac{x(x+2)}{(x+3)^2}\right] = \log\left[x(x+2)\right] - \log(x+3)^2 = \log x + \log(x+2) - 2\log(x+3)$$

41.
$$\ln\left[\frac{x^2-x-2}{(x+4)^2}\right]^{1/3} = \frac{1}{3}\ln\left[\frac{x^2-x-2}{(x+4)^2}\right] = \frac{1}{3}\left[\ln(x^2-x-2) - \ln(x+4)^2\right]$$
$$= \frac{1}{3}\left[\ln\left[(x-2)(x+1)\right] - \ln(x+4)^2\right]$$
$$= \frac{1}{3}\left[\ln(x-2) + \ln(x+1) - 2\ln(x+4)\right]$$
$$= \frac{1}{3}\ln(x-2) + \frac{1}{3}\ln(x+1) - \frac{2}{3}\ln(x+4)$$

43.
$$\ln\frac{5x\sqrt{1+3x}}{(x-4)^3} = \ln\left[5x\sqrt{1+3x}\right] - \ln(x-4)^3 = \ln 5 + \ln x + \frac{1}{2}\ln(1+3x) - 3\ln(x-4)$$

45. $3\log_5 u + 4\log_5 v = \log_5 u^3 + \log_5 v^4 = \log_5\left(u^3 v^4\right)$

47.
$$\log_3\sqrt{x} - \log_3 x^3 = \log_3\left(\frac{\sqrt{x}}{x^3}\right) = \log_3 x^{-5/2} = -\log_3 x^{5/2} \text{ or } -\frac{5}{2}\log_3 x$$

49.
$$\log_4(x^2-1) - 5\log_4(x+1) = \log_4(x^2-1) - \log_4(x+1)^5 = \log_4\left[\frac{x^2-1}{(x+1)^5}\right]$$
$$= \log_4\left[\frac{(x-1)\,(x+1)}{(x+1)^{\cancel{5}\,4}}\right] = \log_4\left[\frac{x-1}{(x+1)^4}\right]$$

51.
$$\ln\left(\frac{x}{x-1}\right) + \ln\left(\frac{x+1}{x}\right) - \ln(x^2-1) = \ln\left[\frac{\cancel{x}}{x-1} \cdot \frac{x+1}{\cancel{x}} \cdot \frac{1}{(x-1)(x+1)}\right]$$
$$= \ln\left[\frac{1}{(x-1)^2}\right] = \ln(x-1)^{-2} = -2\ln(x-1)$$

53.
$$8\log_2\sqrt{3x-2} - \log_2\left(\frac{4}{x}\right) + \log_2 4 = \log_2\left(\sqrt{3x-2}\right)^8 - \log_2\left(\frac{4}{x}\right) + \log_2 4$$
$$= \log_2(3x-2)^4 - \log_2\left(\frac{4}{x}\right) + \log_2 4$$

$$= \log_2 \left[\frac{(3x-2)^4 \cdot \cancel{A}}{\frac{\cancel{A}}{x}} \right] = \log_2 \left[x(3x-2)^4 \right]$$

55.
$$2 \log_a (5x^3) - \frac{1}{2} \log_a (2x+3) = \log_a (5x^3)^2 - \log_a \sqrt{2x+3}$$

$$= \log_a \frac{(5x^3)^2}{\sqrt{2x+3}} = \log_a \frac{25x^6}{\sqrt{2x+3}}$$

57.
$$2 \log_2 (x+1) - \log_2 (x+3) - \log_2 (x-1) = \log_2 (x+1)^2 - \log_2 (x+3) - \log_2 (x-1)$$

$$= \log_2 \frac{(x+1)^2}{(x+3)(x-1)} = \log_2 \frac{(x+1)^2}{x^2 + 2x - 3}$$

59.
$$\log_3 21 = \frac{\ln 21}{\ln 3} = 2.771$$

61.
$$\log_{1/3} 71 = \frac{\ln 71}{\ln \frac{1}{3}} = -3.880$$

63.
$$\log_{\sqrt{2}} 7 = \frac{\ln 7}{\ln \sqrt{2}} = 5.615$$

65.
$$\log_\pi e = \frac{\ln e}{\ln \pi} = 0.874$$

67.
$$y = \log_4 x = \frac{\ln x}{\ln 4}$$

69.
$$y = \log_2 (x+2) = \frac{\ln (x+2)}{\ln 2}$$

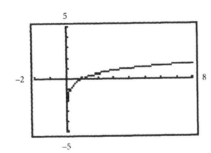

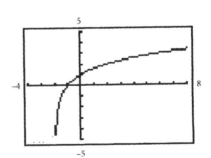

71.
$$y = \log_{x-1} (x+1) = \frac{\ln (x+1)}{\ln (x-1)}$$

73. $\ln y = \ln x + \ln C$

$\ln y = \ln (Cx)$

$y = Cx$

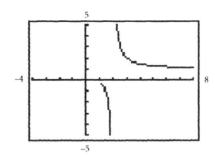

75. $\ln y = \ln x + \ln(x+1) + \ln C$

$\ln y = \ln\left[x(x+1)C\right]$

$y = x(x+1)C$

77. $\ln y = 3x + \ln C$

$\ln y = \ln e^{3x} + \ln C$

$\ln y = \ln Ce^{3x}$

$y = Ce^{3x}$

79. $\ln(y-3) = -4x + \ln C$

$\ln(y-3) = \ln e^{-4x} + \ln C$

$\ln(y-3) = \ln\left(Ce^{-4x}\right)$

$y - 3 = Ce^{-4x}$

$y = Ce^{-4x} + 3$

81.

$3\ln y = \dfrac{1}{2}\ln(2x+1) - \dfrac{1}{3}\ln(x+4) + \ln C$

$\ln y^3 = \ln(2x+1)^{1/2} - \ln(x+4)^{1/3} + \ln C$

$\ln y^3 = \ln\left[C(2x+1)^{1/2}\right] - \ln(x+4)^{1/3}$

$\ln y^3 = \ln\left[\dfrac{C(2x+1)^{1/2}}{(x+4)^{1/3}}\right]$

$y^3 = \dfrac{C(2x+1)^{1/2}}{(x+4)^{1/3}}$

$y = \left[\dfrac{C(2x+1)^{1/2}}{(x+4)^{1/3}}\right]^{1/3} = \dfrac{C(2x+1)^{1/6}}{(x+4)^{1/9}}$ $C^{1/3}$ is still a positive constant. We write C.

83. We use the change of base formula and simplify.

$\log_2 3 \cdot \log_3 4 \cdot \log_4 5 \cdot \log_5 6 \cdot \log_6 7 \cdot \log_7 8$

$= \dfrac{\log 3}{\log 2} \cdot \dfrac{\log 4}{\log 3} \cdot \dfrac{\log 5}{\log 4} \cdot \dfrac{\log 6}{\log 5} \cdot \dfrac{\log 7}{\log 6} \cdot \dfrac{\log 8}{\log 7}$

$= \dfrac{\log 8}{\log 2} = \dfrac{\log 2^3}{\log 2}$

$$= \frac{3 \log 2}{\log 2} = 3$$

85. We use the change of base formula and simplify.

$$\log_2 3 \cdot \log_3 4 \cdot \log_4 5 \cdot \log_5 6 \cdot \log_6 7 \cdot \log_7 8 \cdot \ldots \cdot \log_n (n+1) \cdot \log_{n+1} 2$$

Noticing that the first $n-1$ factors follow the pattern of problem 83, we get

$$\left[\log_2 3 \cdot \log_3 4 \cdot \log_4 5 \cdot \log_5 6 \cdot \log_6 7 \cdot \log_7 8 \cdot \ldots \cdot \log_n (n+1) \right] \cdot \log_{n+1} 2$$

$$= \left[\frac{\log 3}{\log 2} \cdot \frac{\log 4}{\log 3} \cdot \frac{\log 5}{\log 4} \cdot \ldots \cdot \frac{\log (n+1)}{\log n} \right] \cdot \frac{\log 2}{\log (n+1)}$$

$$= 1$$

87.
$$\log_a \left(x + \sqrt{x^2 - 1} \right) + \log_a \left(x - \sqrt{x^2 - 1} \right) = \log_a \left[\left(x + \sqrt{x^2 - 1} \right) \left(x - \sqrt{x^2 - 1} \right) \right]$$

$$= \log_a \left(x^2 - \left(\sqrt{x^2 - 1} \right)^2 \right)$$

$$= \log_a \left(x^2 - \left| x^2 - 1 \right| \right)$$

$$= \log_a \left(x^2 - x^2 + 1 \right)$$

$$= \log_a 1 = 0$$

89.
$$\ln \left(1 + e^{2x} \right) = 2x + \ln \left(1 + e^{-2x} \right)$$

We work from the complicated side (the right) and simplify to get to the left side of the equation.

$$2x + \ln \left(1 + e^{-2x} \right) = \ln e^{2x} + \ln \left(1 + e^{-2x} \right)$$

$$= \ln \left[e^{2x} \cdot \left(1 + e^{-2x} \right) \right]$$

$$= \ln \left[e^{2x} + e^{2x} \cdot e^{-2x} \right]$$

$$= \ln \left[e^{2x} + 1 \right] = \ln \left(1 + e^{2x} \right)$$

91. We show that $-f(x) = \log_{1/a} x$ by using the Change of Base Formula.

$$f(x) = \log_a x \quad \text{so} \quad -f(x) = -\log_a x = -\frac{\ln x}{\ln a}$$

$$= \frac{\ln x}{-\ln a}$$

$$= \frac{\ln x}{\ln a^{-1}} = \frac{\ln x}{\ln \dfrac{1}{a}}$$

Using the Change of base formula in reverse, the last expression becomes $\ln_{1/a} x$. So
$$-f(x) = \log_{1/a} x$$

93. If $f(x) = \log_a x$, then
$$f\left(\frac{1}{x}\right) = \log_a\left(\frac{1}{x}\right) = \log_a x^{-1} = -\log_a x = -f(x)$$

95. If $A = \log_a M$ and $B = \log_a N$, then (writing each expression as an exponential) we get
$$a^A = M \quad \text{and} \quad a^B = N.$$
$$\log_a\left(\frac{M}{N}\right) = \log_a\left(\frac{a^A}{a^B}\right)$$
$$= \log_a a^{A-B}$$
$$= A - B$$
$$= \log_a M - \log_a N$$

97.

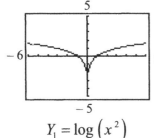

$$Y_1 = \log\left(x^2\right) \qquad Y_2 = 2\log(x)$$

Explanations may vary.

2.6 Continuously Compounded Interest

1. If \$1000 is invested at 4% compounded continuously, the amount A after 3 years is
$$A = Pe^{rt} = 1000e^{(0.04)(3)} = 1000e^{0.12} = \$1127.50$$

3. If \$500 is invested at 5% compounded continuously, the amount A after 3 years is
$$A = Pe^{rt} = 500e^{(0.05)(3)} = 500e^{0.15} = \$580.92$$

5. The present value of \$100 that will be invested at 4% compounded continuously for 6 months is
$$P = Ae^{-rt} = 100e^{-(0.04)(0.5)} = 100e^{-0.02} = \$98.02$$

7. The present value of \$500 that will be invested at 7% compounded continuously for 1 year is
$$P = Ae^{-rt} = 500e^{-(0.07)(1)} = 500e^{-0.07} = \$466.20$$

9. If $1000 is invested at 2% compounded continuously, the amount A after 1 year is
$$A = Pe^{rt} = 1000e^{(0.02)(1)} = 1000e^{0.02} = \$1020.20$$

$A - P = \$20.20$ interest was earned.

11. We need to deposit the present value of $5000 that will be invested at 3% compounded continuously for 4 years.
$$P = Ae^{-rt} = 5000e^{-(0.03)(4)} = 5000e^{-0.12} = \$4434.60$$

We need to deposit the present value of $5000 that will be invested at 3% compounded continuously for 8 years.
$$P = Ae^{-rt} = 5000e^{-(0.03)(8)} = 5000e^{-0.24} = \$3933.14$$

13. If P is invested, it will double when amount $A = 2P$.
$$A = 2P = Pe^{rt}$$
$$2 = e^{3r}$$
$$3r = \ln 2$$
$$r = \frac{\ln 2}{3} \approx 0.2310$$
It will require an interest rate of about 23.1% compounded continuously to double an investment in 3 years.

15. If P is invested, it will triple when amount $A = 3P$.
$$A = 3P = Pe^{rt}$$
$$3 = e^{0.10t}$$
$$0.10t = \ln 3$$
$$t = \frac{\ln 3}{0.10} \approx 10.986$$
It will take about 11 years for an investment to triple at 10 % compounded continuously.

17. We need the present value of $1000 that will be invested at 9% compounded continuously for 1 year.
$$P = Ae^{-rt} = 1000e^{-(0.09)(1)} = 1000e^{-0.09} = \$913.93$$

We need the present value of $1000 that will be invested at 9% compounded continuously for 2 years.
$$P = Ae^{-rt} = 1000e^{-(0.09)(2)} = 1000e^{-0.18} = \$835.27$$

19. Tami and Todd need to invest the present value of $40,000 that will earn 3% compounded continuously for 4 years.
$$P = Ae^{-rt} = 40,000e^{-(0.03)(4)} = 40,000e^{-0.12} = \$35,476.82$$

21. If P is invested, it will triple when amount $A = 3P$.
$$A = 3P = Pe^{rt}$$
$$3 = e^{5r}$$

$$5r = \ln 3$$
$$r = \frac{\ln 3}{5} \approx 0.2197$$

It will require an interest rate of about 22% compounded continuously to triple an investment in 5 years.

23. The rule of 70:
    (a) The actual time it takes to double an investment if $r = 1\%$ is
    $$t = \frac{\ln 2}{0.01} = 69.3147 \text{ years.}$$
    The estimated time it takes to double an investment using the Rule of 70 if $r = 1\%$ is
    $$t = \frac{0.70}{0.01} = 70.0 \text{ years.}$$
    The Rule of 70 overestimates the time to double money by 0.6853 year, about 8 months.

    (b) The actual time it takes to double an investment if $r = 5\%$ is
    $$t = \frac{\ln 2}{0.05} = 13.8629 \text{ years.}$$
    The estimated time it takes to double an investment using the Rule of 70 if $r = 5\%$ is
    $$t = \frac{0.70}{0.05} = 14 \text{ years.}$$

    The Rule of 70 overestimates the time to double money by 0.1371 year, about 1.6 months.

    (c) The actual time it takes to double an investment if $r = 10\%$ is
    $$t = \frac{\ln 2}{0.10} = 6.9315 \text{ years.}$$
    The estimated time it takes to double an investment using the Rule of 70 if $r = 10\%$ is
    $$t = \frac{0.70}{0.10} = 7.0 \text{ years.}$$

    The Rule of 70 overestimates the time to double money by 0.0685 year, almost 1 month.

Chapter 2 Review

TRUE-FALSE ITEMS

1. True   3. True   5. False

7. False   9. False

FILL IN THE BLANKS

1. parabola

3. $x = -\dfrac{b}{2a}$

5. one

7. $(0, \infty)$

9. one

REVIEW EXERCISES

1. First we expand the function.
$$f(x) = (x-2)^2 + 2 = x^2 - 4x + 4 + 2 = x^2 - 4x + 6$$
and find that $a = 1$, $b = -4$, $c = 6$.

$a > 0$; the parabola opens up.

The x-coordinate of the vertex is $-\dfrac{b}{2a} = -\dfrac{(-4)}{2(1)} = 2$, and the y-coordinate is
$$f(2) = (2)^2 - 4(2) + 6 = 2.$$
So the vertex is $(2, 2)$ and the axis of symmetry is the line $x = 2$.

Since $f(0) = c = 6$, the y-intercept is $(0, 6)$.
The x-intercepts are found by solving $f(x) = 0$. Since the discriminant of f
$$b^2 - 4ac = (-4)^2 - 4(1)(6) = -8$$
is negative, the equation $f(x) = 0$ has no real solution, and therefore, the parabola has no x-intercept.

To graph the function, we use symmetry. If we choose the y-intercept, we obtain its symmetric point $(4, 6)$.

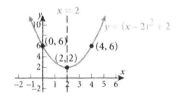

3. $a = \dfrac{1}{4}$, $b = 0$, $c = -16$

$a > 0$, the parabola opens up.

The x-coordinate of the vertex is $-\dfrac{b}{2a} = 0$, and the y-coordinate is
$$f(0) = \dfrac{1}{4}(0)^2 - 16 = -16.$$
So the vertex is $(0, -16)$, and the axis of symmetry is the line $x = 0$.

Since $f(0) = c = -16$, the y-intercept is also $(0, -16)$
The x-intercepts are found by solving $f(x) = 0$.

$$\frac{1}{4}x^2 - 16 = 0$$
$$x^2 - 64 = 0$$
$$x^2 = 64$$
$$x = \pm 8$$

The x-intercepts are $(8, 0)$ and $(-8, 0)$.

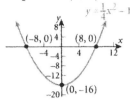

5. $a = -4$, $b = 4$, $c = 0$

$a < 0$, the parabola opens down.

The x-coordinate of the vertex is $-\dfrac{b}{2a} = -\dfrac{4}{2(-4)} = \dfrac{1}{2}$, and the y-coordinate is

$$f\left(\frac{1}{2}\right) = -4\left(\frac{1}{2}\right)^2 + 4x = 1$$

So the vertex is $\left(\dfrac{1}{2}, 1\right)$, and the axis of symmetry is the line $x = \dfrac{1}{2}$.

Since $f(0) = c = 0$, the y-intercept is $(0, 0)$.
The x-intercepts are found by solving $f(x) = 0$.

$$-4x^2 + 4x = 0$$
$$-4x(x - 1) = 0$$
$$x = 0 \quad \text{or} \quad x = 1$$

The x-intercepts are $(0, 0)$ and $(1, 0)$.

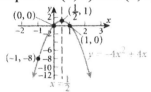

7. $a = \dfrac{9}{2}$, $b = 3$, $c = 1$

$a > 0$; the parabola opens up.

The x-coordinate of the vertex is $-\dfrac{b}{2a} = -\dfrac{3}{2\left(\dfrac{9}{2}\right)} = -\dfrac{1}{3}$, and the y-coordinate is

$$f\left(-\frac{1}{3}\right) = \frac{9}{2}\left(-\frac{1}{3}\right)^2 + 3\left(-\frac{1}{3}\right) + 1 = \frac{1}{2}$$

So the vertex is $\left(-\dfrac{1}{3}, \dfrac{1}{2}\right)$, and the axis of symmetry is the line $x = -\dfrac{1}{3}$.

Since $f(0) = c = 1$, the y-intercept is $(0, 1)$.
The x-intercepts are found by solving $f(x) = 0$. Since the discriminant of f

$$b^2 - 4ac = 3^2 - (4)\left(\frac{9}{2}\right)(1) = -9$$

is negative, the equation $f(x) = 0$ has no real solution, and therefore, the parabola has no x-intercept.

To graph the function, we use symmetry. If we choose the y-intercept, we obtain its symmetric point $\left(-\frac{2}{3}, 1\right)$. Because this point is so close to the vertex, we also use the point $\left(-1, \frac{5}{2}\right)$.

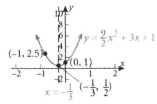

9. $a = 3, b = 4, c = -1$

$a > 0$; the parabola opens up.

The x-coordinate of the vertex is $-\dfrac{b}{2a} = -\dfrac{4}{2(3)} = -\dfrac{2}{3}$, and the y-coordinate is

$$f\left(-\frac{2}{3}\right) = 3\left(-\frac{2}{3}\right)^2 + 4\left(-\frac{2}{3}\right) - 1 = -\frac{7}{3}$$

So the vertex is $\left(-\dfrac{2}{3}, -\dfrac{7}{3}\right)$ and the axis of symmetry is the line $x = -\dfrac{2}{3}$.

Since $f(0) = c = -1$, the y-intercept is $(0, -1)$.
The x-intercepts are found by solving $f(x) = 0$.

$$3x^2 + 4x - 1 = 0$$

$$x = \frac{-4 \pm \sqrt{4^2 - 4(3)(-1)}}{2(3)} = \frac{-4 \pm \sqrt{28}}{6}$$

$$x = \frac{-4 \pm 2\sqrt{7}}{6} = \frac{-2 \pm \sqrt{7}}{3}$$

The x-intercepts are difficult to graph, so we choose another point and use symmetry.
We chose $(-2, 3)$ and by symmetry $\left(\dfrac{2}{3}, 3\right)$.

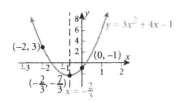

11. $a = 3, b = -6, c = 4$

The quadratic function has a minimum since $a = 3$ is greater than zero.

$$x = -\frac{b}{2a} = -\frac{-6}{2(3)} = 1$$

$$f(1) = 3(1)^2 - 6(1) + 4 = 1$$

The minimum value is 1, and it occurs at $x = 1$.

13. $a = -1, b = 8, c = -4$

The quadratic function has a maximum since $a = -1$ is less than zero.

$$x = -\frac{b}{2a} = -\frac{8}{2(-1)} = 4$$

$$f(4) = -(4)^2 + 8(4) - 4 = 12$$

The maximum value is 12, and it occurs at $x = 4$.

15. $a = -3, b = 12, c = 4$

The quadratic function has a maximum since $a = -3$ is less than zero.

$$x = -\frac{b}{2a} = -\frac{12}{2(-3)} = 2$$

$$f(2) = -3(2)^2 + 12(2) + 4 = 16$$

The maximum value is 16, and it occurs at $x = 2$.

17. Answers will vary.

19.

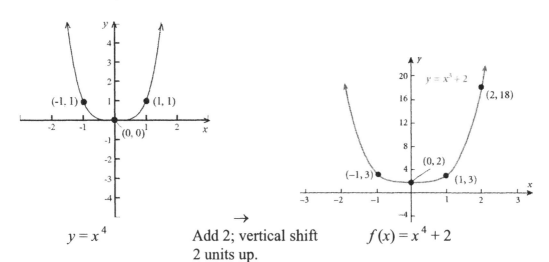

$y = x^4$ Add 2; vertical shift 2 units up. $f(x) = x^4 + 2$

21.

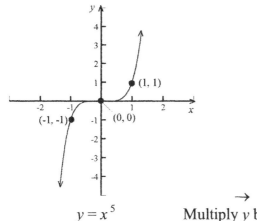

$$y = x^5$$

$$\text{Multiply } y \text{ by } -1;$$
$$\text{reflect about } x\text{-axis.}$$

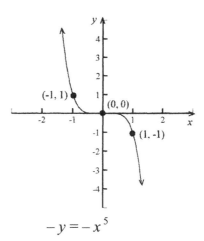

$$-y = -x^5$$

Add 1; vertical
shift up 1 unit.

$$f(x) = -x^5 + 1$$

23. f is a polynomial function. Its degree is 5.

25. f is not a polynomial function. The exponent on the middle term is not a positive integer.

27. The power function that models the end behavior of the function f is $p(x) = -2x^4$.

29. The domain of a rational function is all real numbers except those that make the denominator zero. The denominator of $R(x)$, $x^2 - 9 = 0$ when $x = -3$ or when $x = 3$. The domain of R is $\{x \mid x \neq -3 \text{ or } x \neq 3\}$.

31. The domain of a rational function is all real numbers except those that make the denominator zero. The denominator of $R(x)$, $(x + 2)^2 = 0$, when $x = -2$. The domain of R is $\{x \mid x \neq -2\}$.

33. (a) $f(4) = 3^4 = 81$

(b) $g(9) = \log_3 9 = \log_3 3^2 = 2$

(c) $f(-2) = 3^{-2} = \dfrac{1}{9}$

(d) $g\left(\dfrac{1}{27}\right) = \log_3\left(\dfrac{1}{27}\right) = \log_3\left(\dfrac{1}{3^3}\right)$
$= \log_3 3^{-3} = -3$

35. $\quad 5^2 = z$
$\log_5 z = 2$

37. $\quad \log_5 u = 13$
$5^{13} = u$

39.

The domain of f consists of all x for which $3x - 2 > 0$, that is for all $x > \dfrac{2}{3}$, or using

interval notation, $\left(\dfrac{2}{3}, \infty \right)$.

41.

The domain of H consists of all x for which $-3x + 2 > 0$, that is for all $x < \dfrac{2}{3}$, or using

interval notation, $\left(-\infty, \dfrac{2}{3} \right)$.

43. $\log_2 \left(\dfrac{1}{8} \right) = \log_2 \left(2^{-3} \right) = -3 \log_2 2 = -3$

45. $\log_3 81 = \log_3 \left(3^4 \right) = 4 \log_3 3 = 4$

47. $\ln e^2 = 2 \ln e = 2$

49. $\ln e^{\sqrt{2}} = \sqrt{2} \ln e = \sqrt{2}$

51. $2^{\log_2 0.4} = 0.4$

53.

$$\log_3 \left(\dfrac{uv^2}{w} \right) = \log_3 \left(uv^2 \right) - \log_3 w$$

$$= \log_3 u + \log_3 v^2 - \log_3 w$$

$$= \log_3 u + 2 \log_3 v - \log_3 w$$

55. $\log \left(x^2 \sqrt{x^3 + 1} \right) = \log x^2 + \log \sqrt{x^3 + 1}$

$$= 2 \log x + \dfrac{1}{2} \log \left(x^3 + 1 \right)$$

57.

$$\ln \left(\dfrac{x \sqrt[3]{x^2 + 1}}{x - 3} \right) = \ln \left(x \sqrt[3]{x^2 + 1} \right) - \ln (x - 3)$$

$$= \ln x + \ln \sqrt[3]{x^2 + 1} - \ln (x - 3)$$

$$= \ln x + \dfrac{1}{3} \ln \left(x^2 + 1 \right) - \ln (x - 3)$$

59. $3 \log_4 x^2 + \dfrac{1}{2} \log_4 \sqrt{x} = \log_4 \left(x^2 \right)^3 + \log_4 \left(\sqrt{x} \right)^{1/2}$

$$= \log_4 x^6 + \log_4 x^{1/4}$$

$$= \log_4 \left(x^6 \cdot x^{1/4} \right)$$

$$= \log_4 x^{25/4}$$

61.

$$\ln\left(\frac{x-1}{x}\right) + \ln\left(\frac{x}{x+1}\right) - \ln\left(x^2-1\right) = \ln\left[\left(\frac{x-1}{\cancel{x}}\right)\left(\frac{\cancel{x}}{x+1}\right)\right] - \ln\left(x^2-1\right)$$

$$= \ln\left[\frac{\left(\dfrac{x-1}{x+1}\right)}{\left(x^2-1\right)}\right]$$

$$= \ln\left[\left(\frac{x-1}{x+1}\right)\left(\frac{1}{x^2-1}\right)\right]$$

$$= \ln\left[\left(\frac{\cancel{x-1}}{x+1}\right)\left(\frac{1}{\cancel{(x-1)}(x+1)}\right)\right]$$

$$= \ln\left(\frac{1}{(x+1)^2}\right) = \ln(x+1)^{-2}$$

$$= -2\ln(x+1)$$

63.

$$2\log 2 + 3\log x - \frac{1}{2}\left[\log(x+3) + \log(x-2)\right] = 2\log 2 + 3\log x - \frac{1}{2}\left[\log(x+3)(x-2)\right]$$

$$= \log 2^2 + \log x^3 - \log\left[(x+3)(x-2)\right]^{1/2}$$

$$= \log\left(4x^3\right) - \log\left[(x+3)(x-2)\right]^{1/2}$$

$$= \log\left(\frac{4x^3}{\left[(x+3)(x-2)\right]^{1/2}}\right)$$

$$= \log\left(\frac{4x^3}{\sqrt{x^2+x-6}}\right)$$

65.

$$\log_4 19 = \frac{\ln 19}{\ln 4} = 2.124$$

67.

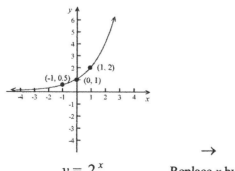

$$y = 2^x$$

→ Replace x by $x-3$;
horizontal shift right 3 units.

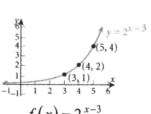

$$f(x) = 2^{x-3}$$

The domain of the function f is all real numbers, or in interval notation $(-\infty, \infty)$; the range is all $\{y \mid y > 0\}$. The x-axis is a horizontal asymptote as f becomes unbounded in the negative direction.

69.

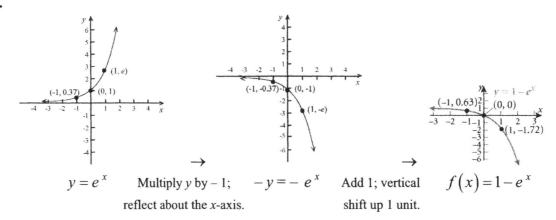

$$y = e^x \qquad \text{Multiply } y \text{ by} -1; \qquad -y = -e^x \qquad \text{Add 1; vertical} \qquad f(x) = 1 - e^x$$
$$\text{reflect about the } x\text{-axis.} \qquad\qquad\qquad \text{shift up 1 unit.}$$

The domain of the function f is all real numbers, or in interval notation $(-\infty, \infty)$; the range is all $\{y \mid y < 1\}$. The line $y = 1$ is a horizontal asymptote as f becomes unbounded in the negative direction.

71.

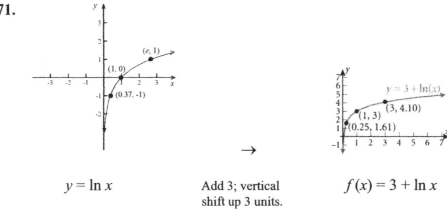

$$y = \ln x \qquad\qquad \text{Add 3; vertical} \qquad f(x) = 3 + \ln x$$
$$\text{shift up 3 units.}$$

The domain of the function f is $\{x \mid x > 0\}$, or in interval notation $(0, \infty)$; the range is all real numbers. The y-axis is a vertical asymptote.

73.
$$4^{1-2x} = 2$$
$$\left(2^2\right)^{1-2x} = 2$$
$$2^{2(1-2x)} = 2^1$$
$$2 - 4x = 1 \qquad \text{Set exponents equal.}$$
$$4x = 1$$
$$x = \frac{1}{4}$$

75.
$$3^{x^2+x} = \sqrt{3} = 3^{1/2}$$
$$x^2 + x = \frac{1}{2}$$

$$x^2 + x - \frac{1}{2} = 0$$
$$2x^2 + 2x - 1 = 0$$

$$x = \frac{-2 \pm \sqrt{2^2 - 4(2)(-1)}}{2(2)} = \frac{-2 \pm \sqrt{4+8}}{4} = \frac{-2 \pm \sqrt{12}}{4} = \frac{-\cancel{2} \pm \cancel{2}\sqrt{3}}{\cancel{4}2} = \frac{-1 \pm \sqrt{3}}{2}$$

$$x = \left\{ \frac{-1+\sqrt{3}}{2}, \frac{-1-\sqrt{3}}{2} \right\}$$

77. $\log_x 64 = -3$

$$x^{-3} = 64 = 4^3$$

$$x^{-3} = \left[\left(\frac{1}{4}\right)^{-1} \right]^3 = \left(\frac{1}{4}\right)^{-3}$$

$$x = \frac{1}{4}$$

79. $9^{2x} = 27^{3x-4}$

$$\left(3^2\right)^{2x} = \left(3^3\right)^{3x-4}$$

$$3^{4x} = 3^{3(3x-4)}$$

$$4x = 3(3x-4)$$

$$4x = 9x - 12$$

$$5x = 12$$

$$x = \frac{12}{5}$$

81. $\log_3 (x-2) = 2$

$$3^2 = x - 2$$

$$9 = x - 2$$

$$x = 11$$

83.   If $100 is invested at 10% compounded continuously, the amount A after 2.25 years (2 years and 3 months) is

$$A = Pe^{rt} = 100e^{(0.1)(2.25)} = 100e^{0.225} = \$125.23$$

85.   I need to invest the present value of $1000 at 4% compounded continuously for 2 years.

$$P = Ae^{-rt} = 1000e^{-(0.04)(2)} = 1000e^{-0.08} = \$923.12$$

87.   If P is invested, it will double when amount $A = 2P$.

$$A = 2P = Pe^{rt}$$

$$2(220,000) = 220,000e^{0.06t}$$

$$2 = e^{0.06t}$$

$$0.06t = \ln 2$$

$$t = \frac{\ln 2}{0.06} \approx 11.55$$

It will take about 11 and one half years for an investment to double at 6 % compounded continuously.

89. When $T = 0°$ and $P_0 = 760$ mm. of mercury,
$$h(x) = 8000\log\left(\frac{760}{x}\right)$$
$$h(300) = 8000\log\left(\frac{760}{300}\right) = 3229.54$$

The Piper Cub is flying at an altitude of approximately 3230 meters above sea level.

91. The perimeter of the pond is $P = 200 = 2l + 2w$
$$l = \frac{200 - 2w}{2} = 100 - w$$

If x denotes the width of the pond, then
$$A = lw = (100 - x)x = 100x - x^2$$

$A = A(x)$ is a quadratic function with $a = -1$, so the maximum value of A is found at the vertex of A.
$$x = -\frac{b}{2a} = -\frac{100}{2(-1)} = 50$$
$$A(50) = 100(50) - 50^2 = 2500 \text{ square feet.}$$

The dimensions should be 50 feet by 50 feet.

93. The perimeter of the window is 100 feet and can be expressed as
$$P = 2l + \pi d = 2l + \pi x = 100$$

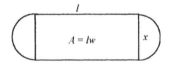

Solving for l in terms of x, we find
$$l = \frac{100 - \pi x}{2} = 50 - \frac{\pi x}{2}$$
The area A of the rectangle is a quadratic function given by
$$A(x) = \left(50 - \frac{\pi x}{2}\right)x = 50x - \frac{\pi x^2}{2}$$
The maximum area is found at the vertex of A. The width of the rectangle with maximum area is
$$x = -\frac{b}{2a} = -\frac{50}{2\left(-\dfrac{\pi}{2}\right)} = \frac{50}{\pi} \text{ feet}$$

The length of the rectangle is
$$l = 50 - \frac{\pi x}{2} = 50 - \frac{\cancel{\pi}\left(\dfrac{50}{\cancel{\pi}}\right)}{2} = 50 - 25 = 25 \text{ feet}$$

95. (a) A 3.5 inch telescope has a limiting magnitude of
$$L = 9 + 5.1\log 3.5 = 11.77$$

(b) If the star's magnitude is 14, the telescope must have a lens with a diameter of

$$9 + 5.1 \log d = 14$$
$$5.1 \log d = 5$$
$$\log d = \frac{5}{5.1}$$
$$d = 10^{5/5.1} = 9.56 \text{ inches.}$$

97. (a) If $620.17 grows to $5000 in 20 years when interest is compounded continuously, the interest rate is 10.436%

$$A = Pe^{rt}$$
$$5000 = 620.17\,e^{20r}$$
$$e^{20r} = \frac{5000}{620.17}$$
$$20r = \ln\left(\frac{5000}{620.17}\right)$$
$$r = \frac{1}{20}\ln\left(\frac{5000}{620.17}\right) = 0.10436$$

(b) An investment of $4000 will have a value A in 20 years if it is invested at 10.436% compounded continuously.

$$A = Pe^{rt} = 4000\,e^{(0.10436)(20)} = \$32,249.24$$

99. (a) The Calloway Company will minimize marginal cost if it produces

$$x = -\frac{b}{2a} = -\frac{(-617.4)}{2(4.9)} = 63 \text{ golf clubs.}$$

(b) The marginal cost of making the 64 golf club is

$$C(63) = 4.9(63)^2 - 617.4(63) + 19,600 = \$151.90$$

CHAPTER 2 PROJECT

1.

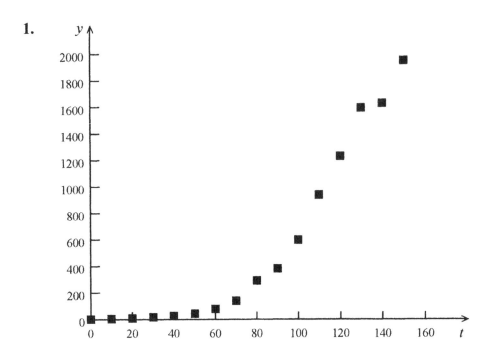

3. (a) $t = 150$ represents the year 2000. The projected population at $t = 150$ is
$$P(150) = 2.4 \cdot 1.59^{150/10} = 2.4 \cdot 1.59^{15} = 2518.7 \text{ thousand persons.}$$

(b) The projected population overestimates the actual population by 565.1 thousand persons, or by 28.9%.

(c) Explanations will vary.

(d) $t = 160$ represents the year 2010. Using the exponential growth function, the predicted population is
$$P(160) = 2.4 \cdot 1.59^{160/10} = 2.4 \cdot 1.59^{16} = 4004.7 \text{ thousand persons.}$$

(e) $t = 200$ represents the year 2050. Using the exponential growth function, the predicted population is
$$P(200) = 2.4 \cdot 1.59^{200/10} = 2.4 \cdot 1.59^{20} = 25,594$$
The predicted population is 25.594 million people.

5. The growth rate for Houston is $P(t) = 2.4 \, e^{\ln 1.59 (t/10)}$

7. Answers will vary.

9. Answers will vary.

11. 3. To do this problem with the result from above, use TBLSET Indpnt: Ask, TABLE.
(a) The population of Houston in 2000 ($t = 50$) is predicted to be 2159.1 thousand.

(b) The prediction overestimates the actual population by 205.5 thousand, or 10.5%.

(c) Answers will vary.

(d) The population of Houston in 2010 ($t = 60$) is predicted to be 2702.0 thousand.

(e) The population of Houston in 2050 ($t = 100$) is predicted to be 6627.0 thousand.

4. If $P = P_0\, a^{t/10} = P_0\, e^{kt/10}$, where $P_0 = 703.4597$ and $a = 1.0227$ then

$$a^{t/10} = e^{kt/10}$$
$$k = \ln a = \ln 1.0227 = 0.022446$$

5. The growth rate for Houston is $P(t) = 703.4597\, e^{\ln 1.0227 (t/10)}$

Explanations will vary.

MATHEMATICAL QUESTIONS FROM PROFESSIONAL EXAMS

1.
(a) $4 \cdot 27 = 2^2 \cdot 3^3 = 6^2 \cdot 3 = \dfrac{6^2 \cdot 3 \cdot 2}{2} = \dfrac{6^2 \cdot 6}{2} = \dfrac{6^3}{2}$

$$\log_6 (4 \cdot 27) = \log_6 \left(\frac{6^3}{2} \right) = \log_6 \left(6^3 \right) - \log_6 (2) = 3 - b$$

3.
(c)
$$y = \frac{e^x - e^{-x}}{2}$$
$$2y = e^x - e^{-x}$$
$$2ye^x = e^x \cdot e^x - e^{-x} \cdot e^x \qquad \text{Multiply both sides by } e^x.$$
$$2ye^x = e^{2x} - 1$$
$$e^{2x} - 2ye^x - 1 = 0 \qquad\qquad a = 1,\, b = -2y,\, c = -1$$
$$e^x = \frac{2y \pm \sqrt{4y^2 + 4}}{2}$$
$$= \frac{2y \pm 2\sqrt{y^2 + 1}}{2}$$
$$= y \pm \sqrt{y^2 + 1}$$

Since $\dfrac{e^x - e^{-x}}{2} > 0$, $y > 0$, so $e^x = y + \sqrt{y^2 + 1}$.

$$x = \ln \left(y + \sqrt{y^2 + 1} \right)$$

5.
(b) $\left(\log_a b \right)\left(\log_b a \right) = \dfrac{\log b}{\log a} \cdot \dfrac{\log a}{\log b} = 1$

7. (a)

$$e^{2\ln(x-1)} = 4$$

$$e^{\ln(x-1)^2} = 4$$

$$(x-1)^2 = 4$$

$$x - 1 = \pm 2$$

$$x = 3$$

Chapter 3
The Limit of a Function

3.1 Finding Limits Using Tables and Graphs

1. Here $f(x) = 2x$, and $c = 1$. We complete the table by evaluating the function f at each value of x.

x	0.9	0.99	0.999
$f(x) = 2x$	1.8	1.98	1.998
x	1.1	1.01	1.001
$f(x) = 2x$	2.2	2.02	2.002

We infer from the table that $\lim\limits_{x \to 1} f(x) = \lim\limits_{x \to 1} 2x = 2$.

3. Here $f(x) = x^2 + 2$, and $c = 0$. We complete the table by evaluating the function f at each value of x.

x	−0.1	−0.01	−0.001
$f(x) = x^2 + 2$	2.01	2.0001	2.0000
x	0.1	0.01	0.001
$f(x) = x^2 + 2$	2.01	2.0001	2.0000

We infer from the table that $\lim\limits_{x \to 0} f(x) = \lim\limits_{x \to 0}(x^2 + 2) = 2$.

5. Here $f(x) = \dfrac{x^2 - 4}{x + 2}$, and $c = -2$. We complete the table by evaluating the function f at each value of x.

x	−2.1	−2.01	−2.001
$f(x) = \dfrac{x^2 - 4}{x + 2}$	−4.1	−4.01	−4.001
x	−1.9	−1.99	−1.999
$f(x) = \dfrac{x^2 - 4}{x + 2}$	−3.9	−3.99	−3.999

We infer from the table that $\lim\limits_{x \to -2} f(x) = \lim\limits_{x \to -2}\left(\dfrac{x^2 - 4}{x + 2}\right) = -4$.

7. Here $f(x) = \dfrac{x^3 + 1}{x + 1}$, and $c = -1$. We complete the table by evaluating the function f at each value of x.

x	-1.1	-1.01	-1.001
$f(x) = \dfrac{x^3 + 1}{x + 1}$	3.31	3.0301	3.0030

x	-0.9	-0.99	-0.999
$f(x) = \dfrac{x^3 + 1}{x + 1}$	2.71	2.9701	2.9970

We infer from the table that $\lim\limits_{x \to -1} f(x) = \lim\limits_{x \to -1}\left(\dfrac{x^3 + 1}{x + 1}\right) = 3$.

9. Here $f(x) = 4x^3$, and $c = 2$. We choose values of x close to 2, starting at 1.99. Then we select additional numbers that get closer to 2, but remain less than 2. Next we choose values of x greater than 2, starting with 2.01, that get closer to 2. Finally we evaluate the function f at each choice to obtain the table:

x	1.99	1.999	1.9999	$\to \leftarrow$	2.0001	2.001	2.01
$f(x) = 4x^3$	31.522	31.952	31.995	$\to \leftarrow$	32.005	32.048	32.482

We infer that as x gets closer to 2, f gets closer to 32. That is,
$$\lim_{x \to 2} f(x) = \lim_{x \to 2}\left(4x^3\right) = 32$$

11. Here $f(x) = \dfrac{x + 1}{x^2 + 1}$, and $c = 0$. We choose values of x close to 0, starting at -0.01. Then

we select additional numbers that get closer to 0, but remain less than 0. Next we choose values of x greater than 0, starting with 0.01, that get closer to 0. Finally we evaluate the function f at each choice to obtain the table:

x	-0.01	-0.001	-0.0001	$\to \leftarrow$	0.0001	0.001	0.01
$f(x) = \dfrac{x + 1}{x^2 + 1}$	0.9899	0.9990	0.9999	$\to \leftarrow$	1.0001	1.001	1.0099

We infer that as x gets closer to 0, f gets closer to 1. That is,
$$\lim_{x \to 0} f(x) = \lim_{x \to 0}\left(\dfrac{x + 1}{x^2 + 1}\right) = 1$$

13. Here $f(x) = \dfrac{x^2 - 4x}{x - 4}$, and $c = 4$. We choose values of x close to 4, starting at 3.99. Then

we select additional numbers that get closer to 4, but remain less than 4. Next we choose values of x greater than 4, starting with 4.01, that get closer to 4. Finally we evaluate the function f at each choice to obtain the table:

x	3.99	3.999	3.9999	$\to \leftarrow$	4.0001	4.001	4.01
$f(x) = \dfrac{x^2 - 4x}{x - 4}$	3.99	3.999	3.9999	$\to \leftarrow$	4.0001	4.001	4.01

We infer that as x gets closer to 4, f gets closer to 4. That is,

$$\lim_{x \to 4} f(x) = \lim_{x \to 4}\left(\frac{x^2 - 4x}{x - 4}\right) = 4$$

15. Here $f(x) = e^x + 1$, and $c = 0$. We choose values of x close to 0, starting at -0.01. Then we select additional numbers that get closer to 0, but remain less than 0. Next we choose values of x greater than 0, starting with 0.01, that get closer to 0. Finally we evaluate the function f at each choice to obtain the table:

x	-0.01	-0.001	-0.0001	$\to \leftarrow$	0.0001	0.001	0.01
$f(x) = e^x + 1$	1.9900	1.9990	1.9999	$\to \leftarrow$	2.0001	2.0010	2.0101

We infer that as x gets closer to 0, f gets closer to 2. That is,

$$\lim_{x \to 0} f(x) = \lim_{x \to 0}\left(e^x + 1\right) = 2$$

17. To determine the $\lim\limits_{x \to 2} f(x)$ we observe that as x gets closer to 2, $f(x)$ gets closer to 3. So We conclude that

$$\lim_{x \to 2} f(x) = 3$$

19. To determine the $\lim\limits_{x \to 2} f(x)$ we observe that as x gets closer to 2, $f(x)$ gets closer to 4. We conclude that

$$\lim_{x \to 2} f(x) = 4$$

21. To determine the $\lim\limits_{x \to 3} f(x)$ we observe that as x gets closer to 3, but remains less than 3, the value of f gets closer to 3. However, we see that as x gets closer to 3, but remains greater than 3, the value of f gets closer to 6. Since there is no single number that the values of f are close to when x is close to 3, we conclude that $\lim\limits_{x \to 3} f(x)$ does not exist.

23. We conclude from the graph that
$$\lim_{x \to 4} f(x) = \lim_{x \to 4}(3x + 1) = 13$$

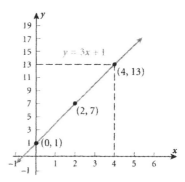

25.   We conclude from the graph that
$$\lim_{x\to 2} f(x)= \lim_{x\to 2}\left(1-x^2\right)=-3$$

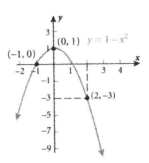

27.   We conclude from the graph that

$$\lim_{x\to -3} f(x)= \lim_{x\to -3}\left(|x|-2\right)=1$$

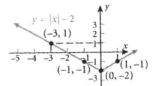

29.   We conclude from the graph that

$$\lim_{x\to 0} f(x)= \lim_{x\to 0} e^x =1$$

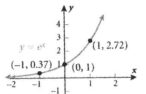

31.   We conclude from the graph that
$$\lim_{x\to -1} f(x)= \lim_{x\to -1}\frac{1}{x}=-1$$

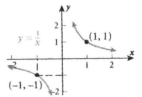

33.   We conclude from the graph that

$$\lim_{x\to 0} f(x)=0$$

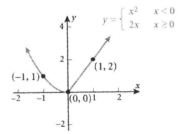

Notice that
$$\lim_{x\to 0^-} f(x)= \lim_{x\to 0^-} x^2=0$$
and
$$\lim_{x\to 0^+} f(x)= \lim_{x\to 0^+}(2x)=0$$

35.   We conclude from the graph that
$\lim_{x\to 1} f(x)$ does not exist.

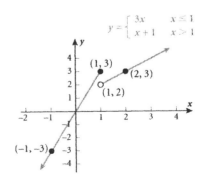

Notice that
$$\lim_{x\to 1^-} f(x)= \lim_{x\to 1^-} 3x=3$$
but
$$\lim_{x\to 1^+} f(x)= \lim_{x\to 1^+}(x+1)=2$$

37. We conclude from the graph that
$$\lim_{x \to 0} f(x) = 0$$

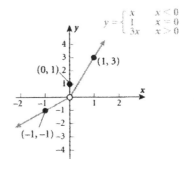

Notice that
$$\lim_{x \to 0^-} f(x) = \lim_{x \to 0^-} x = 0$$
and
$$\lim_{x \to 0^+} f(x) = \lim_{x \to 0^+} (3x) = 0$$

39. We conclude from the graph that
$$\lim_{x \to 0} f(x) = 0$$

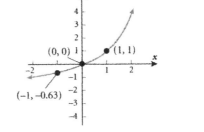

Notice that
$$\lim_{x \to 0^-} f(x) = \lim_{x \to 0^-} (e^x - 1) = 1 - 1 = 0$$
and
$$\lim_{x \to 0^+} f(x) = \lim_{x \to 0^+} x^2 = 0$$

41. To find the limit we create the tables shown below and conclude
$$\lim_{x \to 1} \frac{x^3 - x^2 + x - 1}{x^4 - x^3 + 2x - 2} = \frac{2}{3}$$

X	Y₁
0	.5
.5	.58824
.8	.65287
.9	.66325
.99	.66663
.999	.66667
.9999	.66667
X=.9999	

X	Y₁
2	.5
1.5	.60465
1.2	.65451
1.1	.66346
1.01	.66663
1.001	.66667
1.0001	.66667
X=1.0001	

43. To find the limit we create the tables shown below and conclude
$$\lim_{x \to -2} \frac{x^3 - 2x^2 + 4x - 8}{x^2 + x - 6} = 1.6$$

X	Y₁
1	1.25
1.5	1.3889
1.8	1.5083
1.9	1.5531
1.99	1.5952
1.999	1.5995
1.9999	1.6
X=1.9999	

X	Y₁
3	2.1667
2.5	1.8636
2.2	1.7
2.1	1.649
2.01	1.6048
2.001	1.6005
2.0001	1.6
X=2.0001	

45. To find the limit we create the tables shown below and conclude
$$\lim_{x \to -1} \frac{x^3 + 2x^2 + x}{x^4 + x^3 + 2x + 2} = 0$$

X	Y1
-2	-.3333
-1.5	-.5455
-1.2	.88235
-1.1	.16442
-1.01	.01042
-1.001	.001
-1.0001	.001
X=-1.001	

X	Y1
0	0
.5	-.1333
.8	-.1075
.9	-.0708
.99	-.0096
.999	-1E-3
-.9999	-1E-4
X=-.9999	

3.2 Techniques for Finding Limits of Functions

1. Using formula (1) (p. 243), we find
$$\lim_{x \to 1} 5 = 5$$

3. Using formula (2) (p. 243), we find
$$\lim_{x \to 4} x = 4$$

5.
$$\lim_{x \to 2}(3x + 2) = \lim_{x \to 2}(3x) + \lim_{x \to 2} 2 = \left[\lim_{x \to 2} 3\right]\left[\lim_{x \to 2} x\right] + \lim_{x \to 2} 2$$
$$= (3) \cdot (2) + 2 = 6 + 2 = 8$$

7.
$$\lim_{x \to -1}\left(3x^2 - 5x\right) = 3 \cdot (-1)^2 - 5 \cdot (-1) = 3 \cdot 1 + 5 = 3 + 5 = 8$$

9.
$$\lim_{x \to 1}\left(5x^4 - 3x^2 + 6x - 9\right) = 5 \cdot 1^4 - 3 \cdot 1^2 + 6 \cdot 1 - 9$$
$$= 5 \cdot 1 - 3 \cdot 1 + 6 - 9$$
$$= 5 - 3 - 3 = -1$$

11.
$$\lim_{x \to 1}\left(x^2 + 1\right)^3 = \left[\lim_{x \to 1}\left(x^2 + 1\right)\right]^3 = (2)^3 = 8$$

13.
$$\lim_{x \to 1}\sqrt{5x + 4} = \sqrt{\lim_{x \to 1}(5x + 4)} = \sqrt{9} = 3$$

15. The limit we seek is the limit of a rational function whose domain is $\{x \mid x \neq -2, x \neq 2\}$. Since 0 is in the domain, we use formula (12).
$$\lim_{x \to 0}\frac{x^2 - 4}{x^2 + 4} = \frac{0^2 - 4}{0^2 + 4} = \frac{-4}{4} = -1$$

17.
$$\lim_{x \to 2}(3x - 2)^{\frac{5}{2}} = \left[\lim_{x \to 2}(3x - 2)\right]^{\frac{5}{2}} = (4)^{\frac{5}{2}} = (2)^5 = 32$$

19. The domain of the rational function $R(x) = \dfrac{x^2 - 4}{x^2 - 2x}$ is $\{x \mid x \neq 0, x \neq 2\}$. Since 2 is not in the domain, we cannot evaluate $R(2)$, but we notice that the function can be factored as
$$\frac{x^2 - 4}{x^2 - 2x} = \frac{(x + 2)(x - 2)}{x(x - 2)}$$

Since x is near 2, but $x \neq 2$, we can cancel the $(x - 2)$'s. Formula (11), can then be used to find the limit of the function as x approaches 2.

$$\lim_{x \to 2} \frac{x^2 - 4}{x^2 - 2x} = \lim_{x \to 2} \frac{(x+2)(x-2)}{x(x-2)} = \frac{\lim_{x \to 2}(x+2)}{\lim_{x \to 2}(x)} = \frac{4}{2} = 2$$

21. The domain of the rational function $R(x) = \dfrac{x^2 - x - 12}{x^2 - 9}$ is $\{x \mid x \neq -3, x \neq 3\}$. Since -3 is not in the domain, we cannot evaluate $R(-3)$, but we notice that the function can be factored as

$$\frac{x^2 - x - 12}{x^2 - 9} = \frac{(x-4)(x+3)}{(x-3)(x+3)}$$

Since x is near -3, but $x \neq -3$, we can cancel the $(x + 3)$'s. Formula (11), can then be used to find the limit of the function as x approaches -3.

$$\lim_{x \to -3} \frac{x^2 - x - 12}{x^2 - 9} = \lim_{x \to -3} \frac{(x-4)(x+3)}{(x-3)(x+3)} = \frac{\lim_{x \to -3}(x-4)}{\lim_{x \to -3}(x-3)} = \frac{(-3)-4}{(-3)-3} = \frac{-7}{-6} = \frac{7}{6}$$

23. The domain of the rational function $R(x) = \dfrac{x^3 - 1}{x - 1}$ is $\{x \mid x \neq 1\}$. Since 1 is not in the domain, we cannot evaluate $R(1)$, but we notice that the function can be factored as

$$\frac{x^3 - 1}{x - 1} = \frac{(x-1)(x^2 + x + 1)}{x - 1}$$

Since x is near 1, but $x \neq 1$, we can cancel the $(x - 1)$'s. Formula (11), can then be used to find the limit of the function as x approaches 1.

$$\lim_{x \to 1} \frac{x^3 - 1}{x - 1} = \lim_{x \to 1} \frac{(x-1)(x^2 + x + 1)}{x-1} = \frac{\lim_{x \to 1}(x^2 + x + 1)}{\lim_{x \to 1}(1)} = \frac{1^2 + 1 + 1}{1} = 3$$

25. The domain of the rational function $R(x) = \dfrac{(x+1)^2}{x^2 - 1}$ is $\{x \mid x \neq -1, x \neq 1\}$. Since -1 is not in the domain, we cannot evaluate $R(-1)$, but we notice that the function can be factored as

$$\frac{(x+1)^2}{x^2 - 1} = \frac{(x+1)^2}{(x-1)(x+1)}$$

Since x is near -1, but $x \neq -1$, we can cancel $(x + 1)$'s. Formula (11), can then be used to find the limit of the function as x approaches -1.

$$\lim_{x \to -1} \frac{(x+1)^2}{x^2 - 1} = \lim_{x \to -1} \frac{(x+1)^2}{(x-1)(x+1)} = \lim_{x \to -1} \frac{(x+1)}{(x-1)} = \frac{(-1)+1}{(-1)-1} = \frac{0}{-2} = 0$$

27. The limit of the denominator of this function as x approaches 1 is zero, so formula (11), cannot be used directly. We first factor the function by grouping.

$$\frac{x^3 - x^2 + x - 1}{x^4 - x^3 + 2x - 2} = \frac{x^2(x-1) + 1 \cdot (x-1)}{x^3(x-1) + 2 \cdot (x-1)} = \frac{(x-1)(x^2+1)}{(x-1)(x^3+2)}$$

Since x is near 1, but $x \neq 1$, we can cancel $(x-1)$'s. Then using formula (11), we get

$$\lim_{x \to 1} \frac{x^3 - x^2 + x - 1}{x^4 - x^3 + 2x - 2} = \lim_{x \to 1} \frac{\cancel{(x-1)}(x^2+1)}{\cancel{(x-1)}(x^3+2)} = \frac{\lim\limits_{x \to 1}(x^2+1)}{\lim\limits_{x \to 1}(x^3+2)} = \frac{2}{3}$$

29. The limit of the denominator of this function as x approaches 2 is zero, so formula (11), cannot be used directly. We first factor the function, using grouping to factor the numerator.

$$\frac{x^3 - 2x^2 + 4x - 8}{x^2 + x - 6} = \frac{x^2(x-2) + 4 \cdot (x-2)}{(x+3)(x-2)} = \frac{(x^2+4)(x-2)}{(x+3)(x-2)}$$

Since x is near 2, but $x \neq 2$, we can cancel $(x-2)$'s. Then using formula (11), we get

$$\lim_{x \to 2} \frac{x^3 - 2x^2 + 4x - 8}{x^2 + x - 6} = \lim_{x \to 2} \frac{(x^2+4)\cancel{(x-2)}}{(x+3)\cancel{(x-2)}} = \frac{\lim\limits_{x \to 2}(x^2+4)}{\lim\limits_{x \to 2}(x+3)} = \frac{2^2+4}{2+3} = \frac{8}{5}$$

31. The limit of the denominator of this function as x approaches -1 is zero, so formula (11), cannot be used directly. We first factor the function, using grouping to factor the denominator.

$$\frac{x^3 + 2x^2 + x}{x^4 + x^3 + 2x + 2} = \frac{x(x^2 + 2x + 1)}{x^3(x+1) + 2(x+1)} = \frac{x(x+1)(x+1)}{(x^3+2)(x+1)}$$

Since x is near -1, but $x \neq -1$, we can cancel $(x+1)$'s. Then using formula (11), we get

$$\lim_{x \to -1} \frac{x^3 + 2x^2 + x}{x^4 + x^3 + 2x + 2} = \lim_{x \to -1} \frac{x(x+1)\cancel{(x+1)}}{(x^3+2)\cancel{(x+1)}} = \frac{\lim\limits_{x \to -1} x(x+1)}{\lim\limits_{x \to -1}(x^3+1)} = \frac{(-1)\cdot[(-1)+1]}{(-1)^3+2} = \frac{(-1)\cdot(0)}{1} = 0$$

33. The average rate of change of f from 2 to x is

$$\frac{\Delta y}{\Delta x} = \frac{f(x) - f(2)}{x-2} = \frac{(5x-3) - (5 \cdot 2 - 3)}{x-2} = \frac{5x - 3 - 7}{x-2} = \frac{5x-10}{x-2} = \frac{5(x-2)}{x-2}$$

The limit of the average rate of change as x approaches 2 is

$$\lim_{x \to 2} \frac{f(x) - f(2)}{x-2} = \lim_{x \to 2} \frac{(5x-3) - 7}{x-2} = \lim_{x \to 2} \frac{5x-10}{x-2} = \lim_{x \to 2} \frac{5\cancel{(x-2)}}{\cancel{x-2}} = 5$$

35. The average rate of change of f from 3 to x is

$$\frac{\Delta y}{\Delta x} = \frac{f(x) - f(3)}{x-3} = \frac{x^2 - 3^2}{x-3} = \frac{(x+3)(x-3)}{x-3}$$

The limit of the average rate of change as x approaches 3 is

$$\lim_{x\to 3}\frac{f(x)-f(3)}{x-3}=\lim_{x\to 3}\frac{x^2-3^2}{x-3}=\lim_{x\to 3}\frac{(x+3)(x-3)}{x-3}=6$$

37. The average rate of change of f from -1 to x is

$$\frac{\Delta y}{\Delta x}=\frac{f(x)-f(-1)}{x-(-1)}=\frac{\left(x^2+2x\right)-\left[(-1)^2+2\cdot(-1)\right]}{x+1}=\frac{\left(x^2+2x\right)-(-1)}{x+1}=\frac{x^2+2x+1}{x+1}=\frac{(x+1)^2}{x+1}$$

The limit of the average rate of change as x approaches -1 is

$$\lim_{x\to -1}\frac{f(x)-f(-1)}{x-(-1)}=\lim_{x\to -1}\frac{\left(x^2+2x\right)-(-1)}{x+1}=\lim_{x\to -1}\frac{x^2+2x+1}{x+1}=\lim_{x\to -1}\frac{(x+1)^2}{x+1}=(-1)+1=0$$

39. The average rate of change of f from 0 to x is

$$\frac{\Delta y}{\Delta x}=\frac{f(x)-f(0)}{x-0}=\frac{\left(3x^3-2x^2+4\right)-(4)}{x}=\frac{3x^3-2x^2}{x}=\frac{x^2(3x-2)}{x}$$

The limit of the average rate of change as x approaches 0 is

$$\lim_{x\to 0}\frac{f(x)-f(0)}{x-0}=\lim_{x\to 0}\frac{3x^3-2x^2}{x}=\lim_{x\to 0}\frac{x^2(3x-2)}{x}=0$$

41. The average rate of change of f from 1 to x is

$$\frac{\Delta y}{\Delta x}=\frac{f(x)-f(1)}{x-1}=\frac{\dfrac{1}{x}-\dfrac{1}{1}}{x-1}=\frac{\dfrac{1}{x}-\dfrac{x}{x}}{x-1}=\frac{\dfrac{1-x}{x}}{x-1}=\frac{1-x}{x(x-1)}=\frac{(-1)\cdot(x-1)}{x(x-1)}$$

$$\uparrow\qquad\qquad\qquad\uparrow\qquad\qquad\uparrow$$

Find a common denominator   Simplify   Factor out (-1)

The limit of the average rate of change as x approaches 1 is

$$\lim_{x\to 1}\frac{f(x)-f(1)}{x-1}=\lim_{x\to 1}\frac{1-x}{x(x-1)}=\lim_{x\to 1}\frac{(-1)\cdot(x-1)}{x(x-1)}=-1$$

43. The average rate of change of f from 4 to x is

$$\frac{\Delta y}{\Delta x}=\frac{f(x)-f(4)}{x-4}=\frac{\sqrt{x}-\sqrt{4}}{x-4}=\frac{\sqrt{x}-2}{\left(\sqrt{x}+2\right)\left(\sqrt{x}-2\right)}$$

$$\uparrow$$

Factor the denominator

The limit of the average rate of change as x approaches 4 is

$$\lim_{x\to 4}\frac{f(x)-f(4)}{x-4}=\lim_{x\to 4}\frac{\sqrt{x}-\sqrt{4}}{x-4}=\lim_{x\to 4}\frac{\sqrt{x}-2}{\left(\sqrt{x}+2\right)\left(\sqrt{x}-2\right)}=\frac{1}{\sqrt{4}+2}=\frac{1}{4}$$

45. Since $\lim_{x \to c} f(x) = 5$ and $\lim_{x \to c} g(x) = 2$, $\lim_{x \to c} \left[2f(x) \right] = \left(\lim_{x \to c} 2 \right)\left(\lim_{x \to c} f(x) \right) = 2 \cdot 5 = 10$

47. Since $\lim_{x \to c} f(x) = 5$ and $\lim_{x \to c} g(x) = 2$, $\lim_{x \to c} \left[g(x) \right]^3 = \left[\lim_{x \to c} g(x) \right]^3 = 2^3 = 8$

49. Since $\lim_{x \to c} f(x) = 5$ and $\lim_{x \to c} g(x) = 2$, $\lim_{x \to c} \dfrac{4}{f(x)} = \dfrac{\lim_{x \to c} 4}{\lim_{x \to c} f(x)} = \dfrac{4}{5}$

51. Since $\lim_{x \to c} f(x) = 5$ and $\lim_{x \to c} g(x) = 2$,

$$\lim_{x \to c} \left[4f(x) - 5g(x) \right] = \lim_{x \to c} \left[4f(x) \right] - \lim_{x \to c} \left[5g(x) \right]$$
$$= \left(\lim_{x \to c} 4 \right) \cdot \left(\lim_{x \to c} f(x) \right) - \left(\lim_{x \to c} 5 \right) \cdot \left(\lim_{x \to c} g(x) \right)$$
$$= 4 \cdot 5 - 5 \cdot 2 = 20 - 10 = 10$$

3.3 One-Sided Limits; Continuous Functions

1. The domain of f is $\{x \mid -8 \le x < -3 \text{ or } -3 < x < 4 \text{ or } 4 < x \le 6\}$ or the intervals $[8, -3)$ or $(-3, 4)$ or $(4, 6]$.

3. The x-intercepts of the graph of f are $(-8, 0)$ and $(-5, 0)$. At these points the graph of f either crosses or touches the x-axis.

5. $f(-8) = 0$ and $f(-4) = 2$

7. To find $\lim_{x \to -6^-} f(x)$, we look at the values of f when x is close to -6, but less than -6. Since the graph of f is approaching $y = 3$ for these values, we have $\lim_{x \to -6^-} f(x) = 3$.

9. To find $\lim_{x \to -4^-} f(x)$, we look at the values of f when x is close to -4, but less than -4. Since the graph of f is approaching $y = 2$ for these values, we have $\lim_{x \to -4^-} f(x) = 2$.

11. To find $\lim_{x \to 2^-} f(x)$, we look at the values of f when x is close to 2, but less than 2. Since the graph of f is approaching $y = 1$ for these values, we have $\lim_{x \to 2^-} f(x) = 1$.

13. The $\lim_{x\to 4} f(x)$ exists because both $\lim_{x\to 4^-} f(x) = 0$ and $\lim_{x\to 4^+} f(x) = 0$. Since both one-sided limits exist and are equal, the limit of f as x approaches 4 exists and is equal to the one-sided limits. That is, $\lim_{x\to 4} f(x) = 0$.

15. The function f is not continuous at $x = -6$, because $\lim_{x\to -6} f(x)$ does not exist. (The one-sided limits are not equal. See Problems 7 and 8.)

17. The function f is continuous at $x = 0$. The function is defined at zero, $f(0) = 3$, and $\lim_{x\to 0} f(x) = f(0) = 3$. (See Problem 14.)

19. The function f is not continuous at $x = 4$. The function is not defined at 4. That is, 4 is not part of the domain of f.

21. To find the one-sided limit we look at values of x close to 1, but greater than 1. Since $f(x) = 2x + 3$ for such numbers, we conclude that
$$\lim_{x\to 1^+} f(x) = \lim_{x\to 1^+} (2x + 3) = 5$$

23. To find the one-sided limit we look at values of x close to 1, but less than 1. Since $f(x) = 2x^3 + 5x$ for such numbers, we conclude that
$$\lim_{x\to 1^-} f(x) = \lim_{x\to 1^-} (2x^3 + 5x) = 7$$

25. To find the one-sided limit we look at values of x close to 0, but less than 0. Since $f(x) = e^x$ for such numbers, we conclude that
$$\lim_{x\to 0^-} f(x) = \lim_{x\to 0^-} (e^x) = 1$$

27. To find the one-sided limit we look at values of x close to 2, but greater than 2. Since $f(x) = \dfrac{x^2 - 4}{x - 2}$ for such numbers, we first factor the function.
$$f(x) = \frac{x^2 - 4}{x - 2} = \frac{(x+2)(x-2)}{x-2}$$
We can then conclude that
$$\lim_{x\to 2^+} f(x) = \lim_{x\to 2^+} \frac{x^2 - 4}{x - 2} = \lim_{x\to 2^+} \frac{(x+2)\cancel{(x-2)}}{\cancel{x-2}} = 4$$

29. To find the one-sided limit we look at values of x close to -1, but less than -1. Since $f(x) = \dfrac{x^2 - 1}{x^3 + 1}$ for such numbers, we first factor the function.
$$f(x) = \frac{x^2 - 1}{x^3 + 1} = \frac{(x+1)(x-1)}{(x+1)(x^2 - x + 1)}$$

We can then conclude that

$$\lim_{x \to -1^-} f(x) = \lim_{x \to -1^-} \frac{x^2 - 1}{x^3 + 1} = \lim_{x \to -1^-} \frac{(x+1)(x-1)}{(x+1)(x^2 - x + 1)} = \frac{-2}{3} = -\frac{2}{3}$$

31.   To find the one-sided limit we look at values of x close to -2, but greater than -2. Since $f(x) = \dfrac{x^2 + x - 2}{x^2 + 2x}$ for such numbers, we first factor the function.

$$f(x) = \frac{x^2 + x - 2}{x^2 + 2x} = \frac{(x+2)(x-1)}{x(x+2)}$$

We can then conclude that

$$\lim_{x \to 2^+} f(x) = \lim_{x \to 2^+} \frac{x^2 + x - 2}{x^2 + 2x} = \lim_{x \to 2^+} \frac{(x+2)(x-1)}{x(x+2)} = \lim_{x \to 2^+} \frac{x-1}{x} = \frac{1}{2}$$

33. $f(x) = x^3 - 3x^2 + 2x - 6$ is continuous at $c = 2$ because f is a polynomial, and polynomials are continuous at every number.

35. $f(x) = \dfrac{x^2 + 5}{x - 6}$ is a rational function whose domain is $\{x \mid x \neq 6\}$. f is continuous at $c = 3$ since f is defined at 3.

37. $f(x) = \dfrac{x+3}{x-3}$ is a rational function whose domain is $\{x \mid x \neq 3\}$. f is not continuous at $c = 3$ since f is not defined at 3.

39. $f(x) = \dfrac{x^3 + 3x}{x^2 - 3x}$ is a rational function whose domain is $\{x \mid x \neq 0, x \neq 3\}$. f is not continuous at $c = 0$ since f is not defined at 0.

41.   To determine whether $f(x) = \begin{cases} \dfrac{x^3 + 3x}{x^2 - 3x} & \text{if } x \neq 0 \\ 1 & \text{if } x = 0 \end{cases}$ is continuous at $c = 0$, we investigate f when $x = 0$.

$$f(0) = 1$$

$$\lim_{x \to 0^-} f(x) = \lim_{x \to 0^-} \frac{x^3 + 3x}{x^2 - 3x} = \lim_{x \to 0^-} \frac{x(x^2 + 3)}{x(x-3)} = \frac{3}{-3} = -1$$

$$\lim_{x \to 0^+} f(x) = \lim_{x \to 0^+} \frac{x^3 + 3x}{x^2 - 3x} = \lim_{x \to 0^+} \frac{x(x^2 + 3)}{x(x-3)} = \frac{3}{-3} = -1$$

Since $\lim_{x \to 0} f(x) = -1 \neq f(0) = 1$, the function f is not continuous at $c = 0$.

43. To determine whether $f(x) = \begin{cases} \dfrac{x^3 + 3x}{x^2 - 3x} & \text{if } x \neq 0 \\ -1 & \text{if } x = 0 \end{cases}$ is continuous at $c = 0$, we investigate

f when $x = 0$.

$$f(0) = -1$$

$$\lim_{x \to 0^-} f(x) = \lim_{x \to 0^-} \frac{x^3 + 3x}{x^2 - 3x} = \lim_{x \to 0^-} \frac{\cancel{x}\left(x^2 + 3\right)}{\cancel{x}(x - 3)} = \frac{3}{-3} = -1$$

$$\lim_{x \to 0^+} f(x) = \lim_{x \to 0^+} \frac{x^3 + 3x}{x^2 - 3x} = \lim_{x \to 0^+} \frac{\cancel{x}\left(x^2 + 3\right)}{\cancel{x}(x - 3)} = \frac{3}{-3} = -1$$

Since $\lim_{x \to 0} f(x) = -1 = f(0)$, the function f is continuous at $c = 0$.

45. To determine whether $f(x) = \begin{cases} \dfrac{x^3 - 1}{x^2 - 1} & \text{if } x < 1 \\ 2 & \text{if } x = 1 \\ \dfrac{3}{x + 1} & \text{if } x > 1 \end{cases}$ is continuous at $c = 1$, we investigate f

when $x = 1$.

$$f(1) = 2$$

$$\lim_{x \to 1^-} f(x) = \lim_{x \to 1^-} \frac{x^3 - 1}{x^2 - 1} = \lim_{x \to 1^-} \frac{\cancel{(x - 1)}\left(x^2 + x + 1\right)}{\cancel{(x - 1)}(x + 1)} = \frac{1 + 1 + 1}{1 + 1} = \frac{3}{2}$$

$$\lim_{x \to 1^+} f(x) = \lim_{x \to 1^+} \frac{3}{x + 1} = \frac{3}{2}$$

Since $\lim_{x \to 1} f(x) = \dfrac{3}{2} \neq f(1)$, the function f is not continuous at $c = 1$.

47. To determine whether $f(x) = \begin{cases} 2e^x & \text{if } x < 0 \\ 2 & \text{if } x = 0 \\ \dfrac{x^3 + 2x^2}{x^2} & \text{if } x > 0 \end{cases}$ is continuous at $c = 0$, we

investigate f when $x = 0$.

$$f(0) = 2$$

$$\lim_{x \to 0^-} f(x) = \lim_{x \to 0^-} 2e^x = 2$$

$$\lim_{x \to 0^+} f(x) = \lim_{x \to 0^+} \frac{x^3 + 2x^2}{x^2} = \lim_{x \to 0^+} \frac{\cancel{x^2}(x + 2)}{\cancel{x^2}} = 2$$

The $\lim_{x \to 0} f(x)$ exists, and $\lim_{x \to 0} f(x) = f(0) = 2$. So we conclude that the function f is

continuous at $c = 0$.

49. $f(x) = 2x + 3$ is a first degree polynomial function. Polynomial functions are continuous at all real numbers.

51. $f(x) = 3x^2 + x$ is a second degree polynomial function. Polynomial functions are continuous at all real numbers.

53. $f(x) = 4 \ln x$ is the product of a constant function $h(x) = 4$, which is continuous at every number, and the logarithmic function $g(x) = \ln x$, which is continuous for every number in the domain $(0, \infty)$. So $f(x) = 4 \ln x$ is continuous for all values $x > 0$.

55. $f(x) = 3e^x$ is the product of a constant function $h(x) = 3$, which is continuous at every number, and the exponential function $g(x) = e^x$, which is continuous for every number in the domain $(-\infty, \infty)$. So $f(x) = 3e^x$ is continuous for all real numbers.

57. $f(x) = \dfrac{2x+5}{x^2 - 4}$ is a rational function. Rational functions are continuous at every number in the domain. The domain of f is $\{x \mid x \neq -2,\ x \neq 2\}$, and f is continuous at all those numbers. f is discontinuous at $x = -2$ and $x = 2$.

59. $f(x) = \dfrac{x-3}{\ln x}$ is the quotient of a polynomial function, which is continuous at all real numbers and the logarithmic function, which is continuous at all numbers in the domain $(0, \infty)$. So f is continuous at all positive numbers or for $x > 0$.

61. The "pieces" of f, that is, $y = 3x + 1$, $y = -x^2$, and $y = \dfrac{1}{2}x - 5$, are each continuous for every number since they are polynomials. So we only need to investigate $x = 0$ and $x = 2$, the two points at which the pieces change.

For $x = 0$: $\qquad\qquad\qquad f(0) = 3(0) + 1 = 1$

$$\lim_{x \to 0^-} f(x) = \lim_{x \to 0^-} (3x+1) = 1$$

$$\lim_{x \to 0^+} f(x) = \lim_{x \to 0^+} (-x^2) = 0$$

Since $\lim_{x \to 0^+} f(x) \neq f(0)$, we conclude that the function f is discontinuous at $x = 0$.

For $x = 2$: $\qquad\qquad\qquad f(2) = -2^2 = -4$

$$\lim_{x \to 2^-} f(x) = \lim_{x \to 2^-} (-x^2) = -4$$

$$\lim_{x \to 2^+} f(x) = \lim_{x \to 2^+} \left(\dfrac{1}{2}x - 5\right) = -4$$

Since $\lim_{x \to 2} f(x) = f(2) = -4$, we conclude that f is continuous at $x = 2$.

63. The cost function is $C(x) = \begin{cases} 39.99 & \text{if } 0 < x \le 350 \\ 0.25x - 47.51 & \text{if } x > 350 \end{cases}$.

(a) $\lim\limits_{x \to 350^-} C(x) = \lim\limits_{x \to 350^-} (39.99) = 39.99$

(b) $\lim\limits_{x \to 350^+} C(x) = \lim\limits_{x \to 350^+} (0.25x - 47.51) = 39.99$

(c) The left limit equals the right limit, so $\lim\limits_{x \to 350} C(x)$ exists. C is continuous at $x = 350$

since $\lim\limits_{x \to 350} C(x) = C(350) = 39.99$.

(d) Answers will vary.

65. (a) If $t = 10°$ then $W(v) = \begin{cases} 10 & 0 \le v < 1.79 \\ 33 - \dfrac{23(10.45 + 10\sqrt{v} - v)}{22.04} & 1.79 \le v \le 20 \\ -3.7034 & v > 20 \end{cases}$.

(b) $\lim\limits_{v \to 0^+} W(v) = \lim\limits_{v \to 0^+} (10) = 10$

(c) $\lim\limits_{v \to 1.79^-} W(v) = \lim\limits_{v \to 1.79^-} (10) = 10$

(d) $\lim\limits_{v \to 1.79^+} W(v) = \lim\limits_{v \to 1.79^+} \left[33 - \dfrac{23(10.45 + 10\sqrt{1.79} - 1.79)}{22.04} \right] = 10.00095$

(e) $W(1.79) = 10.00095$

(f) W is not continuous at $v = 1.79$ since $\lim\limits_{v \to 1.79^-} W(v) \ne \lim\limits_{v \to 1.79^+} W(v)$, and therefore,

$\lim\limits_{v \to 1.79} W(v)$ does not exist. In order to be continuous at 1.79, $\lim\limits_{v \to 1.79} W(v)$ must exist.

(g) Rounded to two decimal places,

$\lim\limits_{v \to 1.79^-} W(v) = 10.00$ $\qquad$ $\lim\limits_{v \to 1.79^+} W(v) = 10.00$ $\qquad$ $W(1.79) = 10.00$

Now the function W is continuous at 1.79.

(h) Answers will vary.

(i) $\lim\limits_{v \to 20^-} W(v) = \lim\limits_{v \to 20^-} \left[33 - \dfrac{23(10.45 + 10\sqrt{20} - 20)}{22.04} \right] = -3.7033$

(j) $\lim\limits_{v \to 20^+} W(v) = \lim\limits_{v \to 20^+} (-3.7034) = -3.7034$

(k) $W(20) = -3.7033$

(l) W is not continuous at 20. The right limit is not equal to $W(20)$.

(m) Rounded to two decimal places,
$$\lim_{v \to 20^-} W(v) = -3.70 \qquad \lim_{v \to 20^+} W(v) = -3.70 \qquad W(20) = -3.70$$
Now the function W is continuous at 20.

(n) Answers will vary.

3.4 Limits at Infinity; Infinite Limits; End Behavior; Asymptotes

1. As $x \to \infty$, $x^3 + x^2 + 2x - 1 = x^3$, and $x^3 + x + 1 = x^3$, so
$$\lim_{x \to \infty} \frac{x^3 + x^2 + 2x - 1}{x^3 + x + 1} = \lim_{x \to \infty} \frac{x^3}{x^3} = 1$$

3. As $x \to \infty$, $2x + 4 = 2x$, and $x - 1 = x$, so
$$\lim_{x \to \infty} \frac{2x + 4}{x - 1} = \lim_{x \to \infty} \frac{2x}{x} = 2$$

5. As $x \to \infty$, $3x^2 - 1 = 3x^2$, and $x^2 + 4 = x^2$, so
$$\lim_{x \to \infty} \frac{3x^2 - 1}{x^2 + 4} = \lim_{x \to \infty} \frac{3x^2}{x^2} = 3$$

7. As $x \to -\infty$, $5x^3 - 1 = 5x^3$, and $x^4 + 1 = x^4$, so
$$\lim_{x \to -\infty} \frac{5x^3 - 1}{x^4 + 1} = \lim_{x \to -\infty} \frac{5x^3}{x^4} = \lim_{x \to -\infty} \frac{5}{x} = 5 \lim_{x \to -\infty} \frac{1}{x} = 0$$

9. As $x \to \infty$, $5x^3 + 3 = 5x^3$, and $x^2 + 1 = x^2$, so
$$\lim_{x \to \infty} \frac{5x^3 + 3}{x^2 + 1} = \lim_{x \to \infty} \frac{5x^3}{x^2} = \lim_{x \to \infty} 5x = \infty$$

11. As $x \to -\infty$, $4x^5 = 4x^5$, and $x^2 + 1 = x^2$, so
$$\lim_{x \to -\infty} \frac{4x^5}{x^2 + 1} = \lim_{x \to -\infty} \frac{4x^5}{x^2} = \lim_{x \to -\infty} 4x^3 = 4 \cdot \lim_{x \to -\infty} x^3 = -\infty$$

13. Here $f(x) = \dfrac{1}{x - 2}$, $x \neq 2$. To determine $\lim\limits_{x \to 2^+} \dfrac{1}{x - 2}$, we examine the values of f that are close to 2, but remain greater than 2.

x	2.1	2.01	2.001	2.0001	2.00001
$f(x) = \dfrac{1}{x - 2}$	10	100	1000	10,000	100,000

We see that as x gets closer to 2 from the right, the value of $f(x) = \dfrac{1}{x-2}$ becomes unbounded in the positive direction, and we write

$$\lim_{x \to 2^+} \frac{1}{x-2} = \infty$$

15. Here $f(x) = \dfrac{x}{(x-1)^2}$, $x \neq 1$. To determine $\displaystyle\lim_{x \to 1^-} \dfrac{x}{(x-1)^2}$, we examine the values of f that are close to 1, but remain smaller than 1.

x	0.9	0.99	0.999	0.9999
$f(x) = \dfrac{x}{(x-1)^2}$	90	9900	999,000	99,990,000

We see that as x gets closer to 1 from the left, the value of $f(x) = \dfrac{x}{(x-1)^2}$ becomes unbounded in the positive direction, and we write

$$\lim_{x \to 1^-} \frac{x}{(x-1)^2} = \infty$$

17. Here $f(x) = \dfrac{x^2+1}{x^3-1}$, $x \neq 1$. To determine $\displaystyle\lim_{x \to 1^+} \dfrac{x^2+1}{x^3-1}$, we examine the values of f that are close to 1, but remain larger than 1.

x	1.1	1.01	1.001	1.0001	1.00001
$f(x) = \dfrac{x^2+1}{x^3-1}$	6.6767	66.66777	666.66678	6666.667	66,666.667

We see that as x gets closer to 1 from the right, the value of $f(x) = \dfrac{x^2+1}{x^3-1}$ becomes unbounded in the positive direction, and we write

$$\lim_{x \to 1^+} \frac{x^2+1}{x^3-1} = \infty$$

19. Here $f(x) = \dfrac{1-x}{3x-6}$, $x \neq 2$. To determine $\displaystyle\lim_{x \to 2^-} \dfrac{1-x}{3x-6}$, we examine the values of f that are close to 2, but remain smaller than 2.

x	1.9	1.99	1.999	1.9999	1.99999
$f(x) = \dfrac{1-x}{3x-6}$	3	33	333	3333	33,333

We see that as x gets closer to 2 from the left, the value of $f(x) = \dfrac{1-x}{3x-6}$ becomes unbounded in the positive direction, and we write

$$\lim_{x \to 2^-} \frac{1-x}{3x-6} = \infty$$

21.   To find horizontal asymptotes, we need to find two limits, $\lim_{x \to \infty} f(x)$ and $\lim_{x \to -\infty} f(x)$.

$$\lim_{x \to \infty} f(x) = \lim_{x \to \infty} \left(3 + \frac{1}{x^2} \right) = \lim_{x \to \infty} 3 + \lim_{x \to \infty} \frac{1}{x^2} = 3 + 0 = 3$$

We conclude that the line $y = 3$ is a horizontal asymptote of the graph when x becomes unbounded in the positive direction.

$$\lim_{x \to -\infty} f(x) = \lim_{x \to -\infty} \left(3 + \frac{1}{x^2} \right) = \lim_{x \to -\infty} 3 + \lim_{x \to -\infty} \frac{1}{x^2} = 3 + 0 = 3$$

We conclude that the line $y = 3$ is a horizontal asymptote of the graph when x becomes unbounded in the negative direction.

To find vertical asymptotes, we need to examine the behavior of the graph of f when x is near 0, the point where f is not defined. This will require looking at the one-sided limits of f at 0.

$\lim_{x \to 0^-} f(x)$:   Since $x \to 0^-$, we know $x < 0$, but $x^2 > 0$. It follows that the expression $\frac{1}{x^2}$ is positive and becomes unbounded as $x \to 0^-$.

$$\lim_{x \to 0^-} f(x) = \lim_{x \to 0^-} \left(3 + \frac{1}{x^2} \right) = \infty$$

$\lim_{x \to 0^+} f(x)$:   Since $x \to 0^+$, we know $x > 0$, and $x^2 > 0$. It follows that the expression $\frac{1}{x^2}$ is positive and becomes unbounded as $x \to 0^+$.

$$\lim_{x \to 0^+} f(x) = \lim_{x \to 0^+} \left(3 + \frac{1}{x^2} \right) = \infty$$

We conclude that the graph of f has a vertical asymptote at $x = 0$.

23.   To find horizontal asymptotes, we need to find two limits, $\lim_{x \to \infty} f(x)$ and $\lim_{x \to -\infty} f(x)$.

$$\lim_{x \to \infty} f(x) = \lim_{x \to \infty} \frac{2x^2}{(x-1)^2} = \lim_{x \to \infty} \frac{2x^2}{x^2 - 2x + 1} = \lim_{x \to \infty} \frac{2x^2}{x^2} = 2$$

We conclude that the line $y = 2$ is a horizontal asymptote of the graph when x becomes unbounded in the positive direction.

$$\lim_{x \to -\infty} f(x) = \lim_{x \to -\infty} \frac{2x^2}{(x-1)^2} = \lim_{x \to -\infty} \frac{2x^2}{x^2 - 2x + 1} = \lim_{x \to -\infty} \frac{2x^2}{x^2} = 2$$

We conclude that the line $y = 2$ is a horizontal asymptote of the graph when x becomes unbounded in the negative direction.

To find vertical asymptotes, we need to examine the behavior of the graph of f when x is near 1, the point where f is not defined. This will require looking at the one-sided limits of f at 1.

$\lim\limits_{x \to 1^-} f(x)$: Since $x \to 1^-$, we know $x < 1$, so $x - 1 < 0$, but $(x-1)^2 > 0$. It follows

that the expression $\dfrac{2x^2}{(x-1)^2}$ is positive and becomes unbounded as $x \to 1^-$.

$$\lim\limits_{x \to 1^-} f(x) = \lim\limits_{x \to 1^-} \frac{2x^2}{(x-1)^2} = \infty$$

$\lim\limits_{x \to 1^+} f(x)$: Since $x \to 1^+$, we know $x > 1$, so both $x - 1 > 0$ and $(x-1)^2 > 0$. It

follows that the expression $\dfrac{2x^2}{(x-1)^2}$ is positive and becomes unbounded as $x \to 1^+$.

$$\lim\limits_{x \to 1^+} f(x) = \lim\limits_{x \to 1^+} \frac{2x^2}{(x-1)^2} = \infty$$

We conclude that the graph of f has a vertical asymptote at $x = 1$.

25. To find horizontal asymptotes, we need to find two limits, $\lim\limits_{x \to \infty} f(x)$ and $\lim\limits_{x \to -\infty} f(x)$.

$$\lim\limits_{x \to \infty} f(x) = \lim\limits_{x \to \infty} \frac{x^2}{x^2 - 4} = \lim\limits_{x \to \infty} \frac{x^2}{x^2} = \lim\limits_{x \to \infty} 1 = 1$$

We conclude that the line $y = 1$ is a horizontal asymptote of the graph when x becomes unbounded in the positive direction.

$$\lim\limits_{x \to -\infty} f(x) = \lim\limits_{x \to -\infty} \frac{x^2}{x^2 - 4} = \lim\limits_{x \to -\infty} \frac{x^2}{x^2} = \lim\limits_{x \to -\infty} 1 = 1$$

We conclude that the line $y = 1$ is a horizontal asymptote of the graph when x becomes unbounded in the negative direction.

To find vertical asymptotes, we need to examine the behavior of the graph of f when x is near -2 and 2, the points where f is not defined. This will require looking at the one-sided limits of f.

$\lim\limits_{x \to -2^-} f(x)$: Since $x \to -2^-$, we know $x < -2$ and $x^2 > 4$, so $x^2 - 4 > 0$. It follows

that the expression $\dfrac{x^2}{x^2 - 4}$ is positive and becomes unbounded as $x \to -2^-$.

$$\lim\limits_{x \to -2^-} f(x) = \lim\limits_{x \to -2^-} \frac{x^2}{x^2 - 4} = \infty$$

$\lim\limits_{x \to -2^+} f(x)$: Since $x \to -2^+$, we know $x > -2$ and $x^2 < 4$, so $x^2 - 4 < 0$. It

follows that the expression $\dfrac{x^2}{x^2 - 4}$ is negative and becomes unbounded as $x \to -2^+$.

$$\lim\limits_{x \to -2^+} f(x) = \lim\limits_{x \to -2^+} \frac{x^2}{x^2 - 4} = -\infty$$

We now examine the limits as $x \rightarrow 2$.

$\lim\limits_{x \rightarrow 2^-} f(x)$: Since $x \rightarrow 2^-$, we know $x < 2$ and $x^2 < 4$, so $x^2 - 4 < 0$. It follows that

the expression $\dfrac{x^2}{x^2 - 4}$ is negative and becomes unbounded as $x \rightarrow 2^-$.

$$\lim\limits_{x \rightarrow 2^-} f(x) = \lim\limits_{x \rightarrow 2^-} \dfrac{x^2}{x^2 - 4} = -\infty$$

$\lim\limits_{x \rightarrow 2^+} f(x)$: Since $x \rightarrow 2^+$, we know $x > 2$ and $x^2 > 4$, so $x^2 - 4 > 0$. It follows that

the expression $\dfrac{x^2}{x^2 - 4}$ is positive and becomes unbounded as $x \rightarrow 2^+$.

$$\lim\limits_{x \rightarrow 2^+} f(x) = \lim\limits_{x \rightarrow 2^+} \dfrac{x^2}{x^2 - 4} = \infty$$

We conclude that the graph of f has a vertical asymptotes at $x = -2$ and at $x = 2$.

27. (a) We observe from the graph that the domain continues indefinitely toward the infinities, since there are arrows on both ends. We also observe a vertical asymptote at $x = 6$. We conclude that the domain of f is $\{x \mid x \neq 6\}$ or all real numbers except 6.

(b) The arrows pointing upward as x approaches 6 (the vertical asymptote) indicate that the range of f is the set of positive numbers or $\{y \mid y \geq 0\}$ or the interval $[0, \infty)$.

(c) (-4, 0) and (0, 0) are the x-intercepts; (0, 0) is also the y-intercept.

(d) Since $f(x) = y$, $f(-2) = 2$.

(e) If $f(x) = 4$, then $x = 8$ or $x = 4$.

(f) f is discontinuous at $x = 6$; 6 is not in the domain of f.

(g) The vertical asymptote is $x = 6$.

(h) $y = 4$ is a horizontal asymptote of the graph when x becomes unbounded in the negative direction.

(i) There is only one local maximum. It occurs at $(-2, 2)$ where the local maximum is $y = 2$.

(j) There are 3 local minima. They occur at $(-4, 0)$ where the local minimum is $y = 0$; at $(0, 0)$ where the local minimum is $y = 0$; and at $(8, 4)$ where the local minimum is $y = 4$.

(k) The function f is increasing on the intervals $(-4, -2)$, $(0, 6)$, and $(8, \infty)$.

(l) The function f is decreasing on the intervals $(-\infty, -4)$, $(-2, 0)$, and $(6, 8)$.

(m) As x approaches $-\infty$, y approaches 4, so $\lim\limits_{x \to -\infty} f(x) = 4$.

(n) As x approaches ∞, y becomes unbounded in the positive direction, so $\lim\limits_{x \to \infty} f(x) = \infty$.

(o) As x approaches 6 from the left, we see that y becomes unbounded in the positive direction, so $\lim\limits_{x \to 6^-} f(x) = \infty$.

(p) As x approaches 6 from the right, we see that y becomes unbounded in the positive direction, so $\lim\limits_{x \to 6^+} f(x) = \infty$.

29. $R(x) = \dfrac{x-1}{x^2-1} = \dfrac{x-1}{(x-1)(x+1)}$. To determine the behavior of the graph near -1 and 1, we look at $\lim\limits_{x \to -1} R(x)$ and $\lim\limits_{x \to 1} R(x)$.

For $\lim\limits_{x \to -1} R(x)$, we have

$$\lim_{x \to -1} R(x) = \lim_{x \to -1} \frac{x-1}{x^2-1} = \lim_{x \to -1} \frac{\cancel{x-1}}{\cancel{(x-1)}(x+1)} = \lim_{x \to -1} \frac{1}{x+1}$$

If $x < -1$ and x is getting closer to -1, the value of $\dfrac{1}{x+1} < 0$ and is becoming unbounded; that is, $\lim\limits_{x \to -1^-} R(x) = -\infty$.

If $x > -1$ and x is getting closer to -1, the value of $\dfrac{1}{x+1} > 0$ and is becoming unbounded; that is, $\lim\limits_{x \to -1^+} R(x) = \infty$.

The graph of R will have a vertical asymptote at $x = -1$.

For $\lim\limits_{x \to 1} R(x)$, we have

$$\lim_{x \to 1} R(x) = \lim_{x \to 1} \frac{x-1}{x^2-1} = \lim_{x \to 1} \frac{\cancel{x-1}}{\cancel{(x-1)}(x+1)} = \lim_{x \to 1} \frac{1}{x+1} = \frac{1}{2}$$

As x gets closer to 1, the graph of R gets closer to $\dfrac{1}{2}$. Since R is not defined at 1, the graph will have a hole at $\left(1, \dfrac{1}{2}\right)$.

31. $R(x) = \dfrac{x^2+x}{x^2-1} = \dfrac{x(x+1)}{(x-1)(x+1)}$. To determine the behavior of the graph near -1 and 1, look at $\lim\limits_{x \to -1} R(x)$ and $\lim\limits_{x \to 1} R(x)$.

For $\lim\limits_{x \to -1} R(x)$, we have

$$\lim\limits_{x \to -1} R(x) = \lim\limits_{x \to -1} \frac{x^2 + x}{x^2 - 1} = \lim\limits_{x \to -1} \frac{x(x+1)}{(x-1)(x+1)} = \lim\limits_{x \to -1} \frac{x}{x-1} = \frac{-1}{-2} = \frac{1}{2}$$

As x gets closer to -1, the graph of R gets closer to $\dfrac{1}{2}$. Since R is not defined at -1, the graph will have a hole at $\left(-1, \dfrac{1}{2}\right)$.

For $\lim\limits_{x \to 1} R(x)$, we have

$$\lim\limits_{x \to 1} R(x) = \lim\limits_{x \to 1} \frac{x^2 + x}{x^2 - 1} = \lim\limits_{x \to 1} \frac{x(x+1)}{(x-1)(x+1)} = \lim\limits_{x \to 1} \frac{x}{x-1}$$

If $x < 1$ and x is getting closer to 1, the value of $\dfrac{x}{x-1} < 0$ and is becoming unbounded; that is, $\lim\limits_{x \to 1^-} R(x) = -\infty$.

If $x > 1$ and x is getting closer to 1, the value of $\dfrac{x}{x-1} > 0$ and is becoming unbounded; that is, $\lim\limits_{x \to 1^+} R(x) = \infty$.

The graph of R will have a vertical asymptote at $x = 1$.

33. A rational function is undefined at every number that makes the denominator zero. So we solve

$$\begin{aligned}
x^4 - x^3 + 8x - 8 &= 0 && \text{Set the denominator} = 0. \\
x^3(x-1) + 8(x-1) &= 0 && \text{Factor by grouping.} \\
(x^3 + 8)(x-1) &= 0 && \\
x^3 + 8 = 0 \quad \text{or} \quad x - 1 &= 0 && \text{Apply the Zero-Product Property.} \\
x = -2 \quad \text{or} \quad x &= 1 && \text{Solve for } x.
\end{aligned}$$

To determine the behavior of the graph near -2 and near 1, we look at $\lim\limits_{x \to -2} R(x)$ and $\lim\limits_{x \to 1} R(x)$.

For $\lim\limits_{x \to -2} R(x)$, we have

$$\lim\limits_{x \to -2} R(x) = \lim\limits_{x \to -2} \frac{x^3 - x^2 + x - 1}{x^4 - x^3 + 8x - 8} = \lim\limits_{x \to -2} \frac{x^2(x-1) + (x-1)}{x^3(x-1) + 8(x-1)}$$

$$= \lim\limits_{x \to -2} \frac{(x^2 + 1)(x-1)}{(x^3 + 8)(x-1)} = \lim\limits_{x \to -2} \frac{x^2 + 1}{x^3 + 8}$$

Since the limit of the denominator is 0, we use one-sided limits. If $x < -2$ and x is getting closer to -2, the value of $x^3 + 8 < 0$, so the quotient $\dfrac{x^2+1}{x^3+8} < 0$ and is becoming unbounded; that is, $\lim\limits_{x \to -2^-} R(x) = -\infty$.

If $x > -2$ and x is getting closer to -2, the value of $x^3 + 8 > 0$, so the quotient $\dfrac{x^2+1}{x^3+8} > 0$ and is becoming unbounded; that is, $\lim\limits_{x \to -2^+} R(x) = \infty$.

The graph of R will have a vertical asymptote at $x = -2$.

For $\lim\limits_{x \to 1} R(x)$, we have

$$\lim_{x \to 1} R(x) = \lim_{x \to 1} \frac{x^3 - x^2 + x - 1}{x^4 - x^3 + 8x - 8} = \lim_{x \to 1} \frac{x^2+1}{x^3+8} = \frac{2}{9}$$

As x gets closer to 1, the graph of R gets closer to $\dfrac{2}{9}$. Since R is not defined at 1, the graph will have a hole at $\left(1, \dfrac{2}{9}\right)$.

35. A rational function is undefined at every number that makes the denominator zero. So we solve

$$\begin{aligned}
x^2 + x - 6 &= 0 && \text{Set the denominator} = 0. \\
(x+3)(x-2) &= 0 && \text{Factor.} \\
x + 3 = 0 \quad \text{or} \quad x - 2 &= 0 && \text{Apply the Zero-Product Property.} \\
x = -3 \quad \text{or} \quad x &= 2 && \text{Solve for } x.
\end{aligned}$$

To determine the behavior of the graph near -3 and near 2, we look at $\lim\limits_{x \to -3} R(x)$ and $\lim\limits_{x \to 2} R(x)$.

For $\lim\limits_{x \to -3} R(x)$, we have

$$\begin{aligned}
\lim_{x \to -3} R(x) &= \lim_{x \to -3} \frac{x^3 - 2x^2 + 4x - 8}{x^2 + x - 6} = \lim_{x \to -3} \frac{x^2(x-2) + 4(x-2)}{(x+3)(x-2)} \\
&= \lim_{x \to -3} \frac{(x^2+4)\cancel{(x-2)}}{(x+3)\cancel{(x-2)}} = \lim_{x \to -3} \frac{x^2+4}{x+3}
\end{aligned}$$

Since the limit of the denominator is 0, we use one-sided limits. If $x < -3$ and x is getting closer to -3, the value of $x + 3 < 0$, so the quotient $\dfrac{x^2+4}{x+3} < 0$ and is becoming unbounded; that is, $\lim\limits_{x \to -3^-} R(x) = -\infty$.

If $x > -3$ and x is getting closer to -3, the value of $x + 3 > 0$, so the quotient $\dfrac{x^2+4}{x+3} > 0$ and is becoming unbounded; that is, $\lim\limits_{x \to -3^+} R(x) = \infty$.

The graph of R will have a vertical asymptote at $x = -3$.

For $\lim_{x \to 2} R(x)$, we have

$$\lim_{x \to 2} R(x) = \lim_{x \to 2} \frac{x^3 - 2x^2 + 4x - 8}{x^2 + x - 6} = \lim_{x \to 2} \frac{x^2 + 4}{x + 3} = \frac{8}{5}$$

As x gets closer to 2, the graph of R gets closer to $\frac{8}{5}$. Since R is not defined at 2, the

graph will have a hole at $\left(2, \frac{8}{5}\right)$.

37. A rational function is undefined at every number that makes the denominator zero. So
   we solve

$$x^4 + x^3 + x + 1 = 0 \qquad \text{Set the denominator} = 0.$$
$$x^3(x + 1) + (x + 1) = 0 \qquad \text{Factor by grouping.}$$
$$\left(x^3 + 1\right)(x + 1) = 0$$
$$x^3 + 1 = 0 \quad \text{or} \quad x + 1 = 0 \qquad \text{Apply the Zero-Product Property.}$$
$$x = -1 \quad \text{or} \qquad x = -1 \qquad \text{Solve for } x.$$

To determine the behavior of the graph near -1, we look at $\lim_{x \to -1} R(x)$.

$$\lim_{x \to -1} R(x) = \lim_{x \to -1} \frac{x^3 + 2x^2 + x}{x^4 + x^3 + x + 1} = \lim_{x \to -1} \frac{x\left(x^2 + 2x + 1\right)}{\left(x^3 + 1\right)(x + 1)}$$

$$= \lim_{x \to -1} \frac{x\cancel{(x+1)}\cancel{(x+1)}}{\cancel{(x+1)}\left(x^2 - x + 1\right)\cancel{(x+1)}} = \lim_{x \to -1} \frac{x}{x^2 - x + 1} = -\frac{1}{3}$$

As x gets closer to -1, the graph of R gets closer to $-\frac{1}{3}$. Since R is not defined at 1,

the graph will have a hole at $\left(-1, -\frac{1}{3}\right)$.

39. (a) Production costs are the sum of fixed costs and variable costs. So the cost function C
       of producing x calculators is
$$C = C(x) = 10x + 79{,}000$$
   (b) The domain of C is $\{x \mid x \geq 0\}$.
   (c) The average cost per calculator, when x calculators are produced is given by the
   function $\overline{C}(x) = \frac{C(x)}{x} = \frac{10x + 79{,}000}{x} = 10 + \frac{79{,}000}{x}$.
   (d) The domain of $\overline{C}$ is $\{x \mid x > 0\}$.
   (e) $\lim_{x \to 0^+} \overline{C} = \lim_{x \to 0^+} \left(10 + \frac{79{,}000}{x}\right) = \infty$

   The average cost of making nearly 0 calculators becomes unbounded.
   (f) $\lim_{x \to \infty} \overline{C}(x) = \lim_{x \to \infty} \frac{10x + 79{,}000}{x} = \lim_{x \to \infty} \frac{10\cancel{x}}{\cancel{x}} = 10$

The average cost of producing a calculator when a very large number of calculators are produced is $10.

41. (a) $\lim\limits_{x\to 100^-} C(x) = \lim\limits_{x\to 100^-} \dfrac{5x}{100-x}$ Since x is approaching 100, but is remaining less than

100, $100 - x > 0$, and the quotient $\dfrac{5x}{100-x} > 0$ and is becoming unbounded; so

$\lim\limits_{x\to 100^-} C(x) = \infty$.

(b) It is not possible to remove 100% of the pollutant. Explanations will vary.

43. Graphs will vary.

Chapter 3 Review

TRUE-FALSE ITEMS

1. True   **3.** True

5. True   **7.** True

FILL-IN-THE-BLANKS

1. $\lim\limits_{x\to c} f(x) = N$ **3.** not exist

5. is not equal to   **7.** $y = 2$ ... horizontal

REVIEW EXERCISES

1. Here $f(x) = \dfrac{x^3 - 8}{x-2}$, and $c = 2$. We find the limit by evaluating the function f at values of x close to 2.

x	1.9	1.99	1.999	1.9999
$f(x) = \dfrac{x^3-8}{x-2}$	11.41	11.94	11.9940	11.9994
x	2.1	2.01	2.001	2.0001
$f(x) = \dfrac{x^3-8}{x-2}$	12.61	12.0601	2.0060	12.0006

We infer from the table that $\lim\limits_{x\to 2} f(x) = \lim\limits_{x\to 2} \dfrac{x^3-8}{x-2} = 12$.

3.

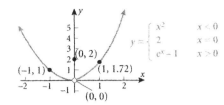

$$\lim_{x \to 0} f(x) = 0$$

5. $f(x) = 3x^2 - 2x + 1$ is a polynomial. We know that for polynomials, $\lim_{x \to c} f(x) = f(c)$.

$$\lim_{x \to 2} \left(3x^2 - 2x + 1\right) = 3 \cdot 2^2 - 2 \cdot 2 + 1 = 12 - 4 + 1 = 9 \qquad \text{Limit of a polynomial.}$$

7. $f(x) = x^2 + 1$ is a polynomial. We know that for polynomials, $\lim_{x \to c} f(x) = f(c)$.

$$\lim_{x \to -2} \left(x^2 + 1\right)^2 = \left(\lim_{x \to -2} \left(x^2 + 1\right)\right)^2 = \left[(-2)^2 + 1\right]^2 = (5)^2 = 25$$

$\qquad\qquad\qquad\qquad\quad\uparrow\qquad\qquad\qquad\qquad\uparrow$

$\qquad\qquad\qquad$ Limit of a Power $\qquad$ Limit of a polynomial

9. $f(x) = \sqrt{x^2 + 7}$; its domain is all real numbers.

$$\lim_{x \to 3} \sqrt{x^2 + 7} = \sqrt{\lim_{x \to 3}\left(x^2 + 7\right)} = \sqrt{\left(3^2 + 7\right)} = \sqrt{16} = 4$$

$\qquad\quad\uparrow\qquad\qquad\qquad\uparrow$

$\quad$ Limit of a Root $\quad$ Limit of a polynomial

11. $f(x) = \sqrt{1 - x^2}$. Its domain is the set of numbers that keeps $1 - x^2 \geq 0$.

$$1 - x^2 \geq 0 \quad \text{or} \quad x^2 \leq 1 \text{ or } x \geq -1 \text{ and } x \leq 1$$

So the domain of f is $\{x \mid -1 \leq x \leq 1\}$ or x in the interval $[-1, 1]$.

As $x \to 1^-$, x gets closer to 1, but remains less than 1; x is in the domain of f. So we need only to consider x as it approaches 1.

$$\lim_{x \to 1^-} \sqrt{1 - x^2} = \lim_{x \to 1} \sqrt{1 - x^2} = \sqrt{\lim_{x \to 1^-}\left(1 - x^2\right)} = \sqrt{1 - 1} = 0$$

$\qquad\qquad\qquad\qquad\qquad\qquad\quad\uparrow\qquad\qquad\qquad\uparrow$

$\qquad\qquad\qquad\qquad\qquad$ Limit of a Root $\qquad$ Limit of a Polynomial

13. $f(x) = 5x + 6$ is a polynomial, so as x approaches c, $f(x)$ approaches $f(c)$.

$$\lim_{x \to 2}(5x + 6)^{3/2} = \left[\lim_{x \to 2}(5x + 6)\right]^{3/2} = (5 \cdot 2 + 6)^{3/2} = 16^{3/2} = 64$$

$\qquad\qquad\qquad\quad\uparrow\qquad\qquad\qquad\qquad\uparrow$

$\qquad\qquad$ Limit of a Power $\qquad$ Limit of a Polynomial

15.   Here $f(x) = x^2 + x + 2$ and $g(x) = x^2 - 9$ are both polynomials. So,

$$\lim_{x \to -1} \left(x^2 + x + 2\right)\left(x^2 - 9\right) = \left[\lim_{x \to -1}\left(x^2 + x + 2\right)\right]\left[\lim_{x \to -1}\left(x^2 - 9\right)\right] = \left[(-1)^2 + (-1) + 2\right]\left[(-1)^2 - 9\right] = -16$$

↑   ↑
Limit of a Product   Limits of Polynomials

17.   Here $f(x) = \dfrac{x-1}{x^3-1}$. As x approaches 1, the limit of the denominator equals zero, so Formula (11) cannot be used directly. We factor the expression first.

$$\lim_{x \to 1} \frac{x-1}{x^3-1} = \lim_{x \to 1} \frac{x-1}{(x-1)(x^2+x+1)} = \lim_{x \to 1} \frac{1}{x^2+x+1} = \frac{\lim_{x \to 1} 1}{\lim_{x \to 1} x^2+x+1} = \frac{1}{3}$$

↑   ↑
Factor.   Limit of a Quotient

19.   Here $f(x) = \dfrac{x^2-9}{x^2-x-12}$. As x approaches -3, the limit of the denominator equals zero, so Formula (11) cannot be used directly. We factor the expression first.

$$\lim_{x \to -3} \frac{x^2-9}{x^2-x-12} = \lim_{x \to -3} \frac{(x-3)(x+3)}{(x-4)(x+3)} = \lim_{x \to -3} \frac{x-3}{x-4} = \frac{\lim_{x \to -3}(x-3)}{\lim_{x \to -3}(x-4)} = \frac{-6}{-7} = \frac{6}{7}$$

↑   ↑
Factor.   Limit of a Quotient

21.

Here $f(x) = \dfrac{x^2-1}{x^3-1}$. As x approaches -1 from the left, the limit of the denominator equals zero, so Formula (11) cannot be used directly. We factor the expression first.

$$\lim_{x \to -1^-} \frac{x^2-1}{x^3-1} = \lim_{x \to -1^-} \frac{(x-1)(x+1)}{(x-1)(x^2+x+1)} = \lim_{x \to -1^-} \frac{x+1}{x^2+x+1} = \frac{\lim_{x \to -1^-}(x+1)}{\lim_{x \to -1^-}(x^2+x+1)} = \frac{0}{1} = 0$$

↑   ↑
Factor.   Limit of a Quotient

23.   Here $f(x) = \dfrac{x^3-8}{x^3-2x^2+4x-8}$. As x approaches 2, the limit of the denominator equals zero, so Formula (11) cannot be used directly. We factor the expression first.

$$\lim_{x \to 2} \frac{x^3-8}{x^3-2x^2+4x-8} = \lim_{x \to 2} \frac{(x-2)(x^2+2x+4)}{x^2(x-2)+4(x-2)} = \lim_{x \to 2} \frac{(x-2)(x^2+2x+4)}{(x^2+4)(x-2)}$$

$$=$$

$$\lim_{x \to 2} \frac{x^2+2x+4}{x^2+4} = \frac{\lim_{x \to 2}\left(x^2+2x+4\right)}{\lim_{x \to 2}\left(x^2+4\right)} = \frac{2^2+2 \cdot 2+4}{2^2+4} = \frac{12}{8} = \frac{3}{2}$$

↑
Limit of a Quotient

25. Here $f(x) = \dfrac{x^4 - 3x^3 + x - 3}{x^3 - 3x^2 + 2x - 6}$. As x approaches 3, the limit of the denominator equals zero, so Formula (11) cannot be used directly. We factor the expression first.

$$\frac{x^4 - 3x^3 + x - 3}{x^3 - 3x^2 + 2x - 6} = \frac{x^3(x-3) + 1(x-3)}{x^2(x-3) + 2(x-3)} = \frac{(x^3 + 1)(x - 3)}{(x^2 + 2)(x - 3)} \qquad \text{Factor both the numerator and the denominator by grouping.}$$

$$\lim_{x \to 3} \frac{x^4 - 3x^3 + x - 3}{x^3 - 3x^2 + 2x - 6} = \lim_{x \to 3} \frac{(x^3 + 1)(\cancel{x - 3})}{(x^2 + 2)(\cancel{x - 3})} = \frac{\lim\limits_{x \to 3}(x^3 + 1)}{\lim\limits_{x \to 3}(x^2 + 2)} = \frac{28}{11}$$

$$\uparrow$$
$$\text{Limit of a Quotient}$$

27. $\displaystyle\lim_{x \to \infty} \frac{5x^4 - 8x^3 + x}{3x^4 + x^2 + 5} = \lim_{x \to \infty} \frac{5x^4}{3x^4}$ \qquad As $x \to \infty$, $5x^4 - 8x^3 + x = 5x^4$ and $3x^4 + x^2 + 5 = 3x^4$.

$$= \lim_{x \to \infty} \frac{5}{3} = \frac{5}{3}$$

29. $f(x) = \dfrac{x^2}{x - 3}$ is not defined at $x = 3$. When $x \to 3^-$, $x - 3 < 0$. Since $x^2 \geq 0$, it follows

that the expression $\dfrac{x^2}{x - 3}$ is negative and becomes unbounded as $x \to 3^-$.

$$\lim_{x \to 3^-} \frac{x^2}{x - 3} = -\infty$$

31. $\displaystyle\lim_{x \to \infty} \frac{8x^4 - x^2 + 2}{-4x^3 + 1} = \lim_{x \to \infty} \frac{8x^4}{-4x^3}$ \qquad As $x \to \infty$, $8x^4 - x^2 + 2 = 8x^4$ and $-4x^3 + 1 = -4x^3$.

$$= \lim_{x \to \infty} \frac{8x}{-4} = -\infty$$

33. $f(x) = \dfrac{1 - 9x^2}{x^2 - 9}$ is not defined at $x = -3$. When $x \to -3^+$, $x > -3$ and $x^2 - 9 < 0$. Since

$1 - 9x^2 < 0$, it follows that $\dfrac{1 - 9x^2}{x^2 - 9}$ is positive and as becomes unbounded $x \to -3^+$.

$$\lim_{x \to -3^+} \frac{1 - 9x^2}{x^2 - 9} = \infty$$

35. $f(x) = 3x^4 - x^2 + 2$ is a polynomial function, and polynomial functions are continuous at all values of x. So $f(x)$ is continuous at $c = 5$.

37. $f(x) = \dfrac{x^4 - 4}{x + 2}$ is a rational function which is continuous at all values of x in its domain.

Since $x = -2$ is not in the domain of f, the function f is not continuous at $c = -2$.

39. The function f is defined at $c = -2$; $f(-2) = 4$.

The $\lim\limits_{x \to -2} f(x) = \lim\limits_{x \to -2} \dfrac{x^2 - 4}{x + 2} = \lim\limits_{x \to -2} (x - 2) = -4$

Since the limit as x approaches -2 does not equal $f(-2)$, the function is not continuous at $c = -2$.

41. The function f is defined at $c = -2$; $f(-2) = -4$.

The $\lim\limits_{x \to -2} f(x) = \lim\limits_{x \to -2} \dfrac{x^2 - 4}{x + 2} = \lim\limits_{x \to -2} (x - 2) = -4$

Since the limit as x approaches -2 equals $f(-2)$, the function is continuous at $c = -2$

43. To find any horizontal asymptotes we need to find $\lim\limits_{x \to \infty} f(x)$ and $\lim\limits_{x \to -\infty} f(x)$.

$$\lim\limits_{x \to \infty} f(x) = \lim\limits_{x \to \infty} \frac{3x}{x^2 - 1} = \lim\limits_{x \to \infty} \frac{3x}{x^2} = \lim\limits_{x \to \infty} \frac{3}{x} = 0$$

The line $y = 0$ is a horizontal asymptote of the graph when x is sufficiently positive.

$$\lim\limits_{x \to -\infty} f(x) = \lim\limits_{x \to -\infty} \frac{3x}{x^2 - 1} = \lim\limits_{x \to -\infty} \frac{3x}{x^2} = \lim\limits_{x \to -\infty} \frac{3}{x} = 0$$

The line $y = 0$ is a horizontal asymptote of the graph when x is sufficiently negative.

The domain of f is $\{x \mid x \neq -1, x \neq 1\}$. To locate any vertical asymptotes we look at $\lim\limits_{x \to -1} f(x)$ and $\lim\limits_{x \to 1} f(x)$.

Looking at one-sided limits of f at -1, we find

$\lim\limits_{x \to -1^-} f(x)$: When $x \to -1$ from the left, $x < -1$ and $x^2 > 1$ or $x^2 - 1 > 0$. So, the expression $\dfrac{3x}{x^2 - 1}$ is negative and becomes unbounded.

$$\lim\limits_{x \to -1^-} f(x) = \lim\limits_{x \to -1^-} \frac{3x}{x^2 - 1} = -\infty$$

$\lim\limits_{x \to -1^+} f(x)$: When $x \to -1$ from the right, $x > -1$ and $x^2 < 1$ or $x^2 - 1 < 0$. So, the expression $\dfrac{3x}{x^2 - 1}$ is positive and becomes unbounded.

$$\lim\limits_{x \to -1^+} f(x) = \lim\limits_{x \to -1^+} \frac{3x}{x^2 - 1} = \infty$$

We conclude f has a vertical asymptote at $x = -1$.

$\lim\limits_{x \to 1^-} f(x)$: When $x \to 1$ from the left, $x < 1$ and $x^2 < 1$ or $x^2 - 1 < 0$. So, the expression $\dfrac{3x}{x^2-1}$ is negative and becomes unbounded.

$$\lim_{x\to 1^-} f(x) = \lim_{x\to 1^-} \frac{3x}{x^2-1} = -\infty$$

$\lim\limits_{x\to 1^+} f(x)$: When $x\to 1$ from the right, $x>1$ and $x^2-1>0$. So, the expression

$\dfrac{3x}{x^2-1}$ is positive and becomes unbounded.

$$\lim_{x\to 1^+} f(x) = \lim_{x\to 1^+} \frac{3x}{x^2-1} = \infty$$

We conclude f has a vertical asymptote at $x=1$.

45. To find any horizontal asymptotes we need to find $\lim\limits_{x\to\infty} f(x)$ and $\lim\limits_{x\to-\infty} f(x)$.

$$\lim_{x\to\infty} f(x) = \lim_{x\to\infty} \frac{5x}{x+2} = \lim_{x\to\infty} \frac{5x}{x} = \lim_{x\to\infty} \frac{5}{1} = 5$$

The line $y=5$ is a horizontal asymptote of the graph when x is sufficiently positive.

$$\lim_{x\to-\infty} f(x) = \lim_{x\to-\infty} \frac{5x}{x+2} = \lim_{x\to-\infty} \frac{5x}{x} = \lim_{x\to-\infty} \frac{5}{1} = 5$$

The line $y=5$ is a horizontal asymptote of the graph when x is sufficiently negative.

The domain of f is $\{x \mid x \neq -2\}$. To locate any vertical asymptotes we look at $\lim\limits_{x\to-2} f(x)$.

Looking at one-sided limits of f at -2, we find
$\lim\limits_{x\to-2^-} f(x)$: When $x\to-2$ from the left, $x<-2$ and $x+2<0$. So, the expression

$\dfrac{5x}{x+2}$ is positive and becomes unbounded.

$$\lim_{x\to-2^-} f(x) = \lim_{x\to-2^-} \frac{5x}{x+2} = \infty$$

$\lim\limits_{x\to-2^+} f(x)$: When $x\to-2$ from the right, $x>-2$ and $x+2>0$. So, the expression

$\dfrac{5x}{x+2}$ is negative and becomes unbounded.

$$\lim_{x\to-2^+} f(x) = \lim_{x\to-2^+} \frac{5x}{x+2} = -\infty$$

We conclude f has a vertical asymptote at $x=-2$.

47. (a) There is a vertical asymptote at $x = 2$ and f is not defined at 2, so the domain of f is the intervals $(-\infty, 2)$ or $(2, 5)$ or $(5, \infty)$.

(b) The range of f is the set of all real numbers, that is all y in the interval $(-\infty, \infty)$.

(c) The x-intercepts are the points at which the graph crosses or touches the x-axis. The x-intercepts are $(-2, 0)$, $(0, 0)$, $(1, 0)$, and $(6, 0)$.

(d) The y-intercept is $(0, 0)$.

(e) $f(-6) = 2$ and $f(-4) = 1$

(f) $f(-2) = 0$ and $f(6) = 0$

(g) $\lim\limits_{x \to -4^-} f(x) = 4$; $\lim\limits_{x \to -4^+} f(x) = -2$

(h) $\lim\limits_{x \to -2^-} f(x) = -2$; $\lim\limits_{x \to -2^+} f(x) = 2$

(i) $\lim\limits_{x \to 5^-} f(x) = 2$; $\lim\limits_{x \to 5^+} f(x) = 2$

(j) The $\lim\limits_{x \to 0} f(x)$ does not exist since $\lim\limits_{x \to 0^-} f(x) = 4$ and $\lim\limits_{x \to 0^+} f(x) = 1$ are not equal.

(k) The $\lim\limits_{x \to 2} f(x)$ does not exist since $\lim\limits_{x \to 2^-} f(x) = -\infty$ and $\lim\limits_{x \to 2^+} f(x) = \infty$.

(l) f is not continuous at -2 since $\lim\limits_{x \to -2} f(x)$ does not exist.

(m) f is not continuous at -4 since $\lim\limits_{x \to -4} f(x)$ does not exist.

(n) f is not continuous at 0 since $\lim\limits_{x \to 0} f(x)$ does not exist.

(o) f is not continuous at 2; there is a vertical asymptote at 2.

(p) f is continuous at 4.

(q) f is not continuous at 5 since f is not defined at $x = 5$.

(r) f is increasing on the open intervals $(-6, -4)$, $(-2, 0)$, and $(6, \infty)$.

(s) f is decreasing on the open intervals $(-\infty, -6)$, $(0, 2)$, $(2, 5)$, and $(5, 6)$.

(t) $\lim\limits_{x \to -\infty} f(x) = \infty$ and $\lim\limits_{x \to \infty} f(x) = 2$

(u) There are no local maxima. There is a local minimum of 2 at $x = -6$, a local minimum of 0 at $x = 0$, and a local minimum of 0 at $x = 6$.

(v) There is a horizontal asymptote of $y = 2$ as x becomes unbounded in the positive direction, and a vertical asymptote at $x = 2$.

49. The average rate of change of $f(x)$ from -2 to x is

$$\frac{\Delta y}{\Delta x} = \frac{f(x) - f(-2)}{x - (-2)} = \frac{\left(2x^2 - 3x\right) - \left(2(-2)^2 - 3(-2)\right)}{x - (-2)}$$

$$= \frac{2x^2 - 3x - 8 - 6}{x + 2} \qquad \text{Remove parentheses.}$$

$$= \frac{(2x - 7)(x + 2)}{x + 2} \qquad \text{Factor.}$$

The limit as $x \to -2$ is

$$\lim_{x \to -2} \frac{(2x - 7)\cancel{(x + 2)}}{\cancel{x + 2}} = \lim_{x \to -2} (2x - 7) = -11$$

51. The average rate of change of $f(x)$ from 3 to x is

$$\frac{\Delta y}{\Delta x} = \frac{f(x) - f(3)}{x - 3} = \frac{\dfrac{x}{x - 1} - \dfrac{3}{3 - 1}}{x - 3} = \frac{\dfrac{x}{x - 1} - \dfrac{3}{2}}{x - 3}$$

$$= \frac{2x - 3(x - 1)}{2(x - 1)(x - 3)} \qquad \text{Write as a single fraction.}$$

$$= \frac{-x + 3}{2(x - 1)(x - 3)}$$

The limit as $x \to 3$ is

$$\lim_{x \to 3} \frac{\overset{-1}{\cancel{-x + 3}}}{2(x - 1)\cancel{(x - 3)}} = \lim_{x \to 3} \frac{-1}{2(x - 1)} = -\frac{1}{4}$$

53. $R(x) = \dfrac{x + 4}{x^2 - 16}$. To determine the behavior of the graph near -4 and 4, we look at $\lim_{x \to -4} R(x)$ and $\lim_{x \to 4} R(x)$.

For $\lim_{x \to -4} R(x)$, we have

$$\lim_{x \to -4} R(x) = \lim_{x \to -4} \frac{x + 4}{x^2 - 16} = \lim_{x \to -4} \frac{\cancel{x + 4}}{\cancel{(x + 4)}(x - 4)} = \lim_{x \to -4} \frac{1}{x - 4} = -\frac{1}{8}$$

As x gets closer to -4, the graph of R gets closer to $-\dfrac{1}{8}$. Since R is not defined at -4, the graph will have a hole at $\left(-4, -\dfrac{1}{8}\right)$.

For $\lim_{x \to 4} R(x)$, we have

$$\lim_{x \to 4} R(x) = \lim_{x \to 4} \frac{x+4}{x^2-16} = \lim_{x \to 4} \frac{\cancel{x+4}}{\cancel{(x+4)}(x-4)} = \lim_{x \to 4} \frac{1}{x-4}$$

Since the limit of the denominator is 0, we use one-sided limits to investigate $\lim_{x \to 4} \dfrac{1}{x-4}$.

If $x < 4$ and x is getting closer to 4, the value of $\dfrac{1}{x-4} < 0$ and is becoming unbounded;

that is, $\lim_{x \to 4^-} \dfrac{1}{x-4} = -\infty$.

If $x > 4$ and x is getting closer to 4, the value of $\dfrac{1}{x-4} > 0$ and is becoming unbounded;

that is, $\lim_{x \to 4^+} \dfrac{1}{x-4} = \infty$.

The graph of R will have a vertical asymptote at $x = 4$.

55. Rational functions are undefined at values of x that would make the denominator of the function equal zero. Solving $x^2 - 11x + 18 = 0$ or $(x-9)(x-2) = 0$ we get $x = 9$ or $x = 2$. So R is undefined at $x = 2$ and $x = 9$.

To analyze the behavior of the graph near 2 and 9, we look at $\lim_{x \to 2} R(x)$ and $\lim_{x \to 9} R(x)$.

For $\lim_{x \to 2} R(x)$, we have

$$\lim_{x \to 2} R(x) = \lim_{x \to 2} \frac{x^3 - 2x^2 + 4x - 8}{x^2 - 11x + 18} = \lim_{x \to 2} \frac{x^2(x-2) + 4(x-2)}{(x-2)(x-9)}$$

$$= \lim_{x \to 2} \frac{(x^2+4)\cancel{(x-2)}}{\cancel{(x-2)}(x-9)} = \lim_{x \to 2} \frac{x^2+4}{x-9} = -\frac{8}{7}$$

As x gets closer to 4, the graph of R gets closer to $-\dfrac{8}{7}$. Since R is not defined at 2, the

graph will have a hole at $\left(2, -\dfrac{8}{7}\right)$.

For $\lim_{x \to 9} R(x)$, we have

$$\lim_{x \to 9} R(x) = \lim_{x \to 9} \frac{x^3 - 2x^2 + 4x - 8}{x^2 - 11x + 18} = \lim_{x \to 9} \frac{(x^2+4)\cancel{(x-2)}}{\cancel{(x-2)}(x-9)} = \lim_{x \to 9} \frac{x^2+4}{x-9}$$

Since the limit of the denominator is 0, we use one-sided limits to investigate

$\lim_{x \to 9} \dfrac{x^2+4}{x-9}$.

If $x < 9$ and x is getting closer to 9, the value of $\dfrac{x^2+4}{x-9} < 0$ and is becoming unbounded;

that is, $\lim\limits_{x \to 9^-} \dfrac{x^2+4}{x-9} = -\infty$.

If $x > 9$ and x is getting closer to 9, the value of $\dfrac{x^2+4}{x-9} > 0$ and is becoming unbounded;

that is, $\lim\limits_{x \to 9^+} \dfrac{x^2+4}{x-9} = \infty$.

The graph of R will have a vertical asymptote at $x = 9$.

57. Answers will vary.

59. (a) $\lim\limits_{x \to \infty} S(x) = \lim\limits_{x \to \infty} \dfrac{2000x^2}{3.5x^2+1000} = \lim\limits_{x \to \infty} \dfrac{2000x^2}{3.5x^2} = \lim\limits_{x \to \infty} \dfrac{2000}{3.5} = 571.43$

CHAPTER 3 PROJECT

1.
$$R(x) = \begin{cases} 0.10 & 0 < x \le 7000 \\ 0.15 & 7000 < x \le 28,400 \\ 0.25 & 28,400 < x \le 68,800 \\ 0.28 & 68,800 < x \le 143,500 \\ 0.33 & 143,500 < x \le 311,950 \\ 0.35 & x > 311,950 \end{cases}$$

3. The function R is not continuous. It is discontinuous at the endpoints of each tax bracket.

5.
$$A(x) = \begin{cases} 0.10x & 1 < x \le 7000 \\ 700+0.15(x-7000) & 7000 < x \le 28,400 \\ 3910+0.25(x-28,400) & 28,400 < x \le 68,800 \\ 14,010+0.28(x-68,800) & 68,800 < x \le 143,500 \\ 34,926+0.33(x-143,500) & 143,500 < x \le 311,950 \\ 90,514+0.35(x-311,950) & x > 311,950 \end{cases}$$

7. The function A is not continuous if your income is $311,950.

9. To compute column 3, we find the amount of tax paid if a person earns the highest dollar amount allowable in the previous row. That is

Row 2, Column 3: A couple earning $14,000 pays
$$0.10(14,000) = \$1400$$
So the entry will be $1400.

Row 3, Column 3: We calculate the taxes paid by a couple earning $56,800.

$$\$1400 + 0.15(\$56,800 - \$14,000) = \$9710$$

So the entry will be $9710.

Row 4, Column 3: We calculate the taxes paid by a couple earning $114,650.

$$\$9710 + 0.25(\$114,650 - \$56,800) = \$24,172.50$$

So the entry will be $24,172.50.

Row 5, Column 3: We calculate the taxes paid by a couple earning $174,700.

$$\$24,172.50 + 0.28(\$174,700 - \$114,650) = \$40,986.50$$

So the entry will be $40,986.50.

Row 6, Column 3: We calculate the taxes paid by a couple earning $311,950.

$$\$40,986.50 + 0.33(\$311,950 - \$174,700) = \$86,279.00$$

So the entry will be $86,279.00.

MATHEMATICAL QUESTIONS FROM PROFESSIONAL EXAMS

1. (b) $\dfrac{5}{6}$
$$\lim_{x \to 3} \frac{x^2 - x - 6}{x^2 - 9} = \lim_{x \to 3} \frac{\cancel{(x-3)}(x+2)}{\cancel{(x-3)}(x+3)} = \lim_{x \to 3} \frac{(x+2)}{(x+3)} = \frac{5}{6}$$

3. (d) $\dfrac{\sqrt{2}}{4}$
$$\lim_{h \to 0} \frac{\sqrt{2+h} - \sqrt{2}}{h} = \lim_{h \to 0} \left(\frac{\sqrt{2+h} - \sqrt{2}}{h} \cdot \frac{\sqrt{2+h} + \sqrt{2}}{\sqrt{2+h} + \sqrt{2}} \right)$$

$$= \lim_{h \to 0} \left(\frac{\left(\sqrt{2+h}\right)^2 - \left(\sqrt{2}\right)^2}{h\left(\sqrt{2+h} + \sqrt{2}\right)} \right) = \lim_{h \to 0} \left(\frac{2 + h - 2}{h\left(\sqrt{2+h} + \sqrt{2}\right)} \right)$$

$$= \lim_{h \to 0} \frac{\cancel{h}}{\cancel{h}\left(\sqrt{2+h} + \sqrt{2}\right)} = \lim_{h \to 0} \frac{1}{\left(\sqrt{2+h} + \sqrt{2}\right)}$$

$$= \frac{1}{2\sqrt{2}} = \frac{\sqrt{2}}{4}$$

Chapter 4
The Derivative of a Function

4.1 The Definition of a Derivative

1. The slope of the tangent line to the graph of $f(x) = 3x + 5$ at the point $(1, 8)$ is

$$m_{\tan} = \lim_{x \to 1} \frac{f(x) - f(1)}{x - 1} = \lim_{x \to 1} \frac{(3x + 5) - 8}{x - 1} = \lim_{x \to 1} \frac{3x - 3}{x - 1} = \lim_{x \to 1} \frac{3(x - 1)}{x - 1} = \lim_{x \to 1} 3 = 3$$

An equation of the tangent line is

$$\begin{aligned} y - 8 &= 3(x - 1) \qquad & y - f(c) = m_{\tan}(x - c) \\ y &= 3x + 5 & \text{Simplify.} \end{aligned}$$

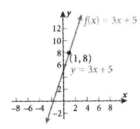

3. The slope of the tangent line to the graph of $f(x) = x^2 + 2$ at the point $(-1, 3)$ is

$$m_{\tan} = \lim_{x \to -1} \frac{f(x) - f(-1)}{x - (-1)} = \lim_{x \to -1} \frac{(x^2 + 2) - (3)}{x - (-1)} = \lim_{x \to -1} \frac{x^2 - 1}{x + 1} = \lim_{x \to -1} \frac{(x - 1)(x + 1)}{(x + 1)}$$

$$= \lim_{x \to -1} (x - 1) = -2$$

An equation of the tangent line is

$$\begin{aligned} y - 3 &= (-2)[x - (-1)] \qquad & y - f(c) = m_{\tan}(x - c) \\ y - 3 &= -2x - 2 & \text{Simplify.} \\ y &= -2x + 1 & \text{Add 3 to both sides.} \end{aligned}$$

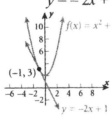

5. The slope of the tangent line to the graph of $f(x) = 3x^2$ at the point $(2, 12)$ is

$$m_{\tan} = \lim_{x \to 2} \frac{f(x) - f(2)}{x - 2} = \lim_{x \to 2} \frac{3x^2 - 12}{x - 2} = \lim_{x \to 2} \frac{3(x^2 - 4)}{x - 2} = \lim_{x \to 2} \frac{3(x - 2)(x + 2)}{x - 2}$$

$$= \lim_{x \to 2} [3(x + 2)] = 12$$

An equation of the tangent line is

$$y - 12 = 12(x - 2) \qquad y - f(c) = m_{\tan}(x - c)$$

$$y - 12 = 12x - 24 \qquad \text{Simplify.}$$
$$y = 12x - 12 \qquad \text{Add 12 to both sides.}$$

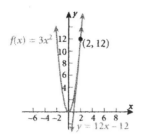

7. The slope of the tangent line to the graph of $f(x) = 2x^2 + x$ at the point $(1, 3)$ is

$$m_{\tan} = \lim_{x \to 1} \frac{f(x) - f(1)}{x - 1} = \lim_{x \to 1} \frac{(2x^2 + x) - 3}{x - 1} = \lim_{x \to 1} \frac{(2x + 3)(x - 1)}{x - 1} = \lim_{x \to 1} (2x + 3) = 5$$

An equation of the tangent line is

$$y - 3 = 5(x - 1) \qquad y - f(c) = m_{\tan}(x - c)$$
$$y = 5x - 2 \qquad \text{Simplify.}$$

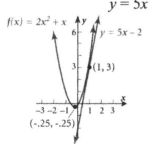

9. The slope of the tangent line to the graph of $f(x) = x^2 - 2x + 3$ at the point $(-1, 6)$ is

$$m_{\tan} = \lim_{x \to -1} \frac{f(x) - f(-1)}{x - (-1)} = \lim_{x \to -1} \frac{(x^2 - 2x + 3) - 6}{x + 1} = \lim_{x \to -1} \frac{x^2 - 2x - 3}{x + 1}$$

$$= \lim_{x \to -1} \frac{(x - 3)(x + 1)}{x + 1} = \lim_{x \to -1} (x - 3) = -4$$

An equation of the tangent line is

$$y - 6 = (-4)[x - (-1)] \qquad y - f(c) = m_{\tan}(x - c)$$
$$y - 6 = -4x - 4 \qquad \text{Simplify.}$$
$$y = -4x + 2 \qquad \text{Add 6 to both sides.}$$

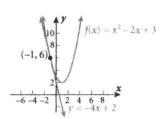

11. The slope of the tangent line to the graph of $f(x) = x^3 + x^2$ at the point $(-1, 0)$ is

$$m_{\tan} = \lim_{x \to -1} \frac{(x^3 + x^2) - 0}{x - (-1)} = \lim_{x \to -1} \frac{x^2(x + 1)}{(x + 1)} = \lim_{x \to -1} x^2 = 1$$

An equation of the tangent line is

$$y - 0 = 1[x - (-1)] \qquad\qquad y - f(c) = m_{tan}(x - c)$$
$$y = x + 1 \qquad\qquad\qquad \text{Simplify.}$$

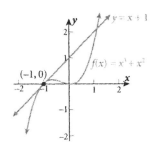

13. To find $f'(3)$, we follow the three steps outlined in the text.
    Step 1: $f(3) = -4(3) + 5 = -12 + 5 = -7$
    Step 2: $\dfrac{f(x) - f(3)}{x - 3} = \dfrac{(-4x + 5) - (-7)}{x - 3} = \dfrac{-4x + 12}{x - 3} = \dfrac{(-4)(x - 3)}{x - 3}$
    Step 3: The derivative of f at 3 is

    $$f'(3) = \lim_{x \to 3} \frac{f(x) - f(3)}{x - 3} = \lim_{x \to 3} \frac{(-4)(x - 3)}{x - 3} = -4$$

15. To find $f'(0)$, we follow the three steps outlined in the text.
    Step 1: $f(0) = (0)^2 - 3 = -3$
    Step 2: $\dfrac{f(x) - f(0)}{x - 0} = \dfrac{(x^2 - 3) - (-3)}{x} = \dfrac{x^2}{x}$
    Step 3: The derivative of f at 0 is

    $$f'(0) = \lim_{x \to 0} \frac{f(x) - f(0)}{x - 0} = \lim_{x \to 0} \frac{x^2}{x} = \lim_{x \to 0} x = 0$$

17. To find $f'(1)$, we follow the three steps outlined in the text.
    Step 1: $f(1) = 2 \cdot 1^2 + 3 \cdot 1 = 5$
    Step 2: $\dfrac{f(x) - f(1)}{x - 1} = \dfrac{(2x^2 + 3x) - (5)}{x - 1} = \dfrac{2x^2 + 3x - 5}{x - 1} = \dfrac{(2x + 5)(x - 1)}{x - 1}$
    Step 3: The derivative of f at 1 is

    $$f'(1) = \lim_{x \to 1} \frac{f(x) - f(1)}{x - 1} = \lim_{x \to 1} \frac{(2x + 5)(x - 1)}{x - 1} = \lim_{x \to 1} (2x + 5) = 7$$

19. To find $f'(0)$, we follow the three steps outlined in the text.
    Step 1: $f(0) = 0^3 + 4 \cdot 0 = 0$
    Step 2: $\dfrac{f(x) - f(0)}{x - 0} = \dfrac{(x^3 + 4x) - (0)}{x} = \dfrac{x(x^2 + 4)}{x}$
    Step 3: The derivative of f at 0 is

    $$f'(0) = \lim_{x \to 0} \frac{f(x) - f(0)}{x - 0} = \lim_{x \to 0} \frac{x(x^2 + 4)}{x} = \lim_{x \to 0} (x^2 + 4) = 4$$

21. To find $f'(1)$, we follow the three steps outlined in the text.

Step 1: $f(1) = 1^3 + 1^2 - 2 \cdot 1 = 0$

Step 2: $\dfrac{f(x) - f(1)}{x - 1} = \dfrac{(x^3 + x^2 - 2x) - 0}{x - 1} = \dfrac{x(x^2 + x - 2)}{x - 1} = \dfrac{x(x+2)(x-1)}{x-1}$

Step 3: The derivative of f at 1 is

$$f'(1) = \lim_{x \to 1} \frac{f(x) - f(1)}{x - 1} = \lim_{x \to 1} \frac{x(x+2)\,\cancel{(x-1)}}{\cancel{x-1}}$$

$$= \lim_{x \to 1}\left[x(x+2)\right] = \lim_{x \to 1} x \cdot \lim_{x \to 1}(x+2) = 1 \cdot 3 = 3$$

23. To find $f'(1)$, we follow the three steps outlined in the text.

Step 1: $f(1) = \dfrac{1}{1} = 1$

Step 2: $\dfrac{f(x) - f(1)}{x - 1} = \dfrac{\left(\dfrac{1}{x}\right) - (1)}{x - 1} = \dfrac{\dfrac{1 - x}{x}}{x - 1} = \dfrac{(-1)(x-1)}{x(x-1)}$

Step 3: The derivative of f at 1 is

$$f'(1) = \lim_{x \to 1} \frac{f(x) - f(1)}{x - 1} = \lim_{x \to 1} \frac{(-1)\,\cancel{(x-1)}}{x\,\cancel{(x-1)}} = \lim_{x \to 1} \frac{-1}{x} = -1$$

25. First we find the difference quotient of $f(x) = 2x$.

$$\frac{f(x+h) - f(x)}{h} = \frac{2(x+h) - 2x}{h} = \frac{2x + 2h - 2x}{h} = \frac{2h}{h} = 2$$

$$\uparrow \qquad\qquad\qquad\qquad \uparrow$$

$$\text{Simplify} \qquad\qquad \text{Cancel the } h\text{'s.}$$

The derivative of f is the limit of the difference quotient as $h \to 0$, that is,

$$f'(x) = \lim_{h \to 0} \frac{f(x+h) - f(x)}{h} = \lim_{h \to 0} 2 = 2$$

27. First we find the difference quotient of $f(x) = 1 - 2x$.

$$\frac{f(x+h) - f(x)}{h} = \frac{[1 - 2(x+h)] - [1 - 2x]}{h} = \frac{1 - 2x - 2h - 1 + 2x}{h} = \frac{-2h}{h} = -2$$

$$\uparrow \qquad\qquad\qquad\qquad\qquad \uparrow$$

$$\text{Simplify} \qquad\qquad\qquad \text{Cancel the } h\text{'s.}$$

The derivative of f is the limit of the difference quotient as $h \to 0$, that is,

$$f'(x) = \lim_{h \to 0} \frac{f(x+h) - f(x)}{h} = \lim_{h \to 0} (-2) = -2$$

29. First we find the difference quotient of $f(x) = x^2 + 2$.

$$\frac{f(x+h) - f(x)}{h} = \frac{\left[(x+h)^2 + 2\right] - \left[x^2 + 2\right]}{h}$$

$$= \frac{x^2 + 2xh + h^2 + 2 - x^2 - 2}{h}$$

$$= \frac{2xh + h^2}{h} \qquad \text{Simplify.}$$

$$= \frac{h(2x + h)}{h} \qquad \text{Factor out } h.$$

$$= 2x + h \qquad \text{Cancel the } h\text{'s.}$$

The derivative of f is the limit of the difference quotient as $h \to 0$, that is,

$$f'(x) = \lim_{h \to 0} \frac{f(x+h) - f(x)}{h} = \lim_{h \to 0} (2x + h) = 2x$$

31. First we find the difference quotient of $f(x) = 3x^2 - 2x + 1$.

$$\frac{f(x+h) - f(x)}{h} = \frac{\left[3(x+h)^2 - 2(x+h) + 1\right] - \left[3x^2 - 2x + 1\right]}{h}$$

$$= \frac{3x^2 + 6xh + 3h^2 - 2x - 2h + 1 - 3x^2 + 2x - 1}{h}$$

$$= \frac{6xh + 3h^2 - 2h}{h} \qquad \text{Simplify.}$$

$$= \frac{h(6x + 3h - 2)}{h} \qquad \text{Factor out } h.$$

$$= 6x + 3h - 2 \qquad \text{Cancel the } h\text{'s.}$$

The derivative of f is the limit of the difference quotient as $h \to 0$, that is,

$$f'(x) = \lim_{h \to 0} \frac{f(x+h) - f(x)}{h} = \lim_{h \to 0} (6x + 3h - 2) = 6x - 2$$

33. First we find the difference quotient of $f(x) = x^3$.

$$\frac{f(x+h) - f(x)}{h} = \frac{(x+h)^3 - x^3}{h}$$

$$= \frac{x^3 + 3x^2 h + 3xh^2 + h^3 - x^3}{h}$$

$$= \frac{3x^2 h + 3xh^2 + h^3}{h} \qquad \text{Simplify.}$$

$$= \frac{h(3x^2 + 3xh + h^2)}{h} \qquad \text{Factor out } h.$$

$$= 3x^2 + 3xh + h^2 \qquad \text{Cancel the } h\text{'s.}$$

The derivative of f is the limit of the difference quotient as $h \to 0$, that is,

$$f'(x) = \lim_{h \to 0} \frac{f(x+h) - f(x)}{h} = \lim_{h \to 0} (3x^2 + 3xh + h^2) = 3x^2$$

35. First we find the difference quotient of $f(x) = mx + b$.

$$\frac{f(x+h) - f(x)}{h} = \frac{\left[m(x+h) + b\right] - \left[mx + b\right]}{h}$$

$$= \frac{mx + mh + b - mx - b}{h}$$

$$= \frac{mh}{h} \qquad \text{Simplify}$$

$$= m \qquad \text{Cancel the } h\text{'s.}$$

The derivative of f is the limit of the difference quotient as $h \to 0$, that is,

$$f'(x) = \lim_{h \to 0} \frac{f(x+h) - f(x)}{h} = \lim_{h \to 0} m = m$$

37. (a) The average rate of change of $f(x) = 3x + 4$ as x changes from 1 to 3 is

$$\frac{\Delta f}{\Delta x} = \frac{f(3) - f(1)}{3 - 1} = \frac{(3 \cdot 3 + 4) - (3 \cdot 1 + 4)}{3 - 1} = \frac{13 - 7}{2} = \frac{6}{2} = 3$$

(b) The instantaneous rate of change at $x = 1$ is the derivative of f at 1.

$$f'(1) = \lim_{x \to 1} \frac{f(x) - f(1)}{x - 1} = \lim_{x \to 1} \frac{(3x + 4) - 7}{x - 1} = \lim_{x \to 1} \frac{3x - 3}{x - 1} = \lim_{x \to 1} \frac{3(x - 1)}{x - 1} = 3$$

The instantaneous rate of change of f at 1 is 3.

39. (a) The average rate of change of $f(x) = 3x^2 + 1$ as x changes from 1 to 3 is

$$\frac{\Delta f}{\Delta x} = \frac{f(3) - f(1)}{3 - 1} = \frac{(3 \cdot 3^2 + 1) - (3 \cdot 1^2 + 1)}{2} = \frac{28 - 4}{2} = \frac{24}{2} = 12$$

(b) The instantaneous rate of change at $x = 1$ is the derivative of f at 1.

$$f'(1) = \lim_{x \to 1} \frac{f(x) - f(1)}{x - 1} = \lim_{x \to 1} \frac{(3x^2 + 1) - (4)}{x - 1} = \lim_{x \to 1} \frac{3x^2 - 3}{x - 1}$$

$$= \lim_{x \to 1} \frac{3(x^2 - 1)}{x - 1} = \lim_{x \to 1} \frac{3(x - 1)(x + 1)}{x - 1} = \lim_{x \to 1} 3(x + 1) = 6$$

The instantaneous rate of change of f at 1 is 6.

41. (a) The average rate of change of $f(x) = x^2 + 2x$ as x changes from 1 to 3 is

$$\frac{\Delta f}{\Delta x} = \frac{f(3) - f(1)}{3 - 1} = \frac{(3^2 + 2 \cdot 3) - (1^2 + 2 \cdot 1)}{2} = \frac{15 - 3}{2} = \frac{12}{2} = 6$$

(b) The instantaneous rate of change at $x = 1$ is the derivative of f at 1.

$$f'(1) = \lim_{x \to 1} \frac{f(x) - f(1)}{x - 1} = \lim_{x \to 1} \frac{(x^2 + 2x) - (3)}{x - 1} = \lim_{x \to 1} \frac{x^2 + 2x - 3}{x - 1}$$

$$= \lim_{x \to 1} \frac{(x + 3)(x - 1)}{x - 1} = \lim_{x \to 1} (x + 3) = 4$$

The instantaneous rate of change of f at 1 is 4.

43.   (a) The average rate of change of $f(x) = 2x^2 - x + 1$ as x changes from 1 to 3 is

$$\frac{\Delta f}{\Delta x} = \frac{f(3) - f(1)}{3 - 1} = \frac{\left(2 \cdot 3^2 - 3 + 1\right) - \left(2 \cdot 1^2 - 1 + 1\right)}{2} = \frac{16 - 2}{2} = \frac{14}{2} = 7$$

(b) The instantaneous rate of change at $x = 1$ is the derivative of f at 1.

$$f'(1) = \lim_{x \to 1} \frac{f(x) - f(1)}{x - 1} = \lim_{x \to 1} \frac{\left(2x^2 - x + 1\right) - (2)}{x - 1} = \lim_{x \to 1} \frac{2x^2 - x - 1}{x - 1}$$

$$= \lim_{x \to 1} \frac{(2x + 1)(x - 1)}{x - 1} = \lim_{x \to 1} (2x + 1) = 3$$

The instantaneous rate of change of f at 1 is 3.

45.   The display below is from a TI-83 Plus graphing calculator.

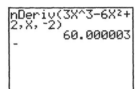

So $f'(-2) = 60$.

47.   The display below is from a TI-83 Plus graphing calculator.

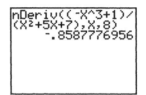

So $f'(8) = -0.85878$.

49.   The display below is from a TI-83 Plus graphing calculator.

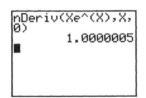

So $f'(0) = 1$.

51.   The display below is from a TI-83 Plus graphing calculator.

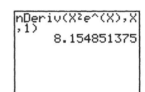

So $f'(1) = 8.15485$.

53.   The display below is from a TI-83 Plus graphing calculator.

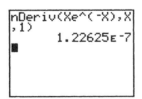

So $f'(1) = 0$.

55.   We first find an equation of the tangent line to the graph of $y = x^2$ at $(1, 1)$.

The slope of the tangent line at $(1, 1)$ is

$$m_{\tan} = \lim_{x \to 1} \frac{f(x) - f(1)}{x - 1} = \lim_{x \to 1} \frac{x^2 - 1}{x - 1} = \lim_{x \to 1} \frac{(x+1)\,(x-1)}{(x-1)} = \lim_{x \to 1} (x+1) = 2$$

An equation of the tangent line is

$y - 1 = 2(x - 1)$	$y - f(c) = m_{\tan}\,(x - c)$
$y - 1 = 2x - 2$	Simplify.
$y = 2x - 1$	Add 1 to both sides.

Now we see if the point (2, 5) satisfies the equation of the tangent line.

$$2 \cdot 2 - 1 = 3 \qquad\qquad y = 2x - 1;\ x = 2,\ y = 5.$$
$$5 \neq 3$$

So the graph of the tangent line does not pass through the point (2, 5).

57. For the rocket bomb to hit its target, the point (1, 0) must be on the graph of the tangent line to the graph of $y = x^2$ at some point $(c,\ c^2)$.

The slope of the tangent line at $(c,\ c^2)$ is

$$\lim_{x \to c} \frac{x^2 - c^2}{x - c} = \lim_{x \to c} \frac{(x-c)(x+c)}{x-c} = \lim_{x \to c} (x+c) = 2c$$

An equation of the tangent line is

$y - c^2 = 2c(x - c)$	$y - f(c) = m_{\tan}\,(x - c)$
$y - c^2 = 2cx - 2c^2$	Simplify.
$y = 2cx - c^2$	Add c^2 to both sides.

The point (1, 0) satisfies the equation of the tangent line, so

$0 = 2c(1) - c^2$	$y = 2cx - c^2;\ x = 1;\ y = 0$
$c^2 - 2c = 0$	Simplify.
$c(c - 2) = 0$	Factor.
$c = 0 \qquad c = 2$	Apply the Zero-Product Property.

Since the dive bomber is flying from right to left, the bomber reaches $c = 2$ first and should release the bomb at point (2, 4).

59. (a) The average rate of change in sales S from day $x = 1$ to day $x = 5$ is

$$\frac{\Delta S}{\Delta x} = \frac{S(5) - S(1)}{5 - 1} = \frac{\left(4(5)^2 + 50(5) + 5000\right) - \left(4(1)^2 + 50(1) + 5000\right)}{4}$$
$$= \frac{5350 - 5054}{4} = \frac{296}{4} = 74 \text{ tickets per day.}$$

(b) The average rate of change in sales S from day $x = 1$ to day $x = 10$ is

$$\frac{\Delta S}{\Delta x} = \frac{S(10) - S(1)}{10 - 1} = \frac{\left(4(10)^2 + 50(10) + 5000\right) - \left(4(1)^2 + 50(1) + 5000\right)}{9}$$
$$= \frac{5900 - 5054}{9} = \frac{846}{9} = 94 \text{ tickets per day.}$$

(c) The average rate of change in sales S from day $x = 5$ to day $x = 10$ is

$$\frac{\Delta S}{\Delta x} = \frac{S(10) - S(5)}{10 - 5} = \frac{\left(4(10)^2 + 50(10) + 5000\right) - \left(4(5)^2 + 50(5) + 5000\right)}{5}$$

$$= \frac{5900-5350}{5} = \frac{550}{5} = 110 \text{ tickets per day.}$$

(d) The instantaneous rate of change in sales on day 5 is the derivative of S at $x = 5$.

$$S'(5) = \lim_{x \to 5} \frac{S(x)-S(5)}{x-5} = \lim_{x \to 5} \frac{(4x^2+50x+5000)-(5350)}{x-5} = \lim_{x \to 5} \frac{4x^2+50x-350}{x-5}$$

$$= \lim_{x \to 5} \frac{2(2x+35)(x-5)}{x-5} = \lim_{x \to 5} \left[2(2x+35)\right] = 2 \lim_{x \to 5} (2x+35) = 2 \cdot (10+35) = 90$$

The instantaneous rate of change of S on day 5 is 90 ticket sales per day.

(e) The instantaneous rate of change in sales on day 10 is the derivative of S at $x = 10$.

$$S'(10) = \lim_{x \to 10} \frac{S(x)-S(10)}{x-10} = \lim_{x \to 10} \frac{(4x^2+50x+5000)-(5900)}{x-10} = \lim_{x \to 10} \frac{4x^2+50x-900}{x-10}$$

$$= \lim_{x \to 10} \frac{2(2x+45)(x-10)}{x-10} = \lim_{x \to 10} \left[2(2x+45)\right] = 2 \lim_{x \to 10} (2x+45) = 2 \cdot 65 = 130$$

The instantaneous rate of change of S on day 10 is 130 ticket sales per day.

61.   (a) At $x = \$10$ per crate, the farmer is willing to supply
$$S(10) = 50 \cdot 10^2 - 50 \cdot 10 = 4500 \text{ crates of grapefruits.}$$
$$\uparrow$$
$$S(x) = 50x^2 - 50x$$

(b) At $x = \$13$ per crate, the farmer is willing to supply
$$S(13) = 50 \cdot 13^2 - 50 \cdot 13 = 7800 \text{ crates of grapefruits.}$$

(c) The average rate of change in supply from $\$10$ to $\$13$ is
$$\frac{\Delta S}{\Delta x} = \frac{S(13)-S(10)}{13-10} = \frac{7800-4500}{3} = \frac{3300}{3} = 1100$$

The average rate of change in crates of grapefruit supplied is 1100 crates per dollar increase in price.

(d) The instantaneous rate of change in supply at $x = 10$ is the derivative $S'(10)$.

$$S'(10) = \lim_{x \to 10} \frac{S(x)-S(10)}{x-10} = \lim_{x \to 10} \frac{[50x^2-50x]-[4500]}{x-10} = \lim_{x \to 10} \frac{50(x^2-x-90)}{x-10}$$

$$= \lim_{x \to 10} \frac{50(x-10)(x+9)}{x-10} = \lim_{x \to 10} \left[50(x+9)\right] = 50 \lim_{x \to 10} (x+9) = 50 \cdot 19 = 950$$

The instantaneous rate of change in supply at $x = \$10$ is 950 crates.

(e) The average rate of change in supply over the price interval from $\$10$ to $\$13$ is 1100 crates of grapefruit per $\$1.00$ change in price.

The instantaneous rate of change in supply of 950 crates is the increase in supply of grapefruit as the price changes from $\$10$ to $\$11$.

63.   (a) The marginal revenue is the derivative $R'(x)$.

$$R'(x) = \lim_{h \to 0} \frac{R(x+h) - R(x)}{h} = \lim_{h \to 0} \frac{\left[8(x+h) - (x+h)^2\right] - \left[8x - x^2\right]}{h}$$

$$= \lim_{h \to 0} \frac{8x + 8h - x^2 - 2xh - h^2 - 8x + x^2}{h} \qquad \text{Simplify.}$$

$$= \lim_{h \to 0} \frac{8h - 2xh - h^2}{h} \qquad \text{Simplify.}$$

$$= \lim_{h \to 0} \frac{\cancel{h}(8 - 2x - h)}{\cancel{h}} \qquad \text{Factor out the } h.$$

$$= \lim_{h \to 0} (8 - 2x - h) = 8 - 2x \qquad \text{Cancel the } h\text{'s. Go to the limit.}$$

The marginal revenue is $R'(x) = 8 - 2x$.

(b) The marginal cost is the derivative $C'(x)$.

$$C'(x) = \lim_{h \to 0} \frac{C(x+h) - C(x)}{h} = \lim_{h \to 0} \frac{\left[2(x+h) + 5\right] - \left[2x + 5\right]}{h}$$

$$= \lim_{h \to 0} \frac{2x + 2h + 5 - 2x - 5}{h} = \lim_{h \to 0} \frac{2\cancel{h}}{\cancel{h}} = \lim_{h \to 0} 2 = 2$$

(c) To find the break-even point we solve the equation $R(x) = C(x)$.

$$8x - x^2 = 2x + 5$$
$$x^2 - 6x + 5 = 0 \qquad \text{Put the quadratic equation in standard form.}$$
$$(x - 5)(x - 1) = 0 \qquad \text{Factor.}$$
$$x - 5 = 0 \quad x - 1 = 0 \qquad \text{Apply the Zero-Product Property.}$$
$$x = 5 \qquad x = 1 \qquad \text{Solve.}$$

There are two break-even points. One is when 1000 units are produced, and the other is when 5000 units are produced.

(d) To find the number x for which marginal revenue equals marginal cost, we solve the equation $R'(x) = C'(x)$.

$$8 - 2x = 2$$
$$2x = 6$$
$$x = 3$$

Marginal revenue equals marginal cost when 3000 units are produced and sold.

(e)

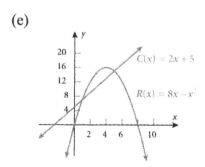

65. (a) The revenue function $R(x) = xp = x(-10x + 2000) = -10x^2 + 2000x$.

(b) The marginal revenue is the derivative $R'(x)$.

$$R'(x) = \lim_{h \to 0} \frac{R(x+h) - R(x)}{h} = \lim_{h \to 0} \frac{\left[-10(x+h)^2 + 2000(x+h)\right] - \left[-10x^2 + 2000x\right]}{h}$$

$$= \lim_{h \to 0} \frac{-10x^2 - 20xh - 10h^2 + 2000x + 2000h + 10x^2 - 2000x}{h} \qquad \text{Simplify.}$$

$$= \lim_{h \to 0} \frac{-20xh - 10h^2 + 2000h}{h} \qquad \text{Simplify.}$$

$$= \lim_{h \to 0} \frac{h(-20x - 10h + 2000)}{h} \qquad \text{Factor out an } h.$$

$$= \lim_{h \to 0} (-20x - 10h + 2000) \qquad \text{Cancel out the } h\text{'s.}$$

$$= -20x + 2000 \qquad \text{Go to the limit.}$$

(c) The marginal revenue at $x = 100$ tons is $R'(100) = (-20) \cdot 100 + 2000 = 0$ dollars.

(d) The average rate in change in revenue from $x = 100$ to $x = 101$ tons is

$$\frac{\Delta R}{\Delta x} = \frac{R(101) - R(100)}{101 - 100} = \frac{\left[(-10)(101^2) + 2000(101)\right] - \left[(-10)(100^2) + 2000(1000)\right]}{1}$$

$$= 99{,}990 - 100{,}000 = -10$$

(e) $R'(100) = 0$ indicates that there is no additional revenue gained by selling the 101st ton of cement.

The average rate of change in revenue from selling the 101st ton of cement represents a decrease in revenue of $10.

67. (a) The revenue function $R(x) = xp$, where p is the unit price and x is the number of units sold. $R(x) = xp = x(90 - 0.02x) = 90x - 0.02x^2$

(b) The marginal revenue is the derivative $R'(x)$.

$$R'(x) = \lim_{h \to 0} \frac{R(x+h) - R(x)}{h} = \lim_{h \to 0} \frac{\left[90(x+h) - 0.02(x+h)^2\right] - \left[90x - 0.02x^2\right]}{h}$$

$$= \lim_{h \to 0} \frac{90x + 90h - 0.02x^2 - 0.04xh - 0.02h^2 - 90x + 0.02x^2}{h} \qquad \text{Simplify.}$$

$$= \lim_{h \to 0} \frac{90h - 0.04xh - 0.02h^2}{h} \qquad \text{Simplify.}$$

$$= \lim_{h \to 0} \frac{\cancel{h}(90 - 0.04x - 0.02h)}{\cancel{h}} \qquad \text{Factor out an } h.$$

$$= \lim_{h \to 0} (90 - 0.04x - 0.02h) = 90 - 0.04x \qquad \text{Cancel the } h\text{'s; take the limit.}$$

(c) It costs $10 per unit to produce the product, so the cost function $C = C(x) = 10x$. The marginal cost is the derivative $C'(x)$.

$$C'(x) = \lim_{h \to 0} \frac{C(x+h) - C(x)}{h} = \lim_{h \to 0} \frac{\left[10(x+h)\right] - \left[10x\right]}{h}$$

$$= \lim_{h \to 0} \frac{10x + 10h - 10x}{h} = \lim_{h \to 0} \frac{10\cancel{h}}{\cancel{h}} = \lim_{h \to 0} 10 = 10$$

(d) A break-even point is a number x for which $R(x) = C(x)$. We solve the equation

$R(x) = C(x)$	
$90x - 0.02x^2 = 10x$	
$0.02x^2 - 80x = 0$	Put the quadratic equation in standard form.
$x^2 - 4000x = 0$	Multiply both sides by 50.
$x(x - 4000) = 0$	Factor.
$x = 0 \quad x - 4000 = 0$	Apply the Zero-Product Property.
$x = 0 \quad\quad x = 4000$	Solve for x.

There are two break-even points. One is when no units are produced and sold; the other is when $x = 4000$ units are produced and sold.

(e) The marginal revenue equals marginal cost when $R'(x) = C'(x)$.

$$R'(x) = C'(x)$$
$$90 - 0.04x = 10$$
$$80 = 0.04x$$
$$x = 2000$$

The marginal revenue equals marginal cost when 2000 units are produced and sold.

69. The instantaneous rate of change of the volume of the cylinder with respect to the radius r at $r = 3$ is the derivative $V'(3)$.

$$V'(3) = \lim_{r \to 3} \frac{V(r) - V(3)}{r - 3} = \lim_{r \to 3} \frac{3\pi r^2 - 3\pi(3^2)}{r - 3} = \lim_{r \to 3} \frac{3\pi r^2 - 27\pi}{r - 3} = \lim_{r \to 3} \frac{3\pi(r^2 - 9)}{r - 3}$$

$$= \lim_{r \to 3} \frac{3\pi \,(\cancel{r - 3})(r + 3)}{\cancel{r - 3}} = \lim_{r \to 3} \left[3\pi(r + 3)\right] = 18\pi \approx 56.55$$

4.2 The Derivative of a Power Function; Sum and Difference Formulas

1. The function $f(x) = 4$ is a constant. $f'(x) = 0$

3. The function $f(x) = x^5$ is a power function. $f'(x) = 5x^{5-1} = 5x^4$

5. $f'(x) = \dfrac{d}{dx}(6x^2) = 6\dfrac{d}{dx}x^2 = 6 \cdot 2x = 12x$

7. $f'(t) = \dfrac{d}{dt}\left(\dfrac{t^4}{4}\right) = \dfrac{1}{4}\dfrac{d}{dt}t^4 = \dfrac{1}{4} \cdot 4t^{4-1} = t^3$

9. $f'(x) = \dfrac{d}{dx}(x^2 + x) = \dfrac{d}{dx}x^2 + \dfrac{d}{dx}x$ Use the derivative of a sum formula (Formula (5)).

$$= 2x + 1 \qquad\qquad\qquad \frac{d}{dx}x = 1$$

11. $f'(x) = \dfrac{d}{dx}(x^3 - x^2 + 1) = \dfrac{d}{dx}x^3 - \dfrac{d}{dx}x^2 + \dfrac{d}{dx}1$ Use Formulas (6) and (5).

$$= 3x^2 - 2x \qquad\qquad \dfrac{d}{dx}1 = 0$$

13. $f'(t) = \dfrac{d}{dt}(2t^2 - t + 4) = \dfrac{d}{dt}(2t^2) - \dfrac{d}{dt}t + \dfrac{d}{dt}4$ Use Formulas (5) and (6).

$$= 2\dfrac{d}{dt}t^2 - \dfrac{d}{dt}t + 0 \qquad \text{Use Formulas(4) and (2).}$$

$$= 4t - 1 \qquad\qquad \text{Differentiate.}$$

15.
$$f'(x) = \dfrac{d}{dx}\left(\dfrac{1}{2}x^8 + 3x + \dfrac{2}{3}\right) = \dfrac{d}{dx}\left(\dfrac{1}{2}x^8\right) + \dfrac{d}{dx}(3x) + \dfrac{d}{dx}\dfrac{2}{3} \qquad \text{Use Formula (5).}$$

$$= \dfrac{1}{2}\dfrac{d}{dx}x^8 + 3\dfrac{d}{dx}x + 0 \qquad \text{Use Formula (4); } \dfrac{d}{dx}\dfrac{2}{3} = 0.$$

$$= \dfrac{1}{2}\cdot 8x^7 + 3 = 4x^7 + 3 \qquad \text{Differentiate and simplify.}$$

17. $f'(x) = \dfrac{d}{dx}\left[\dfrac{1}{3}(x^5 - 8)\right] = \dfrac{1}{3}\dfrac{d}{dx}(x^5 - 8) = \dfrac{1}{3}\left[\dfrac{d}{dx}x^5 - \dfrac{d}{dx}8\right] = \dfrac{1}{3}[5x^4 - 0] = \dfrac{5}{3}x^4$

                  ↑   ↑   ↑   ↑

    Use Formula (4).   Use Formula (6).   Differentiate.   Simplify.

19. $f'(x) = \dfrac{d}{dx}[ax^2 + bx + c] = \dfrac{d}{dx}(ax^2) + \dfrac{d}{dx}(bx) + \dfrac{d}{dx}c = a\dfrac{d}{dx}x^2 + b\dfrac{d}{dx}x + 0$

                   ↑   ↑

        Use Formula (5).   Use Formula (4); $\dfrac{d}{dx}c = 0$

$$= a\cdot 2x + b\cdot 1 = 2ax + b$$

         ↑   ↑

    Differentiate.   Simplify.

21. $\dfrac{d}{dx}(-6x^2 + x + 4) = \dfrac{d}{dx}(-6x^2) + \dfrac{d}{dx}x + \dfrac{d}{dx}4 = (-6)\dfrac{d}{dx}x^2 + 1 + 0 = (-6)\cdot 2x + 1 = -12x + 1$

23. $\dfrac{d}{dt}(-16t^2 + 80t) = \dfrac{d}{dt}(-16t^2) + \dfrac{d}{dt}(80t) = (-16)\dfrac{d}{dt}t^2 + 80\dfrac{d}{dt}t = (-16)\cdot 2t + 80\cdot 1 = -32t + 80$

25. $\dfrac{dA}{dr} = \dfrac{d}{dr}(\pi r^2) = \pi\dfrac{d}{dr}r^2 = \pi\cdot 2r = 2\pi r$

27. $\dfrac{dV}{dr} = \dfrac{d}{dr}\left(\dfrac{4}{3}\pi r^3\right) = \dfrac{4}{3}\pi\dfrac{d}{dr}r^3 = \dfrac{4}{3}\pi\cdot 3r^2 = 4\pi r^2$

29. To find $f'(-3)$, first we find the derivative of the function f.

$$f'(x) = \frac{d}{dx}(4x^2) = 8x$$

Then we substitute -3 for x.

$$f'(-3) = 8 \cdot (-3) = -24$$

31. To find $f'(4)$, first we find the derivative of the function f.

$$f'(x) = \frac{d}{dx}(2x^2 - x) = \frac{d}{dx}(2x^2) - \frac{d}{dx}x = 2\frac{d}{dx}x^2 - 1 = 4x - 1$$

Then we substitute 4 for x.

$$f'(4) = 4 \cdot 4 - 1 = 15$$

33. To find $f'(3)$, first we find the derivative of the function f.

$$f'(x) = \frac{d}{dt}\left(-\frac{1}{3}t^3 + 5t\right) = \frac{d}{dt}\left(-\frac{1}{3}t^3\right) + \frac{d}{dt}(5t) = -\frac{1}{3}\frac{d}{dt}t^3 + 5\frac{d}{dt}t = -\frac{1}{3} \cdot 3t^2 + 5 = -t^2 + 5$$

Then we substitute 3 for t.

$$f'(3) = -3^2 + 5 = -4$$

35. To find $f'(1)$, first we find the derivative of the function f.

$$f'(x) = \frac{d}{dx}\left[\frac{1}{2}(x^6 - x^4)\right] = \frac{1}{2}\frac{d}{dx}(x^6 - x^4) = \frac{1}{2}\left[\frac{d}{dx}x^6 - \frac{d}{dx}x^4\right]$$

$$= \frac{1}{2}(6x^5 - 4x^3) = 3x^5 - 2x^3$$

Then we substitute 1 for x.

$$f'(1) = 3 \cdot 1^5 - 2 \cdot 1^3 = 1$$

37. First we find the derivative of the function f. In Problem 19 we found $f'(x) = 2ax + b$.

So we now substitute $-\dfrac{b}{2a}$ for x.

$$f'\left(-\frac{b}{2a}\right) = 2a\left(-\frac{b}{2a}\right) + b = -b + b = 0$$

39. First we find the derivative $\dfrac{dy}{dx} = 4x^3$

Then we evaluate the derivative at the point $(1, 1)$ by substituting 1 for x.

$$\frac{dy}{dx} = 4 \cdot 1^3 = 4$$

41. First we find the derivative $\dfrac{dy}{dx} = 2x - 0 = 2x$

Then we evaluate the derivative at the point $(4, 2)$ by substituting 4 for x.

$$\frac{dy}{dx} = 2 \cdot 4 = 8$$

43. First we find the derivative $\dfrac{dy}{dx} = 6x - 1$

Then we evaluate the derivative at the point $(-1, 4)$ by substituting -1 for x.

$$\frac{dy}{dx} = 6 \cdot (-1) - 1 = -7$$

45. First we find the derivative $\dfrac{dy}{dx} = \dfrac{1}{2} \cdot 2x = x$

Then we evaluate the derivative at the point $\left(1, \dfrac{1}{2}\right)$ by substituting 1 for x.

$$\frac{dy}{dx} = 1$$

47. First we find the derivative $\dfrac{dy}{dx} = 0 - 2 + 3x^2 = -2 + 3x^2$

Then we evaluate the derivative at the point $(2, 6)$ by substituting 2 for x.

$$\frac{dy}{dx} = -2 + 3 \cdot 2^2 = 10$$

49. The slope of the tangent line to the graph of $f(x) = x^3 + 3x - 1$ at the point $(0, -1)$ is the derivative of the function f evaluated at the point $(0, -1)$. The derivative of f is

$$f'(x) = 3x^2 + 3$$
$$m_{\tan} = f'(0) = 3 \cdot (0)^2 + 3 = 3$$

An equation of the tangent line is

$$
\begin{array}{ll}
y - (-1) = 3(x - 0) & y - f(c) = m_{\tan}(x - c) \\
y + 1 = 3x & \text{Simplify.} \\
y = 3x - 1 & \text{Subtract 1 from both sides.}
\end{array}
$$

51. We first find the derivative $f'(x)$.

$$f'(x) = 6x - 12 + 0 = 6x - 12$$

We then solve the equation $f'(x) = 0$.

$$
\begin{array}{c}
6x - 12 = 0 \\
6x = 12 \\
x = 2
\end{array}
$$

53. We first find the derivative $f'(x)$.

$$f'(x) = 3x^2 - 3 + 0 = 3x^2 - 3$$

We then solve the equation $f'(x) = 0$.

$$
\begin{array}{ll}
3x^2 - 3 = 0 & f'(x) = 3x^2 - 3 \\
3(x^2 - 1) = 0 & \text{Factor out the 3.} \\
x^2 - 1 = 0 & \text{Divide both sides by 3.} \\
x = 1 \quad x = -1 & \text{Solve using the Square Root Method.}
\end{array}
$$

55. We first find the derivative $f'(x)$.
$$f'(x) = 3x^2 + 1$$

We then solve the equation $f'(x) = 0$.
$$3x^2 + 1 = 0 \qquad\qquad f'(x) = 3x^2 + 1$$
has no solutions.

57. The slope of the tangent line to the function $f(x) = 9x^3$ is the derivative of f.
$$f'(x) = 27x^2$$
To find the slope of the line $3x - y + 2 = 0$, we put the equation in slope-intercept form.
$$y = 3x + 2$$
The slope of the line is $m = 3$. For the tangent line to the graph of f to be parallel to the line y, $m_{\tan} = 3$. We solve the equation $f'(x) = 3$.

$$27x^2 = 3 \qquad\qquad f'(x) = 27x^2$$

$$x^2 = \frac{1}{9} \qquad\qquad \text{Divide both sides by 27.}$$

$$x = \frac{1}{3} \quad x = -\frac{1}{3} \qquad \text{Solve using the Square Root Method.}$$

There are two points on the graph of the function $y = 9x^3$ for which the slope of the tangent line is parallel to the graph of the line $3x - y + 2 = 0$. They are

$$\left(\frac{1}{3}, f\left(\frac{1}{3}\right)\right) = \left(\frac{1}{3}, \frac{1}{3}\right) \quad \text{and} \quad \left(-\frac{1}{3}, f\left(-\frac{1}{3}\right)\right) = \left(-\frac{1}{3}, -\frac{1}{3}\right)$$

59. If $(x, y) = (x, 2x^2 - 4x + 1)$ is a point on the graph of the function $y = 2x^2 - 4x + 1$ for which the tangent line to the graph of y passes through the point $(1, -3)$, then the slope of the tangent line is given by

$$m_{\tan} = \frac{\Delta y}{\Delta x} = \frac{y_2 - y_1}{x_2 - x_1} = \frac{(2x^2 - 4x + 1) - (-3)}{x - 1} = \frac{2x^2 - 4x + 4}{x - 1}$$

The slope of the tangent line to the graph of the function y is also given by the derivative of y.

$$m_{\tan} = \frac{d}{dx}(2x^2 - 4x + 1) = 4x - 4$$

Since both these expressions for the slope must be same, we will set them equal to each other and solve for x.

$$\frac{2x^2 - 4x + 4}{x - 1} = 4x - 4 \qquad \text{Set the slope equal to the derivative.}$$
$$2x^2 - 4x + 4 = (4x - 4)(x - 1) \qquad \text{Multiply both sides by } x - 1.$$
$$2x^2 - 4x + 4 = 4x^2 - 8x + 4 \qquad \text{Multiply out the right side.}$$
$$2x^2 - 4x = 0 \qquad \text{Put the quadratic equation in standard form.}$$
$$2x(x - 2) = 0 \qquad \text{Factor.}$$
$$2x = 0 \quad x - 2 = 0 \qquad \text{Apply the Zero-Product Property.}$$
$$x = 0 \quad x = 2 \qquad \text{Solve for } x.$$

We evaluate the tangent lines to the graph of y using each of these two values of x.

If $x = 0$ then $y = f(0) = 1$. The point on the graph of y is $(0, 1)$. The slope of the tangent line to the graph of y at $(0, 1)$ is $f'(0) = 4 \cdot 0 - 4 = -4$. The equation of the tangent line through $(0, 1)$ and $(1, -3)$ is

$$y - 1 = -4(x - 0)$$
$$y = -4x + 1$$

If $x = 2$ then $y = f(2) = 2 \cdot 2^2 - 4 \cdot 2 + 1 = 1$. The point on the graph of y is (2, 1). The slope of the tangent line to the graph of y at (2, 1) is $f'(2) = 4 \cdot 2 - 4 = 4$. The equation of the tangent line through (2, 1) and (1, –3) is

$$y - 1 = 4(x - 2)$$

$$y - 1 = 4x - 8$$
$$y = 4x - 7$$

61. (a) The average cost of producing 10 additional pairs of eyeglasses is

$$\frac{\Delta y}{\Delta x} = \frac{C(110) - C(100)}{110 - 100} = \frac{\left[0.2 \cdot 110^2 + 3 \cdot 110 + 1000\right] - \left[0.2 \cdot 100^2 + 3 \cdot 100 + 1000\right]}{10}$$

$$= \frac{3750 - 3300}{10} = \frac{450}{10} = 45 \text{ dollars per pair.}$$

(b) The marginal cost of producing an additional pair of eyeglasses is
$$C'(x) = 0.4x + 3$$

(c) The marginal cost at $x = 100$ is $C'(100) = 0.4(100) + 3 = 43$ dollars.

(d) The marginal cost at $x = 100$ is the cost of producing one additional pair of eyeglasses when 100 pairs have already been produced.

63. (a) The derivative $V'(R) = 4kR^3$.

(b) $V'(0.3) = 4k \cdot 0.3^3 = 0.108k$ centimeters cubed

(c) $V'(0.4) = 4k \cdot 0.4^3 = 0.256k$ centimeters cubed

65. (a) It costs $C(40) = 2000 + 50 \cdot 40 - 0.05 \cdot 40^2 = 3920$ dollars to produce 40 microwave ovens.

(b) The marginal cost function is the derivative of $C(x)$.
$$C'(x) = 50 - 0.10x$$

(c) $C'(40) = 50 - 0.10 \cdot 40 = 46$. The marginal cost at $x = 40$ indicates that the cost of producing the 41st microwave oven is $46.00.

(d) An estimate of the cost of producing 41 microwave ovens can be obtained by adding $C(40)$ and $C'(40)$.
$$C(40) + C'(40) = 3920 + 46 = 3966 \text{ dollars to produce 41 microwaves.}$$

(e) The actual cost of producing 41 microwave ovens is
$$C(41) = 2000 + 50 \cdot 41 - 0.05 \cdot 41^2 = 3965.95 \text{ dollars.}$$
The actual cost of producing 41 microwave ovens is $0.05 less than the estimated cost.

(f) The actual cost of producing the 41st microwave oven is $45.95.
$$C(41) - C(40) = 3965.95 - 3920 = \$45.95$$

(g) The average cost function for producing x microwave ovens is
$$\overline{C}(x) = \frac{2000 + 50x - 0.05x^2}{x} = \frac{2000}{x} + 50 - 0.05x$$

(h) The average cost of producing 41 microwave ovens is $\overline{C}(41) = \$96.73$
$$\overline{C}(41) = \frac{2000 + 50 \cdot 41 - 0.05 \cdot 41^2}{41} = 96.73$$

67. The marginal price of beans in year t is the derivative of $p(t)$.
$$p'(t) = 0.021t^2 - 1.26t + 0.005$$
(a) The marginal price of beans in 1995 is $p'(2)$, since $t = 0$ represents 1993, $t = 2$ represents 1995.
$$p'(2) = 0.021 \cdot 2^2 - 1.26 \cdot 2 + 0.005 = -2.431$$

(b) The marginal price of beans in 2002 is $p'(9)$, since $t = 0$ represents 1993, $t = 9$ represents 2002.
$$p'(9) = 0.021 \cdot 9^2 - 1.26 \cdot 9 + 0.005 = -9.634$$

(c) Answers will vary.

69. The instantaneous rate of change of the volume V of a sphere with respect to its radius r when $r = 2$ feet is $V'(2)$.
$$V(r) = \frac{4}{3}\pi r^3$$
$$V'(r) = 4\pi r^2$$
$$V'(2) = 4\pi \cdot 2^2 = 16\pi \text{ feet cubed}$$

71. The instantaneous rate of change of work output at time t is the derivative of $A(t)$.
$$A'(t) = 3a_3 t^2 + 2a_2 t + a_1$$

73. Formula (3) from Section 2 of the text is $\dfrac{d}{dx}x^n = nx^{n-1}$

To prove this formula we begin with the difference quotient.
$$\frac{d}{dx}x^n = f'(x) = \lim_{h \to 0} \frac{f(x+h) - f(x)}{h}$$
$$= \lim_{h \to 0} \frac{(x+h)^n - x^n}{h} \qquad \text{Use the difference quotient.}$$
$$= \lim_{h \to 0} \frac{\left[x^n + nx^{n-1}h + \dfrac{n(n-1)}{2}x^{n-2}h^2 + \ldots + h^n \right] - \left[x^n \right]}{h} \qquad \text{Use the hint provided.}$$

$$= \lim_{h \to 0} \frac{nx^{n-1}h + \frac{n(n-1)}{2}x^{n-2}h^2 + \ldots + h^n}{h}$$

Simplify.

$$= \lim_{h \to 0} \frac{h\left(nx^{n-1} + \frac{n(n-1)}{2}x^{n-2}h + \ldots + h^{n-1}\right)}{h}$$

Factor out an h.

$$= \lim_{h \to 0} \left(nx^{n-1} + \frac{n(n-1)}{2}x^{n-2}h + \ldots + h^{n-1}\right)$$

Cancel the h's.

$$= nx^{n-1}$$

Find the limit.

4.3 Product and Quotient Formulas

1. The function f is the product of two functions $g(x) = 2x + 1$ and $h(x) = 4x - 3$ so that using the formula for the derivative of a product, we have

$$f'(x) = (2x+1)\left[\frac{d}{dx}(4x-3)\right] + (4x-3)\left[\frac{d}{dx}(2x+1)\right] \quad \text{Derivative of a product formula.}$$

$$= (2x + 1)(4) + (4x - 3)(2) \qquad\qquad \text{Differentiate.}$$
$$= 8x + 4 + 8x - 6 \qquad\qquad\qquad \text{Simplify.}$$
$$= 16x - 2 \qquad\qquad\qquad\qquad \text{Simplify.}$$

3. The function f is the product of two functions $g(t) = t^2 + 1$ and $h(t) = t^2 - 4$ so that using the formula for the derivative of a product, we have

$$f'(t) = (t^2+1)\left[\frac{d}{dt}(t^2-4)\right] + (t^2-4)\left[\frac{d}{dt}(t^2+1)\right] \quad \text{Derivative of a product formula.}$$

$$= (t^2 + 1)(2t) + (t^2 - 4)(2t) \qquad\qquad \text{Differentiate.}$$
$$= 2t^3 + 2t + 2t^3 - 8t \qquad\qquad\qquad \text{Simplify.}$$
$$= 4t^3 - 6t \qquad\qquad\qquad\qquad\quad \text{Simplify.}$$

5. The function f is the product of two functions $g(x) = 3x - 5$ and $h(x) = 2x^2 + 1$ so that using the formula for the derivative of a product, we have

$$f'(x) = (3x-5)\left[\frac{d}{dx}(2x^2+1)\right] + (2x^2+1)\left[\frac{d}{dx}(3x-5)\right] \quad \text{Derivative of a product formula.}$$

$$= (3x - 5)(4x) + (2x^2 + 1)(3) \qquad\qquad \text{Differentiate.}$$
$$= 12x^2 - 20x + 6x^2 + 3 \qquad\qquad\qquad \text{Simplify.}$$
$$= 18x^2 - 20x + 3 \qquad\qquad\qquad\qquad \text{Simplify.}$$

7. The function f is the product of two functions $g(x) = x^5 + 1$ and $h(x) = 3x^3 + 8$ so that using the formula for the derivative of a product, we have

$$f'(x) = (x^5+1)\left[\frac{d}{dx}(3x^3+8)\right] + (3x^3+8)\left[\frac{d}{dx}(x^5+1)\right] \quad \text{Derivative of a product formula.}$$

$$= (x^5 + 1)(9x^2) + (3x^3 + 8)(5x^4) \qquad\qquad \text{Differentiate.}$$
$$= 9x^7 + 9x^2 + 15x^7 + 40x^4 \qquad\qquad\qquad \text{Simplify.}$$
$$= 24x^7 + 40x^4 + 9x^2 \qquad\qquad\qquad\qquad \text{Simplify.}$$

9. The function f is the quotient of two functions $g(x) = x$ and $h(x) = x + 1$. We use the formula for the derivative of a quotient to get

$$\frac{d}{dx}\left(\frac{x}{x+1}\right) = \frac{(x+1)\frac{d}{dx}x - x\frac{d}{dx}(x+1)}{(x+1)^2} \qquad \text{Derivative of a quotient formula.}$$

$$= \frac{(x+1)(1) - x(1)}{(x+1)^2} \qquad \text{Differentiate.}$$

$$= \frac{x+1-x}{(x+1)^2} \qquad \text{Simplify.}$$

$$= \frac{1}{(x+1)^2} \qquad \text{Simplify.}$$

11. The function f is the quotient of two functions $g(x) = 3x + 4$ and $h(x) = 2x - 1$. We use the formula for the derivative of a quotient to get

$$\frac{d}{dx}\left(\frac{3x+4}{2x-1}\right) = \frac{(2x-1)\frac{d}{dx}(3x+4) - (3x+4)\frac{d}{dx}(2x-1)}{(2x-1)^2} \qquad \text{Derivative of a quotient formula.}$$

$$= \frac{(2x-1)(3) - (3x+4)(2)}{(2x-1)^2} \qquad \text{Differentiate.}$$

$$= \frac{6x-3-6x-8}{(2x-1)^2} \qquad \text{Simplify.}$$

$$= -\frac{11}{(2x-1)^2} \qquad \text{Simplify.}$$

13. The function f is the quotient of two functions $g(x) = x^2$ and $h(x) = x - 4$. We use the formula for the derivative of a quotient to get

$$\frac{d}{dx}\left(\frac{x^2}{x-4}\right) = \frac{(x-4)\frac{d}{dx}x^2 - x^2\frac{d}{dx}(x-4)}{(x-4)^2} \qquad \text{Derivative of a quotient formula.}$$

$$= \frac{(x-4)(2x) - x^2(1)}{(x-4)^2} \qquad \text{Differentiate.}$$

$$= \frac{2x^2 - 8x - x^2}{(x-4)^2} \qquad \text{Simplify.}$$

$$= \frac{x^2 - 8x}{(x-4)^2} \qquad \text{Simplify.}$$

15. The function f is the quotient of two functions $g(x) = 2x + 1$ and $h(x) = 3x^2 + 4$. We use the formula for the derivative of a quotient to get

$$\frac{d}{dx}\left(\frac{2x+1}{3x^2+4}\right) = \frac{(3x^2+4)\frac{d}{dx}(2x+1) - (2x+1)\frac{d}{dx}(3x^2+4)}{(3x^2+4)^2} \qquad \text{Derivative of a quotient formula.}$$

$$= \frac{(3x^2+4)(2)-(2x+1)(6x)}{(3x^2+4)^2} \qquad \text{Differentiate.}$$

$$= \frac{6x^2+8-12x^2-6x}{(3x^2+4)^2} \qquad \text{Simplify.}$$

$$= -\frac{6x^2+6x-8}{(3x^2+4)^2} \qquad \text{Simplify.}$$

17. $\dfrac{d}{dx}\dfrac{-2}{t^2} = \dfrac{d}{dx}\left(-2t^{-2}\right) = -2\dfrac{d}{dx}t^{-2} = -2\left(-2t^{-3}\right) = \dfrac{4}{t^3}$

19. $\dfrac{d}{dx}\left(1+\dfrac{1}{x}+\dfrac{1}{x^2}\right) = \dfrac{d}{dx}\left(1+x^{-1}+x^{-2}\right) = \dfrac{d}{dx}1+\dfrac{d}{dx}x^{-1}+\dfrac{d}{dx}x^{-2} = 0-1x^{-2}-2x^{-3} = -\dfrac{1}{x^2}-\dfrac{2}{x^3}$

21. The slope of the tangent line to the function f at $(1, 2)$ is the derivative of f at $x = 1$.

$$f'(x) = \frac{d}{dx}\left(x^3-2x+2\right)(x+1)$$

$$= \left(x^3-2x+2\right)\frac{d}{dx}(x+1)+(x+1)\frac{d}{dx}\left(x^3-2x+2\right) \quad \text{Derivative of a product.}$$

$$= \left(x^3-2x+2\right)(1)+(x+1)\left(3x^2-2\right) \qquad \text{Differentiate.}$$

$$= x^3-2x+2+3x^3+3x^2-2x-2 \qquad \text{Simplify.}$$

$$= 4x^3+3x^2-4x \qquad \text{Simplify.}$$

$$m_{\tan} = f'(1) = 3$$

An equation of the tangent line is
$$\begin{array}{ll} y-2 = 3(x-1) & y-y_1 = m(x-x_1) \\ y-2 = 3x-3 & \text{Simplify.} \\ y = 3x-1 & \text{Add 2 to both sides.} \end{array}$$

23. The slope of the tangent line to the function f at $\left(1, \dfrac{1}{2}\right)$ is the derivative of f at $x = 1$.

$$f'(x) = \frac{d}{dx}\frac{x^3}{x+1} = \frac{(x+1)\dfrac{d}{dx}x^3 - x^3\dfrac{d}{dx}(x+1)}{(x+1)^2} \qquad \text{Derivative of a quotient.}$$

$$= \frac{(x+1)\left(3x^2\right)-x^3(1)}{(x+1)^2} \qquad \text{Differentiate.}$$

$$= \frac{3x^3+3x^2-x^3}{(x+1)^2} \qquad \text{Simplify.}$$

$$= \frac{2x^3+3x^2}{(x+1)^2} \qquad \text{Simplify.}$$

$$m_{\tan} = f'(1) = \frac{2\left(1^3\right)+3\left(1^2\right)}{(1+1)^2} = \frac{5}{4}$$

An equation of the tangent line is

$$y - \frac{1}{2} = \frac{5}{4}(x-1) \qquad y - y_1 = m(x - x_1)$$

$$y - \frac{1}{2} = \frac{5}{4}x - \frac{5}{4} \qquad \text{Simplify.}$$

$$y = \frac{5}{4}x - \frac{3}{4} \qquad \text{Add } \frac{1}{2} \text{ to both sides.}$$

25. We first find the derivative $f'(x)$.

$$f'(x) = \frac{d}{dx}\left[(x^2-2)(2x-1)\right] = (x^2-2)\frac{d}{dx}(2x-1)+(2x-1)\frac{d}{dx}(x^2-2)$$

$$= (x^2-2)(2)+(2x-1)(2x)$$

$$= 2x^2-4+4x^2-2x = 6x^2-2x-4$$

We then solve the equation $f'(x) = 0$.

$$6x^2-2x-4 = 0 \qquad f'(x) = 6x^2-2x-4$$
$$3x^2-1x-2 = 0$$
$$(3x+2)(x-1) = 0 \qquad \text{Factor.}$$
$$x = \frac{2}{3} \quad x = 1 \qquad \text{Apply the Zero-Product Property.}$$

27. We first find the derivative $f'(x)$.

$$f'(x) = \frac{d}{dx}\frac{x^2}{x+1} = \frac{(x+1)\frac{d}{dx}x^2 - x^2\frac{d}{dx}(x+1)}{(x+1)^2}$$

$$= \frac{(x+1)(2x) - x^2(1)}{(x+1)^2}$$

$$= \frac{2x^2+2x-x^2}{(x+1)^2} = \frac{x^2+2x}{(x+1)^2}$$

We then solve the equation $f'(x) = 0$.

$$\frac{x^2+2x}{(x+1)^2} = 0 \qquad f'(x) = \frac{x^2+2x}{(x+1)^2}$$
$$x^2+2x = 0 \qquad \text{Multiply both sides by } (x+2)^2; x \ne -2.$$
$$x(x+2) = 0 \qquad \text{Factor.}$$
$$x = 0 \quad x+2 = 0 \qquad \text{Apply the Zero-Product Property.}$$
$$x = -2 \qquad \text{Solve for } x.$$

29. y is the product of two functions, we will use the formula for the derivative of a product.

$$y' = \frac{d}{dx}\left[x^2(3x-2)\right] = x^2\frac{d}{dx}(3x-2)+(3x-2)\frac{d}{dx}x^2$$

$$= x^2(3)+(3x-2)(2x) \qquad \text{Differentiate.}$$

$$= 3x^2+6x^2-4x \qquad \text{Simplify.}$$

$$= 9x^2-4x \qquad \text{Simplify.}$$

31. y is the product of two functions, we will use the formula for the derivative of a product.

$$y' = \frac{d}{dx}\left[(x^2+4)(4x^2+3)\right] = (x^2+4)\frac{d}{dx}(4x^2+3)+(4x^2+3)\frac{d}{dx}(x^2+4)$$

$$= (x^2+4)(8x)+(4x^2+3)(2x) \qquad \text{Differentiate.}$$
$$= 8x^3+32x+8x^3+6x \qquad \text{Simplify.}$$
$$= 16x^3+38x \qquad \text{Simplify.}$$

33. y is the quotient of two functions, we will use the formula for the derivative of a quotient.

$$y' = \frac{d}{dx}\frac{2x+3}{3x+5} = \frac{(3x+5)\frac{d}{dx}(2x+3)-(2x+3)\frac{d}{dx}(3x+5)}{(3x+5)^2}$$

$$= \frac{(3x+5)(2)-(2x+3)(3)}{(3x+5)^2} \qquad \text{Differentiate.}$$

$$= \frac{6x+10-6x-9}{(3x+5)^2} \qquad \text{Simplify.}$$

$$= \frac{1}{(3x+5)^2} \qquad \text{Simplify.}$$

35. y is the quotient of two functions, so we will use the formula for the derivative of a quotient.

$$y' = \frac{d}{dx}\frac{x^2}{x^2-4} = \frac{(x^2-4)\frac{d}{dx}x^2-x^2\frac{d}{dx}(x^2-4)}{(x^2-4)^2}$$

$$= \frac{(x^2-4)(2x)-x^2(2x)}{(x^2-4)^2} \qquad \text{Differentiate.}$$

$$= \frac{2x^3-8x-2x^3}{(x^2-4)^2} \qquad \text{Simplify.}$$

$$= -\frac{8x}{(x^2-4)^2} \qquad \text{Simplify.}$$

37. y is the quotient of two functions, but its numerator is the product of two more functions. So we will use the formula for the derivative of a quotient and when differentiating the numerator the formula for the derivative of a product.

$$y' = \frac{d}{dx}\frac{(3x+4)(2x-3)}{2x+1} = \frac{(2x+1)\frac{d}{dx}\left[(3x+4)(2x-3)\right]-\left[(3x+4)(2x-3)\right]\frac{d}{dx}(2x+1)}{(2x+1)^2}$$

$$= \frac{(2x+1)\left[(3x+4)\frac{d}{dx}(2x-3)+(2x-3)\frac{d}{dx}(3x+4)\right]-(3x+4)(2x-3)\frac{d}{dx}(2x+1)}{(2x+1)^2}$$

$$= \frac{(2x+1)\left[(3x+4)(2)+(2x-3)(3)\right]-(3x+4)(2x-3)(2)}{(2x+1)^2} \qquad \text{Differentiate.}$$

$$= \frac{(2x+1)\left[6x+8+6x-9\right]-\left(6x^2+8x-9x-12\right)(2)}{(2x+1)^2} \qquad \text{Simplify.}$$

$$= \frac{(2x+1)(12x-1)-\left(6x^2-x-12\right)(2)}{(2x+1)^2} \qquad \text{Add like terms.}$$

$$= \frac{24x^2-2x+12x-1-12x^2+2x+24}{(2x+1)^2} \qquad \text{Multiply.}$$

$$= \frac{12x^2+12x+23}{(2x+1)^2} \qquad \text{Simplify.}$$

39. y is the quotient of two functions, so we will use the formula for the derivative of a quotient.

$$y' = \frac{d}{dx}\frac{4x^3}{x^2+4} = \frac{(x^2+4)\dfrac{d}{dx}(4x^3)-4x^3\dfrac{d}{dx}(x^2+4)}{(x^2+4)^2}$$

$$= \frac{(x^2+4)(12x^2)-4x^3(2x)}{(x^2+4)^2} \qquad \text{Differentiate.}$$

$$= \frac{12x^4+48x^2-8x^4}{(x^2+4)^2} \qquad \text{Simplify.}$$

$$= \frac{4x^4+48x^2}{(x^2+4)^2} \qquad \text{Simplify.}$$

41.   (a) The average change in value from $t=2$ to $t=5$ is

$$\frac{\Delta V}{\Delta t} = \frac{V(5)-V(2)}{5-2} = \frac{\left[\dfrac{10{,}000}{5}+6000\right]-\left[\dfrac{10{,}000}{2}+6000\right]}{3}$$

$$= \frac{2000-5000}{3} = \frac{-3000}{3} = -1000$$

   (b) The instantaneous rate of change in value is the derivative of function V.

$$V(t) = \frac{10{,}000}{t}+6000 = 10{,}000t^{-1}+6000$$

$$V'(t) = -10{,}000t^{-2} = -\frac{10{,}000}{t^2}$$

   (c) The instantaneous rate of change after 2 years is $V'(2)$.

$$V'(2) = -\frac{10{,}000}{2^2} = -\frac{10{,}000}{4} = -2500$$

(d) The instantaneous rate of change after 5 years is $V'(5)$.
$$V'(5) = -\frac{10,000}{5^2} = -\frac{10,000}{25} = -400$$

(e) Answers will vary.

43. (a) The revenue function R is the product of the unit price and the number of units sold.
$$R = R(x) = px = \left(10 + \frac{40}{x}\right)x = 10x + 40$$

(b) The marginal revenue is the derivative of the revenue function R.
$$R'(x) = 10$$

(c) The marginal revenue when $x = 4$, is $R'(4) = 10$.

(d) The marginal revenue when $x = 6$, is $R'(6) = 10$

45. (a) $D'(p) = \dfrac{d}{dp}\dfrac{100,000}{p^2 + 10p + 50} = \dfrac{(p^2 + 10p + 50)\dfrac{d}{dp}(100,000) - 100,000\dfrac{d}{dp}(p^2 + 10p + 50)}{(p^2 + 10p + 50)^2}$

$$= \frac{(p^2 + 10p + 50)(0) - 100,000(2p + 10)}{(p^2 + 10p + 50)^2} \qquad \frac{d}{dp}100,000 = 0$$

$$= \frac{-200,000p - 1,000,000}{(p^2 + 10p + 50)^2}$$

(b) $D'(5) = \dfrac{-200,000(5) - 1,000,000}{(5^2 + 10(5) + 50)^2} = -128$

$D'(10) = \dfrac{-200,000(10) - 1,000,000}{(10^2 + 10(10) + 50)^2} = -48$

$D'(15) = \dfrac{-200,000(15) - 1,000,000}{(15^2 + 10(15) + 50)^2} = -22.145$

47. The rate at which the population is growing is given by the derivative of the population function P.

$$P'(t) = \frac{d}{dt}1000\left(1 + \frac{4t}{100 + t^2}\right) = 1000\left[\frac{d}{dt}1 + \frac{d}{dt}\frac{4t}{100 + t^2}\right]$$

$\dfrac{d}{dx}(cf(x)) = c\dfrac{d}{dx}f(x)$;
The derivative of a sum is the sum of the derivatives

$$= 1000\left[\frac{(100 + t^2)\dfrac{d}{dt}(4t) - 4t\dfrac{d}{dt}(100 + t^2)}{(100 + t^2)^2}\right]$$

Derivative of a quotient formula.

$$= 1000\left[\frac{(100 + t^2)(4) - 4t(2t)}{(100 + t^2)^2}\right]$$

Differentiate.

$$= 1000\left[\frac{400+4t^2-8t^2}{\left(100+t^2\right)^2}\right] = \frac{4000\left(100-t^2\right)}{\left(100+t^2\right)^2} \qquad \text{Simplify.}$$

In Parts (a) – (d), we evaluate the derivative $P'(t)$ at the indicated time.

(a) $t = 1$ hour, $P'(1) = \dfrac{4000\left(100-1^2\right)}{\left(100+1^2\right)^2} = 38.820$

(b) $t = 2$ hours, $P'(2) = \dfrac{4000\left(100-2^2\right)}{\left(100+2^2\right)^2} = 35.503$

(c) $t = 3$ hours, $P'(3) = \dfrac{4000\left(100-3^2\right)}{\left(100+3^2\right)^2} = 30.637$

(d) $t = 4$ hours, $P'(4) = \dfrac{4000\left(100-4^2\right)}{\left(100+4^2\right)^2} = 24.970$

49. First we must find the function I that describes the intensity of light with respect to the distance r of the object from the source of the light. Since I is inversely related to the square of distance, and we are told that $I = 1000$ units when $r = 1$ meter, we solve for the constant of proportionality k.

$$I(1) = \frac{k}{1^2} = 1000 \qquad\qquad I(r) = \frac{k}{r^2}$$
$$k = 1000 \qquad\qquad \text{Solve for } k.$$

So, we have

$$I = I(r) = \frac{1000}{r^2}$$

The rate of change of intensity with respect to distance r is the derivative $I'(r)$ of the function I.

$$I'(r) = \frac{dI}{dr} = \frac{d}{dr}\frac{1000}{r^2} = \frac{d}{dr}\left(1000r^{-2}\right) = 1000\frac{d}{dr}r^{-2} = 1000\left(-2r^{-3}\right) = -\frac{2000}{r^3}$$

When $r = 10$ meters the rate of change of the intensity of the light is

$$I'(10) = -\frac{2000}{10^3} = -\frac{2000}{1000} = -2 \text{ units per meter.}$$

51. (a) The marginal cost is the derivative of the cost function C.

$$C'(x) = \frac{d}{dx}\left(100+\frac{x}{10}+\frac{36{,}000}{x}\right) = \frac{d}{dx}\left(100+\frac{x}{10}+36{,}000x^{-1}\right)$$

$$= \frac{d}{dx}100 + \frac{1}{10}\cdot\frac{d}{dx}x + 36{,}000\frac{d}{dx}x^{-1}$$

The derivative of a sum is the sum of the derivatives;

$$\frac{d}{dx}\left(cf(x)\right) = c\frac{d}{dx}f(x)$$

$$= 0 + \frac{1}{10} + 36{,}000 \left(-1x^{-2}\right) \qquad \text{Differentiate.}$$

$$= \frac{1}{10} - \frac{36{,}000}{x^2}$$

In Parts (b) – (d), the derivative $C'(x)$ is evaluated at the indicated ground speeds.

(b) ground speed $x = 500$ mph $\qquad C'(500) = \dfrac{1}{10} - \dfrac{36{,}000}{500^2} = -0.044$

(c) ground speed $x = 550$ mph $\qquad C'(550) = \dfrac{1}{10} - \dfrac{36{,}000}{550^2} = -0.019$

(d) ground speed $x = 450$ mph $\qquad C'(450) = \dfrac{1}{10} - \dfrac{36{,}000}{450^2} = -0.078$

53. (a) First we find the derivative of the function S with respect to reward r.

$$S'(r) = \frac{d}{dr}\frac{ar}{g-r} = \frac{(g-r)\dfrac{d}{dr}(ar) - ar\dfrac{d}{dr}(g-r)}{(g-r)^2} \qquad \text{Use the derivative of a quotient formula.}$$

$$= \frac{(g-r)(a) - ar(-1)}{(g-r)^2} \qquad \text{Differentiate.}$$

$$= \frac{ag - ar + ar}{(g-r)^2} = \frac{ag}{(g-r)^2} \qquad \text{Simplify.}$$

Since both a and g are constants for a given individual $k = ag$, and $S'(r)$ is of the form,

$S'(r) = \dfrac{k}{(g-r)^2}$ which is inversely proportional to the square of the difference between

the personal goal of the individual and the amount of reward received.

(b) Answers may vary.

4.4 The Power Rule

1. $f'(x) = \dfrac{d}{dx} f(x) = \dfrac{d}{dx}(2x-3)^4 = 4(2x-3)^3 \dfrac{d}{dx}(2x-3) = 4(2x-3)^3 (2) = 8(2x-3)^3$

$\qquad\qquad\qquad\qquad\qquad\quad \uparrow \qquad\qquad\qquad\qquad\quad \uparrow \qquad\qquad\quad \uparrow$

$\qquad\qquad\qquad\qquad$ Use the Power Rule. $\qquad$ Differentiate. $\qquad$ Simplify.

3. $f'(x) = \dfrac{d}{dx} f(x) = \dfrac{d}{dx}(x^2+4)^3 = 3(x^2+4)^2 \dfrac{d}{dx}(x^2+4) = 3(x^2+4)^2 (2x) = 6x(x^2+4)^2$

$\qquad\qquad\qquad\qquad\qquad\quad \uparrow \qquad\qquad\qquad\qquad\quad \uparrow \qquad\qquad\quad \uparrow$

$\qquad\qquad\qquad\qquad$ Use the Power Rule. $\qquad$ Differentiate. $\qquad$ Simplify.

5. $f'(x) = \dfrac{d}{dx} f(x) = \dfrac{d}{dx}(3x^2+4)^2 = 2(3x^2+4)\dfrac{d}{dx}(3x^2+4) = 2(3x^2+4)(6x)$

$\qquad\qquad\qquad\qquad\qquad\quad \uparrow \qquad\qquad\qquad\qquad\quad \uparrow$

$\qquad\qquad\qquad\qquad$ Use the Power Rule. $\qquad$ Differentiate.

$$= 12x(3x^2 + 4) = 36x^3 + 48x$$

↑   ↑
Simplify.   Simplify.

7. The function f is the product of x and $(x + 1)^3$. We begin by using the formula for the derivative of a product. That is,

$$f'(x) = \frac{d}{dx}f(x) = x\frac{d}{dx}(x+1)^3 + (x+1)^3\frac{d}{dx}x$$

We continue by using the Power Rule:

$$f'(x) = x\left[3(x+1)^2\frac{d}{dx}(x+1)\right] + (x+1)^3\frac{d}{dx}x$$

$$= x\left[3(x+1)^2 \cdot 1\right] + (x+1)^3 \cdot 1 \qquad \text{Differentiate.}$$

$$= 3x(x+1)^2 + (x+1)^3 \qquad \text{Simplify.}$$

$$= (x+1)^2\left[3x + (x+1)\right] \qquad \text{Factor.}$$

$$= (x+1)^2(4x+1) \qquad \text{Simplify.}$$

9. The function f is the product of $4x^2$ and $(2x+1)^4$. We begin by using the formula for the derivative of a product. That is,

$$f'(x) = 4x^2\frac{d}{dx}(2x+1)^4 + (2x+1)^4\frac{d}{dx}(4x^2) = 4x^2\frac{d}{dx}(2x+1)^4 + (2x+1)^4 \cdot 4 \cdot \frac{d}{dx}x^2$$

We continue by using the Power Rule:

$$f'(x) = 4x^2\left[4(2x+1)^3\frac{d}{dx}(2x+1)\right] + (2x+1)^4 \cdot 4 \cdot \frac{d}{dx}x^2$$

$$= 4x^2\left[4(2x+1)^3(2)\right] + (2x+1)^4 \cdot 4 \cdot 2x \qquad \text{Differentiate.}$$

$$= 32x^2(2x+1)^3 + 8x(2x+1)^4 \qquad \text{Simplify.}$$

$$= 8x(2x+1)^3\left[4x + (2x+1)\right] \qquad \text{Factor.}$$

$$= 8x(2x+1)^3(6x+1) \qquad \text{Simplify.}$$

11. Before differentiating the function f, we simplify it, then we use the Power Rule.

$$f(x) = [x(x-1)]^3 = (x^2 - x)^3$$

$$f'(x) = \frac{d}{dx}(x^2-x)^3 = 3(x^2-x)^2\frac{d}{dx}(x^2-x) \qquad \text{Use the Power Rule.}$$

$$= 3(x^2-x)^2(2x-1) \qquad \text{Differentiate.}$$

$$= 3x^2(x-1)^2(2x-1) \qquad \text{Factor.}$$

13. $$f'(x) = \frac{d}{dx}(3x-1)^{-2} = -2(3x-1)^{-3}\frac{d}{dx}(3x-1) \qquad \text{Use the Power Rule.}$$

$$= -2(3x-1)^{-3}(3) \qquad \text{Differentiate.}$$

$$= -\frac{6}{(3x-1)^3} \qquad \text{Simplify.}$$

15. We rewrite $f(x)$ as $f(x) = 4(x^2 + 4)^{-1}$. Then we use the Power Rule.

$$f'(x) = \frac{d}{dx}\left[4(x^2+4)^{-1}\right] = 4\frac{d}{dx}(x^2+4)^{-1} = 4 \cdot (-1)(x^2+4)^{-2}\frac{d}{dx}(x^2+4)$$

$$\uparrow \qquad\qquad\qquad \uparrow$$
$$f'(cx) = cf'(x) \qquad \text{Use the Power Rule.}$$

$$= -4(x^2+4)^{-2}(2x) = -\frac{8x}{(x^2+4)^2}$$

$$\uparrow \qquad\qquad\qquad \uparrow$$
$$\text{Differentiate.} \qquad\qquad \text{Simplify.}$$

17. We rewrite $f(x)$ as $f(x) = -4(x^2 - 9)^{-3}$. Then we use the Power Rule.

$$f'(x) = \frac{d}{dx}\left[-4(x^2-9)^{-3}\right] = -4 \cdot \frac{d}{dx}(x^2-9)^{-3} = -4 \cdot (-3)(x^2-9)^{-4}\frac{d}{dx}(x^2-9)$$

$$\uparrow \qquad\qquad\qquad \uparrow$$
$$f'(cx) = cf'(x) \qquad \text{Use the Power Rule.}$$

$$= 12(x^2-9)^{-4}(2x) = \frac{24x}{(x^2-9)^4}$$

$$\uparrow \qquad\qquad\qquad \uparrow$$
$$\text{Differentiate.} \qquad\qquad \text{Simplify.}$$

19. In this problem the function f is a quotient raised to the power 3. We begin with the Power Rule and then use the formula for the derivative of a quotient.

$$f'(x) = \frac{d}{dx}\left(\frac{x}{x+1}\right)^3 = 3\left(\frac{x}{x+1}\right)^2\left[\frac{d}{dx}\left(\frac{x}{x+1}\right)\right] \qquad \text{The Power Rule.}$$

$$= 3\left(\frac{x}{x+1}\right)^2\left[\frac{(x+1)\frac{d}{dx}(x) - x\frac{dy}{dx}(x+1)}{(x+1)^2}\right] \qquad \text{The derivative of a quotient.}$$

$$= 3\left(\frac{x}{x+1}\right)^2\left[\frac{(x+1)(1) - x(1)}{(x+1)^2}\right] \qquad \text{Differentiate.}$$

$$= 3\left(\frac{x}{x+1}\right)^2\left[\frac{1}{(x+1)^2}\right] = \frac{3x^2}{(x+1)^4} \qquad \text{Simplify.}$$

21. Here the function f is the quotient of two functions, the numerator of which is raised to a power. We will first use the formula for the derivative of a quotient and then use the power rule when differentiating the numerator.

$$f'(x) = \frac{d}{dx}\left[\frac{(2x+1)^4}{3x^2}\right] = \frac{3x^2\frac{d}{dx}(2x+1)^4 - (2x+1)^4\frac{d}{dx}(3x^2)}{(3x^2)^2}$$

$$= \frac{3x^2\left[4(2x+1)^3\frac{d}{dx}(2x+1)\right] - (2x+1)^4\frac{d}{dx}(3x^2)}{(3x^2)^2} \qquad \text{The Power Rule.}$$

$$= \frac{3x^2 \left[4(2x+1)^3 (2) \right] - (2x+1)^4 (6x)}{(3x^2)^2}$$ Differentiate.

$$= \frac{24x^2 (2x+1)^3 - 6x(2x+1)^4}{9x^4}$$ Simplify.

$$= \frac{6x(2x+1)^3 \left[4x - (2x+1) \right]}{9x^4}$$ Factor.

$$= \frac{\cancel{6}^2 \cancel{x} (2x+1)^3 (2x-1)}{\cancel{9}^3 x^{\cancel{4}3}}$$ Simplify.

$$= \frac{2(2x+1)^3 (2x-1)}{3x^3}$$ Cancel.

23. Here the function f is the quotient of two functions, the numerator of which is raised to a power. We will first use the formula for the derivative of a quotient and then use the power rule when differentiating the numerator.

$$f'(x) = \frac{d}{dx} \left[\frac{(x^2+1)^3}{x} \right] = \frac{x \dfrac{d}{dx}(x^2+1)^3 - (x^2+1)^3 \dfrac{d}{dx} x}{x^2}$$

$$= \frac{x \left[3(x^2+1)^2 \dfrac{d}{dx}(x^2+1) \right] - (x^2+1)^3 \dfrac{d}{dx} x}{x^2}$$ The Power Rule.

$$= \frac{x \left[3(x^2+1)^2 (2x) \right] - (x^2+1)^3 (1)}{x^2}$$ Differentiate.

$$= \frac{6x^2 (x^2+1)^2 - (x^2+1)^3}{x^2}$$ Simplify.

$$= \frac{(x^2+1)^2 \left[6x^2 - (x^2+1) \right]}{x^2}$$ Factor.

$$= \frac{(x^2+1)^2 (5x^2-1)}{x^2}$$ Simplify.

25. We rewrite f as $f(x) = (x + x^{-1})^3$ and use the Power Rule.

$$f'(x) = \frac{d}{dx} f(x) = \frac{d}{dx}(x + x^{-1})^3 = 3(x + x^{-1})^2 \frac{d}{dx}(x + x^{-1}) = 3(x + x^{-1})^2 (1 - x^{-2})$$

          ↑   ↑

    Use the Power Rule.   Differentiate.

$$= 3 \left(x + \frac{1}{x} \right)^2 \left(1 - \frac{1}{x^2} \right) = 3 \left(\frac{x^2+1}{x} \right)^2 \left(\frac{x^2-1}{x^2} \right) = \frac{3(x^2+1)^2 (x^2-1)}{x^4}$$

    ↑   ↑   ↑

Rewrite with positive   Write with a single   Simplify.
   exponents.   denominator.

27. Here the function f is the quotient of two functions, the denominator of which is raised to a power. We will first use the formula for the derivative of a quotient and then use the Power Rule when differentiating the denominator.

$$f'(x) = \frac{d}{dx}\frac{3x^2}{(x^2+1)^2} = \frac{(x^2+1)^2\frac{d}{dx}(3x^2) - 3x^2\frac{d}{dx}(x^2+1)^2}{\left[(x^2+1)^2\right]^2}$$ The derivative of a quotient.

$$= \frac{(x^2+1)^2\frac{d}{dx}(3x^2) - 3x^2 \cdot 2(x^2+1)\frac{d}{dx}(x^2+1)}{(x^2+1)^4}$$ Use the Power Rule.

$$= \frac{(x^2+1)^2(6x) - 3x^2 \cdot 2(x^2+1)(2x)}{(x^2+1)^4}$$ Differentiate.

$$= \frac{(x^2+1)^2(6x) - 12x^3(x^2+1)}{(x^2+1)^4}$$ Simplify.

$$= \frac{6x(x^2+1)\left[(x^2+1) - 2x^2\right]}{(x^2+1)^{\cancel{4}3}}$$ Factor.

$$= \frac{6x(1-x^2)}{(x^2+1)^3}$$ Simplify.

29. The rate at which the car is depreciating is the derivative $V'(t)$.

$$V'(t) = \frac{d}{dt}\frac{29{,}000}{1+0.4t+0.1t^2} = \frac{(1+0.4t+0.1t^2)\frac{d}{dt}29{,}000 - 29{,}000\frac{d}{dt}(1+0.4t+0.1t^2)}{(1+0.4t+0.1t^2)^2}$$

$$= \frac{-29{,}000(0.4+0.2t)}{(1+0.4t+0.1t^2)^2}$$

(a) The rate of depreciation 1 year after purchase is $V'(1)$

$$V'(1) = \frac{-29{,}000(0.4+0.2)}{(1+0.4+0.1)^2} = \frac{-29{,}000(0.6)}{1.5^2} = -7733.33$$

The car is depreciating at a rate of $7733.33 per year when it is one year old.

(b) The rate of depreciation 2 years after purchase is $V'(2)$

$$V'(2) = \frac{-29{,}000(0.4+0.2 \cdot 2)}{(1+0.4 \cdot 2+0.1 \cdot 2^2)^2} = \frac{-29{,}000(0.8)}{2.2^2} = -4793.39$$

The car is depreciating at a rate of $4793.39 per year when it is two years old.

(c) The rate of depreciation 3 years after purchase is $V'(3)$

$$V'(3) = \frac{-29,000(0.4 + 0.2 \cdot 3)}{\left(1 + 0.4 \cdot 3 + 0.1 \cdot 3^2\right)^2} = \frac{-29,000(1.0)}{3.1^2} = -3017.69$$

The car is depreciating at a rate of $3017.69 per year when it is three years old.

(d) The rate of depreciation 4 years after purchase is $V'(4)$

$$V'(4) = \frac{-29,000(0.4 + 0.2 \cdot 4)}{\left(1 + 0.4 \cdot 4 + 0.1 \cdot 4^2\right)^2} = \frac{-29,000(1.2)}{4.2^2} = -1972.79$$

The car is depreciating at a rate of $1972.79 per year when it is four years old.

31. (a) The rate of change is given by the derivative.

$$\frac{dp}{dx} = \frac{d}{dx}\left(\frac{10,000}{5x+100} - 5\right) = \frac{d}{dx}\frac{10,000}{5x+100} - \frac{d}{dx}5 = \frac{d}{dx}\left[10,000\,(5x+100)^{-1}\right]$$

          ↑   ↑

    Derivative of a difference. $\dfrac{d}{dx}5 = 0$

$$= 10,000 \cdot \frac{d}{dx}(5x+100)^{-1} = 10,000 \cdot (-1)(5x+100)^{-2}(5) = -\frac{50,000}{(5x+100)^2} = -\frac{2000}{(x+20)^2}$$

↑   ↑   ↑

$f'(cx) = cf'(x)$ Differentiate.   Simplify.

(b) The revenue function $R = px$.

$$R = R(x) = \left(\frac{10,000}{5x+100} - 5\right)x = \frac{10,000x}{5x+100} - 5x$$

(c) The marginal revenue is the derivative $R'(x)$.

$$R'(x) = \frac{d}{dx}\left(\frac{10,000x}{5x+100} - 5x\right) = \frac{d}{dx}\frac{10,000x}{5x+100} - \frac{d}{dx}(5x)$$

$$= \frac{(5x+100)(10,000) - 10,000x(5)}{(5x+100)^2} - 5 = \frac{1,000,000}{(5x+100)^2} - 5 = \frac{40,000}{(x+20)^2} - 5$$

(d) $R'(10) = \dfrac{1,000,000}{(5 \cdot 10 + 100)^2} - 5 = \dfrac{1,000,000}{150^2} - 5 = 39.44$ dollars

$$R'(40) = \frac{1,000,000}{(5 \cdot 40 + 100)^2} - 5 = \frac{1,000,000}{300^2} - 5 = 6.11 \text{ dollars}$$

33. (a) The average rate of change in the mass of the protein is

$$\frac{\Delta M}{\Delta t} = \frac{M(2) - M(0)}{2 - 0} = \frac{\dfrac{28}{2+2} - \dfrac{28}{0+2}}{2} = \frac{7 - 14}{2} = -\frac{7}{2}$$

grams per hour.

(b) $M'(t) = \dfrac{d}{dt}\dfrac{28}{t+2} = 28\dfrac{d}{dt}(t+2)^{-1} = 28\cdot(-1)(t+2)^{-2}\cdot 1 = -\dfrac{28}{(t+2)^2}$

$\quad M'(0) = -\dfrac{28}{(0+2)^2} = -\dfrac{28}{4} = -7$

4.5 The Derivatives of the Exponential and Logarithmic Functions; The Chain Rule

1. $f'(x) = \dfrac{d}{dx}(x^3 - e^x) = \dfrac{d}{dx}x^3 - \dfrac{d}{dx}e^x = 3x^2 - e^x$

3. Using the formula for the derivative of a product,

$$f'(x) = \dfrac{d}{dx}(x^2 e^x) = x^2\dfrac{d}{dx}e^x + e^x\dfrac{d}{dx}x^2 = x^2 e^x + 2xe^x = x(x+2)e^x$$

5. Using the formula for the derivative of a quotient,

$$f'(x) = \dfrac{d}{dx}\dfrac{e^x}{x^2} = \dfrac{x^2\dfrac{d}{dx}e^x - e^x\dfrac{d}{dx}x^2}{(x^2)^2} = \dfrac{x^2 e^x - e^x\cdot 2x}{x^4} = \dfrac{x(x-2)e^x}{x^4} = \dfrac{(x-2)e^x}{x^3}$$

$$\qquad\qquad\qquad\qquad\uparrow\qquad\qquad\qquad\uparrow\qquad\qquad\uparrow$$
$$\qquad\qquad\qquad\text{Differentiate.}\qquad\text{Factor.}\qquad\text{Simplify.}$$

7. Using the formula for the derivative of a quotient,

$$f'(x) = \dfrac{d}{dx}\dfrac{4x^2}{e^x} = \dfrac{e^x\dfrac{d}{dx}(4x^2) - 4x^2\dfrac{d}{dx}e^x}{(e^x)^2} = \dfrac{e^x\cdot 8x - 4x^2 e^x}{e^{2x}} = \dfrac{4x(2-x)e^x}{e^{2x}} = \dfrac{4x(2-x)}{e^x}$$

$$\qquad\qquad\qquad\qquad\uparrow\qquad\qquad\qquad\uparrow\qquad\qquad\uparrow$$
$$\qquad\qquad\qquad\text{Differentiate.}\qquad\text{Factor.}\qquad\text{Simplify.}$$

$$\qquad\quad = \dfrac{8x - 4x^2}{e^x}$$

9. $y = f(u) = u^5 = (x^3 + 1)^5 = f(x)$

$$\dfrac{dy}{dx} = \dfrac{d}{dx}(x^3 + 1)^5 = 5(x^3 + 1)^4\dfrac{d}{dx}(x^3 + 1) = 5(x^3 + 1)^4(3x^2) = 15x^2(x^3 + 1)^4$$

$$\qquad\qquad\uparrow\qquad\qquad\qquad\uparrow\qquad\qquad\qquad\uparrow$$
$$\qquad\text{Chain Rule.}\qquad\text{Differentiate.}\qquad\text{Simplify.}$$

11. $y = f(u) = \dfrac{u}{u+1} = \dfrac{x^2 + 1}{(x^2 + 1) + 1} = \dfrac{x^2 + 1}{x^2 + 2} = f(x)$

$$\frac{dy}{dx} = \frac{d}{dx}\frac{x^2+1}{x^2+2} = \frac{(x^2+2)\dfrac{d}{dx}(x^2+1)-(x^2+1)\dfrac{d}{dx}(x^2+2)}{(x^2+2)^2}$$

$\uparrow$
Derivative of a quotient

$$= \frac{(x^2+2)(2x)-(x^2+1)(2x)}{(x^2+2)^2} = \frac{2x\big[(x^2+2)-(x^2+1)\big]}{(x^2+2)^2} = \frac{2x}{(x^2+2)^2}$$

$\qquad\qquad\uparrow$ $\qquad\qquad\qquad\qquad\uparrow$ $\qquad\qquad\qquad\uparrow$
Differentiate. $\qquad\qquad\qquad\qquad$ Factor. $\qquad\qquad\qquad$ Simplify.

13. $y = f(u) = (u+1)^2 = \left(\dfrac{1}{x}+1\right)^2 = (x^{-1}+1)^2 = f(x)$

$$\frac{dy}{dx} = \frac{d}{dx}(x^{-1}+1)^2 = 2(x^{-1}+1)\frac{d}{dx}(x^{-1}+1) = 2(x^{-1}+1)\cdot(-1)x^{-2} = \frac{-2\left(\dfrac{1}{x}+1\right)}{x^2} = -\frac{2(1+x)}{x^3}$$

$\qquad\qquad\uparrow$ $\qquad\qquad\qquad\uparrow$ $\qquad\qquad\qquad\uparrow$ $\qquad\uparrow$
Use the Chain Rule. $\quad$ Differentiate. $\quad$ Write with positive $\quad$ Simplify.
$\qquad\qquad\qquad\qquad\qquad\qquad\qquad\qquad\qquad\qquad$ exponents.

15. $y = f(u) = (u^3-1)^5 = \big[(x^{-2})^3-1\big]^5 = (x^{-6}-1)^5 = f(x)$

$$\frac{dy}{dx} = \frac{d}{dx}(x^{-6}-1)^5 = 5(x^{-6}-1)^4\frac{d}{dx}(x^{-6}-1) = 5(x^{-6}-1)^4\cdot(-6)x^{-7}$$

$\qquad\qquad\uparrow$ $\qquad\qquad\qquad\uparrow$
Use the Chain Rule. $\qquad$ Differentiate.

$$= \frac{-30(x^{-6}-1)^4}{x^7} = -\frac{30\left(\dfrac{1}{x^6}-1\right)^4}{x^7} = -\frac{30\left(\dfrac{1-x^6}{x^6}\right)^4}{x^7} = -\frac{30(1-x^6)^4}{(x^6)^4\,x^7} = -\frac{30(1-x^6)^4}{x^{31}}$$

$\quad\uparrow$ $\qquad\qquad\uparrow$ $\qquad\qquad\uparrow$ $\qquad\qquad\qquad\qquad\uparrow$
Simplify. $\quad$ Write with positive $\quad$ Write the numerator $\qquad\qquad$ Simplify.
$\qquad\qquad$ exponent. $\qquad\quad$ as a single quotient.

17. $y = f(u) = u^3 = (e^x)^3 = e^{3x} = f(x)$

$$\frac{dy}{dx} = \frac{d}{dx}e^{3x} = e^{3x}\cdot\frac{d}{dx}(3x) = e^{3x}\cdot 3 = 3e^{3x}$$

$\qquad\uparrow$ $\qquad\uparrow$ $\qquad\qquad\uparrow$
Chain Rule $\quad$ Differentiate. $\quad$ Simplify.

19. $y = f(u) = e^u = e^{x^3} = f(x)$

$$\frac{dy}{dx} = \frac{d}{dx}e^{x^3} = e^{x^3}\cdot\frac{d}{dx}x^3 = e^{x^3}\cdot 3x^2 = 3x^2 e^{x^3}$$

$\qquad\uparrow$ $\qquad\qquad\uparrow$
Chain Rule $\quad$ Differentiate.

21. (a) Using the Chain Rule, $y = (x^3 + 1)^2$ is thought of as $y = u^2$ and $u = x^3 + 1$.

$$\frac{dy}{du} = 2u \quad \text{and} \quad \frac{du}{dx} = 3x^2$$

$$\frac{dy}{dx} = \frac{dy}{du} \cdot \frac{du}{dx} = 2u \cdot 3x^2 = 2(x^3 + 1) \cdot 3x^2 = 6x^2(x^3 + 1) = 6x^5 + 6x^2$$

$$\uparrow \qquad\qquad\qquad \uparrow$$

Substitute $u = x^3 + 1$ \qquad Simplify.

(b) Using the Power Rule,

$$\frac{dy}{dx} = \frac{d}{dx}(x^3 + 1)^2 = 2(x^3 + 1) \cdot \frac{d}{dx}(x^3 + 1) = 2(x^3 + 1) \cdot (3x^2) = 6x^2(x^3 + 1) = 6x^5 + 6x^2$$

$$\uparrow \qquad\qquad\qquad \uparrow \qquad\qquad\qquad \uparrow$$

Power Rule \qquad\qquad Differentiate. \qquad\qquad Simplify.

(c) Expanding, $y = (x^3 + 1)^2 = x^6 + 2x^3 + 1$, and

$$\frac{dy}{dx} = \frac{d}{dx}(x^6 + 2x^3 + 1) = \frac{d}{dx}x^6 + 2\frac{d}{dx}x^3 + \frac{d}{dx}1 = 6x^5 + 6x^2 = 6x^2(x^3 + 1)$$

$$\uparrow \qquad\qquad \uparrow$$

Differentiate. \qquad Factor.

23. $f'(x) = \dfrac{d}{dx}e^{5x} = e^{5x} \cdot \dfrac{d}{dx}(5x) = e^{5x} \cdot 5 = 5e^{5x}$

25. $f'(x) = \dfrac{d}{dx}8e^{-x^2} = 8 \cdot \dfrac{d}{dx}e^{-x^2} = 8e^{-x^2}\dfrac{d}{dx}(-x^2) = 8e^{-x^2} \cdot (-2x) = -16xe^{-x^2}$

27. $f'(x) = \dfrac{d}{dx}x^2 e^{x^2} = x^2\dfrac{d}{dx}e^{x^2} + e^{x^2}\dfrac{d}{dx}x^2 = x^2\left(e^{x^2}\dfrac{d}{dx}x^2\right) + e^{x^2} \cdot 2x$

$$\uparrow \qquad\qquad\qquad \uparrow$$

The derivative of a product \qquad The Chain Rule

$$= x^2 e^{x^2} \cdot 2x + e^{x^2} \cdot 2x = e^{x^2}(2x^3 + 2x) = 2xe^{x^2}(x^2 + 1)$$

$$\uparrow \qquad\qquad\qquad \uparrow$$

Differentiate. \qquad\qquad Factor.

29. $f'(x) = \dfrac{d}{dx}\left[5(e^x)^3\right] = 5\dfrac{d}{dx}(e^x)^3 = 5 \cdot 3(e^x)^2 \cdot \dfrac{d}{dx}e^x = 15(e^x)^2 e^x = 15e^{3x}$

$$\uparrow \qquad\qquad\qquad \uparrow \qquad\qquad \uparrow$$

Use the Power Rule. \qquad Differentiate. \qquad Simplify.

31. $f(x) = \dfrac{x^2}{e^x} = x^2 e^{-x}$

$$f'(x) = \dfrac{d}{dx}\left(x^2 e^{-x}\right) = x^2\dfrac{d}{dx}e^{-x} + e^{-x}\dfrac{d}{dx}x^2 = x^2\left(e^{-x} \cdot \dfrac{d}{dx}(-x)\right) + e^{-x} \cdot 2x$$

$$\uparrow \qquad\qquad\qquad \uparrow$$

Derivative of a product \qquad\qquad Chain Rule

$$= x^2 \cdot e^{-x} \cdot (-1) + 2x e^{-x} = -x^2 e^{-x} + 2x e^{-x} = x e^{-x}(2-x) = \frac{2x - x^2}{e^x}$$

↑   ↑   ↑

Differentiate.   Simplify.   Factor.

33. $f(x) = \dfrac{\left(e^x\right)^2}{x} = x^{-1} e^{2x}$

$$f'(x) = \frac{d}{dx}\left(x^{-1} e^{2x}\right) = x^{-1} \frac{d}{dx} e^{2x} + e^{2x} \frac{d}{dx} x^{-1} = x^{-1}\left(e^{2x} \frac{d}{dx}(2x)\right) + e^{2x} \cdot (-1) x^{-2}$$

↑   ↑

The derivative of a product   Apply the Chain Rule.

$$= 2x^{-1} e^{2x} - x^{-2} e^{2x} = e^{2x}\left(2x^{-1} - x^{-2}\right) = \frac{e^{2x}(2x-1)}{x^2}$$

↑   ↑   ↑

Differentiate.   Factor.   Simplify.

35. $f'(x) = \dfrac{d}{dx}\left(x^2 - 3\ln x\right) = \dfrac{d}{dx} x^2 - 3\dfrac{d}{dx}\ln x = 2x - 3 \cdot \dfrac{1}{x} = 2x - \dfrac{3}{x}$

↑   ↑   ↑

The derivative of a difference   Differentiate.   Simplify.

37. $f'(x) = \dfrac{d}{dx}\left(x^2 \ln x\right) = x^2 \dfrac{d}{dx}\ln x + \ln x \cdot \dfrac{d}{dx} x^2 = x^2 \cdot \dfrac{1}{x} + \ln x \cdot 2x$

↑   ↑

The derivative of a product   Differentiate.

$$= x + 2x \ln x = x(1 + 2\ln x) = x\left(1 + \ln x^2\right)$$

↑   ↑   ↑

Simplify.   Factor.   Alternate form of the answer

39. $f'(x) = \dfrac{d}{dx}\left[3\ln(5x)\right] = 3\dfrac{d}{dx}\ln(5x) = 3 \cdot \dfrac{\dfrac{d}{dx}(5x)}{5x} = 3 \cdot \dfrac{5}{5x} = \dfrac{3}{x}$

↑

Derivative of $\ln g(x)$

41. Here the f is the product of two functions, we use the product rule first and then the Chain Rule when differentiating $\ln(x^2 + 1)$.

$$f'(x) = \frac{d}{dx}\left[x \ln\left(x^2 + 1\right)\right] = x \cdot \frac{d}{dx}\ln\left(x^2 + 1\right) + \ln\left(x^2 + 1\right) \cdot \frac{d}{dx} x \qquad \text{The derivative of a product.}$$

$$= x \cdot \frac{\dfrac{d}{dx}\left(x^2 + 1\right)}{x^2 + 1} + \ln\left(x^2 + 1\right) \cdot 1 \qquad \text{Use the Chain Rule; } \frac{d}{dx} x = 1.$$

$$= x \cdot \frac{2x}{x^2 + 1} + \ln\left(x^2 + 1\right) \qquad \text{Differentiate.}$$

$$= \frac{2x^2}{x^2+1} + \ln(x^2+1) \qquad \text{Simplify.}$$

43. $f'(x) = \dfrac{d}{dx}\left[x + 8\ln(3x)\right] = \dfrac{d}{dx}x + 8\dfrac{d}{dx}\ln(3x) = \dfrac{d}{dx}x + 8 \cdot \dfrac{\dfrac{d}{dx}(3x)}{3x}$

$$\underset{\substack{\uparrow \\ \text{The derivative of a sum}}}{} \qquad \underset{\substack{\uparrow \\ \text{The Chain Rule}}}{}$$

$$= 1 + 8 \cdot \frac{3}{3x} = 1 + \frac{8}{x} = \frac{x+8}{x}$$

$$\underset{\substack{\uparrow \\ \text{Differentiate.}}}{} \qquad \underset{\substack{\uparrow \\ \text{Simplify.}}}{} \quad \underset{\substack{\uparrow \\ \text{Alternate form of the answer.}}}{}$$

45. $f'(x) = \dfrac{d}{dx}\left[8(\ln x)^3\right] = 8\dfrac{d}{dx}(\ln x)^3 = 8 \cdot 3(\ln x)^2 \cdot \dfrac{d}{dx}\ln x = 24\,(\ln x)^2 \cdot \dfrac{1}{x} = \dfrac{24(\ln x)^2}{x}$

$$\underset{\substack{\uparrow \\ \text{Use the Power Rule.}}}{} \qquad \underset{\substack{\uparrow \\ \text{Differentiate.}}}{} \qquad \underset{\substack{\uparrow \\ \text{Simplify.}}}{} \qquad \bullet$$

47. $f'(x) = \dfrac{d}{dx}\log_3 x = \dfrac{1}{x\ln 3} \qquad\qquad \dfrac{d}{dx}\log_a x = \dfrac{1}{x\ln a}$

49. $f'(x) = \dfrac{d}{dx}\left(x^2\log_2 x\right) = x^2 \cdot \dfrac{d}{dx}\log_2 x + \log_2 x \cdot \dfrac{d}{dx}x^2$

$$= x^2 \cdot \frac{1}{x\ln 2} + \log_2 x \cdot 2x$$

$$= \frac{x}{\ln 2} + 2x\log_2 x = \frac{x}{\ln 2} + 2x\frac{\ln x}{\ln 2} \qquad \text{Use the Change of Base Formula.}$$

$$= \frac{x + 2x\ln x}{\ln 2}$$

51. $f'(x) = \dfrac{d}{dx}3^x = 3^x\ln 3 \qquad\qquad \dfrac{d}{dx}a^x = a^x\ln a$

53. $f'(x) = \dfrac{d}{dx}\left(x^2 \cdot 2^x\right) = x^2 \cdot \dfrac{d}{dx}2^x + 2^x \cdot \dfrac{d}{dx}x^2 = x^2 \cdot 2^x\ln 2 + 2^x \cdot 2x = 2^x\left(x^2\ln 2 + 2x\right)$

$$\underset{\substack{\uparrow \\ \text{The derivative of a product}}}{} \qquad \underset{\substack{\uparrow \\ \text{Differentiate;}\ \dfrac{d}{dx}a^x = a^x\ln a\,.}}{} \qquad \underset{\substack{\uparrow \\ \text{Factor.}}}{}$$

55. The slope of the tangent line to the graph of $f(x) = e^{3x}$ at the point $(0, 1)$ is the derivative of the function f evaluated at the point $(0, 1)$. The derivative of f is
$$f'(x) = 3e^{3x}$$
The slope of the tangent line is $m_{\tan} = f'(0) = 3 \cdot e^0 = 3.$

An equation of the tangent line is $y - 1 = 3(x - 0)$ or $y = 3x + 1$.

57. The slope of the tangent line to the graph of $f(x) = \ln x$ at the point $(1, 0)$ is the derivative of the function f evaluated at the point $(1, 0)$. The derivative of f is
$$f'(x) = \frac{1}{x}$$
The slope of the tangent line is $m_{\tan} = f'(1) = 1$.

An equation of the tangent line is $y - 0 = 1(x - 1)$ or $y = x - 1$.

59. The slope of the tangent line to the graph of $f(x) = e^{3x-2}$ at the point $\left(\frac{2}{3}, 1\right)$ is the

derivative of the function f evaluated at the point $\left(\frac{2}{3}, 1\right)$. The derivative of f is
$$f'(x) = 3e^{3x-2}$$

The slope of the tangent line is $m_{\tan} = 3e^{3 \cdot 2/3 - 2} = 3e^{0} = 3$

An equation of the tangent line is $y - 1 = 3\left(x - \frac{2}{3}\right)$ or $y = 3x - 1$.

61. The slope of the tangent line to the graph of $f(x) = x \ln x$ at the point $(1, 0)$ is the derivative of the function f evaluated at the point $(1, 0)$. The derivative of f is
$$f'(x) = x \cdot \frac{d}{dx} \ln x + \ln x \cdot \frac{d}{dx} x = x \cdot \frac{1}{x} + \ln x \cdot 1 = 1 + \ln x$$

    ↑   ↑   ↑

The derivative of a product   Differentiate.   Simplify.

The slope of the tangent line is $m_{\tan} = f'(1) = 1 + \ln 1 = 1 + 0 = 1$

An equation of the tangent line is $y - 0 = 1(x - 1)$ or $y = x - 1$.

63. Parallel lines have the same slope. Since the slope of the line $y = x$ is 1, the slope of the tangent line we seek is also 1. Moreover, the slope of a tangent line is given by the derivative of the function. So we need $m_{\tan} = f'(x) = 1$.
    $f'(x) = e^{x} = 1$ when $x = 0$, and
    $f(0) = e^{0} = 1$
which means an equation of a tangent line to the function is $y - 1 = 1(x - 0)$ or $y = x + 1$.

65. (a) The reaction rate for a dose of 5 units is given by the derivative of R, $R'(x)$ evaluated at $x = 5$ units.
$$R'(x) = \frac{d}{dx}(5.5 \ln x + 10) = \frac{5.5}{x}$$
$$R'(5) = \frac{5.5}{5} = 1.1$$

(b) The reaction rate for a dose of 10 units is given by $R'(10)$.
$$R'(10) = \frac{5.5}{10} = 0.55$$

67. The rate of change in atmospheric pressure is given by the derivative of the atmospheric pressure P.
$$P'(x) = \frac{d}{dx}\left(10^4 e^{-0.00012x}\right) = 10^4 \frac{d}{dx} e^{-0.00012x} = 10^4 \cdot (-0.00012) e^{-0.00012x} = -1.2 e^{-0.00012x}$$
The rate of change of atmospheric pressure at $x = 500$ meters is
$$P'(500) = -1.2 e^{(-0.00012)(500)} = -1.2 e^{-0.060} = -1.130 \text{ kilograms per square meter.}$$

The rate of change of atmospheric pressure at $x = 700$ meters is
$$P'(700) = -1.2 e^{(-0.00012)(700)} = -1.2 e^{-0.060} = -1.103 \text{ kilograms per square meter.}$$

69. (a) The rate of change of A with respect to time is the derivative $A'(t)$.
$$A'(t) = \frac{d}{dx}\left(102 - 90 e^{-0.21t}\right) = \frac{d}{dx}102 - 90 \cdot \frac{d}{dx} e^{-0.21t} = -90 \cdot (-0.21) e^{-0.21t} = 18.9 e^{-0.21t}$$

(b) At $t = 5$, $A'(t) = A'(5) = 18.9 e^{-0.21(5)} = 18.9 e^{-1.05} = 6.614$

(c) At $t = 10$, $A'(t) = A'(10) = 18.9 e^{-0.21(10)} = 18.9 e^{-2.1} = 2.314$

(d) At $t = 30$, $A'(t) = A'(30) = 18.9 e^{-0.21(30)} = 18.9 e^{-6.3} = 0.035$

71. (a) The rate of change of S with respect to x is the derivative $S'(x)$.
$$S'(x) = \frac{d}{dx}(100,000 + 400,000 \ln x) = \frac{d}{dx}100,000 + 400,000 \frac{d}{dx}\ln x$$
$$= 400,000 \cdot \frac{1}{x} = \frac{400,000}{x}$$

(b) $S'(10) = \dfrac{400,000}{10} = 40,000$

(c) $S'(20) = \dfrac{400,000}{20} = 20,000$

73. (a) For $x = 1000$, the price p is
$$p = 50 - 4\ln\left(\frac{1000}{100} + 1\right) = 50 - 4\ln 11 = \$40.41$$

(b) For $x = 5000$, the price p is
$$p = 50 - 4\ln\left(\frac{5000}{100} + 1\right) = 50 - 4\ln 51 = \$34.27$$

(c) The marginal demand is the derivative of the function p.

$$\frac{d}{dx}p = \frac{d}{dx}\left[50 - 4\ln\left(\frac{x}{100}+1\right)\right] = \frac{d}{dx}50 - 4\frac{d}{dx}\ln\left(\frac{x}{100}+1\right)$$

↑
The derivative of a difference.

$$= \frac{d}{dx}50 - 4\frac{d}{dx}\ln\left(\frac{x+100}{100}\right) = -4\frac{d}{dx}\left[\ln(x+100)-\ln100\right]$$

↑   ↑

Write $\frac{x}{100}+1$ as a single fraction.   The logarithm of a quotient is the difference of the logarithms.

$$= -4\left[\frac{d}{dx}\ln(x+100) - \frac{d}{dx}\ln100\right] = -4\cdot\frac{1}{x+100} = -\frac{4}{x+100}$$

↑   ↑   ↑
The derivative of a difference.   Differentiate.   Simplify.

The marginal demand for 1000 t-shirts is

$$p'(1000) = -\frac{4}{1000+100} = -\frac{4}{1100} = -0.0036$$

(d) The marginal demand for 5000 t-shirts is

$$p'(5000) = -\frac{4}{5000+100} = -\frac{4}{5100} = -0.00078$$

(e) The revenue function $R(x) = p\cdot x = \left[50 - 4\ln\left(\frac{x}{100}+1\right)\right]\cdot x = 50x - 4x\ln\left(\frac{x}{100}+1\right)$.

(f) The marginal revenue is the derivative of the function R.

$$R'(x) = \frac{d}{dx}\left[50x - 4x\ln\left(\frac{x}{100}+1\right)\right] = \frac{d}{dx}(50x) - \frac{d}{dx}\left[4x\ln\left(\frac{x}{100}+1\right)\right]$$

$$= 50\frac{d}{dx}x - 4\left[x\frac{d}{dx}\ln\left(\frac{x}{100}+1\right) + \ln\left(\frac{x}{100}+1\right)\frac{d}{dx}x\right]$$

$$= 50 - 4\left[x\cdot\frac{1}{x+100} + \ln\left(\frac{x}{100}+1\right)\right] = 50 - 4\left[\frac{x}{x+100} + \ln\left(\frac{x}{100}+1\right)\right]$$

When $x = 1000$ t-shirts are sold the marginal revenue is

$$R'(1000) = 50 - 4\left[\frac{1000}{1000+100} + \ln\left(\frac{1000}{100}+1\right)\right] = 50 - 4\left[\frac{10}{11} + \ln11\right] = \$36.77$$

(g) When $x = 5000$ t-shirts are sold the marginal revenue is

$$R'(5000) = 50 - 4\left[\frac{5000}{5000+100} + \ln\left(\frac{5000}{100}+1\right)\right] = 50 - 4\left[\frac{50}{51} + \ln51\right] = \$30.35$$

(h) Profit is the difference between revenue and cost.

$$P(x) = R(x) - C(x)$$

$$= \left[50x - 4x\ln\left(\frac{x}{100} + 1\right) \right] - [4x] = 46x - 4x\ln\left(\frac{x}{100} + 1\right)$$

(i) If 1000 t-shirts are sold the profit is

$$P(1000) = 46(1000) - 4(1000)\ln\left(\frac{1000}{100} + 1\right) = 46{,}000 - 4000\ln 11 = \$36{,}408.42$$

(j) If 5000 t-shirts are sold the profit is

$$P(5000) = 46(5000) - 4(5000)\ln\left(\frac{5000}{100} + 1\right) = 230{,}000 - 20{,}000\ln 51 = \$151{,}363.49$$

(k) To use TABLE to find the quantity x that maximizes profit, we enter the profit function into Y_1, then in TBLSET we select a large value for x, choose ΔTbl, and select the Auto option.

X	Y1
3.1E6	1.44E7
3.2E6	1.44E7
3.3E6	1.45E7
3.4E6	1.45E7
3.5E6	1.45E7
3.6E6	**1.45E7**
3.7E6	1.45E7

Y1=14525251.2745

We chose $x = 3{,}000{,}000$ and ΔTbl $= 1000$. Then using TABLE, we increased x until the profit function stopped increasing and began decreasing in magnitude. The quantity (to the nearest thousand) that maximizes profit is 3,632,000 t-shirts.
The maximum profit is $14,525,251.

(l) $p(3{,}632{,}000) = p = 50 - 4\ln\left(\dfrac{3{,}632{,}000}{100} + 1\right) = 50 - 4\ln(36{,}321) = 8.00$

To maximize profit, the t-shirts should be sold for $8.00 each.

75. (a) The rate of change of p with respect to t is the derivative of p.

$$\frac{d}{dt}p = \frac{d}{dt}(0.470 + 0.026\ln t) = \frac{d}{dt}0.470 + 0.026\frac{d}{dt}\ln t = 0.026 \cdot \frac{1}{t} = \frac{0.026}{t}$$

(b) In 2002 the rate of change of p was

$$p'(5) = \frac{0.026}{5} = 0.052$$

(c) In 2007 the rate of change of p was

$$p'(10) = \frac{0.026}{10} = 0.026$$

77. Prove: $\dfrac{d}{dx}\ln g(x) = \dfrac{\dfrac{d}{dx}g(x)}{g(x)} = \dfrac{g'(x)}{g(x)}$

Proof: Let $y = f(u) = \ln u$ and $u = g(x)$.

Then $y' = f'(u) = \dfrac{dy}{du} = \dfrac{d}{du} \ln u = \dfrac{1}{u}$ and $u' = \dfrac{du}{dx} = g'(x)$

According to the Chain Rule if both f and g are differentiable functions, then

$$\frac{dy}{dx} = \frac{dy}{du} \cdot \frac{du}{dx} = \frac{1}{u} g'(x) = \frac{1}{g(x)} \cdot g'(x) = \frac{g'(x)}{g(x)}$$

4.6 Higher-Order Derivatives

1. $f(x) = 2x + 5$
$f'(x) = 2$
$f''(x) = 0$

3. $f(x) = 3x^2 + x - 2$
$f'(x) = 6x + 1$
$f''(x) = 6$

5. $f(x) = -3x^4 + 2x^2$
$f'(x) = -12x^3 + 4x$
$f''(x) = -36x^2 + 4$

7. $f(x) = \dfrac{1}{x} = x^{-1}$

$f'(x) = -x^{-2} = -\dfrac{1}{x^2}$

$f''(x) = 2x^{-3} = \dfrac{2}{x^3}$

9. $f(x) = x + \dfrac{1}{x} = x + x^{-1}$

$f'(x) = 1 - x^{-2} = 1 - \dfrac{1}{x^2}$

$f''(x) = 2x^{-3} = \dfrac{2}{x^3}$

11. $f(x) = \dfrac{x}{x+1}$

$f'(x) = \dfrac{d}{dx}\left(\dfrac{x}{x+1}\right) = \dfrac{(x+1)\dfrac{d}{dx}x - x\dfrac{d}{dx}(x+1)}{(x+1)^2} = \dfrac{(x+1)\cdot 1 - x\cdot 1}{(x+1)^2} = \dfrac{1}{(x+1)^2} = (x+1)^{-2}$

$\qquad\qquad\uparrow\qquad\qquad\qquad\qquad\qquad\uparrow\qquad\qquad\qquad\quad\uparrow\qquad\quad\uparrow$

$\qquad$ The derivative of a quotient.$\qquad$ Differentiate.$\qquad\qquad$ Simplify.$\quad$ Write with a
$\qquad\qquad\qquad\qquad\qquad\qquad\qquad\qquad\qquad\qquad\qquad\qquad\qquad\qquad\qquad$ negative exponent.

$f''(x) = -2(x+1)^{-3}\dfrac{d}{dx}(x+1) = -2(x+1)^{-3}\cdot 1 = -\dfrac{2}{(x+1)^3}$

$\qquad\uparrow\qquad\qquad\qquad\qquad\qquad\qquad\uparrow$

$\quad$ Use the Power Rule.$\qquad\qquad$ Simplify.

13. $f(x) = e^x$
$f'(x) = e^x$
$f''(x) = e^x$

15. $f(x) = \left(x^2 + 4\right)^3$

$f'(x) = 3\left(x^2 + 4\right)^2 \dfrac{d}{dx}\left(x^2 + 4\right) = 3\left(x^2 + 4\right)^2 \cdot 2x = 6x\left(x^2 + 4\right)^2$

                    ↑   ↑
Use the Power Rule.   Simplify.

$$f''(x) = \dfrac{d}{dx}\left[6x\left(x^2 + 4\right)^2\right] = 6x\dfrac{d}{dx}\left(x^2 + 4\right)^2 + \left(x^2 + 4\right)^2\dfrac{d}{dx}(6x) \qquad \text{Derivative of a product.}$$

$$= 6x \cdot 2\left(x^2 + 4\right)\dfrac{d}{dx}\left(x^2 + 4\right) + \left(x^2 + 4\right)^2 \cdot 6 \qquad \text{Use the Power Rule.}$$

$$= 6x \cdot 2\left(x^2 + 4\right) \cdot 2x + \left(x^2 + 4\right)^2 \cdot 6 \qquad \text{Differentiate.}$$

$$= 24x^2\left(x^2 + 4\right) + 6\left(x^2 + 4\right)^2 \qquad \text{Simplify.}$$

$$= 6\left(x^2 + 4\right)\left(4x^2 + x^2 + 4\right) \qquad \text{Factor.}$$

$$= 6\left(x^2 + 4\right)\left(5x^2 + 4\right) \qquad \text{Simplify.}$$

17. $f(x) = \ln x$

$f'(x) = \dfrac{1}{x} = x^{-1}$

$f''(x) = -x^{-2} = -\dfrac{1}{x^2}$

19. $f(x) = xe^x$

$f'(x) = x\dfrac{d}{dx}e^x + e^x\dfrac{d}{dx}x = xe^x + e^x = e^x(x + 1)$

        ↑   ↑   ↑
Derivative of a product.   Differentiate. Factor.

$$f''(x) = \dfrac{d}{dx}\left(xe^x + e^x\right) = \dfrac{d}{dx}xe^x + \dfrac{d}{dx}e^x = xe^x + e^x + e^x = e^x(x + 2)$$

                         ↑   ↑   ↑
Derivative of a sum.   Differentiate; use $f'(x)$.   Factor.

21. $f(x) = (e^x)^2$

$f'(x) = 2\left(e^x\right)\dfrac{d}{dx}e^x = 2\left(e^x\right)^2 = 2e^{2x} \qquad \text{Use the Chain Rule.}$

$f''(x) = 4\left(e^x\right)\dfrac{d}{dx}e^x = 4\left(e^x\right)^2 = 4e^{2x} \qquad \text{Use the Chain Rule.}$

23. $f(x) = \dfrac{1}{\ln x} = (\ln x)^{-1}$

$$f'(x) = -\left(\ln x\right)^{-2}\frac{d}{dx}\ln x = -\left(\ln x\right)^{-2}\cdot\frac{1}{x} = -\frac{1}{x\left(\ln x\right)^2}$$

↑   ↑

Use the Chain Rule.   Simplify.

$$f''(x) = -\frac{x\left(\ln x\right)^2\frac{d}{dx}1 - \frac{d}{dx}\left[x\left(\ln x\right)^2\right]}{\left[x\left(\ln x\right)^2\right]^2} = \frac{x\frac{d}{dx}\left(\ln x\right)^2 + \left(\ln x\right)^2\frac{d}{dx}x}{x^2\left(\ln x\right)^4}$$

↑   ↑

The derivative of a quotient. $\frac{d}{dx}1 = 0$; the derivative of a product.

$$= \frac{2x\ln x\cdot\frac{1}{x} + \left(\ln x\right)^2}{x^2\left(\ln x\right)^4} = \frac{2 + \ln x}{x^2\left(\ln x\right)^3}$$

↑   ↑

Use the Power Rule.   Simplify.

25. (a) The function f is a polynomial, so the domain of f is all real numbers.

(b) $f'(x) = 2x$

(c) $f'(x)$ is a monomial, so the domain of f' is all real numbers.

(d) $f'(x) = 0$ when $2x = 0$, or when $x = 0$.

(e) There are no numbers in the domain of f for which $f'(x)$ does not exist.

(f) $f''(x) = 2$

(g) The domain of f'' is all real numbers.

27. (a) The function f is a polynomial, so the domain of f is all real numbers.

(b) $f'(x) = 3x^2 - 18x + 27$

(c) $f'(x)$ is a polynomial, so the domain of f' is all real numbers.

(d) $f'(x) = 0$ when $3x^2 - 18x + 27 = 0$, or when $x = 3$.

$$\begin{array}{ll} 3x^2 - 18x + 27 = 0 & f'(x) = 0 \\ x^2 - 6x + 9 = 0 & \text{Divide both sides by 3.} \\ (x - 3)^2 = 0 & \text{Factor.} \\ x = 3 & \text{Use the square root method.} \end{array}$$

(e) There are no numbers in the domain of f for which $f'(x)$ does not exist.

(f) $f''(x) = 6x - 18$

(g) The domain of f'' is all real numbers.

29. (a) The function f is a polynomial, so the domain of f is all real numbers.

(b) $f'(x) = 12x^3 - 36x^2$

(c) $f'(x)$ is a polynomial, so the domain of f' is all real numbers.

(d) $f'(x) = 0$ when $12x^3 - 36x^2 = 0$, or when $x = 0$ or $x = 3$.

$$\begin{array}{ll} 12x^3 - 36x^2 = 0 & f'(x) = 0 \\ x^3 - 3x^2 = 0 & \text{Divide both sides by 12.} \\ x^2(x-3)^2 = 0 & \text{Factor.} \\ x = 0 \quad \text{or} \quad x = 3 & \text{Use the square root method.} \end{array}$$

(e) There are no numbers in the domain of f for which $f'(x)$ does not exist.

(f) $f''(x) = 36x^2 - 72x$

(g) f'' is a polynomial, so the domain of f'' is all real numbers.

31. (a) The domain of the function f is all real numbers except $x = 2$ and $x = -2$.

(b) $f'(x) = \dfrac{(x^2-4)\dfrac{d}{dx}x - x\dfrac{d}{dx}(x^2-4)}{(x^2-4)^2} = \dfrac{(x^2-4) - x \cdot 2x}{(x^2-4)^2} = -\dfrac{x^2+4}{(x^2-4)^2}$

(c) The domain of the function $f'(x)$ is all real numbers except $x = 2$ and $x = -2$.

(d) $f'(x)$ is never equal to zero.

(e) There are no numbers in the domain of f for which $f'(x)$ does not exist.

(f) $f''(x) = -\left[\dfrac{(x^2-4)^2\dfrac{d}{dx}(x^2+4) - (x^2+4)\dfrac{d}{dx}(x^2-4)^2}{(x^2-4)^4} \right]$

$= -\left[\dfrac{(x^2-4)^2 \cdot (2x) - (x^2+4) \cdot 2(x^2-4)(2x)}{(x^2-4)^4} \right]$

$= -\left[\dfrac{2x(x^2-4)^2 - 4x(x^2+4)(x^2-4)}{(x^2-4)^3} \right]$

$$= -\left[\frac{2x^3 - 8x - 4x^3 - 16x}{\left(x^2 - 4\right)^3}\right]$$

$$= -\left[\frac{-2x^3 - 24x}{\left(x^2 - 4\right)^3}\right] = \frac{2x^3 + 24x}{\left(x^2 - 4\right)^3}$$

(g) The domain of the function $f''(x)$ is all real numbers except $x = 2$ and $x = -2$.

33.   The function f is a polynomial of degree 3, so the fourth derivative is zero.

35.   The function f is a polynomial of degree 19, so the twentieth derivative is zero.

37.   The function f is a polynomial of degree 8, so the eighth derivative is equal to the constant

$$8! \cdot \frac{1}{8} = 7! = 5040$$

39. $v = s'(t) = 32t + 20$
     $a = v'(t) = s''(t) = 32$

41. $v = s'(t) = 9.8\,t + 4$
     $a = v'(t) = s''(t) = 9.8$

43.   To find a formula for the nth derivative, we take successive derivatives until we see a pattern.

$$f(x) = e^x$$
$$f'(x) = e^x$$
$$f''(x) = e^x$$

We see that each order derivative is e^x, so we conclude that a formula for $f^{(n)}$ is
$$f^{(n)}(x) = e^x$$

45.   To find a formula for the nth derivative, we take successive derivatives until we see a pattern.

$$f(x) = \ln x$$
$$f'(x) = \frac{1}{x} = x^{-1}$$
$$f''(x) = -x^{-2} = -\frac{1}{x^2}$$
$$f'''(x) = 2x^{-3} = \frac{2}{x^3}$$
$$f^{(4)}(x) = -3 \cdot 2x^{-4} = -\frac{3!}{x^4}$$
$$f^{(5)}(x) = (-4) \cdot \left(-3!x^{-5}\right) = \frac{4!}{x^5}$$

$$f^{(6)}(x) = (-5) \cdot \left(4! x^{-5} \right) = -\frac{5!}{x^6}$$

We see a pattern, noticing that the sign of the derivative alternates from positive to negative and conclude the formula for $f^{(n)}$ is

$$f^{(n)}(x) = (-1)^{(n-1)} \cdot \frac{(n-1)!}{x^n}$$

47. To find a formula for the nth derivative, we take successive derivatives until we see a pattern.

$$f(x) = x \ln x$$

$$f'(x) = x \cdot \frac{d}{dx} \ln x + \ln x \cdot \frac{d}{dx} x = x \cdot \frac{1}{x} + \ln x \cdot 1$$

$$= 1 + \ln x$$

$$f''(x) = 0 + \frac{1}{x} = \frac{1}{x} = x^{-1}$$

$$f'''(x) = -x^{-2} = -\frac{1}{x^2}$$

$$f^{(4)}(x) = 2x^{-3} = \frac{2}{x^3}$$

$$f^{(5)}(x) = -3 \cdot 2x^{-4} = -\frac{3!}{x^4}$$

$$f^{(6)}(x) = (-4) \cdot \left(-3! x^{-5} \right) = \frac{4!}{x^5}$$

$$f^{(7)}(x) = (-5) \cdot \left(4! x^{-5} \right) = -\frac{5!}{x^6}$$

We see a pattern, noticing that the sign of the derivative alternates from positive to negative and conclude the formula for $f^{(n)}$ is

$$f^{(n)}(x) = (-1)^{(n)} \cdot \frac{(n-2)!}{x^{n-1}} \text{ provided } n > 1.$$

49. To find a formula for the nth derivative, we take successive derivatives until we see a pattern.

$$f(x) = (2x + 3)^n$$

$$f'(x) = n(2x+3)^{n-1} \frac{d}{dx}(2x+3) = 2n(2x+3)^{n-1} \qquad \text{Use the Power Rule.}$$

$$f''(x) = 2n \cdot (n-1)(2x+3)^{n-2} \frac{d}{dx}(2x+3) = 4n(n-1)(2x+3)^{n-2} \quad \text{Use the Power Rule.}$$

$$= 2^2 n(n-1)(2x+3)^{n-2}$$

$$f'''(x) = 2^2 n(n-1) \cdot (n-2)(2x+3)^{n-3} \frac{d}{dx}(2x+3)$$

$$= 2^3 n(n-1)(n-2)(2x+3)^{n-3}$$
$$f^{(4)}(x) = 2^4 n(n-1)(n-2)(n-3)(2x+3)^{n-4}$$

We see a pattern and conclude the formula for $f^{(n)}$ is
$$f^{(n)}(x) = 2^n \cdot n!$$

51. To find a formula for the nth derivative, we take successive derivatives until we see a pattern.
$$f(x) = e^{ax}$$
$$f'(x) = e^{ax} \frac{d}{dx}(ax) = ae^{ax}$$
$$f''(x) = ae^{ax} \frac{d}{dx}(ax) = a^2 e^{ax}$$
$$f'''(x) = a^2 e^{ax} \frac{d}{dx}(ax) = a^3 e^{ax}$$
We see a pattern and conclude the formula for $f^{(n)}$ is
$$f^{(n)}(x) = a^n \cdot e^{ax}$$

53. To find a formula for the nth derivative, we take successive derivatives until we see a pattern.
$$f(x) = \ln(ax)$$
$$f'(x) = \frac{a}{ax} = \frac{1}{x} = \frac{0!}{x^1}$$
$$f''(x) = -\frac{1}{x^2} = -\frac{1!}{x^2}$$
$$f'''(x) = \frac{2}{x^3} = \frac{2!}{x^3}$$
$$f^{(4)}(x) = -\frac{6}{x^4} = -\frac{3!}{x^4}$$
$$f^{(5)}(x) = \frac{24}{x^5} = \frac{4!}{x^5}$$
We see a pattern and conclude the formula for $f^{(n)}$ is
$$f^{(n)}(x) = (-1)^{n-1} \cdot \frac{(n-1)!}{x^n}$$

55. $y = e^{2x}$
$$y' = \frac{d}{dx}e^{2x} = e^{2x}\frac{d}{dx}(2x) = e^{2x} \cdot 2 = 2e^{2x}$$
$$y'' = \frac{d}{dx}y' = \frac{d}{dx}(2e^{2x}) = 2e^{2x}\frac{d}{dx}(2x) = 2e^{2x} \cdot 2 = 4e^{2x}$$
So, $y'' - 4y = 4e^{2x} - 4e^{2x} = 0$

57. $f(x) = x^2 g(x)$

$$f'(x) = x^2 \frac{d}{dx} g(x) + g(x) \frac{d}{dx} x^2 = x^2 g'(x) + 2x g(x)$$

    ↑   ↑

  Derivative of a product   Differentiate.

$$f''(x) = \frac{d}{dx} x^2 g'(x) + \frac{d}{dx} 2x g(x)$$
$$= \left[x^2 \frac{d}{dx} g'(x) + g'(x) \frac{d}{dx} x^2 \right] + \left[2x \frac{d}{dx} g(x) + g(x) \frac{d}{dx} (2x) \right]$$
$$= \left[x^2 g''(x) + 2x g'(x) \right] + \left[2x g'(x) + 2g(x) \right]$$
$$= x^2 g''(x) + 4x g'(x) + 2g(x)$$

59. (a) The velocity is
$$v = s'(t) = \frac{d}{dx} \left(6 + 80t - 16t^2 \right) = 80 - 32t$$
At $t = 2$ seconds the velocity is $v(2) = 80 - 32(2) = 16$ feet per second.

(b) The ball reaches its maximum height when $v = 0$.
$$v = 80 - 32t = 0 \text{ when } t = \frac{80}{32} = 2.5 \text{ seconds}$$
The ball reaches its maximum height 2.5 seconds after it is thrown.

(c) At $t = 2.5$, $s(2.5) = 6 + 80(2.5) - 16(2.5^2) = 106$ feet. The ball reaches a maximum height of 106 feet.

(d) The acceleration is $a = v'(t) = -32$ feet per second per second.

(e) The ball strikes the ground when $s(t) = 0$. That is when
$$6 + 80t - 16t^2 = 0$$
Using the quadratic formula to solve for t, we find
$$t = \frac{-80 \pm \sqrt{(80)^2 - 4(-16)(6)}}{2(-16)}$$
$$t = \frac{80 \pm \sqrt{6784}}{32}$$
We need only the positive answer, since t represents time, and we get $t = 5.0739$. So the ball is in the air for 5.0739 seconds.

(f) The ball hits the ground at $t = 5.0739$ seconds, the velocity is
$$v(5.0739) = 80 - 32(5.0739) = -82.365 \text{ feet per second.}$$
The ball is moving at a speed of 82.365 feet per second in a downward direction.

(g) The total distance traveled by the ball is the distance up plus the distance down or $(106 - 6) + 106 = 206$ feet.

61. The velocity of the bullet is

$$v = s'(t) = \frac{d}{dx}\left[8 - (2-t)^3\right] = \frac{d}{dx}8 - \frac{d}{dx}(2-t)^3 = -3(2-t)^2\frac{d}{dx}(2-t) = 3(2-t)^2$$

meters per second. After 1 second, the bullet is traveling at a velocity
$$v(1) = 3(2-1)^2 = 3 \cdot 1^2 = 3 \text{ meters per second.}$$

The acceleration is $a = v'(t) = -6(2-t) = 6t - 12$ meters per second per second.

63. (a) The rock hits the ground when its height is zero. Since the rock started from a height of 88.2 meters, when it hits the ground the rock has traveled 88.2 meters.

$$4.9t^2 = 88.2$$
$$t^2 = \frac{88.2}{4.9} = 18$$
$$t = 3\sqrt{2} \approx 4.24$$

It takes the rock approximately 4.24 seconds to hit the ground.

(b) The average velocity is

$$\frac{ds}{dt} = \frac{s\left(3\sqrt{2}\right) - s(0)}{3\sqrt{2} - 0} = \frac{\left(88.2 - 4.9 \cdot 3\sqrt{2}^2\right) - \left(88.2 - 4.9 \cdot 0^2\right)}{3\sqrt{2}} = \frac{-20.8}{3\sqrt{2}} = 4.9$$

meters per second. That is, the rock is moving at an average speed of 4.9 meters per second in the downward direction.

(c) The average velocity in the first 3 seconds is 4.9 meters per second in a downward direction. (See part (b).)

(d) The velocity of the rock is $v = s'(t) = -9.8t$ meters per second.
The rock hits the ground at $t = 4.24$ seconds. The velocity is $v(4.24) = -9.8(4.24) = -41.6$ meters per second.

4.7   Implicit Differentiation

1.
$$\frac{d}{dx}\left(x^2 + y^2\right) = \frac{d}{dx}4$$
$$2x + 2y\frac{dy}{dx} = 0$$

This is a linear equation in $\dfrac{dy}{dx}$. Solving for $\dfrac{dy}{dx}$, we have

$$2y\frac{dy}{dx} = -2x$$
$$\frac{dy}{dx} = \frac{-2x}{2y} = -\frac{x}{y} \quad \text{provided } y \neq 0.$$

3.
$$\frac{d}{dx}\left(x^2 y\right) = \frac{d}{dx} 8$$

$$x^2 \frac{d}{dx} y + y \frac{d}{dx} x^2 = 0$$

$$x^2 \frac{dy}{dx} + y \cdot 2x = 0$$

This is a linear equation in $\frac{dy}{dx}$. Solving for $\frac{dy}{dx}$, we have

$$x^2 \frac{dy}{dx} = -2xy$$

$$\frac{dy}{dx} = \frac{-2xy}{x^2} = -\frac{2y}{x} \quad \text{provided } x \neq 0.$$

5.
$$\frac{d}{dx}\left(x^2 + y^2 - xy\right) = \frac{d}{dx} 2$$

$$2x + 2y\frac{dy}{dx} - \left(x\frac{d}{dx} y + y \frac{d}{dx} x\right) = 0$$

$$2x + 2y\frac{dy}{dx} - x\frac{dy}{dx} - y = 0$$

This is a linear equation in $\frac{dy}{dx}$. Solving for $\frac{dy}{dx}$, we have

$$2x + (2y - x)\frac{dy}{dx} - y = 0$$

$$(2y - x)\frac{dy}{dx} = y - 2x$$

$$\frac{dy}{dx} = \frac{y - 2x}{2y - x} \quad \text{provided } 2y - x \neq 0.$$

7.
$$\frac{d}{dx}\left(x^2 + 4xy + y^2\right) = \frac{d}{dx} y$$

$$\frac{d}{dx}x^2 + 4x\frac{d}{dx} y + y\frac{d}{dx}(4x) + \frac{d}{dx} y^2 = \frac{d}{dx} y$$

$$2x + 4x\frac{dy}{dx} + y \cdot 4 + 2y\frac{dy}{dx} = \frac{dy}{dx}$$

This is a linear equation in $\frac{dy}{dx}$. Solving for $\frac{dy}{dx}$, we have

$$2x + 4y + (4x + 2y)\frac{dy}{dx} = \frac{dy}{dx}$$

$$(4x + 2y)\frac{dy}{dx} - \frac{dy}{dx} = -2x - 4y$$

$$(4x + 2y - 1)\frac{dy}{dx} = -2x - 4y$$

$$\frac{dy}{dx} = \frac{-2x - 4y}{4x + 2y - 1} \quad \text{provided } 4x + 2y - 1 \neq 0.$$

9. $\quad \dfrac{d}{dx}\left(3x^2 + y^3\right) = \dfrac{d}{dx}1$

$\quad\quad 6x + 3y^2\dfrac{dy}{dx} = 0$

This is a linear equation in $\dfrac{dy}{dx}$. Solving for $\dfrac{dy}{dx}$, we have

$$3y^2\frac{dy}{dx} = -6x$$

$$\frac{dy}{dx} = \frac{-6x}{3y^2} = -\frac{2x}{y^2} \quad \text{provided } y \neq 0.$$

11. $\quad \dfrac{d}{dx}\left(4x^3 + 2y^3\right) = \dfrac{d}{dx}x^2$

$\quad\quad 12x^2 + 6y^2\dfrac{dy}{dx} = 2x$

This is a linear equation in $\dfrac{dy}{dx}$. Solving for $\dfrac{dy}{dx}$, we have

$$6y^2\frac{dy}{dx} = 2x - 12x^2$$

$$\frac{dy}{dx} = \frac{2x - 12x^2}{6y^2} = \frac{x - 6x^2}{3y^2} \quad \text{provided } y \neq 0.$$

13. $\quad \dfrac{d}{dx}\left(\dfrac{1}{x^2} - \dfrac{1}{y^2}\right) = \dfrac{d}{dx}\left(x^{-2} - y^{-2}\right) = \dfrac{d}{dx}4$

$$-2x^{-3} + 2y^{-3}\frac{dy}{dx} = 0$$

This is a linear equation in $\dfrac{dy}{dx}$. Solving for $\dfrac{dy}{dx}$, we have

$$2y^{-3}\frac{dy}{dx} = 2x^{-3}$$

$$\frac{dy}{dx} = \frac{2x^{-3}}{2y^{-3}} = \frac{y^3}{x^3} \quad \text{provided } x \neq 0.$$

15. $\dfrac{d}{dx}\left(\dfrac{1}{x}+\dfrac{1}{y}\right)=\dfrac{d}{dx}\left(x^{-1}+y^{-1}\right)=\dfrac{d}{dx}2$

$$-x^{-2}-y^{-2}\dfrac{dy}{dx}=0$$

This is a linear equation in $\dfrac{dy}{dx}$. Solving for $\dfrac{dy}{dx}$, we have

$$-y^{-2}\dfrac{dy}{dx}=x^{-2}$$

$$\dfrac{dy}{dx}=\dfrac{x^{-2}}{-y^{-2}}=-\dfrac{y^2}{x^2}\quad\text{provided }x\neq 0.$$

17. $\dfrac{d}{dx}\left(x^2+y^2\right)=\dfrac{d}{dx}\left(ye^x\right)$

$$\dfrac{d}{dx}x^2+\dfrac{d}{dx}y^2=y\dfrac{d}{dx}e^x+e^x\dfrac{d}{dx}y$$

$$2x+2y\dfrac{dy}{dx}=ye^x+e^x\dfrac{dy}{dx}$$

This is a linear equation in $\dfrac{dy}{dx}$. Solving for $\dfrac{dy}{dx}$, we have

$$2y\dfrac{dy}{dx}-e^x\dfrac{dy}{dx}=ye^x-2x$$

$$\left(2y-e^x\right)\dfrac{dy}{dx}=ye^x-2x$$

$$\dfrac{dy}{dx}=\dfrac{ye^x-2x}{2y-e^x}\quad\text{provided }2y-e^x\neq 0.$$

19. $\dfrac{d}{dx}\left(\dfrac{x}{y}+\dfrac{y}{x}\right)=\dfrac{d}{dx}\left(6e^x\right)$

$$\left[\dfrac{y\dfrac{d}{dx}x-x\dfrac{d}{dx}y}{y^2}\right]+\left[\dfrac{x\dfrac{d}{dx}y-y\dfrac{d}{dx}x}{x^2}\right]=6e^x$$

$$\left[\dfrac{y-x\dfrac{dy}{dx}}{y^2}\right]+\left[\dfrac{x\dfrac{dy}{dx}-y}{x^2}\right]=6e^x$$

$$\dfrac{x^2y-x^3\dfrac{dy}{dx}+xy^2\dfrac{dy}{dx}-y^3}{x^2y^2}=6e^x$$

$$x^2y-x^3\dfrac{dy}{dx}+xy^2\dfrac{dy}{dx}-y^3=6x^2y^2e^x$$

This is a linear equation in $\dfrac{dy}{dx}$. Solving for $\dfrac{dy}{dx}$, we have

$$-\left(x^3 - xy^2\right)\frac{dy}{dx} = 6x^2 y^2 e^x - x^2 y + y^3$$

$$\frac{dy}{dx} = \frac{6x^2 y^2 e^x - x^2 y + y^3}{-x^3 + xy^2} \quad \text{provided } -x^3 + xy^2 \neq 0.$$

21. $\dfrac{d}{dx} x^2 = \dfrac{d}{dx}\left(y^2 \ln x\right)$

$2x = y^2 \dfrac{d}{dx} \ln x + \ln x \dfrac{d}{dx} y^2$

$2x = y^2 \cdot \dfrac{1}{x} + \ln x \cdot 2y \dfrac{dy}{dx}$

This is a linear equation in $\dfrac{dy}{dx}$. Solving for $\dfrac{dy}{dx}$, we have

$$2x = \frac{y^2}{x} + 2y \ln x \, \frac{dy}{dx}$$

$$2x - \frac{y^2}{x} = 2y \ln x \, \frac{dy}{dx}$$

$$\frac{dy}{dx} = \frac{2x - \dfrac{y^2}{x}}{2y \ln x} = \frac{2x^2 - y^2}{2xy \ln x} \quad \text{provided } 2xy \ln x \neq 0.$$

23. $\dfrac{d}{dx}\left(2x + 3y\right)^2 = \dfrac{d}{dx}\left(x^2 + y^2\right)$

$2\left(2x + 3y\right)\dfrac{d}{dx}\left(2x + 3y\right) = 2x + 2y\dfrac{dy}{dx}$

$2\left(2x + 3y\right)\left(2 + 3\dfrac{dy}{dx}\right) = 2x + 2y\dfrac{dy}{dx}$

$\cancel{2}\left(2x + 3y\right)\left(2 + 3\dfrac{dy}{dx}\right) = \cancel{2}x + \cancel{2}y\dfrac{dy}{dx}$

$4x + 6x\dfrac{dy}{dx} + 6y + 9y\dfrac{dy}{dx} = x + y\dfrac{dy}{dx}$

This is a linear equation in $\dfrac{dy}{dx}$. Solving for $\dfrac{dy}{dx}$, we have

$$6x\frac{dy}{dx} + 9y\frac{dy}{dx} - y\frac{dy}{dx} = x - 4x - 6y$$

$$\left(6x + 9y - y\right)\frac{dy}{dx} = x - 4x - 6y$$

$$\frac{dy}{dx} = \frac{-3x - 6y}{6x + 8y} \quad \text{provided } 6x + 8y \neq 0.$$

25.
$$\frac{d}{dx}\left(x^2+y^2\right)^2 = \frac{d}{dx}(x-y)^3$$

$$2\left(x^2+y^2\right)\frac{d}{dx}\left(x^2+y^2\right) = 3(x-y)^2\frac{d}{dx}(x-y)$$

$$2\left(x^2+y^2\right)\left(2x+2y\frac{dy}{dx}\right) = 3(x-y)^2\left(1-\frac{dy}{dx}\right)$$

$$4x^3+4x^2y\frac{dy}{dx}+4xy^2+4y^3\frac{dy}{dx} = 3x^2-6xy+3y^2-\left(3x^2-6xy+3y^2\right)\frac{dy}{dx}$$

This is a linear equation in $\dfrac{dy}{dx}$. Solving for $\dfrac{dy}{dx}$, we have

$$4x^2y\frac{dy}{dx}+4y^3\frac{dy}{dx}+\left(3x^2-6xy+3y^2\right)\frac{dy}{dx} = 3x^2-6xy+3y^2-4x^3-4xy^2$$

$$\left(4x^2y+4y^3+3x^2-6xy+3y^2\right)\frac{dy}{dx} = 3x^2-6xy+3y^2-4x^3-4xy^2$$

$$\frac{dy}{dx} = \frac{3x^2-6xy+3y^2-4x^3-4xy^2}{4x^2y+4y^3+3x^2-6xy+3y^2}$$

provided $4x^2y+4y^3+3x^2-6xy+3y^2 \neq 0$.

27.
$$\frac{d}{dx}\left(x^3+y^3\right)^2 = \frac{d}{dx}\left(x^2y^2\right)$$

$$2\left(x^3+y^3\right)\frac{d}{dx}\left(x^3+y^3\right) = x^2\frac{d}{dx}y^2+y^2\frac{d}{dx}x^2$$

$$2\left(x^3+y^3\right)\left(3x^2+3y^2\frac{dy}{dx}\right) = x^2\cdot 2y\frac{dy}{dx}+y^2\cdot 2x$$

$$\cancel{2}\left(x^3+y^3\right)\left(3x^2+3y^2\frac{dy}{dx}\right) = \cancel{2}x^2y\frac{dy}{dx}+\cancel{2}xy^2$$

$$3x^5+3x^3y^2\frac{dy}{dx}+3x^2y^3+3y^5\frac{dy}{dx} = x^2y\frac{dy}{dx}+xy^2$$

This is a linear equation in $\dfrac{dy}{dx}$. Solving for $\dfrac{dy}{dx}$, we have

$$3x^3y^2\frac{dy}{dx}+3y^5\frac{dy}{dx}-x^2y\frac{dy}{dx} = xy^2-3x^5-3x^2y^3$$

$$\left(3x^3y^2+3y^5-x^2y\right)\frac{dy}{dx} = xy^2-3x^5-3x^2y^3$$

$$\frac{dy}{dx} = \frac{xy^2-3x^5-3x^2y^3}{3x^3y^2+3y^5-x^2y} \quad \text{provided } 3x^3y^2+3y^5-x^2y \neq 0.$$

29.
$$\frac{d}{dx}y = \frac{d}{dx}e^{x^2+y^2}$$

$$\frac{dy}{dx} = e^{x^2+y^2}\frac{d}{dx}\left(x^2+y^2\right)$$

$$\frac{dy}{dx} = e^{x^2+y^2}\left(2x + 2y\frac{dy}{dx}\right)$$

$$\frac{dy}{dx} = 2xe^{x^2+y^2} + 2ye^{x^2+y^2}\frac{dy}{dx}$$

This is a linear equation in $\frac{dy}{dx}$. Solving for $\frac{dy}{dx}$, we have

$$\frac{dy}{dx} - 2ye^{x^2+y^2}\frac{dy}{dx} = 2xe^{x^2+y^2}$$

$$\left(1 - 2ye^{x^2+y^2}\right)\frac{dy}{dx} = 2xe^{x^2+y^2}$$

$$\frac{dy}{dx} = \frac{2xe^{x^2+y^2}}{1 - 2ye^{x^2+y^2}} \quad \text{provided } 1 - 2ye^{x^2+y^2} \neq 0.$$

31. The first derivative is $y' = -\dfrac{x}{y}$, provided $y \neq 0$ (from Problem 1).

The second derivative is

$$\frac{d}{dx}\left(2x + 2y\frac{dy}{dx}\right) = \frac{d}{dx}0$$

$$2\frac{d}{dx}x + 2\frac{dy}{dx}(yy') = 0$$

$$2 + 2\left[y\frac{d}{dx}y' + y'\frac{d}{dx}y\right] = 0$$

$$\cancel{2} + \cancel{2}\left[yy'' + y'y'\right] = 0$$

$$1 + yy'' + (y')^2 = 0$$

$$y'' = \frac{-(y')^2 - 1}{y} = \frac{-\left(-\dfrac{x}{y}\right)^2 - 1}{y} = -\frac{\dfrac{x^2}{y^2} + 1}{y} = -\frac{x^2 + y^2}{y^3}$$

provided $y \neq 0$.

33. The first derivative is

$$\frac{d}{dx}\left(xy + yx^2\right) = \frac{d}{dx}2$$

$$\left[x\frac{d}{dx}y + y\frac{d}{dx}x\right] + \left[y\frac{d}{dx}x^2 + x^2\frac{d}{dx}y\right] = 0$$

$$\left[xy' + y\right] + \left[y \cdot 2x + x^2y'\right] = 0$$

$$xy' + y + 2xy + x^2y' = 0$$

$$\left(x + x^2\right)y' + \left(1 + 2x\right)y = 0$$

$$\left(x + x^2\right)y' = -\left(1 + 2x\right)y$$

$$y' = -\frac{(1+2x)y}{x^2+x} \quad \text{provided } x \neq 0, \ x \neq -1.$$

Using the fifth line from above, we find the second derivative is

$$\frac{d}{dx}\left[\left(x+x^2\right)y' + (1+2x)y\right] = \frac{d}{dx}0$$

$$\left[\left(x+x^2\right)\frac{d}{dx}y' + y'\frac{d}{dx}\left(x+x^2\right)\right] + \left[(1+2x)\frac{d}{dx}y + y\frac{d}{dx}(1+2x)\right] = 0$$

$$\left[\left(x+x^2\right)y'' + y'(1+2x)\right] + \left[(1+2x)y' + y\cdot 2\right] = 0$$

$$\left(x+x^2\right)y'' + y'(1+2x) + (1+2x)y' + 2y = 0$$

$$\left(x+x^2\right)y'' + 2(1+2x)y' + 2y = 0$$

$$\left(x+x^2\right)y'' = -2(1+2x)y' - 2y$$

$$y'' = -\frac{2(1+2x)y' + 2y}{x^2+x}$$

$$y'' = -\frac{2(1+2x)\left(\dfrac{y+2xy}{x^2+x}\right) + 2y}{x^2+x}$$

$$y'' = -\frac{2(1+2x)(y+2xy) + 2y\left(x^2+x\right)}{\left(x^2+x\right)^2}$$

$$y'' = -\frac{2y+8xy+8x^2y+2x^2y+2xy}{\left(x^2+x\right)^2}$$

$$y'' = -\frac{2y+10xy+10x^2y}{\left(x^2+x\right)^2}$$

provided $x \neq 0, \ x \neq -1$.

35. The slope of the tangent line is $\dfrac{dy}{dx}$, which is

$$\frac{d}{dx}\left(x^2+y^2\right) = \frac{d}{dx}5$$

$$2x + 2y\frac{dy}{dx} = 0$$

Solving for $\dfrac{dy}{dx}$, we have

$$2y\frac{dy}{dx} = -2x$$

$$\frac{dy}{dx} = \frac{-2x}{2y} = -\frac{x}{y} \quad \text{provided } y \neq 0.$$

The slope of the tangent line at the point $(1, 2)$ is $m_{tan} = -\dfrac{1}{2}$. The equation of the tangent line is

$$y - y_1 = m(x - x_1)$$
$$y - 2 = -\frac{1}{2}(x - 1)$$
$$y - 2 = -\frac{1}{2}x + \frac{1}{2}$$
$$y = -\frac{1}{2}x + \frac{5}{2}$$

37. (Note: The problem as printed in the first printing of the text is incorrect. That equation has no tangent line at $(0, 0)$.) The slope of the tangent line is $\dfrac{dy}{dx}$, which is

$$\frac{d}{dx}e^{xy} = \frac{d}{dx}x$$
$$e^{xy}\frac{d}{dx}(xy) = 1$$
$$e^{xy}\left(x\frac{d}{dx}y + y\frac{d}{dx}\right) = 1$$
$$e^{xy}\left(x\frac{dy}{dx} + y\right) = 1$$
$$e^{xy}x\frac{dy}{dx} + ye^{xy} = 1$$
$$\frac{dy}{dx} = \frac{1 - ye^{xy}}{xe^{xy}}$$

The slope of the tangent line at the point $(1, 0)$ is $m_{tan} = \dfrac{1 - 0 \cdot e^0}{1e^0} = 1$. The equation of the tangent line is

$$y - y_1 = m(x - x_1)$$
$$y - 0 = 1(x - 1)$$
$$y = x - 1$$

39. The tangent line is horizontal when the slope is zero, that is when $\dfrac{dy}{dx} = 0$.

$$\frac{d}{dx}(x^2 + y^2) = \frac{d}{dx}4$$
$$2x + 2y\frac{dy}{dx} = 0$$
$$\frac{dy}{dx} = \frac{-2x}{2y} = -\frac{x}{y} \quad \text{provided } y \neq 0.$$

$\dfrac{dy}{dx} = 0$ when $x = 0$.

When $x = 0$, $y^2 = 4$, or $y = \pm 2$. So there are horizontal tangent lines at the points $(0, 2)$ and $(0, -2)$.

41. The tangent line is horizontal when the slope is zero, that is when $\dfrac{dy}{dx} = 0$.

$$\frac{d}{dx}\left(y^2 + 4x^2\right) = \frac{d}{dx}16$$

$$2y\frac{dy}{dx} + 8x = 0$$

$$\frac{dy}{dx} = \frac{-8x}{2y} = -\frac{4x}{y} \quad \text{provided } y \neq 0.$$

$\dfrac{dy}{dx} = 0$ when $x = 0$.

When $x = 0$, $y^2 = 16$, or $y = \pm 4$. So there are horizontal tangent lines at the points $(0, 4)$ and $(0, -4)$.

43. (a) The slope of the tangent line is $\dfrac{dy}{dx}$, which is

$$\frac{d}{dx}\left(x + xy + 2y^2\right) = \frac{d}{dx}6$$

$$\frac{d}{dx}x + x\frac{d}{dx}y + y\frac{d}{dx}x + \frac{d}{dx}\left(2y^2\right) = 0$$

$$1 + x\frac{dy}{dx} + y + 4y\frac{dy}{dx} = 0$$

$$\left(x + 4y\right)\frac{dy}{dx} = -y - 1$$

$$\frac{dy}{dx} = \frac{-y-1}{x+4y} = -\frac{y+1}{x+4y} \quad \text{provided } x + 4y \neq 0.$$

(b) At the point $(2, 1)$ the slope of the tangent line is $m_{\text{tan}} = -\dfrac{1+1}{2+4 \cdot 1} = -\dfrac{2}{6} = -\dfrac{1}{3}$, and

an equation of the tangent line is

$$y - y_1 = m\left(x - x_1\right)$$

$$y - 1 = -\frac{1}{3}\left(x - 2\right)$$

$$y = -\frac{1}{3}x + \frac{5}{3}$$

(c) To find the coordinates of the points (x, y) at which the slope of the tangent line equals the slope of the tangent line at $(2, 1)$, we need to solve the system of equations

$$x + xy + 2y^2 = 6 \qquad (1)$$

$$-\frac{y+1}{x+4y} = -\frac{1}{3} \qquad (2)$$

Beginning with equation (2), we get $3y + 3 = x + 4y$ or $3 = x + y$ or $x = 3 - y$.
Substituting $x = 3 - y$ into equation (1), we get

$$(3 - y) + (3 - y)y + 2y^2 = 6$$

$$3 - y + 3y - y^2 + 2y^2 = 6$$

$$3 + 2y + y^2 = 6$$

$$y^2 + 2y - 3 = 0$$

$$(y - 1)(y + 3) = 0$$

$$y = 1 \qquad y = -3$$

Back substituting $y = 1$ into equation (2) we get $x = 2$.
Back substituting $y = -3$ into equation (2) we get $x = 6$.

So the coordinates of the point which has the same slope as the tangent line at $(2, 1)$ is $(6, -3)$.

45.

$$\frac{d}{dP}\left(P + \frac{a}{V^2}\right) = \frac{d}{dP}\frac{C}{V - b}$$

$$\frac{d}{dP}P + \frac{d}{dP}aV^{-2} = C\frac{d}{dP}(V - b)^{-1}$$

$$1 - 2aV^{-3}\frac{dV}{dP} = -C(V - b)^{-2}\frac{d}{dP}(V - b)$$

$$1 - \frac{2a}{V^3}\frac{dV}{dP} = -\frac{C}{(V - b)^2}\frac{dV}{dP}$$

$$1 = \left(\frac{2a}{V^3} - \frac{C}{(V - b)^2}\right)\frac{dV}{dP}$$

$$\left(\frac{2a(V - b)^2 - CV^3}{V^3(V - b)^2}\right)\frac{dV}{dP} = 1$$

$$\frac{dV}{dP} = \frac{V^3(V - b)^2}{2a(V - b)^2 - CV^3}$$

47.

(a) $\quad \dfrac{d}{dt}e^{N(t)} = \dfrac{d}{dt}\left(430{,}163t + \dfrac{3t}{t^2 + 2}\right)$

$$e^{N(t)}\frac{d}{dt}N(t) = \frac{d}{dt}430{,}163t + \frac{(t^2 + 2)\dfrac{d}{dt}(3t) - 3t\dfrac{d}{dt}(t^2 + 2)}{(t^2 + 2)^2}$$

$$e^{N(t)}\frac{dN}{dt} = 430{,}163 + \frac{3(t^2 + 2) - 6t^2}{(t^2 + 2)^2}$$

$$e^{N(t)} \frac{dN}{dt} = \frac{430{,}163(t^2+2)^2 + 3(t^2+2) - 6t^2}{(t^2+2)^2}$$

$$\frac{dN}{dt} = \frac{430{,}163(t^2+2)^2 + 3(t^2+2) - 6t^2}{e^{N(t)}(t^2+2)^2} = \frac{430{,}163(t^2+2)^2 + 3(t^2+2) - 6t^2}{\left(430{,}163t + \dfrac{3t}{t^2+2}\right)(t^2+2)^2}$$

$$= \frac{430{,}163(t^2+2)^2 + 3(t^2+2) - 6t^2}{430{,}163t(t^2+2)^2 + 3t(t^2+2)}$$

(b) $N(2) = \dfrac{430{,}163(2^2+2)^2 + 3(2^2+2) - 6 \cdot 2^2}{(430{,}163 \cdot 2)(2^2+2)^2 + (3 \cdot 2)(2^2+2)} = 0.500$

$N(4) = \dfrac{430{,}163(4^2+2)^2 + 3(4^2+2) - 6 \cdot 4^2}{(430{,}163 \cdot 4)(4^2+2)^2 + (3 \cdot 4)(4^2+2)} = 0.250$

4.8   The Derivative of $x^{p/q}$

1. $f'(x) = \dfrac{d}{dx} x^{4/3} = \dfrac{4}{3} x^{(4/3)-1} = \dfrac{4}{3} x^{1/3}$

3. $f'(x) = \dfrac{d}{dx} x^{2/3} = \dfrac{2}{3} x^{(2/3)-1} = \dfrac{2}{3} x^{-1/3} = \dfrac{2}{3x^{1/3}}$

5. $f'(x) = \dfrac{d}{dx} \dfrac{1}{x^{1/2}} = \dfrac{d}{dx} x^{-1/2} = -\dfrac{1}{2} x^{(-1/2)-1} = -\dfrac{1}{2} x^{-3/2} = -\dfrac{1}{2x^{3/2}}$

7. $f'(x) = \dfrac{d}{dx} (2x+3)^{3/2} = \dfrac{3}{2}(2x+3)^{(3/2)-1} \dfrac{d}{dx}(2x+3) = \dfrac{3}{2}(2x+3)^{1/2} \cdot 2 = 3(2x+3)^{1/2}$

9. $f'(x) = \dfrac{d}{dx}(x^2+4)^{3/2} = \dfrac{3}{2}(x^2+4)^{(3/2)-1} \dfrac{d}{dx}(x^2+4) = \dfrac{3}{2}(x^2+4)^{1/2} \cdot 2x = 3x(x^2+4)^{1/2}$

11. We first change each radical to its fractional exponent equivalent.

$f(x) = (2x+3)^{1/2}$

$f'(x) = \dfrac{d}{dx}(2x+3)^{1/2} = \dfrac{1}{2}(2x+3)^{(1/2)-1} \dfrac{d}{dx}(2x+3) = \dfrac{1}{2}(2x+3)^{-1/2} \cdot 2 = \dfrac{1}{(2x+3)^{1/2}}$

$$= \dfrac{1}{\sqrt{2x+3}}$$

13. We first change each radical to its fractional exponent equivalent.

$$f(x) = \left(9x^2 + 1\right)^{1/2}$$

$$f'(x) = \frac{d}{dx}\left(9x^2 + 1\right)^{1/2} = \frac{1}{2}\left(9x^2 + 1\right)^{(1/2)-1}\frac{d}{dx}\left(9x^2 + 1\right) = \frac{1}{2}\left(9x^2 + 1\right)^{-1/2} \cdot 18x$$

$$= \frac{9x}{\left(9x^2 + 1\right)^{1/2}} = \frac{9x}{\sqrt{9x^2 + 1}}$$

15.

$$f'(x) = \frac{d}{dx}\left(3x^{5/3} - 6x^{1/3}\right) = 3\frac{d}{dx}x^{5/3} - 6\frac{d}{dx}x^{1/3}$$

$$= 3 \cdot \frac{5}{3}x^{(5/3)-1} - 6 \cdot \frac{1}{3}x^{(1/3)-1}$$

$$= 5x^{2/3} - 2x^{-2/3} = 5x^{2/3} - \frac{2}{x^{2/3}}$$

17. To find the derivative of function f we can either use the formula for the derivative of a product or multiply the factors and find the derivative of the sum. We chose to multiply first.

$$f(x) = x^{1/3}\left(x^2 - 4\right) = x^{7/3} - 4x^{1/3}$$

$$f'(x) = \frac{d}{dx}\left(x^{7/3} - 4x^{1/3}\right) = \frac{d}{dx}x^{7/3} - 4\frac{d}{dx}x^{1/3} = \frac{7}{3}x^{(7/3)-1} - 4 \cdot \frac{1}{3}x^{(1/3)-1}$$

$$= \frac{7}{3}x^{4/3} - \frac{4}{3}x^{-2/3} = \frac{7}{3}x^{4/3} - \frac{4}{3x^{2/3}}$$

19. We first change each radical to its fractional exponent equivalent.

$$f(x) = \frac{x}{\sqrt{x^2 - 4}} = \frac{x}{\left(x^2 - 4\right)^{1/2}}$$

$$f'(x) = \frac{d}{dx}\frac{x}{\left(x^2 - 4\right)^{1/2}} = \frac{\left(x^2 - 4\right)^{1/2}\frac{d}{dx}x - x\frac{d}{dx}\left(x^2 - 4\right)^{1/2}}{\left[\left(x^2 - 4\right)^{1/2}\right]^2} \qquad \text{Use the formula for the derivative of a quotient.}$$

$$= \frac{\left(x^2 - 4\right)^{1/2} - x \cdot \frac{1}{2}\left(x^2 - 4\right)^{(1/2)-1}\frac{d}{dx}\left(x^2 - 4\right)}{\left(x^2 - 4\right)}$$

$$= \frac{\left(x^2 - 4\right)^{1/2} - x \cdot \frac{1}{2}\left(x^2 - 4\right)^{-1/2} \cdot 2x}{\left(x^2 - 4\right)}$$

$$= \frac{\left(x^2 - 4\right)^{1/2} - x^2\left(x^2 - 4\right)^{-1/2}}{\left(x^2 - 4\right)}$$

$$= \frac{\left(x^2 - 4\right) - x^2}{\left(x^2 - 4\right)^{3/2}} = -\frac{4}{\left(x^2 - 4\right)^{3/2}} \qquad \text{Multiply by } \frac{\left(x^2 - 4\right)^{1/2}}{\left(x^2 - 4\right)^{1/2}} \text{; simplify.}$$

21. We first change each radical to its fractional exponent equivalent.

$$f(x) = \sqrt{e^x} = \left(e^x\right)^{1/2}$$

$$f'(x) = \frac{d}{dx}\left(e^x\right)^{1/2} = \frac{1}{2}\left(e^x\right)^{(1/2)-1}\frac{d}{dx}e^x = \frac{1}{2}\left(e^x\right)^{-1/2}e^x = \frac{1}{2}\left(e^x\right)^{1/2} = \frac{\sqrt{e^x}}{2}$$

23. We first change each radical to its fractional exponent equivalent.

$$f(x) = \sqrt{\ln x} = \left(\ln x\right)^{1/2}$$

$$f'(x) = \frac{d}{dx}\left(\ln x\right)^{1/2} = \frac{1}{2}\left(\ln x\right)^{(1/2)-1}\frac{d}{dx}\ln x = \frac{1}{2}\left(\ln x\right)^{-1/2}\cdot\frac{1}{x} = \frac{1}{2x\left(\ln x\right)^{1/2}} = \frac{1}{2x\sqrt{\ln x}}$$

25. We first change each radical to its fractional exponent equivalent.

$$f(x) = e^{\sqrt[3]{x}} = e^{x^{1/3}}$$

$$f'(x) = \frac{d}{dx}e^{x^{1/3}} = e^{x^{1/3}}\frac{d}{dx}x^{1/3} = e^{x^{1/3}}\cdot\frac{1}{3}x^{(1/3)-1} = \frac{1}{3}x^{-2/3}e^{x^{1/3}} = \frac{e^{x^{1/3}}}{3x^{2/3}} = \frac{e^{\sqrt[3]{x}}}{3x^{2/3}} = \frac{e^{\sqrt[3]{x}}}{3\sqrt[3]{x^2}}$$

27. We first change each radical to its fractional exponent equivalent.

$$f(x) = \sqrt[3]{\ln x} = \left(\ln x\right)^{1/3}$$

$$f'(x) = \frac{d}{dx}\left(\ln x\right)^{1/3} = \frac{1}{3}\left(\ln x\right)^{(1/3)-1}\frac{d}{dx}\ln x = \frac{1}{3}\left(\ln x\right)^{-2/3}\cdot\frac{1}{x} = \frac{1}{3x\left(\ln x\right)^{2/3}} = \frac{1}{3x\sqrt[3]{\left(\ln x\right)^2}}$$

29. We first change each radical to its fractional exponent equivalent.

$$f(x) = \sqrt{x}\,e^x = x^{1/2}e^x$$

$$f'(x) = \frac{d}{dx}\left(x^{1/2}e^x\right) = x^{1/2}\frac{d}{dx}e^x + e^x\frac{d}{dx}x^{1/2} \qquad \text{Use the formula for the derivative of a product.}$$

$$= x^{1/2}e^x + e^x\cdot\frac{1}{2}x^{(1/2)-1}$$

$$= x^{1/2}e^x + \frac{1}{2}x^{-1/2}e^x$$

$$= x^{1/2}e^x + \frac{e^x}{2x^{1/2}}$$

$$= \frac{2xe^x + e^x}{2x^{1/2}} = \frac{e^x\left(2x+1\right)}{2\sqrt{x}}$$

31. We first change each radical to its fractional exponent equivalent.

$$f(x) = e^{2x}\sqrt{x^2+1} = e^{2x}\left(x^2+1\right)^{1/2}$$

$$f'(x) = \frac{d}{dx}\left(e^{2x}\left(x^2+1\right)^{1/2}\right) = e^{2x}\frac{d}{dx}\left(x^2+1\right)^{1/2} + \left(x^2+1\right)^{1/2}\frac{d}{dx}e^{2x} \qquad \begin{array}{l}\text{Use the formula for the}\\\text{derivative of a quotient.}\end{array}$$

$$= e^{2x}\frac{1}{2}\left(x^2+1\right)^{(1/2)-1}\frac{d}{dx}\left(x^2+1\right) + \left(x^2+1\right)^{1/2}e^{2x}\frac{d}{dx}\left(2x\right) \qquad \text{Use the Chain Rule.}$$

$$= e^{2x} \frac{1}{2}\left(x^2+1\right)^{-1/2} \cdot 2x + \left(x^2+1\right)^{1/2} e^{2x} \cdot 2 \qquad \text{Differentiate.}$$

$$= xe^{2x}\left(x^2+1\right)^{-1/2} + 2e^{2x}\left(x^2+1\right)^{1/2} \qquad \text{Simplify.}$$

$$= \frac{xe^{2x}}{\left(x^2+1\right)^{1/2}} + 2e^{2x}\left(x^2+1\right)^{1/2} = \frac{xe^{2x}+2e^{2x}\left(x^2+1\right)}{\left(x^2+1\right)^{1/2}} = \frac{e^{2x}\left(2x^2+x+2\right)}{\sqrt{x^2+1}}$$

33.
$$\frac{d}{dx}\left(\sqrt{x}+\sqrt{y}\right) = \frac{d}{dx}4$$

$$\frac{1}{2\sqrt{x}} + \frac{1}{2\sqrt{y}}\frac{dy}{dx} = 0$$

$$\frac{1}{2\sqrt{y}}\frac{dy}{dx} = -\frac{1}{2\sqrt{x}}$$

$$\frac{dy}{dx} = -\frac{2\sqrt{y}}{2\sqrt{x}} = -\frac{\sqrt{y}}{\sqrt{x}}$$

35. $\sqrt{x^2+y^2} = \left(x^2+y^2\right)^{1/2}$

$$\frac{d}{dx}\left(x^2+y^2\right)^{1/2} = \frac{d}{dx}x$$

$$\frac{1}{2}\left(x^2+y^2\right)^{(1/2)-1}\frac{d}{dx}\left(x^2+y^2\right) = 1$$

$$\frac{1}{2}\left(x^2+y^2\right)^{-1/2}\left(2x+2y\frac{dy}{dx}\right) = 1$$

$$\cancel{2}x+\cancel{2}y\frac{dy}{dx} = \cancel{2}\left(x^2+y^2\right)^{1/2}$$

$$y\frac{dy}{dx} = \left(x^2+y^2\right)^{1/2}-x$$

$$\frac{dy}{dx} = \frac{\left(x^2+y^2\right)^{1/2}-x}{y} = \frac{\sqrt{x^2+y^2}-x}{y}$$

37.
$$\frac{d}{dx}\left(x^{1/3}+y^{1/3}\right) = \frac{d}{dx}1$$

$$\frac{1}{3}x^{(1/3)-1} + \frac{1}{3}y^{(1/3)-1} = 0$$

$$x^{-2/3} + y^{-2/3}\frac{dy}{dx} = 0$$

$$y^{-2/3}\frac{dy}{dx} = -x^{-2/3}$$

$$\frac{dy}{dx} = -x^{-2/3}y^{2/3} = -\frac{y^{2/3}}{x^{2/3}}$$

39.

$$\frac{d}{dx}\left(e^{\sqrt{x}}+e^{\sqrt{y}}\right)=\frac{d}{dx}4$$

$$e^{\sqrt{x}}\frac{d}{dx}\sqrt{x}+e^{\sqrt{y}}\frac{d}{dx}\sqrt{y}=0 \qquad \text{Use the Chain Rule.}$$

$$e^{\sqrt{x}}\cdot\frac{1}{2\sqrt{x}}+e^{\sqrt{y}}\cdot\frac{1}{2\sqrt{y}}\frac{dy}{dx}=0 \qquad \text{Differentiate.}$$

$$\frac{e^{\sqrt{x}}}{\sqrt{x}}+\frac{e^{\sqrt{y}}}{\sqrt{y}}\frac{dy}{dx}=0 \qquad \text{Simplify.}$$

$$\frac{e^{\sqrt{y}}}{\sqrt{y}}\frac{dy}{dx}=-\frac{e^{\sqrt{x}}}{\sqrt{x}}$$

$$\frac{dy}{dx}=-\frac{e^{\sqrt{x}}}{\sqrt{x}}\cdot\frac{\sqrt{y}}{e^{\sqrt{y}}}=-\frac{\sqrt{y}\,e^{\sqrt{x}}}{\sqrt{x}\,e^{\sqrt{y}}}=-\frac{\sqrt{y}\,e^{\sqrt{x}-\sqrt{y}}}{\sqrt{x}}$$

41. (a) The domain of f is $\{x \mid x \geq 0\}$ or on the interval $[0, \infty)$..

(b) $f'(x)=\dfrac{1}{2\sqrt{x}}$

(c) The domain of $f'(x)$ is $\{x \mid x > 0\}$ or on the interval $(0, \infty)$.

(d) $f'(x)$ is never equal to 0.

(e) $x = 0$ is in the domain of f, but not in the domain of $f'(x)$.

(f) $f''(x)=\dfrac{d}{dx}\left(\dfrac{1}{2\sqrt{x}}\right)=\dfrac{1}{2}\dfrac{d}{dx}x^{-1/2}=\dfrac{1}{2}\cdot\left(-\dfrac{1}{2}\right)x^{(-1/2)-1}=-\dfrac{1}{4}x^{-3/2}=-\dfrac{1}{4x^{3/2}}$

(g) The domain of $f''(x)$ is $\{x \mid x > 0\}$ or on the interval $(0, \infty)$.

43. (a) The domain of f is all real numbers or the interval $(-\infty, \infty)$.

(b) $f'(x)=\dfrac{d}{dx}x^{2/3}=\dfrac{2}{3}x^{(2/3)-1}=\dfrac{2}{3}x^{-1/3}=\dfrac{2}{3x^{1/3}}=\dfrac{2}{3\sqrt[3]{x}}$

(c) The domain of $f'(x)$ is all real numbers except $x = 0$, that is the set $\{x \mid x \neq 0\}$.

(d) $f'(x)$ is never equal to 0.

(e) $x = 0$ is in the domain of f, but not in the domain of $f'(x)$.

(f) $f''(x)=\dfrac{d}{dx}\left(\dfrac{2}{3}x^{-1/3}\right)=\dfrac{2}{3}\cdot\left(-\dfrac{1}{3}\right)x^{(-1/3)-1}=-\dfrac{2}{9}x^{-4/3}=-\dfrac{2}{9x^{4/3}}$

(g) The domain of $f''(x)$ is all real numbers except $x = 0$, that is the set $\{x \mid x \neq 0\}$.

45. (a) The domain of f is all real numbers or the interval $(-\infty, \infty)$.

(b) $f'(x) = \dfrac{d}{dx}\left(x^{2/3} + 2x^{1/3}\right) = \dfrac{2}{3}x^{(2/3)-1} + 2 \cdot \dfrac{1}{3}x^{(1/3)-1}$

$\qquad = \dfrac{2}{3}x^{-1/3} + \dfrac{2}{3}x^{-2/3} = \dfrac{2}{3x^{1/3}} + \dfrac{2}{3x^{2/3}}$

(c) The domain of $f'(x)$ is all real numbers except $x = 0$, that is the set $\{x \mid x \neq 0\}$.

(d) $f'(x)$ is never equal to zero.

(e) $x = 0$ is in the domain of f, but not in the domain of $f'(x)$.

(f) $f''(x) = \dfrac{d}{dx}\left(\dfrac{2}{3}x^{-1/3} + \dfrac{2}{3}x^{-2/3}\right) = \dfrac{2}{3}\cdot\left(-\dfrac{1}{3}\right)x^{(-1/3)-1} + \dfrac{2}{3}\cdot\left(-\dfrac{2}{3}\right)x^{(-2/3)-1}$

$\qquad = -\dfrac{2}{9}x^{-4/3} - \dfrac{4}{9}x^{-5/3} = -\dfrac{2}{9x^{4/3}} - \dfrac{4}{9x^{5/3}}$

(g) The domain of $f''(x)$ is all real numbers except $x = 0$, that is the set $\{x \mid x \neq 0\}$.

47. (a) The domain of f is all real numbers or the interval $(-\infty, \infty)$.

(b) $f'(x) = \dfrac{d}{dx}\left(x^2 - 1\right)^{2/3} = \dfrac{2}{3}\left(x^2 - 1\right)^{(2/3)-1}\dfrac{d}{dx}\left(x^2 - 1\right) = \dfrac{2}{3}\left(x^2 - 1\right)^{-1/3} \cdot 2x$

$\qquad\qquad = \dfrac{4x}{3\left(x^2 - 1\right)^{1/3}}$

(c) The domain of $f'(x)$ is all real numbers except $x = 1$ and $x = -1$, that is the set $\{x \mid x \neq 1 \text{ and } x \neq -1\}$.

(d) $f'(x) = 0$ when $x = 0$.

(e) $x = 1$ and $x = -1$ are in the domain of f, but not in the domain of $f'(x)$.

(f) $f''(x) = \dfrac{d}{dx}\dfrac{4x}{3\left(x^2 - 1\right)^{1/3}} = \dfrac{4}{3}\left[\dfrac{\left(x^2 - 1\right)^{1/3}\dfrac{d}{dx}x - x\dfrac{d}{dx}\left(x^2 - 1\right)^{1/3}}{\left(x^2 - 1\right)^{2/3}}\right]$

$\qquad = \dfrac{4}{3}\left[\dfrac{\left(x^2 - 1\right)^{1/3} - x \cdot \dfrac{1}{3}\left(x^2 - 1\right)^{(1/3)-1}\dfrac{d}{dx}\left(x^2 - 1\right)}{\left(x^2 - 1\right)^{2/3}}\right]$

$$= \frac{4}{3} \left[\frac{\left(x^2-1\right)^{1/3} - \frac{x}{3}\left(x^2-1\right)^{-2/3} \cdot 2x}{\left(x^2-1\right)^{2/3}} \right]$$

$$= \frac{4}{3} \left[\frac{\left(x^2-1\right)^{1/3} - \frac{2x^2}{3\left(x^2-1\right)^{2/3}}}{\left(x^2-1\right)^{2/3}} \right]$$

$$= \frac{4}{3} \left[\frac{3\left(x^2-1\right)-2x^2}{3\left(x^2-1\right)^{4/3}} \right] = \frac{4\left(x^2-3\right)}{9\left(x^2-1\right)^{4/3}}$$

(g) The domain of $f''(x)$ is all real numbers except $x=1$ and $x=-1$, that is the set $\{x \mid x \neq 1 \text{ and } x \neq -1\}$.

49. (a) Since $\sqrt{1-x^2} \geq 0$, $1-x^2 \geq 0$ or $x^2 \leq 1$. Solving for x, we get $-1 \leq x \leq 1$, so the domain of f is the interval $[-1, 1]$.

(b) $f'(x) = \frac{d}{dx}\left(x\sqrt{1-x^2}\right) = \frac{d}{dx}\left[x\left(1-x^2\right)^{1/2}\right] = x\frac{d}{dx}\left(1-x^2\right)^{1/2} + \left(1-x^2\right)^{1/2}\frac{d}{dx}x$

$$= x \cdot \frac{1}{2}\left(1-x^2\right)^{(1/2)-1}\frac{d}{dx}\left(1-x^2\right) + \left(1-x^2\right)^{1/2} \cdot 1$$

$$= \frac{x}{2}\left(1-x^2\right)^{-1/2}\left(-2x\right) + \left(1-x^2\right)^{1/2}$$

$$= \frac{-x^2}{\left(1-x^2\right)^{1/2}} + \left(1-x^2\right)^{1/2} = \frac{-x^2+\left(1-x^2\right)}{\left(1-x^2\right)^{1/2}} = \frac{1-2x^2}{\left(1-x^2\right)^{1/2}}$$

(c) Since $\left(1-x^2\right)^{1/2} > 0$, $1-x^2 > 0$ or $x^2 < 1$. Solving for x, we get $-1 < x < 1$, so the domain of $f'(x)$ is all real numbers or the interval $(-1, 1)$.

(d) $f'(x) = 0$ when $1-2x^2 = 0$, or when $x^2 = \frac{1}{2}$, or $x = -\sqrt{\frac{1}{2}}$ or $x = \sqrt{\frac{1}{2}}$.

(e) The points $x = -1$ and $x = 1$ are in the domain of f, but are not part of the domain of f'.

(f) $f''(x) = \frac{d}{dx}\frac{1-2x^2}{\left(1-x^2\right)^{1/2}} = \frac{\left(1-x^2\right)^{1/2}\frac{d}{dx}\left(1-2x^2\right) - \left(1-2x^2\right)\frac{d}{dx}\left(1-x^2\right)^{1/2}}{\left(\left(1-x^2\right)^{1/2}\right)^2}$

$$= \frac{\left(1-x^2\right)^{1/2}(-4x) - \left(1-2x^2\right)\frac{1}{2}\left(1-x^2\right)^{(1/2)-1}\frac{d}{dx}\left(1-x^2\right)}{\left(1-x^2\right)}$$

$$= \frac{-4x\left(1-x^2\right)^{1/2} - \left(1-2x^2\right)\frac{1}{2}\left(1-x^2\right)^{-1/2}(-2x)}{\left(1-x^2\right)}$$

$$= \frac{-4x\left(1-x^2\right)^{1/2} + \left(x-2x^3\right)\left(1-x^2\right)^{-1/2}}{\left(1-x^2\right)}$$

$$= \frac{-4x\left(1-x^2\right) + \left(x-2x^3\right)}{\left(1-x^2\right)^{3/2}}$$

$$= \frac{-4x+4x^3+x-2x^3}{\left(1-x^2\right)^{3/2}} = \frac{2x^3-3x}{\left(1-x^2\right)^{3/2}}$$

(g) Since $\left(1-x^2\right)^{3/2} > 0$, $1-x^2 > 0$ or $x^2 < 1$. Solving for x, we get $-1 < x < 1$, so the domain of $f''(x)$ is all real numbers or the interval $(-1, 1)$.

51.

(a) $N'(t) = \dfrac{d}{dx}\left(-\dfrac{10,000}{\sqrt{1+0.1t}} + 11,000\right) = \dfrac{d}{dx}\left(-10,000(1+0.1t)^{-1/2} + 11,000\right)$

$$= -10,000\frac{d}{dx}(1+0.1t)^{-1/2} + \frac{d}{dx}11,000$$

$$= -10,000 \cdot \left(-\frac{1}{2}\right)(1+0.1t)^{(-1/2)-1}\frac{d}{dx}(1+0.1t)$$

$$= 5,000(1+0.1t)^{-3/2}(0.1) = \frac{500}{(1+0.1t)^{3/2}}$$

(b) In 10 years, $t = 10$.

$$N'(10) = \frac{500}{\left(1+0.1(10)\right)^{3/2}} = 176.777 = 177 \text{ students}$$

53. Since z is a constant, we write $z = K$. Then the production function becomes
$$K = x^{0.5}y^{0.4}$$

We will find $\dfrac{dy}{dx}$ using implicit differentiation.

$$\frac{d}{dx}K = \frac{d}{dx}\left(x^{0.5}y^{0.4}\right)$$

$$0 = x^{0.5}\frac{d}{dx}y^{0.4} + y^{0.4}\frac{d}{dx}x^{0.5}$$

$$0 = x^{0.5} \cdot 0.4y^{0.4-1}\frac{dy}{dx} + y^{0.4} \cdot 0.5x^{0.5-1}$$

$$0 = 0.4x^{0.5}y^{-0.6}\frac{dy}{dx} + 0.5y^{0.4}x^{-0.5}$$

$$0 = \frac{0.4x^{0.5}}{y^{0.6}}\frac{dy}{dx} + \frac{0.5y^{0.4}}{x^{0.5}}$$

$$\frac{0.4x^{0.5}}{y^{0.6}}\frac{dy}{dx} = -\frac{0.5y^{0.4}}{x^{0.5}}$$

$$\frac{dy}{dx} = -\frac{0.5y^{0.4}}{x^{0.5}} \cdot \frac{y^{0.6}}{0.4x^{0.5}} = -\frac{0.5y^{0.4+.06}}{0.4x^{0.5+0.5}} \cdot \frac{10}{10} = -\frac{5y}{4x}$$

55. (a) The instantaneous rate of pollution is the derivative of the function A.

$$A'(t) = \frac{d}{dt}\left(t^{1/4} + 3\right)^3 = 3\left(t^{1/4} + 3\right)^2 \frac{d}{dt}\left(t^{1/4} + 3\right) = 3\left(t^{1/4} + 3\right)^2 \cdot \frac{1}{4}t^{(1/4)-1} = \frac{3\left(t^{1/4} + 3\right)^2}{4t^{3/4}}$$

(b) After 16 years the rate of

$$A'(16) = \frac{3\left(16^{1/4} + 3\right)^2}{4 \cdot 16^{3/4}} = \frac{3 \cdot 5^2}{32} = 2.344 \text{ units per year.}$$

57. (a) Velocity is the derivative of the distance function s.

$$v = s'(t) = \frac{d}{dt}t^{3/2} = \frac{3}{2}t^{(3/2)-1} = \frac{3t^{1/2}}{2} \text{ feet per second.}$$

After 1 second the child has a velocity of $v(1) = \dfrac{3 \cdot 1^{1/2}}{2} = \dfrac{3}{2} = 1.5$ feet per second.

(b) If the slide is 8 feet long, it takes

$$t^{3/2} = 8$$

$$\left(t^{3/2}\right)^{2/3} = (8)^{2/3} = 4 \text{ seconds to get down the slide and strike the ground.}$$

The velocity the child when striking the ground is

$$v(4) = \frac{3 \cdot 4^{1/2}}{2} = 3 \text{ feet per second.}$$

Chapter 4 Review

TRUE-FALSE ITEMS

1. True   **3.** True   **5.** False   **7.** True

FILL-IN-THE-BLANKS

1. tangent   **3.** Power Rule; Chain Rule   **5.** zero

REVIEW EXERCISES

1. $f'(x) = 2;\ f'(2) = 2$

3. $f'(x) = 2x;\ f'(2) = 2 \cdot 2 = 4$

5. $f'(x) = 2x - 2;\ f'(1) = 2 \cdot 1 - 2 = 0$

7. $f'(x) = e^{3x} \dfrac{d}{dx}(3x) = 3e^{3x};\ f'(0) = 3e^{0} = 3$

9. $f'(x) = \lim\limits_{h \to 0} \dfrac{\left[4(x+h)+3\right] - \left[4x+3\right]}{h} = \lim\limits_{h \to 0} \dfrac{4x + 4h + 3 - 4x - 3}{h} = \lim\limits_{h \to 0} \dfrac{4\cancel{h}}{\cancel{h}} = \lim\limits_{h \to 0} 4 = 4$

11. $f'(x) = \lim\limits_{h \to 0} \dfrac{\left[2(x+h)^2 + 1\right] - \left[2x^2 + 1\right]}{h} = \lim\limits_{h \to 0} \dfrac{2x^2 + 4xh + 2h^2 - 2x^2 - 1}{h}$

$= \lim\limits_{h \to 0} \dfrac{\cancel{h}(4x + 2h)}{\cancel{h}} = \lim\limits_{h \to 0}(4x + 2h) = \lim\limits_{h \to 0} 4x + \lim\limits_{h \to 0} 2h = 4x + 0 = 4x$

13. $f'(x) = 5x^4$

15. $f'(x) = \dfrac{1}{4}\dfrac{d}{dx}x^4 = \dfrac{1}{4} \cdot 4x^3 = x^3$

17. $f'(x) = 2 \cdot 2x - 3 = 4x - 3$

19. $f'(x) = 7\dfrac{d}{dx}(x^2 - 4) = 7 \cdot 2x = 14x$

21. $f'(x) = 5\left[(x^2 - 3x)\dfrac{d}{dx}(x-6) + (x-6)\dfrac{d}{dx}(x^2 - 3x)\right] = 5\left[x^2 - 3x + (x-6)(2x-3)\right]$

$= 5(x^2 - 3x + 2x^2 - 15x + 18) = 5(3x^2 - 18x + 18) = 15(x^2 - 6x + 6)$

23. $f'(x) = \dfrac{d}{dx}\left[12x(8x^3 + 2x^2 - 5x + 2)\right] = 12\dfrac{d}{dx}(8x^4 + 2x^3 - 5x^2 + 2x)$

$= 12(32x^3 + 6x^2 - 10x + 2) = 24(16x^3 + 3x^2 - 5x + 1)$

25. $f'(x) = \dfrac{d}{dx}\dfrac{2x+2}{5x-3} = \dfrac{(5x-3)\dfrac{d}{dx}(2x+2) - (2x+2)\dfrac{d}{dx}(5x-3)}{(5x-3)^2}$

$= \dfrac{(5x-3) \cdot 2 - (2x+2) \cdot 5}{(5x-3)^2} = \dfrac{10x - 6 - 10x - 10}{(5x-3)^2} = \dfrac{-16}{(5x-3)^2}$

27. $f'(x) = 2 \cdot (-12)x^{-13} = -24x^{-13} = -\dfrac{24}{x^{13}}$

29. $f'(x) = \dfrac{d}{dx}\left(2 + \dfrac{3}{x} + \dfrac{4}{x^2}\right) = \dfrac{d}{dx}(2 + 3x^{-1} + 4x^{-2}) = -3x^{-2} - 8x^{-3} = -\dfrac{3}{x^2} - \dfrac{8}{x^3}$

31. $f'(x) = \dfrac{d}{dx}\left(\dfrac{3x-2}{x+5}\right) = \dfrac{(x+5)\dfrac{d}{dx}(3x-2)-(3x-2)\dfrac{d}{dx}(x+5)}{(x+5)^2}$ Derivative of a quotient.

$\quad\quad = \dfrac{(x+5)(3)-(3x-2)(1)}{(x+5)^2} = \dfrac{3x+15-3x+2}{(x+5)^2} = \dfrac{17}{(x+5)^2}$

33. $f'(x) = \dfrac{d}{dx}\left(3x^2-2x\right)^5 = 5\left(3x^2-2x\right)^4\dfrac{d}{dx}\left(3x^2-2x\right)$ Use the Power Rule.

$\quad\quad = 5\left(3x^2-2x\right)^4(6x-2) = 10\left(3x^2-2x\right)^4(3x-1)$

35. $f'(x) = \dfrac{d}{dx}\left[7x\left(x^2+2x+1\right)^2\right]$

$\quad\quad = 7x\dfrac{d}{dx}\left(x^2+2x+1\right)^2 + \left(x^2+2x+1\right)^2\dfrac{d}{dx}(7x)$ The derivative of a product.

$\quad\quad = 7x \cdot 2\left(x^2+2x+1\right)\dfrac{d}{dx}\left(x^2+2x+1\right) + \left(x^2+2x+1\right)^2 \cdot 7$ Use the Power Rule.

$\quad\quad = 14x\left(x^2+2x+1\right)(2x+2) + 7\left(x^2+2x+1\right)^2$ Differentiate.

$\quad\quad = 7\left(x^2+2x+1\right)\left[\,2x(2x+2)+\left(x^2+2x+1\right)\right]$ Factor.

$\quad\quad = 7\left(x^2+2x+1\right)\left[\,4x^2+4x+x^2+2x+1\right]$ Simplify in the brackets.

$\quad\quad = 7\left(x^2+2x+1\right)\left[\,5x^2+6x+1\right]$ Simplify in the brackets.

$\quad\quad = 7(x+1)^2\left[(5x+1)(x+1)\right] = 7(x+1)^3(5x+1)$ Factor.

37. $f'(x) = \dfrac{d}{dx}\left(\dfrac{x+1}{3x+2}\right)^2 = 2\left(\dfrac{x+1}{3x+2}\right)\dfrac{d}{dx}\left(\dfrac{x+1}{3x+2}\right)$ Use the Power Rule.

$\quad\quad = 2\left(\dfrac{x+1}{3x+2}\right)\left[\dfrac{(3x+2)\dfrac{d}{dx}(x+1)-(x+1)\dfrac{d}{dx}(3x+2)}{(3x+2)^2}\right]$ Derivative of a quotient.

$\quad\quad = 2\left(\dfrac{x+1}{3x+2}\right)\left[\dfrac{(3x+2)\cdot 1-(x+1)\cdot 3}{(3x+2)^2}\right]$ Differentiate.

$\quad\quad = \dfrac{2(x+1)(3x+2-3x-3)}{(3x+2)^3}$ Simplify.

$\quad\quad = -\dfrac{2(x+1)}{(3x+2)^3}$

39. $f'(x) = \dfrac{d}{dx} \dfrac{7}{\left(x^3 + 4\right)^2} = 7\dfrac{d}{dx}\left(x^3 + 4\right)^{-2}$ Write the function with a negative exponent.

$\qquad = 7 \cdot (-2)\left(x^3 + 4\right)^{-3} \dfrac{d}{dx}\left(x^3 + 4\right)$ Use the Power Rule.

$\qquad = -14\left(x^3 + 4\right)^{-3}\left(3x^2\right) = -\dfrac{42x^2}{\left(x^3 + 4\right)^3}$ Differentiate and simplify.

41. $f'(x) = \dfrac{d}{dx}\left(3x + \dfrac{4}{x}\right)^3 = \dfrac{d}{dx}\left(\dfrac{3x^2 + 4}{x}\right)^3$ Write the function with a common denominator.

$\qquad = 3\left(\dfrac{3x^2 + 4}{x}\right)^2 \dfrac{d}{dx}\left(\dfrac{3x^2 + 4}{x}\right)$ Use the Power Rule.

$\qquad = 3\left(\dfrac{3x^2 + 4}{x}\right)^2 \left[\dfrac{x\dfrac{d}{dx}\left(3x^2 + 4\right) - \left(3x^2 + 4\right)\dfrac{d}{dx}x}{x^2}\right]$ The derivative of a quotient.

$\qquad = 3\left(\dfrac{3x^2 + 4}{x}\right)^2 \left[\dfrac{x\left(6x\right) - \left(3x^2 + 4\right)}{x^2}\right] = 3\left(\dfrac{3x^2 + 4}{x}\right)^2 \left[\dfrac{3x^2 - 4}{x^2}\right]$ Differentiate; simplify.

$\qquad = \dfrac{3\left(3x^2 + 4\right)^2\left(3x^2 - 4\right)}{x^4}$ Simplify.

43. $f'(x) = \dfrac{d}{dx}\left(3e^x + x^2\right) = 3\dfrac{d}{dx}e^x + \dfrac{d}{dx}x^2 = 3e^x + 2x$

45. $f'(x) = \dfrac{d}{dx}e^{3x+1} = e^{3x+1}\dfrac{d}{dx}(3x + 1) = 3e^{3x+1}$ Use the Chain Rule.

47. $f'(x) = \dfrac{d}{dx}\left[e^x\left(2x^2 + 7x\right)\right] = e^x\dfrac{d}{dx}\left(2x^2 + 7x\right) + \left(2x^2 + 7x\right)\dfrac{d}{dx}e^x$ The derivative of a product.

$\qquad = e^x\left(4x + 7\right) + \left(2x^2 + 7x\right)e^x = e^x\left(2x^2 + 11x + 7\right)$

49. $f'(x) = \dfrac{d}{dx}\dfrac{1+x}{e^x} = \dfrac{e^x\dfrac{d}{dx}(1 + x) - (1 + x)\dfrac{d}{dx}e^x}{\left(e^x\right)^2} = \dfrac{e^x - (1 + x)e^x}{e^{2x}} = \dfrac{-xe^x}{e^{2x}} = -\dfrac{x}{e^x}$

$\qquad\qquad\qquad\uparrow \qquad\qquad\qquad\qquad\qquad\qquad\uparrow \qquad\qquad\quad\uparrow$

$\qquad$ Derivative of a quotient. $\qquad\qquad$ Differentiate. $\qquad$ Simplify.

51. $f'(x) = \dfrac{d}{dx}\left(\dfrac{e^x}{3x}\right)^2 = 2\left(\dfrac{e^x}{3x}\right)\dfrac{d}{dx}\left(\dfrac{e^x}{3x}\right) = 2\left(\dfrac{e^x}{3x}\right)\left[\dfrac{1}{3}\cdot\dfrac{x\dfrac{d}{dx}e^x - e^x\dfrac{d}{dx}x}{x^2}\right]$

           ↑   ↑

   Use the Power Rule.   Use the formula for the derivative of a quotient.

$= 2\left(\dfrac{e^x}{3x}\right)\left[\dfrac{1}{3}\cdot\dfrac{xe^x - e^x}{x^2}\right] = \dfrac{2e^x\left(xe^x - e^x\right)}{9x^3} = \dfrac{2e^{2x}\left(x-1\right)}{9x^3}$

    ↑   ↑   ↑

  Differentiate.   Simplify.   Factor.

53. $f'(x) = \dfrac{d}{dx}\ln(4x) = \dfrac{1}{4x}\dfrac{d}{dx}(4x) = \dfrac{4}{4x} = \dfrac{1}{x}$

55. $f'(x) = \dfrac{d}{dx}\left(x^2\ln x\right) = x^2\dfrac{d}{dx}\ln x + \ln x\dfrac{d}{dx}x^2 = x^2\cdot\dfrac{1}{x} + \ln x\cdot 2x = x + 2x\ln x$

              ↑   ↑   ↑

   Use the formula for the   Differentiate.   Simplify.
   derivative of a product.

57. $f'(x) = \dfrac{d}{dx}\ln\left(2x^3 + 1\right) = \dfrac{1}{2x^3 + 1}\dfrac{d}{dx}\left(2x^3 + 1\right) = \dfrac{1}{2x^3 + 1}\cdot\left(6x^2\right) = \dfrac{6x^2}{2x^3 + 1}$

                      ↑

          Use the Chain Rule.

59. $f'(x) = \dfrac{d}{dx}\left[2^x + x^2\right] = \dfrac{d}{dx}2^x + \dfrac{d}{dx}x^2 = 2^x\ln 2 + 2x$

61. $f'(x) = \dfrac{d}{dx}\left(x + \log x\right) = \dfrac{d}{dx}x + \dfrac{d}{dx}\log x = 1 + \dfrac{1}{x\ln 10}$

63. $f'(x) = \dfrac{d}{dx}\sqrt{x} = \dfrac{d}{dx}(x)^{1/2} = \dfrac{1}{2}x^{(1/2)-1} = \dfrac{1}{2}x^{-1/2} = \dfrac{1}{2x^{1/2}} = \dfrac{1}{2\sqrt{x}}$

65. $f'(x) = \dfrac{d}{dx}\left(3x^{5/3} + 5\right) = 3\dfrac{d}{dx}x^{5/3} + \dfrac{d}{dx}5 = 3\cdot\dfrac{5}{3}x^{(5/3)-1} = 5x^{2/3}$

67. $f'(x) = \dfrac{d}{dx}\sqrt{x^2 - 3x} = \dfrac{d}{dx}\left(x^2 - 3x\right)^{1/2} = \dfrac{1}{2}\left(x^2 - 3x\right)^{(1/2)-1}\dfrac{d}{dx}\left(x^2 - 3x\right)$

                  ↑   ↑

  Change the radical to an exponent.   Use the Power Rule.

$$= \frac{1}{2}\left(x^2 - 3x\right)^{-1/2}\left(2x - 3\right) = \frac{2x - 3}{2\left(x^2 - 3x\right)^{1/2}} = \frac{2x - 3}{2\sqrt{x^2 - 3x}}$$

↑	↑	↑
Differentiate.	Simplify.	Alternate form of the solution.

69. $f'(x) = \dfrac{d}{dx}\dfrac{x+1}{\sqrt{x+5}} = \dfrac{d}{dx}\dfrac{x+1}{\left(x+5\right)^{1/2}} = \dfrac{\left(x+5\right)^{1/2}\dfrac{d}{dx}\left(x+1\right) - \left(x+1\right)\dfrac{d}{dx}\left(x+5\right)^{1/2}}{\left[\left(x+5\right)^{1/2}\right]^2}$

$$= \frac{\left(x+5\right)^{1/2} - \left(x+1\right)\cdot\dfrac{1}{2}\left(x+5\right)^{(1/2)-1}\dfrac{d}{dx}\left(x+5\right)}{\left(x+5\right)}$$

$$= \frac{\left(x+5\right)^{1/2} - \left(x+1\right)\cdot\dfrac{1}{2}\left(x+5\right)^{-1/2}}{\left(x+5\right)}$$

$$= \frac{2\left(x+5\right) - \left(x+1\right)}{2\left(x+5\right)^{3/2}} = \frac{2x+10 - x - 1}{2\left(x+5\right)^{3/2}} = \frac{x+9}{2\left(x+5\right)^{3/2}}$$

71. $f'(x) = \dfrac{d}{dx}\left[\left(1+x\right)\sqrt{e^x}\right] = \dfrac{d}{dx}\left[\left(1+x\right)e^{x/2}\right] = \left(1+x\right)\dfrac{d}{dx}e^{x/2} + e^{x/2}\dfrac{d}{dx}\left(1+x\right)$

$$= \left(1+x\right)e^{x/2}\frac{d}{dx}\left(\frac{x}{2}\right) + e^{x/2}\cdot 1$$

$$= \left(1+x\right)e^{x/2}\cdot\frac{1}{2} + e^{x/2}$$

$$= \frac{\left(1+x\right)e^{x/2} + 2e^{x/2}}{2}$$

$$= \frac{e^{x/2}\left(1+x+2\right)}{2} = \frac{e^{x/2}\left(x+3\right)}{2}$$

73. $f'(x) = \dfrac{d}{dx}\left(\sqrt{x}\ln x\right) = \dfrac{d}{dx}\left(x^{1/2}\ln x\right) = x^{1/2}\dfrac{dy}{dx}\ln x + \ln x\dfrac{d}{dx}x^{1/2}$ Derivative of a product.

$$= x^{1/2}\cdot\frac{1}{x} + \ln x\cdot\frac{1}{2}x^{-1/2} \qquad \text{Differentiate.}$$

$$= \frac{1}{x^{1/2}} + \ln x\cdot\frac{1}{2x^{1/2}} = \frac{2 + \ln x}{2x^{1/2}} = \frac{2 + \ln x}{2\sqrt{x}} \qquad \text{Simplify.}$$

75. $f'(x) = \dfrac{d}{dx}\left(x^3 - 8\right) = 3x^2$

$$f''(x) = \frac{d}{dx}\left(3x^2\right) = 6x$$

77. $f'(x) = \dfrac{d}{dx} e^{-3x} = e^{-3x} \dfrac{d}{dx}(-3x) = -3e^{-3x}$

$f''(x) = \dfrac{d}{dx}\left(-3e^{-3x}\right) = -3e^{-3x}\dfrac{d}{dx}(-3x) = -3e^{-3x} \cdot (-3) = 9e^{-3x}$

79. $f'(x) = \dfrac{d}{dx}\dfrac{x}{2x+1} = \dfrac{(2x+1)\dfrac{d}{dx}x - x\dfrac{d}{dx}(2x+1)}{(2x+1)^2}$ Use the formula for the derivative of a quotient.

$\qquad\qquad = \dfrac{(2x+1)\cdot 1 - x(2)}{(2x+1)^2} = \dfrac{(2x+1)-2x}{(2x+1)^2} = \dfrac{1}{(2x+1)^2}$

$f''(x) = \dfrac{d}{dx}\dfrac{1}{(2x+1)^2} = \dfrac{d}{dx}(2x+1)^{-2} = -2(2x+1)^{-3}\dfrac{d}{dx}(2x+1)$ Use the Power Rule.

$\qquad\qquad = -2(2x+1)^{-3}\cdot 2 = -\dfrac{4}{(2x+1)^3}$ Simplify.

81.
$$\dfrac{d}{dx}\left(xy + 3y^2\right) = \dfrac{d}{dx}(10x)$$

$$\left(x\dfrac{d}{dx}y + y\dfrac{d}{dx}x\right) + 3\dfrac{d}{dx}y^2 = 10$$

$$x\dfrac{dy}{dx} + y + 6y\dfrac{dy}{dx} = 10$$

$$(x+6y)\dfrac{dy}{dx} = 10 - y$$

$$\dfrac{dy}{dx} = \dfrac{10-y}{x+6y} \quad \text{provided } x+6y \neq 0$$

83.
$$\dfrac{d}{dx}\left(xe^y\right) = \dfrac{d}{dx}\left(4x^2\right)$$

$$x\dfrac{d}{dx}e^y + e^y\dfrac{d}{dx}x = 4\cdot 2x$$

$$xe^y\dfrac{dy}{dx} + e^y = 8x$$

$$xe^y\dfrac{dy}{dx} = 8x - e^y$$

$$\dfrac{dy}{dx} = \dfrac{8x - e^y}{xe^y} \quad \text{provided } x \neq 0.$$

85. The slope of the tangent line to the graph of $f(x) = 2x^2 + 3x - 7$ at the point $(-1, -8)$ is

$$m_{\tan} = \lim_{x\to -1}\dfrac{f(x) - f(-1)}{x - (-1)} = \lim_{x\to -1}\dfrac{(2x^2 + 3x - 7) - (-8)}{x+1} = \lim_{x\to -1}\dfrac{2x^2 + 3x + 1}{x+1}$$

$$= \lim_{x \to -1} \frac{(2x+1)\,\cancel{(x+1)}}{\cancel{x+1}} = \lim_{x \to -1} (2x+1) = -1$$

An equation of the tangent line is

$$\begin{aligned} y - (-8) &= (-1)[x - (-1)] \\ y + 8 &= -x - 1 \\ y &= -x - 9 \end{aligned} \qquad \begin{aligned} &y - f(c) = m_{\tan}\,(x - c) \\ &\text{Simplify.} \end{aligned}$$

87. The slope of the tangent line to the graph of f is the derivative evaluated at $(0, 1)$.

$$f'(x) = \frac{d}{dx}\left(x^2 + e^x\right) = 2x + e^x$$

$$m_{\tan} = f'(0) = 2 \cdot 0 + e^0 = 1$$

An equation of the tangent line is

$$\begin{aligned} y - 1 &= 1(x - 0) \\ y &= x + 1 \end{aligned}$$

89. (a) The average rate of change of f as x changes from 0 to 2 is

$$\frac{\Delta y}{\Delta x} = \frac{f(2) - f(0)}{2 - 0} = \frac{\left[2^3 + 3 \cdot 2\right] - \left[0^3 + 3 \cdot 0\right]}{2} = \frac{14 - 0}{2} = 7$$

(b) The instantaneous rate of change is the derivative of f evaluated at 2.

$$f'(x) = \frac{d}{dx}\left(x^3 + 3x\right) = 3x^2 + 3$$

$$f'(2) = 3 \cdot 2^2 + 3 = 15$$

91. (a) When the stone hits the water its height is zero. So we solve the equation $h(t) = 0$.

$$\begin{aligned} -16t^2 + 100 &= 0 \\ 16t^2 &= 100 \\ t^2 &= \frac{100}{16} \\ t &= \sqrt{\frac{100}{16}} = \frac{10}{4} = 2.5 \end{aligned}$$

It takes the stone 2.5 seconds to hit the water.

(b) The average velocity during its fall is

$$\frac{\Delta y}{\Delta t} = \frac{h(2.5) - h(0)}{2.5 - 0} = \frac{\left[-16(2.5^2) + 100\right] - \left[-16(0^2) + 100\right]}{2.5} = \frac{-100}{2.5} = -40$$

feet per second in the downward direction.

(c) The instantaneous velocity is the derivative of h.

$$v = h'(t) = \frac{d}{dt}\left(-16t^2 + 100\right) = -32t$$

When it hits the water the stone's velocity is $v(2.5) = (-32)(2.5) = -80$ feet per second.

93. (a) The ball reaches its highest point when its velocity is zero. The velocity is the derivative of the distance function s.

$$v = s'(t) = \frac{d}{dt}\left(-16t^2 + 128t + 6\right) = -32t + 128$$

We set $v = 0$ and solve.

$$-32t + 128 = 0$$
$$32t = 128$$
$$t = 4$$

The ball reach its highest point 4 seconds after it is thrown.

(b) The maximum height of the ball is

$$s(4) = -16(4^2) + 128(4) + 6 = 262$$

feet above the ground.

(c) The ball travels 262 feet downward and $262 - 6$ feet upward for a total of 518 feet.

(d) The velocity of the ball at time t is $v(t) = -32t + 128$ feet per second as we calculated in part (a).

(e) The velocity of the ball is zero at $t = 4$ seconds. At that time it momentarily stops, changing from the upward to the downward direction.

(f) The ball is in the air for 4 seconds on it way up. To find how long it is in the air on the way down, we solve the equation $s(t) = 0$.

$$-16t^2 + 128t + 6 = 0$$

Using the quadratic formula with $a = 16$, $b = -128$, and $c = -6$, we find

$$t = \frac{128 \pm \sqrt{(-128)^2 - 4(16)(-6)}}{2 \cdot 16} = \frac{128 \pm \sqrt{16,768}}{32}$$

We only need the positive answer, and so

$$t = \frac{128 + \sqrt{16,768}}{32} = 8.047 \approx 8.0 \text{ seconds.}$$

(g) The velocity of the ball when it hits the ground is
$v(8.047) = -32(8.047) + 128 = -129.504$ feet per second in the downward direction.

(h) The acceleration a is the derivative of the velocity at time t.

$$a = v'(t) = s''(t) = \frac{d}{dx}\left(-32t + 128\right) = -32 \text{ feet per second per second.}$$

(i) The velocity of the ball is $v(2) = -32(2) + 128 = 64$ feet per second after it is in the air for 2 seconds.

The velocity of the ball is $v(6) = -32(6) + 128 = -64$ feet per second after it is in the air for 6 seconds.

95. (a) Revenue is the product of price and quantity. So the revenue function R is
$$R(x) = px = (-0.50x + 75)x = -0.50x^2 + 75x$$

(b) The marginal revenue function is the derivative of the revenue function R.
$$R'(x) = \frac{d}{dx}(-0.50x^2 + 75x) = -1.00x + 75$$

(c) The marginal cost function is the derivative of the cost function C.
$$C'(x) = \frac{d}{dx}(15x + 550) = 15$$

(d) The break even point is the x-value for which $R(x) = C(x)$.
$$-0.50x^2 + 75x = 15x + 550$$
$$-0.50x^2 + 60x - 550 = 0$$
$$x^2 - 120x + 1100 = 0$$
$$(x - 110)(x - 10) = 0$$
$$(x - 110) = 0 \qquad (x - 10) = 0$$
$$x = 110 \qquad\qquad x = 10$$

(e) The marginal revenue equals the marginal cost when
$$-x + 75 = 15$$
$$x = 60 \text{ units are produced and sold.}$$

CHAPTER 4 PROJECT

1. Since x persons are in the pool, the probability that the pooled test is negative is the same as the probability that each person in the sample tests negative. So, $p_- = q \cdot q \cdot q \cdot \ldots \cdot q = q^x$. On the other hand, the pooled test will be positive if at least 1 person tests positive, so $p_+ = 1 -$ probability that all tests were negative $= 1 - q^x$.

3. Answers will vary. (Note: $\dfrac{N}{x} \cdot x = N$)

5. If $C(x) = KN\left(1 - (0.9944)^x + \dfrac{1}{x}\right)$ then the marginal cost function is the derivative of C.
$$C'(x) = \frac{d}{dx}\left[KN\left(1 - (0.9944)^x + \frac{1}{x}\right)\right] = KN\left[\frac{d}{dx}1 - \frac{d}{dx}(0.9944)^x + \frac{d}{dx}x^{-1}\right]$$
$$= KN\left[-(0.9944)^x \ln 0.9944 - x^{-2}\right] = KN\left[(0.0056)(0.9944)^x - \frac{1}{x^2}\right]$$

7. N will increase (or decrease) the unit marginal cost by a factor of N, the number of persons tested.

9. If each test costs $5.00 and individual tests are performed, the total cost will be $5000.

   If the pooling procedure is used the total cost will be $\$5000\left(1-(0.9944)^x+\dfrac{1}{x}\right)$ where x is the number of persons in the pooled group. The savings will be

   $$\$5000-\$5000\left(1-(0.9944)^x+\frac{1}{x}\right)=\$5000\left[(0.9944)^x-\frac{1}{x}\right]$$

   If $x = 13$ persons are in each group, the savings would be approximately

   $$\$5000\left[(0.9944)^{13}-\frac{1}{13}\right]=\$4263.37$$

   If $x = 14$ persons are in each group, the savings would be approximately

   $$\$5000\left[(0.9944)^{14}-\frac{1}{14}\right]=\$4264.81$$

MATHEMATICAL QUESTIONS FROM PROFESSIONAL EXAMS

1. (e) $\dfrac{d}{dx}\left(x^2e^{x^2}\right)=x^2\dfrac{d}{dx}e^{x^2}+e^{x^2}\dfrac{d}{dx}x^2=x^2e^{x^2}\dfrac{d}{dx}x^2+e^{x^2}(2x)=2x^3e^{x^2}+2xe^{x^2}$

3. (d) $\dfrac{d}{dx}\left(be^{c^2+x^2}\right)=be^{c^2+x^2}\dfrac{d}{dx}\left(c^2+x^2\right)=2bxe^{c^2+x^2}$

5. (e) $f''(x)=\dfrac{d}{dx}f'(x)=\dfrac{dy}{dx}\left(x\,f(x)\right)=x\,\dfrac{d}{dx}\left[f(x)\right]+f(x)\dfrac{d}{dx}x$
   $$=xf'(x)+f(x)=x\left[x\,f(x)\right]+f(x)=\left(x^2+1\right)f(x)$$
   $$f''(-2)=\left[(-2)^2+1\right]f(-2)=5\cdot 3=15$$

Chapter 5
Applications: Graphing Functions; Optimization

5.1 Horizontal and Vertical Tangent Lines; Continuity and Differentiability

1. $f'(x) = 2x - 4$

   Horizontal tangent lines occur where $f'(x) = 0$.
   $$2x - 4 = 0$$
   $$2x = 4$$
   $$x = 2$$
   Evaluate the function f at $x = 2$: $f(2) = (2)^2 - 4(2) = -4$

   The tangent line to the graph of f is horizontal at the point $(2, f(2)) = (2, -4)$. There is no vertical tangent line since $f'(x)$ is never unbounded.

3. $f'(x) = -2x + 8$

   Horizontal tangent lines occur where $f'(x) = 0$.
   $$-2x + 8 = 0$$
   $$-2x = -8$$
   $$x = 4$$
   Evaluate the function f at $x = 4$: $f(4) = -(4)^2 + 8(4) = 16$

   The tangent line to the graph of f is horizontal at the point $(4, f(4)) = (4, 16)$. There is no vertical tangent line since $f'(x)$ is never unbounded.

5. $f'(x) = -4x + 8$

   Horizontal tangent lines occur where $f'(x) = 0$.
   $$-4x + 8 = 0$$
   $$-4x = -8$$
   $$x = 2$$
   Evaluate the function f at $x = 2$: $f(2) = -2(2)^2 + 8(2) + 1 = -8 + 16 + 1 = 9$

   The tangent line to the graph of f is horizontal at the point $(2, f(2)) = (2, 9)$. There is no vertical tangent line since $f'(x)$ is never unbounded.

7. $f'(x) = \dfrac{2}{3} \cdot 3x^{-1/3} = \dfrac{2}{x^{1/3}}$

   $f'(x)$ is never zero, so there is no horizontal tangent line.

$f'(x)$ is unbounded at $x = 0$. We evaluate f at 0: $f(0) = 3 \cdot 0^{2/3} + 1 = 1$.

The tangent line to the graph of f is vertical at the point $(0, 1)$.

9. $f'(x) = -3x^2 + 3$

Horizontal tangent lines occur where $f'(x) = 0$.

$$-3x^2 + 3 = 0$$
$$x^2 - 1 = 0$$
$$(x - 1)(x + 1) = 0$$
$$x - 1 = 0 \quad \text{or} \quad x + 1 = 0$$
$$x = 1 \quad \text{or} \quad x = -1$$

Evaluate f at 1 and -1.

$$f(1) = -1^3 + 3(1) + 1 = 3$$
$$f(-1) = -(-1)^3 + 3(-1) + 1 = -1$$

There are two horizontal lines tangent to the graph of f, one at $(1, 3)$ and the other at $(-1, -1)$. There is no vertical tangent line since $f'(x)$ is never unbounded.

11. $f'(x) = \dfrac{3}{4} \cdot 4x^{-1/4} = \dfrac{3}{x^{1/4}}$

$f'(x) \neq 0$ so there is no horizontal tangent line.

$f'(x)$ is unbounded at $x = 0$. We evaluate $f(0) = 4(0)^{3/4} - 2 = -2$.

The tangent line to the graph of f at $(0, -2)$ is vertical.

13. $f'(x) = 5x^4 - 40x^3$

Horizontal tangents occur where $f'(x) = 0$.

$$5x^4 - 40x^3 = 0$$
$$5x^3(x - 8) = 0$$
$$5x^3 = 0 \quad \text{or} \quad x - 8 = 0$$
$$x = 0 \quad \text{or} \quad x = 8$$

Evaluate f at 0 and 8.

$$f(0) = (0)^5 - 10(0)^4 = 0 \qquad\qquad f(8) = (8)^5 - 10(8)^4 = -8192$$

We conclude that the graph of f has 2 horizontal tangent lines, one at $(0, 0)$, the other at $(8, -8192)$. There is no vertical tangent line.

15. $f'(x) = 15x^4 + 60x^2$

Horizontal tangent lines occur where $f'(x) = 0$.

$$15x^4 + 60x^2 = 0$$

$$15x^2\left(x^2+4\right)=0$$
$$15x^2=0 \quad \text{or} \quad x^2+4=0$$
$$x=0$$

Evaluate f at 0: $\quad f(0)=3(0)^5+20(0)^3-1=-1$

The graph of f has a horizontal tangent line at $(0,-1)$; f' is never unbounded, so f has no vertical tangent.

17.
$$f'(x)=\frac{2}{3}x^{-1/3}+\frac{1}{3}\cdot 2x^{-2/3}=\frac{2}{3x^{1/3}}+\frac{2}{3x^{2/3}}$$
$$=\frac{2}{3x^{1/3}}\cdot\frac{x^{1/3}}{x^{1/3}}+\frac{2}{3x^{2/3}}=\frac{2x^{1/3}}{3x^{2/3}}+\frac{2}{3x^{2/3}}=\frac{2x^{1/3}+2}{3x^{2/3}}$$

Horizontal tangent lines occur where $f'(x)=0$.
$$\frac{2x^{1/3}+2}{3x^{2/3}}=0$$
$$2x^{1/3}+2=0$$
$$x^{1/3}=-1$$
$$x=-1$$

Vertical tangent lines occur where $f'(x)$ is unbounded. $f'(x)$ is unbounded at 0.

Evaluate f at 0 and -1:
$$f(0)=0^{2/3}+2(0)^{1/3}=0$$
$$f(-1)=(-1)^{2/3}+2(-1)^{1/3}=-1$$

We conclude that the graph of f has a horizontal tangent line at $(-1,-1)$ and a vertical tangent line at $(0,0)$.

19. $\quad f(x)=x^{2/3}(x-10)=x^{5/3}-10x^{2/3}$
$$f'(x)=\frac{5}{3}x^{2/3}-\frac{2}{3}\cdot 10x^{-1/3}=\frac{5x^{2/3}}{3}-\frac{20}{3x^{1/3}}$$
$$=\frac{5x^{2/3}}{3}\cdot\frac{x^{1/3}}{x^{1/3}}-\frac{20}{3x^{1/3}}=\frac{5x}{3x^{1/3}}-\frac{20}{3x^{1/3}}=\frac{5x-20}{3x^{1/3}}$$

Horizontal tangent lines occur where $f'(x)=0$.
$$\frac{5x-20}{3x^{1/3}}=0$$
$$5x-20=0$$

$$5x = 20$$
$$x = 4$$

Vertical tangent lines occur where $f'(x)$ is unbounded. $f'(x)$ is unbounded at 0.

Evaluate f at 0 and 4:
$$f(0) = 0^{2/3}(0-10) = 0$$
$$f(4) = 4^{2/3}(4-10) = -6 \cdot 4^{2/3} = -6 \cdot (8 \cdot 2)^{1/3} = -6 \cdot 2 \cdot 2^{1/3} = -12 \cdot 2^{1/3} = -12\sqrt[3]{2}$$

We conclude that the graph of f has a horizontal tangent line at $(4, -12\sqrt[3]{2})$ and a vertical tangent line at $(0, 0)$.

21. $f(x) = x^{2/3}(x^2 - 16) = x^{8/3} - 16x^{2/3}$

$$f'(x) = \frac{8}{3}x^{5/3} - \frac{2}{3} \cdot 16x^{-1/3} = \frac{8x^{5/3}}{3} - \frac{32}{3x^{1/3}}$$

$$= \frac{8x^{5/3}}{3} \cdot \frac{x^{1/3}}{x^{1/3}} - \frac{32}{3x^{1/3}} = \frac{8x^2}{3x^{1/3}} - \frac{32}{3x^{1/3}} = \frac{8x^2 - 32}{3x^{1/3}}$$

Horizontal tangent lines occur where $f'(x) = 0$.
$$\frac{8x^2 - 32}{3x^{1/3}} = 0$$
$$8x^2 - 32 = 0$$
$$8(x^2 - 4) = 0$$
$$(x - 2)(x + 2) = 0$$
$$x - 2 = 0 \quad \text{or} \quad x + 2 = 0$$
$$x = 2 \quad \text{or} \quad x = -2$$

Vertical tangent lines occur where $f'(x)$ is unbounded. $f'(x)$ is unbounded at 0.

Evaluate f at 2, -2, and 0:
$$f(2) = 2^{2/3}(2^2 - 16) = 2^{2/3}(-12) = -12 \cdot 2^{2/3} = -12\sqrt[3]{4}$$
$$f(-2) = (-2)^{2/3}((-2)^2 - 16) = 2^{2/3}(-12) = -12 \cdot 2^{2/3} = -12\sqrt[3]{4}$$
$$f(0) = 0^{2/3}(0^2 - 16) = 0$$

We conclude that the graph of f has horizontal tangent lines at $\left(2, -12\sqrt[3]{4}\right)$ and $\left(-2, -12\sqrt[3]{4}\right)$ and a vertical tangent line at $(0, 0)$.

23. $f(x) = \dfrac{x^{2/3}}{x - 2} \quad x \neq 2$

$$f'(x) = \frac{\frac{2}{3}x^{-1/3}\cdot(x-2)-x^{2/3}\cdot 1}{(x-2)^2} = \frac{1}{(x-2)^2}\left[\frac{2(x-2)}{3x^{1/3}}-x^{2/3}\right]$$

$$= \frac{1}{(x-2)^2}\left[\frac{2(x-2)}{3x^{1/3}}-x^{2/3}\cdot\frac{3x^{1/3}}{3x^{1/3}}\right]$$

$$= \frac{1}{(x-2)^2}\left[\frac{2(x-2)}{3x^{1/3}}-\frac{3x}{3x^{1/3}}\right]$$

$$= \frac{1}{(x-2)^2}\left[\frac{2x-4-3x}{3x^{1/3}}\right] = \frac{-x-4}{3x^{1/3}(x-2)^2}$$

Horizontal tangent lines occur where $f'(x) = 0$.

$$\frac{-x-4}{3x^{1/3}(x-2)^2} = 0$$

$$-x-4 = 0$$

$$x = -4$$

Vertical tangent lines occur where $f'(x)$ is unbounded. $f'(x)$ is unbounded if $x = 0$ or $x = 2$, but $x = 2$ is not part of the domain of f, so we disregard it.

Evaluate f at -4 and 0:

$$f(-4) = \frac{(-4)^{2/3}}{(-4)-2} = \frac{16^{1/3}}{-6} = -\frac{2\cdot 2^{1/3}}{6} = -\frac{2^{1/3}}{3} = -\frac{\sqrt[3]{2}}{3}$$

$$f(0) = \frac{0^{2/3}}{0-2} = 0$$

We conclude that the graph of f has a horizontal tangent line at $\left(-4, -\frac{\sqrt[3]{2}}{3}\right)$ and a vertical tangent line at $(0, 0)$.

25.

$$f(x) = \frac{x^{1/3}}{x-1} \qquad x\neq 1$$

$$f'(x) = \frac{\frac{1}{3}x^{-2/3}(x-1)-x^{1/3}\cdot 1}{(x-1)^2} = \frac{1}{(x-1)^2}\left[\frac{x-1}{3x^{2/3}}-x^{1/3}\right]$$

$$= \frac{1}{(x-1)^2}\left[\frac{x-1}{3x^{2/3}}-x^{1/3}\cdot\frac{3x^{2/3}}{3x^{2/3}}\right]$$

$$= \frac{1}{(x-1)^2}\left[\frac{x-1}{3x^{2/3}}-\frac{3x}{3x^{2/3}}\right]$$

$$= \frac{1}{(x-1)^2}\left[\frac{x-1-3x}{3x^{2/3}}\right] = \frac{-2x-1}{3x^{2/3}(x-1)^2}$$

Horizontal tangent lines occur where $f'(x) = 0$.

$$\frac{-2x-1}{3x^{2/3}(x-1)^2} = 0$$

$$-2x - 1 = 0$$

$$x = -\frac{1}{2}$$

Vertical tangent lines occur where $f'(x)$ is unbounded. $f'(x)$ is unbounded if $x = 0$ or $x = 1$, but $x = 1$ is not part of the domain of f, so we disregard it.

Evaluate f at $-\frac{1}{2}$ and 0:

$$f\left(-\frac{1}{2}\right) = \frac{\left(-\frac{1}{2}\right)^{1/3}}{-\frac{1}{2}-1} = \frac{\left(-\frac{1}{2}\right)^{1/3}}{-\frac{3}{2}} = \left(-\frac{1}{2}\right)^{1/3}\cdot\left(-\frac{2}{3}\right) = \frac{1}{2^{1/3}}\cdot\frac{2}{3} = \frac{2^{2/3}}{3} = \frac{\sqrt[3]{4}}{3}$$

$$f(0) = \frac{0^{1/3}}{0-1} = 0$$

We conclude that the graph of f has a horizontal tangent line at $\left(-\frac{1}{2}, \frac{\sqrt[3]{4}}{3}\right)$ and a vertical tangent line at $(0, 0)$.

27. (a) $f(0) = 0^{2/3} = 0$. The one-sided limits are

$$\lim_{x\to 0^-} f(x) = \lim_{x\to 0^-} x^{2/3} = 0 \qquad \lim_{x\to 0^+} f(x) = \lim_{x\to 0^+} x^{2/3} = 0$$

Since $\lim_{x\to 0} f(x) = f(0)$ the function is continuous at 0.

(b) The derivative of f at 0 is $f'(0) = \lim_{x\to 0}\frac{f(x)-f(0)}{x-0} = \lim_{x\to 0}\frac{x^{2/3}-0}{x-0}$.

We look at the one-sided limits:

$$\lim_{x\to 0^-}\frac{x^{2/3}-0}{x-0} = \lim_{x\to 0^-}\frac{x^{2/3}}{x} = \lim_{x\to 0^-}\frac{1}{x^{1/3}} = -\infty$$

$$\lim_{x\to 0^+}\frac{x^{2/3}-0}{x-0} = \lim_{x\to 0^+}\frac{x^{2/3}}{x} = \lim_{x\to 0^+}\frac{1}{x^{1/3}} = \infty$$

Since the one-sided limits are not equal, $f'(0)$ does not exist.

(c) The derivative is unbounded at $x = 0$, so there is a vertical tangent line at 0.

29. (a) $f(1)$ is not defined, so f is not continuous at $x = 1$.

(b) Since f is not continuous at $x = 1$, $f'(1)$ does not exist.

31. (a) $f(0) = 0^2 = 0$. The one sided limits are
$$\lim_{x \to 0^-} f(x) = \lim_{x \to 0^-} 3x = 0 \qquad \lim_{x \to 0^+} f(x) = \lim_{x \to 0^+} x^2 = 0$$
Since $\lim_{x \to 0} f(x) = f(0)$, the function f is continuous at 0.

(b) The derivative of f at 0 is $f'(0) = \lim_{x \to 0} \dfrac{f(x) - f(0)}{x - 0} = \lim_{x \to 0} \dfrac{f(x) - 0}{x - 0}$.

We look at the one-sided limits:
$$\lim_{x \to 0^-} \frac{f(x) - 0}{x - 0} = \lim_{x \to 0^-} \frac{3x - 0}{x - 0} = \lim_{x \to 0^-} \frac{3x}{x} = \lim_{x \to 0^-} 3 = 3$$
$$\lim_{x \to 0^+} \frac{f(x) - 0}{x - 0} = \lim_{x \to 0^+} \frac{x^2 - 0}{x - 0} = \lim_{x \to 0^+} \frac{x^2}{x} = \lim_{x \to 0^+} x = 0$$

Since the one-sided limits are not equal, $f'(0)$ does not exist.

(c) There is no tangent line at $x = 0$.

(d)

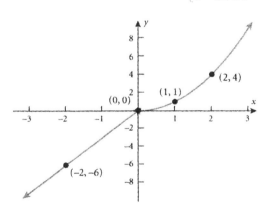

33. (a) $f(2) = 4 \cdot 2 = 8$. The one-sided limits are
$$\lim_{x \to 2^-} f(x) = \lim_{x \to 2^-} 4x = 4 \cdot 2 = 8 \qquad \lim_{x \to 2^+} f(x) = \lim_{x \to 2^+} x^2 = 2^2 = 4$$
Since the one-sided limits are unequal, the $\lim_{x \to 2} f(x)$ doesn't exist and the function is not continuous at $x = 2$.

(b) The function is not continuous at $x = 2$, so $f'(2)$ does not exist.

(c) Does not apply to this problem.

(d)

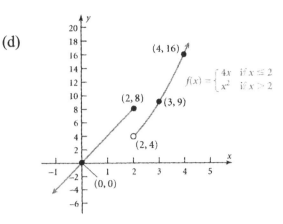

35. (a) $f(0) = 0^3 = 0$. The one-sided limits are:

$$\lim_{x \to 0^-} f(x) = \lim_{x \to 0^-} x^2 = 0^2 = 0 \qquad \lim_{x \to 0^+} f(x) = \lim_{x \to 0^+} x^3 = 0^3 = 0$$

The one-sided limits are equal so $\lim_{x \to 0} f(x)$ exists, and since $\lim_{x \to 0} f(x) = f(0)$, the function is continuous at $x = 0$.

(b) The derivative of f at 0 is $f'(0) = \lim_{x \to 0} \dfrac{f(x) - f(0)}{x - 0} = \lim_{x \to 0} \dfrac{f(x) - 0}{x - 0} = \lim_{x \to 0} \dfrac{f(x)}{x}$.

We look at one-sided limits.

$$\lim_{x \to 0^-} \frac{f(x)}{x} = \lim_{x \to 0^-} \frac{x^2}{x} = \lim_{x \to 0^-} x = 0 \qquad \lim_{x \to 0^+} \frac{f(x)}{x} = \lim_{x \to 0^+} \frac{x^3}{x} = \lim_{x \to 0^+} x^2 = 0$$

We conclude that $f'(0) = 0$, and that there is a horizontal tangent line at 0.

(c) Does not apply to this problem.

(d)

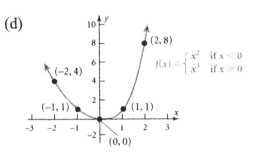

5.2 Increasing and Decreasing Functions; the First Derivative Test

1. The domain of f is $[x_1, x_9]$.

3. The graph of f is increasing on (x_1, x_4), (x_5, x_7), and (x_8, x_9).

5. $f'(x) = 0$ for x_4, x_7, and x_8.

7. f has a local maximum at (x_4, y_4) and at (x_7, y_7).

9. $f(x) = -2x^2 + 4x - 2$
    STEP 1 The domain of f is all real numbers.

    STEP 2 Let $x = 0$. Then $y = f(0) = -2$. The y-intercept is $(0, -2)$. Now let $y = 0$. Then
    $$-2x^2 + 4x - 2 = 0$$
    $$x^2 - 2x + 1 = 0$$
    $$(x - 1)^2 = 0$$
    $$x - 1 = 0$$
    $$x = 1$$
    The x-intercept is $(1, 0)$.

    STEP 3 To find where the graph is increasing or decreasing, we find $f'(x)$:
    $$f'(x) = -4x + 4$$

    The solution to $f'(x) = 0$ is
    $$-4x + 4 = 0$$
    $$x = 1$$
    Use 1 to separate the number line into 2 parts, and use $x = 0$ and $x = 2$ as test numbers.

    $$-\infty < x < 1 \quad \text{and} \quad 1 < x < \infty$$

    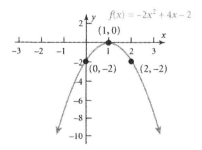

    $$f'(0) = 4 \qquad f'(2) = -4(2) + 4 = -4$$

    We conclude that the graph of f is increasing on the interval $(-\infty, 1)$; f is decreasing on the interval $(1, \infty)$.

    STEP 4 Since f is increasing to the left of 1 and decreasing to the right of 1, we conclude that there is a local maximum at $(1, f(1)) = (1, 0)$.

    STEP 5 We found $f'(1) = 0$, indicating that there is a horizontal tangent at $(1, 0)$. The first derivative is never unbounded, so there is no vertical tangent line.

    STEP 6 Since f is a polynomial function, its end behavior is that of $y = -2x^2$. Polynomial functions have no asymptotes.

11. $f(x) = x^3 - 9x^2 + 27x - 27$

STEP 1 The domain of f is all real numbers.

STEP 2 Let $x = 0$. Then $y = f(0) = -27$. The y-intercept is $(0, -27)$.
Now let $y = 0$. Then $x^3 - 9x^2 + 27x - 27 = 0$
$$(x-3)^3 = 0$$
$$x - 3 = 0$$
$$x = 3$$
The x-intercept is $(3, 0)$.

STEP 3 To find where the graph is increasing or decreasing, we find $f'(x)$:
$$f'(x) = 3x^2 - 18x + 27$$
The solution to $f'(x) = 0$ is
$$3x^2 - 18x + 27 = 0$$
$$3(x^2 - 6x + 9) = 0$$
$$3(x-3)^2 = 0$$
$$x - 3 = 0$$
$$x = 3$$
Use 3 to separate the number line into 2 parts, and use 0 and 4 as test numbers.

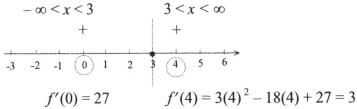

$$-\infty < x < 3 \qquad\qquad 3 < x < \infty$$

$$f'(0) = 27 \qquad\qquad f'(4) = 3(4)^2 - 18(4) + 27 = 3$$
We conclude that the function is always increasing that is on the interval $(-\infty, \infty)$.

STEP 4 There are no local extreme points since the first derivative never changes signs.

STEP 5 We found $f'(3) = 0$, indicating that there is a horizontal tangent at $(3, f(3)) = (3, 0)$.
  The first derivative is never unbounded, so there is no vertical tangent.

STEP 6 Since f is a polynomial function, its end behavior is that of $y = x^3$. Polynomial functions have no asymptotes.

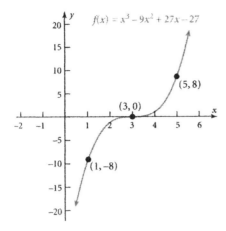

13. $f(x) = 2x^3 - 15x^2 + 36x$

STEP 1 The domain of f is all real numbers.

STEP 2 Let $x = 0$. Then $y = f(0) = 0$. The y-intercept is (0, 0).
Now let $y = 0$. Then $2x^3 - 15x^2 + 36x = 0$
$$x\left(2x^2 - 15x + 36\right) = 0$$
$$x = 0$$
The x-intercept is (0, 0).
($2x^2 - 15x + 36 = 0$ has no real solution; its discriminant, $b^2 - 4ac = -63$, is negative.)

STEP 3 To find where the graph is increasing or decreasing, we find $f'(x)$:
$$f'(x) = 6x^2 - 30x + 36$$
The solutions to $f'(x) = 0$ are
$$6x^2 - 30x + 36 = 0$$
$$6\left(x^2 - 5x + 6\right) = 0$$
$$6(x - 2)(x - 3) = 0$$
$$x - 2 = 0 \quad \text{or} \quad x - 3 = 0$$
$$x = 2 \quad \text{or} \quad x = 3$$
Use the numbers to separate the number line into 3 parts, and use 0, 2.5, and 4 as test numbers.

$$-\infty < x < 2 \qquad 2 < x < 3 \qquad 3 < x < \infty$$

$$+ \qquad\qquad - \qquad\qquad +$$

-1   (0)   1   2 (2.5) 3   (4)

$f'(0) = 36$ $\quad f'(2.5) = 6(2.5)^2 - 30(2.5) + 36 = -1.5$ $\quad f'(4) = 6(4)^2 - 30(4) + 36 = 12$

We conclude that the function is increasing on the intervals $(-\infty, 2)$ and $(3, \infty)$ and is decreasing on the interval (2, 3).

STEP 4 Since the function is increasing to the left of 2 and decreasing to the right of 2, the point $(2, f(2)) = (2, 28)$ is a local maximum.

The function is decreasing to the left of 3 and increasing to the right of 3, so the point $(3, f(3)) = (3, 27)$ is a local minimum.

STEP 5 We found $f'(2) = 0$ and $f'(3) = 0$, indicating that there are horizontal tangent lines at $(2, f(2)) = (2, 28)$ and $(3, f(3)) = (3, 27)$.

The first derivative is never unbounded, so there is no vertical tangent.

STEP 6 Since f is a polynomial function, its end behavior is that of $y = 2x^3$. Polynomial functions have no asymptotes.

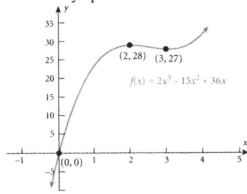

15. $f(x) = -x^3 + 3x - 1$

STEP 1 The domain of f is all real numbers.

STEP 2 Let $x = 0$. Then $y = f(0) = -1$. The y-intercept is $(0, -1)$. The x-intercept is hard to find, so we skip it.

STEP 3 To find where the graph is increasing or decreasing, we find $f'(x)$:
$$f'(x) = -3x^2 + 3$$
The solutions to $f'(x) = 0$ are
$$-3x^2 + 3 = 0$$
$$-3(x^2 - 1) = 0$$
$$-3(x-1)(x+1) = 0$$
$$x - 1 = 0 \quad \text{or} \quad x + 1 = 0$$
$$x = 1 \quad \text{or} \quad x = -1$$

We use the numbers 1 and -1 to separate the number line into three parts:
$$-\infty < x < -1 \qquad -1 < x < 1 \qquad 1 < x < \infty$$
and choose a test point from each interval.

For $x = -2$: $f'(-2) = -3(-2)^2 + 3 = -9$

For $x = 0$: $f'(0) = -3(0)^2 + 3 = 3$

For $x = 2$: $f'(2) = -3(2)^2 + 3 = -9$

We conclude that the graph of f is increasing on the interval $(-1, 1)$; f is decreasing on the intervals $(-\infty, -1)$ and $(1, \infty)$.

STEP 4 Since the graph is decreasing to the left of -1 and increasing to the right of -1, the point $(-1, f(-1)) = (-1, -3)$ is a local minimum.

Since the graph is increasing to the left of 1 and decreasing to the right of 1, the point $(1, f(1)) = (1, 1)$ is a local maximum.

STEP 5 $f'(x) = 0$ for $x = 1$ and $x = -1$. The graph of f has horizontal tangent lines at the points $(1, f(1)) = (1, 1)$ and $(-1, f(-1)) = (-1, -3)$.

The first derivative is never unbounded, so there is no vertical tangent.

STEP 6 Since f is a polynomial function, its end behavior is that of $y = -x^3$. Polynomial functions have no asymptotes.

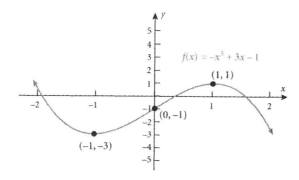

17. $f(x) = 3x^4 - 12x^3 + 2$

STEP 1 The domain of f is all real numbers.

STEP 2 Let $x = 0$. Then $y = f(0) = 2$. The y-intercept is $(0, 2)$.
The x-intercept is hard to find, so we skip it.

STEP 3 To find where the graph is increasing or decreasing, we find $f'(x)$:
$$f'(x) = 12x^3 - 36x^2$$
The solutions to $f'(x) = 0$ are
$$12x^3 - 36x^2 = 0$$
$$12x^2(x - 3) = 0$$
$$12x^2 = 0 \quad \text{or} \quad x - 3 = 0$$
$$x = 0 \quad \text{or} \quad x = 3$$

We use the numbers 0 and 3 to separate the number line into three parts:
$$-\infty < x < 0 \qquad 0 < x < 3 \qquad 3 < x < \infty$$
and choose a test number from each interval.

For $x = -1$: $f'(-1) = 12(-1)^3 - 36(-1)^2 = -48$

For $x = 1$: $f'(1) = 12(1)^3 - 36(1)^2 = -24$

For $x = 4$: $f'(4) = 12(4)^3 - 36(4)^2 = 192$

We conclude that the graph of f is increasing on the interval $(3, \infty)$; f is decreasing on the interval $(-\infty, 3)$.

STEP 4 Since the graph is decreasing to the left of 3 and increasing to the right of 3, the point $(3, f(3)) = (3, -79)$ is a local minimum.
  There is no local maximum.

STEP 5 $f'(x) = 0$ for $x = 0$ and $x = 3$. The graph of f has horizontal tangent lines at the points $(0, f(0)) = (0, 2)$ and $(3, f(3)) = (3, -79)$.
  The first derivative is never unbounded, so there is no vertical tangent.

STEP 6 Since f is a polynomial function, its end behavior is that of $y = 3x^4$. Polynomial functions have no asymptotes.

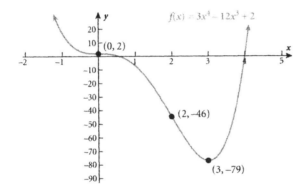

19. $f(x) = x^5 - 5x + 1$

  STEP 1 The domain of f is all real numbers.

  STEP 2 Let $x = 0$. Then $y = f(0) = 1$. The y-intercept is $(0, 1)$.
  The x-intercept is hard to find, so we skip it.

  STEP 3 To find where the graph is increasing or decreasing, we find $f'(x)$:
  $$f'(x) = 5x^4 - 5$$
  The solutions to $f'(x) = 0$ are
  $$5x^4 - 5 = 0$$
  $$5(x^4 - 1) = 0$$
  $$5(x^2 - 1)(x^2 + 1) = 0$$
  $$5(x - 1)(x + 1)(x^2 + 1) = 0$$

$$x - 1 = 0 \quad \text{or} \quad x + 1 = 0 \quad \text{or} \quad x^2 + 1 = 0$$
$$x = 1 \quad \text{or} \quad x = -1$$

The discriminant of $x^2 + 1$ is negative, so $x^2 + 1 = 0$ has no solution.

We use the numbers -1 and 1 to separate the number line into three parts:
$$-\infty < x < -1 \qquad -1 < x < 1 \qquad 1 < x < \infty$$
and choose a test point from each part.

For $x = -2$: $\quad f'(-2) = 5(-2)^4 - 5 = 75$

For $x = 0$: $\quad f'(0) = 5(0)^4 - 5 = -5$

For $x = 2$: $\quad f'(2) = 5(2)^4 - 5 = 75$

We conclude that the graph of f is increasing on the intervals $(-\infty, -1)$ and $(1, \infty)$; f is decreasing on the interval $(-1, 1)$.

STEP 4 Since the graph is increasing to the left of -1 and decreasing to the right of -1, the point $(-1, f(-1)) = (-1, 5)$ is a local maximum.

The graph is decreasing to the left of 1 and increasing to the right of 1, so the point $(1, f(1)) = (1, -3)$ is a local minimum.

STEP 5 $f'(x) = 0$ for $x = -1$ and for $x = 1$. The graph of f has horizontal tangent lines at the points $(-1, f(-1)) = (-1, 5)$ and $(1, f(1)) = (1, -3)$.

The first derivative is never unbounded, so there is no vertical tangent

STEP 6 Since f is a polynomial function, its end behavior is that of $y = x^5$. Polynomial functions have no asymptotes.

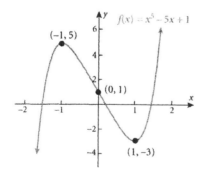

21. $f(x) = 3x^5 - 20x^3 + 1$

STEP 1 The domain of f is all real numbers.

STEP 2 Let $x = 0$. Then $y = f(0) = 1$. The y-intercept is $(0, 1)$.
The x-intercept is hard to find, so we skip it.

STEP 3 To find where the graph is increasing or decreasing, we find $f'(x)$:

$$f'(x) = 15x^4 - 60x^2$$

The solutions to $f'(x) = 0$ are

$$15x^4 - 60x^2 = 0$$
$$15x^2(x^2 - 4) = 0$$
$$15x^4(x - 2)(x + 2) = 0$$

$$15x^4 = 0 \quad \text{or} \quad x - 2 = 0 \quad \text{or} \quad x + 2 = 0$$
$$x = 0 \quad \text{or} \quad x = 2 \quad \text{or} \quad x = -2$$

We use the numbers -2, 0, and 2 to separate the number line into four parts:

$$-\infty < x < -2 \qquad -2 < x < 0 \qquad 0 < x < 2 \qquad 2 < x < \infty$$

and choose a test point from each part.

For $x = -3$: $\quad f'(x)(-3) = 15(-3)^4 - 60(-3)^2 = 675$

For $x = -1$: $\quad f'(x)(-1) = 15(-1)^4 - 60(-1)^2 = -45$

For $x = 1$: $\quad f'(x)(1) = 15(1)^4 - 60(1)^2 = -45$

For $x = 3$: $\quad f'(x)(3) = 15(3)^4 - 60(3)^2 = 675$

We conclude that the graph of f is increasing on the intervals $(-\infty, -2)$ and $(2, \infty)$; f is decreasing on the interval $(-2, 2)$.

STEP 4 Since the graph is increasing to the left of -2 and decreasing to the right of -2, the point $(-2, f(-2)) = (-2, 65)$ is a local maximum.

The graph is decreasing to the left of 2 and increasing to the right of 2, so the point $(2, f(2)) = (2, -63)$ is a local minimum.

STEP 5 $f'(x) = 0$ for $x = -2$, $x = 0$, and $x = 2$. The graph of f has horizontal tangent lines at the points $(-2, f(-2)) = (-2, 65)$, $(0, f(0)) = (0, 1)$, and $(2, f(2)) = (2, -63)$.

The first derivative is never unbounded, so there is no vertical tangent

STEP 6 Since f is a polynomial function, its end behavior is that of $y = 3x^5$. Polynomial functions have no asymptotes.

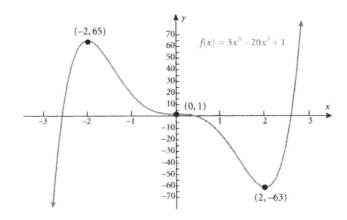

23. $f(x) = x^{2/3} + 2x^{1/3}$

STEP 1 The domain of f is the set of all real numbers.

STEP 2 Let $x = 0$. Then $y = f(0) = 0$. The y-intercept is $(0, 0)$.
Now let $y = 0$. Then $x^{2/3} + 2x^{1/3} = 0$

$$x^{1/3}\left(x^{1/3} + 2\right) = 0$$
$$x^{1/3} = 0 \quad \text{or} \quad x^{1/3} + 2 = 0$$
$$x = 0 \quad \text{or} \quad x^{1/3} = -2$$
$$x = -8$$

The x-intercepts are $(0, 0)$ and $(8, 0)$.

STEP 3 To find where the graph is increasing or decreasing, we find $f'(x)$:

$$f'(x) = \frac{2}{3}x^{-1/3} + \frac{2}{3}x^{-2/3}$$

The solutions to $f'(x) = 0$ are

$$\frac{2}{3}x^{-1/3} + \frac{2}{3}x^{-2/3} = 0$$
$$x^{-1/3} + x^{-2/3} = 0$$
$$\frac{1}{x^{1/3}} + \frac{1}{x^{2/3}} = 0$$
$$\frac{x^{1/3} + 1}{x^{2/3}} = 0$$
$$x^{1/3} = -1$$

$f'(x)$ is not defined when $x = 0$, $f'(x) = 0$ when $x = -1$.
We use these two numbers to separate the number line into three parts:

$$-\infty < x < -1 \qquad -1 < x < 0 \qquad 0 < x < \infty$$

and we choose a test point in each part.

For $x = -8$: $f'(x)(-8) = \frac{2}{3}(-8)^{-1/3} + \frac{2}{3}(-8)^{-2/3} = -\frac{1}{6} \approx -0.167$

For $x = -\frac{1}{8}$: $f'(x)\left(-\frac{1}{8}\right) = \frac{2}{3}\left(-\frac{1}{8}\right)^{-1/3} + \frac{2}{3}\left(-\frac{1}{8}\right)^{-2/3} = \frac{4}{3} \approx 1.333$

For $x = 1$: $f'(x)(1) = \frac{2}{3}(1)^{-1/3} + \frac{2}{3}(1)^{-2/3} = \frac{4}{3} \approx 1.333$

We conclude that the graph of f is increasing on the interval $(-1, \infty)$; f is decreasing on the interval $(-\infty, -1)$.

STEP 4 The graph is decreasing to the left of -1 and increasing to the right of -1, so the point $(-1, f(-1)) = (-1, -1)$ is a local minimum.
There is no local maximum.

STEP 5 $f'(x) = 0$ for $x = -1$. The graph of f has a horizontal tangent line at the point $(-1, f(-1)) = (-1, -1)$.

   The first derivative is unbounded at $x = 0$, so there is a vertical tangent line at the point $(0, f(0)) = (0, 0)$.

STEP 6 For the end behavior of f, we look at the two limits at infinity:

$$\lim_{x \to -\infty} f(x) = \lim_{x \to -\infty} \left(x^{2/3} + 2x^{1/3} \right) = \lim_{x \to -\infty} x^{2/3} = \left(\lim_{x \to -\infty} x^{1/3} \right)^2 = \infty$$

$$\lim_{x \to \infty} f(x) = \lim_{x \to \infty} \left(x^{2/3} + 2x^{1/3} \right) = \lim_{x \to \infty} x^{2/3} = \left(\lim_{x \to \infty} x^{1/3} \right)^2 = \infty$$

The graph of f becomes unbounded in the positive direction as $x \to \pm \infty$.

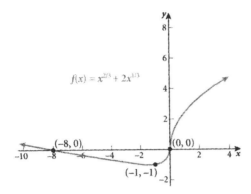

25. $f(x) = \left(x^2 - 1 \right)^{2/3}$

STEP 1 The domain of f is all real numbers.

STEP 2 Let $x = 0$. Then $y = f(0) = (0-1)^{2/3} = (-1)^{2/3} = 1$. The y-intercept is $(0, 1)$.

Now let $y = 0$. Then $\left(x^2 - 1 \right)^{2/3} = 0$

$$x^2 - 1 = 0$$
$$x^2 = 1$$
$$x = \pm 1$$

The x-intercepts are $(-1, 0)$ and $(1, 0)$.

STEP 3 To find where the graph is increasing or decreasing, we find $f'(x)$:

$$f'(x) = \frac{2}{3}\left(x^2 - 1 \right)^{-1/3} \cdot 2x = \frac{4x}{3\left(x^2 - 1 \right)^{1/3}} \qquad x \neq \pm 1$$

The solution to $f'(x) = 0$ is

$$\frac{4x}{3\left(x^2 - 1 \right)^{1/3}} = 0$$
$$4x = 0$$
$$x = 0$$

We use the numbers -1, 0, and 1 to separate the number line into four parts:

$$-\infty < x < -1 \qquad -1 < x < 0 \qquad 0 < x < 1 \qquad 1 < x < \infty$$

and choose a test number from each part.

For $x = -8$: $\quad f'(-8) = \dfrac{4(-8)}{3\left[(-8)^2 - 1\right]^{1/3}} = -2.681$

For $x = -\dfrac{1}{8}$: $\quad f'\left(-\dfrac{1}{8}\right) = \dfrac{4\left(-\dfrac{1}{8}\right)}{3\left[\left(-\dfrac{1}{8}\right)^2 - 1\right]^{1/3}} = 0.168$

For $x = \dfrac{1}{8}$: $\quad f'\left(\dfrac{1}{8}\right) = \dfrac{4\left(\dfrac{1}{8}\right)}{3\left[\left(\dfrac{1}{8}\right)^2 - 1\right]^{1/3}} = -0.168$

For $x = 8$: $\quad f'(8) = \dfrac{4(8)}{3\left[(8)^2 - 1\right]^{1/3}} = 2.681$

We conclude that the graph of f is increasing on the intervals $(-1, 0)$ and $(1, \infty)$; f is decreasing on the intervals $(-\infty, -1)$ and $(0, 1)$.

STEP 4 The graph is decreasing to the left of -1 and increasing to the right of -1, so the point $(-1, f(-1)) = (-1, 0)$ is a local minimum. The graph is also decreasing to the left of 1 and increasing to the right of 1, so the point $(1, f(1)) = (1, 0)$ is another local minimum.

The graph is increasing to the left of 0 and decreasing to the right of 0, so the point $(0, f(0)) = (0, 1)$ is a local maximum.

STEP 5 $f'(x) = 0$ for $x = 0$. The graph of f has a horizontal tangent line at the point $(0, f(0)) = (0, 1)$. The first derivative is unbounded at $x = -1$ and $x = 1$, so there are vertical tangent lines at the points $(-1, 0)$ and $(1, 0)$.

STEP 6 For the end behavior we look at the limits at infinity. Since f is an even function, we need only to consider the limit as $x \to \infty$.

$$\lim_{x \to \infty} \left(x^2 - 1\right)^{2/3} = \lim_{x \to \infty} \left(x^2\right)^{2/3} = \lim_{x \to \infty} x^{4/3} = \infty$$

The graph becomes unbounded at $x \to \pm \infty$.

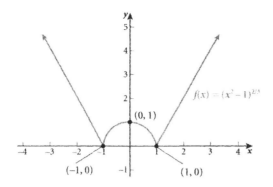

27.
$$f(x) = \frac{8}{x^2 - 16} = \frac{8}{(x-4)(x+4)} \qquad x \neq \pm 4$$

STEP 1 The domain of f is $\{x \mid x \neq -4, x \neq 4\}$.

STEP 2 Let $x = 0$. Then $y = f(0) = \dfrac{8}{0^2 - 16} = \dfrac{8}{-16} = -\dfrac{1}{2}$. The y-intercept is $\left(0, -\dfrac{1}{2} \right)$.

Now let $y = 0$. But $f(x) \neq 0$, so there is no x-intercept.

STEP 3 To find where the graph is increasing or decreasing, we find $f'(x)$:
$$f(x) = \frac{8}{x^2 - 16} = 8\left(x^2 - 16 \right)^{-1}$$
$$f'(x) = -8\left(x^2 - 16 \right)^{-2} \cdot 2x = \frac{-16x}{\left(x^2 - 16 \right)^2} \qquad x \neq \pm 4$$

$f'(x)$ is not defined when $x = -4$ or $x = 4$.
The solution to $f'(x) = 0$ is
$$-16x = 0$$
$$x = 0$$

We use the numbers -4, 0, and 4 to separate the number line into four parts:
$$-\infty < x < -4 \qquad -4 < x < 0 \qquad 0 < x < 4 \qquad 4 < x < \infty$$
and choose a test number from each part.

For $x = -5$: $f'(-5) = -\dfrac{16(-5)}{\left[(-5)^2 - 16 \right]^2} = 0.988$

For $x = -1$: $f'(-1) = -\dfrac{16(-1)}{\left[(-1)^2 - 16 \right]^2} = 0.071$

For $x = 1$: $f'(1) = -\dfrac{16(1)}{\left[(1)^2 - 16 \right]^2} = -0.071$

For $x = 5$: $\quad f'(5) = -\dfrac{16(5)}{\left[(5)^2 - 16\right]^2} = -0.988$

We conclude that the graph of f is increasing on the intervals $(-\infty, -4)$ and $(-4, 0)$; f is decreasing on the intervals $(0, 4)$ and $(4, \infty)$.

STEP 4 The graph is increasing to the left of 0 and decreasing to the right of 0, so the point $(0, f(0)) = \left(0, -\dfrac{1}{2}\right)$ is a local maximum.

There is no local minimum.

STEP 5 $f'(x) = 0$ for $x = 0$. The graph of f has a horizontal tangent line at the point $\left(0, \dfrac{1}{2}\right)$. The first derivative is unbounded at $x = -4$ and $x = 4$, but they are not in the domain of the function so there are no vertical tangent lines.

STEP 6 For the end behavior we look at the limits at infinity. Since f is an even function, we need only to consider the limit as $x \to \infty$.

$$\lim_{x \to \infty} f(x) = \lim_{x \to \infty} \left(\frac{8}{x^2 - 16}\right) = \frac{\lim_{x \to \infty} 8}{\lim_{x \to \infty} (x^2 - 16)} = \frac{8}{\lim_{x \to \infty} x^2} = 0$$

The x-axis ($y = 0$) is a horizontal asymptote of f as x becomes unbounded in the positive and negative directions.

Since f is a rational function and f is unbounded at $x = -4$ and $x = 4$, the graph of f will have vertical asymptotes at $x = -4$ and $x = 4$.

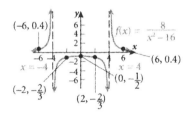

29.

$$f(x) = \frac{x}{x^2 - 9} = \frac{x}{(x-3)(x+3)} \qquad x \neq \pm 3$$

STEP 1 The domain of f is $\{x \mid x \neq -3, x \neq 3\}$.

STEP 2 Let $x = 0$. Then $y = f(0) = \dfrac{0}{0^2 - 9} = 0$. The y-intercept is $(0, 0)$.

Now let $y = 0$. Then $\dfrac{x}{x^2 - 9} = 0$ or $x = 0$. The x-intercept is also $(0, 0)$.

STEP 3 To find where the graph is increasing or decreasing, we find $f'(x)$:

$$f'(x) = \frac{(x^2-9)\cdot 1 - x\cdot 2x}{(x^2-9)^2} = \frac{x^2-9-2x^2}{(x^2-9)^2} = -\frac{x^2+9}{(x^2-9)^2} \qquad x \neq \pm 3$$

$f'(x)$ is not defined when $x = -3$ or $x = 3$.

$f'(x) = 0$ when $-\dfrac{x^2+9}{(x^2-9)^2} = 0$ or when $x^2 + 9 = 0$, which has no solution. So $f'(x) \neq 0$.

We use the numbers -3 and 3 to separate the number line into three parts:
$$-\infty < x < -3 \qquad -3 < x < 3 \qquad 3 < x < \infty$$
and choose a test number from each part.

For $x = -4$: $\quad f'(-4) = -\dfrac{(-4)^2+9}{\left[(-4)^2-9\right]^2} = -0.510$

For $x = 0$: $\quad f'(0) = -\dfrac{0^2+9}{(0^2-9)^2} = -0.111$

For $x = 4$: $\quad f'(4) = -\dfrac{4^2+9}{(4^2-9)^2} = -0.510$

We conclude that the graph of f is decreasing on the intervals $(-\infty, -3)$, $(-3, 3)$, and $(3, \infty)$.

STEP 4 The graph is always decreasing so there is neither a local minimum nor a local maximum.

STEP 5 $f'(x) \neq 0$ so the graph of f has no horizontal tangent line.
     The first derivative is unbounded at $x = -3$ and $x = 3$, but these points are not in the domain of the function so there are no vertical tangent lines.

STEP 6 For the end behavior we look at the limits at infinity. Since f is an odd function, we need only to consider the limit as $x \to \infty$.
$$\lim_{x \to \infty} f(x) = \lim_{x \to \infty}\left(\frac{x}{x^2-9}\right) = \lim_{x \to \infty}\left(\frac{x}{x^2}\right) = \lim_{x \to \infty}\left(\frac{1}{x}\right) = 0$$
The x-axis is a horizontal asymptote to the graph as f becomes unbounded.

Since f is a rational function and f is unbounded at $x = -3$ and $x = 3$, the graph of f will have vertical asymptotes at $x = -3$ and $x = 3$.

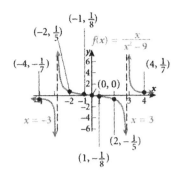

$(-1, \frac{1}{8})$
$(-2, \frac{1}{5})$
$f(x) = \frac{x}{x^2 - 9}$
$(-4, -\frac{1}{7})$ $(4, \frac{1}{7})$
$(0, 0)$
$x = -3$ $x = 3$
$(2, -\frac{1}{5})$
$(1, -\frac{1}{8})$

31.

$$f(x) = \frac{x^2}{x^2 - 4} \qquad x \neq \pm 2$$

STEP 1 The domain of f is $\{x \mid x \neq -2, x \neq 2\}$.

STEP 2 Let $x = 0$. Then $y = f(0) = \dfrac{0^2}{0^2 - 4} = 0$. The y-intercept is $(0, 0)$.

Now let $y = 0$. Then $\dfrac{x^2}{x^2 - 4} = 0$ or $x = 0$. The x-intercept is also $(0, 0)$.

STEP 3 To find where the graph is increasing or decreasing, we find $f'(x)$:

$$f'(x) = \frac{(x^2 - 4) \cdot 2x - x^2 \cdot 2x}{(x^2 - 4)^2} = \frac{2x^3 - 8x - 2x^3}{(x^2 - 4)^2} = -\frac{8x}{(x^2 - 4)^2} \qquad x \neq \pm 2$$

$f'(x)$ is not defined when $x = -2$ or $x = 2$.
The solutions to $f'(x) = 0$ are

$$-\frac{8x}{(x^2 - 4)^2} = 0 \text{ or } x = 0$$

We use the numbers -2, 0, and 2 to separate the number line into four parts:

$$-\infty < x < -2 \qquad -2 < x < 0 \qquad 0 < x < 2 \qquad 2 < x < \infty$$

and choose a test number from each part.

For $x = -3$: $f'(-3) = -\dfrac{8(-3)}{\left[(-3)^2 - 4\right]^2} = 0.96$

For $x = -1$: $f'(-1) = -\dfrac{8(-1)}{\left[(-1)^2 - 4\right]^2} = 0.889$

For $x = 1$: $f'(1) = -\dfrac{8(1)}{(1^2 - 4)^2} = -0.889$

For $x = 3$: $f'(3) = -\dfrac{8(3)}{(3^2 - 4)^2} = -0.96$

We conclude that the graph of f is increasing on the intervals $(-\infty, -2)$ and $(-2, 0)$; f is decreasing on the intervals $(0, 2)$ and $(2, \infty)$.

STEP 4 The graph is increasing to the left of 0 and decreasing to the right of 0, so the point $(0, f(0)) = (0, 0)$ is a local maximum. There is no local minimum.

STEP 5 $f'(x) = 0$ for $x = 0$. The graph of f has a horizontal tangent line at the point $(0, f(0)) = (0, 0)$.

The first derivative is unbounded at $x = -2$ and $x = 2$, but these points are not in the domain of the function so there are no vertical tangent lines.

STEP 6 For the end behavior we look at the limits at infinity. Since f is an even function, we need only to consider the limit as $x \to \infty$.

$$\lim_{x \to \infty} f(x) = \lim_{x \to \infty} \frac{x^2}{x^2 - 4} = \lim_{x \to \infty} \frac{x^2}{x^2} = \lim_{x \to \infty} 1 = 1$$

The line $y = 1$ is a horizontal asymptote to the graph as f becomes unbounded. Since f is a rational function and f is unbounded at $x = -2$ and $x = 2$, the graph of f will have vertical asymptotes at $x = -2$ and $x = 2$.

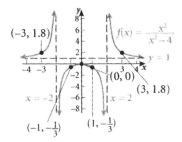

33. $f(x) = x \ln x$

STEP 1 The domain of f is $\{x \mid x > 0\}$.

STEP 2 $x = 0$ is not in the domain of f so there is no y-intercept.
Let $y = 0$. Then $x \ln x = 0$
$$x = 0 \quad \text{or} \quad \ln x = 0$$
$$x = 1$$
We disregard $x = 0$ since it is not in the domain of f. The x-intercept is $(1, 0)$.

STEP 3 To find where the graph is increasing or decreasing, we find $f'(x)$:
$$f'(x) = x \cdot \frac{1}{x} + 1 \cdot \ln x = 1 + \ln x$$
The solution to $f'(x) = 0$ is
$$1 + \ln x = 0$$
$$\ln x = -1$$
$$x = \frac{1}{e} \approx 0.368$$

We use the number 0.368 to separate the positive number line into two parts:

$$0 < x < 0.368 \qquad 0.368 < x < \infty$$

and choose a test number from each part.

For $x = 0.1$: $\quad f'(0.1) = 1 + \ln 0.1 = -1.303$

For $x = 1$: $\quad f'(1) = 1 + \ln 1 = 1$

We conclude that the graph of f is increasing on the interval $\left(\dfrac{1}{e}, \infty \right)$; f is decreasing on

the interval $\left(0, \dfrac{1}{e} \right)$.

STEP 4 The graph is decreasing to the left of $\dfrac{1}{e}$ and increasing to the right of $\dfrac{1}{e}$, so the

point $\left(\dfrac{1}{e}, f\left(\dfrac{1}{e}\right) \right) = \left(\dfrac{1}{e}, -\dfrac{1}{e} \right)$ is a local minimum. There is no local maximum.

STEP 5 $f'(x) = 0$ for $x = \dfrac{1}{e}$. The graph of f has a horizontal tangent line at the point

$\left(\dfrac{1}{e}, -\dfrac{1}{e} \right)$. There are no vertical tangent lines.

STEP 6 For the end behavior we look at the limit at infinity.

$$\lim_{x \to \infty} f(x) = \lim_{x \to \infty} (x \ln x) = \infty$$

The graph of the function becomes unbounded as $x \to \infty$.

Since f becomes unbounded as $x \to 0$, there is a vertical tangent at $x = 0$.

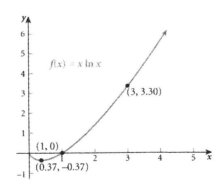

35. S is increasing on the interval where $S'(x) > 0$.

$$S'(x) = 8x + 50$$
$$8x + 50 > 0$$

$$8x > -50$$
$$x > -\frac{50}{8} = -\frac{25}{4} = -6.25$$

Since $S'(x) > 0$ on the domain of S, S is an increasing function.

37. (a) R is increasing on the interval where $R'(x) > 0$ and decreasing where $R'(x) < 0$.
$$R'(x) = -0.010x + 20$$
$$-0.010x + 20 > 0$$

$$-0.010x > -20$$
$$x < 2000$$

The graph of R is increasing on the interval $(0, 2000)$ and it is decreasing on the interval $(2000, \infty)$.

(b) Since the function R is increasing to the left of 2000 and decreasing to the right of 2000, then selling $x = 2000$ trucks will maximize revenue.

(c) The maximum revenue is $R(2000) = -0.005(2000)^2 + 20(2000) = \$20{,}000$.

(d)

39. (a) A is increasing on the interval where $A'(t) > 0$ and decreasing where $A'(t) < 0$.
$$A'(t) = -238.4t + 113.4$$
$$-238.4t + 113.4 > 0$$
$$-238.4t > -113.4$$
$$t < 0.476$$

The function A is increasing on the interval $(0, 0.476)$.

(b) According to this model the acreage of wheat planted will be decreasing from 2004 to 2008.

41. (a) The yield is increasing when $f'(x) > 0$.
$$f'(x) = -0.417 + \frac{1}{2} \cdot 0.852x^{-1/2} = -0.417 + \frac{0.426}{x^{1/2}} = \frac{-0.417x^{1/2} + 0.426}{x^{1/2}} \qquad x > 0$$

$f'(x) = 0$ when

$$-0.417x^{1/2} + 0.426 = 0$$
$$-0.417x^{1/2} = -0.426$$
$$x^{1/2} = 1.022$$
$$x = 1.044$$

We use 1.044 to separate the positive number line into two parts.

$$0 < x < 1.044 \qquad 1.044 < x < \infty$$

Testing a point in each interval we find

when $x = 1$: $f'(1) = \dfrac{-0.417(1)^{1/2} + 0.426}{(1)^{1/2}} = 0.009$

when $x = 4$: $f'(4) = \dfrac{-0.417(4)^{1/2} + 0.426}{4^{1/2}} = \dfrac{-0.417(2) + 0.426}{2} = -0.204$

The yield will be increasing when the amount of nitrogen is in the interval (0, 1.044).

(b) The yield will be decreasing when the amount of nitrogen is greater than 1.044.

43. $f(x) = 2x^2 - 2x$

1. f is a polynomial function. It is continuous everywhere on its domain. So it is continuous on [0, 1].

2. f is a polynomial function. It is differentiable everywhere on its domain. So f is differentiable on (0, 1).

3. $f(0) = 0, \qquad f(1) = 2(1)^2 - 2(1) = 2 - 2 = 0$

$f'(x) = 4x - 2$, $f'(x) = 0$ when

$$4x - 2 = 0$$
$$4x = 2$$
$$x = \frac{1}{2}$$

Since $\dfrac{1}{2}$ is in the interval [0, 1], Rolle's Theorem is verified.

45. $f(x) = x^4 - 1$

1. f is a polynomial function. It is continuous everywhere on its domain. So it is continuous on [− 1, 1].

2. A polynomial function is differentiable everywhere on its domain. So f is differentiable on (− 1, 1).

3. $f(-1) = (-1)^4 - 1 = 1 - 1 = 0$ $\qquad\qquad$ $f(1) = 1^4 - 1 = 1 - 1 = 0$

$f'(x) = 4x^3$, $f'(x) = 0$ when $4x^3 = 0$ or when $x = 0$. Since 0 is in the interval $[-1, 1]$, Rolle's Theorem is verified.

47. $f(x) = x^2$

1. f is a polynomial function. It is continuous everywhere on its domain. So it is continuous on $[0, 3]$.

2. A polynomial function is differentiable everywhere on its domain. So f is differentiable on $(0, 3)$.

3. $f(0) = 0^2 = 0$ $\qquad\qquad$ $f(3) = 3^2 = 9$

$$\frac{f(b) - f(a)}{b - a} = \frac{f(3) - f(0)}{3 - 0} = \frac{9 - 0}{3 - 0} = \frac{9}{3} = 3$$

$f'(x) = 2x$. $f'(x) = 3$ when $2x = 3$ or when $x = \frac{3}{2} = 1.5$. Since $\frac{3}{2}$ is in the interval $[0, 3]$,

the Mean Value Theorem is verified.

49. $f(x) = \frac{1}{x^2}$ $\quad x \neq 0$

1. f is a rational function. It is continuous everywhere on its domain. So it is continuous on $[1, 2]$.

2. A rational function is differentiable everywhere on its domain. So f is differentiable on $(1, 2)$.

3. $f(1) = \frac{1}{1^2} = 1$ $\qquad\qquad$ $f(2) = \frac{1}{2^2} = \frac{1}{4}$

$$\frac{f(b) - f(a)}{b - a} = \frac{f(2) - f(1)}{2 - 1} = \frac{\frac{1}{4} - 1}{1} = -\frac{3}{4}$$

$$f'(x) = -2x^{-3} = -\frac{2}{x^3}. \quad f'(x) = -\frac{3}{4} \text{ when}$$

$$-\frac{2}{x^3} = -\frac{3}{4}$$

$$-3x^3 = -8$$

$$x^3 = \frac{8}{3}$$

$$x = \frac{2}{\sqrt[3]{3}} \approx 1.387$$

Since 1.387 is in the interval $[1, 2]$, the Mean Value Theorem is verified.

5.3 Concavity; the Second Derivative Test

1. The domain of f is $\{x \mid x_1 \le x < x_4 \text{ or } x_4 < x < x_7\}$ or all the x in the interval $[x_1, x_4)$ or (x_4, x_7).

3. The graph of the function is increasing on the intervals and (x_1, x_3), $(0, x_4)$, and (x_4, x_6).

5. $f'(x) = 0$ when $x = 0$ and when $x = x_6$.

7. f has a local maximum at x_3 and x_6.
9. The graph is concave up on the intervals (x_1, x_3) and (x_3, x_4).

11. There is a vertical asymptote at $x = x_4$.

13. $f(x) = x^3 - 6x^2 + 1$
$f'(x) = 3x^2 - 12x$
$f''(x) = 6x - 12$
$f''(x) > 0$ when $6x - 12 > 0$ or when $6x > 12$ or $x > 2$.
The graph of f is concave up on the interval $(2, \infty)$. It is concave down on the interval $(-\infty, 2)$. The inflection point is $(2, f(2)) = (2, -15)$.

15. $f(x) = x^4 - 2x^3 + 6x - 1$
$f'(x) = 4x^3 - 6x^2 + 6$
$f''(x) = 12x^2 - 12x$
We solve $f''(x) = 0$ and use test points to determine the intervals for which f is concave up and concave down.
$$12x^2 - 12x = 0$$
$$12x(x - 1) = 0$$
$$12x = 0 \quad \text{or} \quad x - 1 = 0$$
$$x = 0 \quad \text{or} \quad x = 1$$

We separate the number line into three parts:
$$-\infty < x < 0 \quad 0 < x < 1 \quad 1 < x < \infty$$
For $x = -1$: $\quad f''(-1) = 12(-1)^2 - 12(-1) = 24$
For $x = 0.5$: $\quad f''(0.5) = 12(0.5)^2 - 12(0.5) = -3$
For $x = 2$: $\quad f''(2) = 12(2)^2 - 12(2) = 24$

We conclude that the graph of f is concave up on the intervals $(-\infty, 0)$ and $(1, \infty)$; f is concave down on the interval $(0, 1)$. The inflection points are $(0, f(0)) = (0, -1)$ and $(1, f(1)) = (1, 4)$.

17. $f(x) = 3x^5 - 5x^4 + 60x + 10$

$f'(x) = 15x^4 - 20x^3 + 60$

$f''(x) = 60x^3 - 60x^2$

We solve $f''(x) = 0$ and use test points to determine the intervals for which f is concave up and concave down.

$$60x^3 - 60x^2 = 0$$
$$60x^2(x-1) = 0$$

$$60x^2 = 0 \quad \text{or} \quad x - 1 = 0$$
$$x = 0 \quad \text{or} \quad x = 1$$

We separate the number line into three parts:

$$-\infty < x < 0 \qquad 0 < x < 1 \qquad 1 < x < \infty$$

For $x = -1$: $\qquad f''(-1) = 60(-1)^3 - 60(-1)^3 = -120$

For $x = 0.5$: $\qquad f''(0.5) = 60(0.5)^3 - 60(0.5)^3 = -7.5$

For $x = 2$: $\qquad f''(2) = 60(2)^3 - 60(2)^3 = 240$

We conclude that the graph of f is concave up on the interval $(1, \infty)$; f is concave down on the interval $(-\infty, 1)$. The inflection point is $(1, f(1)) = (1, 68)$.

19. $f(x) = 3x^5 - 10x^3 + 10x + 10$

$f'(x) = 15x^4 - 30x^2 + 10$

$f''(x) = 60x^3 - 60x$

We solve $f''(x) = 0$ and use test points to determine the intervals for which f is concave up and concave down.

$$60x^3 - 60x = 0$$
$$60x(x^2 - 1) = 0$$
$$60x(x - 1)(x + 1) = 0$$
$$60x = 0 \quad \text{or} \quad x - 1 = 0 \quad \text{or} \quad x + 1 = 0$$
$$x = 0 \quad \text{or} \quad x = 1 \quad \text{or} \quad x = -1$$

We separate the number line into four parts:

$$-\infty < x < -1 \qquad -1 < x < 0 \qquad 0 < x < 1 \qquad 1 < x < \infty$$

For $x = -2$: $\quad f''(-2) = 60(-2)^3 - 60(-2) = -360$

For $x = -0.5$: $\quad f''(-0.5) = 60(-0.5)^3 - 60(-0.5) = 22.5$

For $x = 0.5$: $\quad f''(0.5) = 60(0.5)^3 - 60(0.5) = -22.5$

For $x = 2$: $f''(2) = 60(2)^3 - 60(2) = 360$

We conclude that the graph of f is concave up on the intervals $(-1, 0)$ and $(1, \infty)$; f is concave down on the intervals $(-\infty, -1)$ and $(0, 1)$. The inflection points are $(-1, f(-1)) = (-1, 7)$, $(0, f(0)) = (0, 10)$, and $(1, f(1)) = (1, 13)$.

21. $f(x) = x^5 - 10x^2 + 4$

$f'(x) = 5x^4 - 20x$

$f''(x) = 20x^3 - 20 = 20(x^3 - 1)$

We solve $f''(x) = 0$ and use test points to determine the intervals for which f is concave up and concave down.

$$20(x^3 - 1) = 0$$

$$20(x - 1)(x^2 + x + 1) = 0$$

$$x - 1 = 0 \quad \text{or} \quad x^2 + x + 1 = 0$$
$$x = 1$$

The discriminant of $x^2 + x + 1$ is negative, and so $x^2 + x + 1 = 0$ has no real solution. We use 1 to separate the number line into two parts:

$$-\infty < x < 1 \qquad 1 < x < \infty$$

For $x = 0$: $f''(0) = 20(0^3 - 1) = -20$

For $x = 2$: $f''(2) = 20(2^3 - 1) = 140$

We conclude that the graph of f is concave up on the interval $(1, \infty)$; f is concave down on the interval $(-\infty, 1)$. The inflection point is $(1, f(1)) = (1, -5)$.

23. $f(x) = 3x^{1/3} + 9x + 2$

$f'(x) = \dfrac{1}{3} \cdot 3x^{-2/3} + 9 = x^{-2/3} + 9$

$f''(x) = -\dfrac{2}{3} x^{-5/3} = -\dfrac{2}{3x^{5/3}}$

$f''(x)$ is unbounded at $x = 0$. We use 0 to separate the number line into two parts:

$$-\infty < x < 0 \qquad 0 < x < \infty$$

For $x = -1$: $f''(-1) = -\dfrac{2}{3(-1)^{5/3}} = \dfrac{2}{3}$

For $x = 1$: $f''(1) = -\dfrac{2}{3(1)^{5/3}} = -\dfrac{2}{3}$

We conclude that the graph of f is concave up on the interval $(-\infty, 0)$; f is concave down on the interval $(0, \infty)$. The inflection point is $(0, f(0)) = (0, 2)$.

25. $f(x) = x^{2/3}(x - 10) = x^{5/3} - 10x^{2/3}$

$$f'(x) = \frac{5}{3}x^{2/3} - \frac{20}{3}x^{-1/3}$$

$$f''(x) = \frac{10}{9}x^{-1/3} + \frac{20}{9}x^{-4/3} = \frac{10}{9x^{1/3}} + \frac{20}{9x^{4/3}} = \frac{10}{9x^{1/3}} \cdot \frac{x}{x^{3/3}} + \frac{20}{9x^{4/3}} = \frac{10x + 20}{9x^{4/3}}$$

$f''(x)$ is unbounded at $x = 0$. We solve $f''(x) = 0$ and use test points to determine the intervals for which f is concave up and concave down.

$$\frac{10x + 20}{9x^{4/3}} = 0$$
$$10x + 20 = 0$$
$$10x = -20$$
$$x = -2$$

We use 0 and -2 to separate the number line into three parts:

$$-\infty < x < -2 \qquad -2 < x < 0 \qquad 0 < x < \infty$$

For $x = -3$: $f''(-3) = f''(-3) = \dfrac{10(-3) + 20}{9(-3)^{4/3}} = -0.257$

For $x = -1$: $f''(-1) = \dfrac{10(-1) + 20}{9(-1)^{4/3}} = 1.111$

For $x = 1$: $f''(1) = \dfrac{10(1) + 20}{9(1)^{4/3}} = 3.333$

We conclude that the graph of f is concave up on the intervals $(-2, 0)$ and $(0, \infty)$; f is concave down on the interval $(-\infty, -2)$. The inflection point is $(-2, f(-2)) = (-2, -19.049)$.

27. $f(x) = x^{2/3}(x^2 - 16) = x^{8/3} - 16x^{2/3}$

$$f'(x) = \frac{8}{3}x^{5/3} - \frac{32}{3}x^{-1/3}$$

$$f''(x) = \frac{40}{9}x^{2/3} + \frac{32}{9}x^{-4/3} = \frac{40x^{2/3}}{9} + \frac{32}{9x^{4/3}} = \frac{40x^2 + 32}{9x^{4/3}}$$

$f''(x)$ is unbounded at $x = 0$. We solve $f''(x) = 0$ and use test points to determine the intervals for which f is concave up and concave down.

$$\frac{40x^2 + 32}{9x^{4/3}} = 0 \quad \text{or} \quad 40x^2 + 32 = 0$$

The discriminant of $40x^2 + 32$ is negative, so $40x^2 + 32 = 0$ has no real solution.

We use 0 to separate the number line into two parts:
$$-\infty < x < 0 \qquad\qquad 0 < x < \infty$$

For $x = -1$: $f''(-1) = \dfrac{40(-1)^2 + 32}{9(-1)^{4/3}} = 8$

For $x = 1$: $f''(1) = \dfrac{40(1)^2 + 32}{9(1)^{4/3}} = 8$

Since $f''(x)$ is positive for all x, we conclude that the graph of f is concave up on the interval $(-\infty, 0)$ and $(0, \infty)$. There is no inflection point.

29. $f(x) = x^3 - 6x^2 + 1$

STEP 1 Since f is a polynomial, the domain of f is all real numbers.

STEP 2 Let $x = 0$. Then $y = f(0) = 1$. The y-intercept is $(0, 1)$. The x-intercept is hard to find, so we skip it.

STEP 3 To find where the graph is increasing or decreasing, we find $f'(x)$:
$$f'(x) = 3x^2 - 12x$$
The solutions to $f'(x) = 0$ are
$$3x^2 - 12x = 0$$
$$3x(x-4) = 0$$
$$3x = 0 \quad \text{or} \quad x - 4 = 0$$
$$x = 0 \quad \text{or} \quad x = 4$$
We use the numbers to separate the number line into three parts:
$$-\infty < x < 0 \qquad 0 < x < 4 \qquad 4 < x < \infty$$
and choose a test number from each part.

For $x = -2$: $f'(-2) = 3(-2)^2 - 12(-2) = 36$

For $x = 2$: $f'(2) = 3(2)^2 - 12(2) = -12$

For $x = 5$: $f'(5) = 3(5)^2 - 12(5) = 15$

We conclude that the graph of f is increasing on the intervals $(-\infty, 0)$ and $(4, \infty)$; f is decreasing on the interval $(0, 4)$.

STEP 4 The graph is increasing to the left of 0 and decreasing to the right of 0, so the point $(0, f(0)) = (0, 1)$ is a local maximum.

The graph is decreasing to the left of 4 and increasing to the right of 4, so the point $(4, f(4)) = (4, -31)$ is a local minimum.

STEP 5 $f'(x) = 0$ for $x = 0$ and $x = 4$. The graph of f has a horizontal tangent line at the points $(0, 1)$ and $(4, -31)$.

The first derivative is never unbounded, so there is no vertical tangent.

STEP 6 Since f is a polynomial function, its end behavior is that of $y = x^3$. Polynomial functions have no asymptotes.

STEP 7 To identify the inflection point, if any, we find $f''(x)$
$$f''(x) = 6x - 12$$

$f''(x) = 0$ when $6x = 12$, that is when $x = 2$.

We use the number 2 to separate the number line into two parts:
$$-\infty < x < 2 \qquad 2 < x < \infty$$
and choose a test number from each part.

For $x = 0$: $f''(0) = 6(0) - 12 = -12$

For $x = 3$: $f''(3) = 6(3) - 12 = 6$

We conclude that the graph of f is concave up on the interval $(2, \infty)$ and is concave down on the interval $(-\infty, 2)$. Since the concavity changes at the point $(2, -15)$, it is an inflection point.

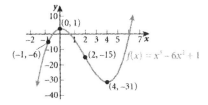

31. $f(x) = x^4 - 2x^2 + 1$

STEP 1 Since f is a polynomial, the domain of f is all real numbers.

STEP 2 Let $x = 0$. Then $y = f(0) = 1$. The y-intercept is $(0, 1)$.
Now let $y = 0$. Then $x^4 - 2x^2 + 1 = 0$
$$\left(x^2 - 1\right)^2 = 0$$
$$x^2 = 1; \quad x = \pm 1$$
The x-intercepts are $(1, 0)$ and $(-1, 0)$.

STEP 3 To find where the graph is increasing or decreasing, we find $f'(x)$:
$$f'(x) = 4x^3 - 4x$$
The solutions to $f'(x) = 0$ are
$$4x^3 - 4x = 0$$
$$4x\left(x^2 - 1\right) = 0$$
$$4x(x-1)(x+1) = 0$$
$$4x = 0 \quad \text{or} \quad x - 1 = 0 \quad \text{or} \quad x + 1 = 0$$
$$x = 0 \quad \text{or} \quad x = 1 \quad \text{or} \quad x = -1$$

We use the numbers to separate the number line into four parts:

$$-\infty < x < -1 \qquad -1 < x < 0 \qquad 0 < x < 1 \qquad 1 < x < \infty$$

and choose a test number from each part.

For $x = -2$ $\qquad f'(-2) = 4(-2)^3 - 4(-2) = -24$

For $x = -0.5$ $\qquad f'(-0.5) = 4(-0.5)^3 - 4(-0.5) = 1.5$

For $x = 0.5$ $\qquad f'(0.5) = 4(0.5)^3 - 4(0.5) = -1.5$

For $x = 2$ $\qquad f'(2) = 4(2)^3 - 4(2) = 24$

We conclude that the graph of f is increasing on the intervals $(-1, 0)$ and $(1, \infty)$; f is decreasing on the intervals $(-\infty, -1)$ and $(0, 1)$.

STEP 4 The graph is decreasing to the left of -1 and increasing to the right of -1, so the point $(-1, f(-1)) = (-1, 0)$ is a local minimum. The graph is also decreasing to the left of 1 and increasing to the right of 1, so the point $(1, f(1)) = (1, 0)$ is another local minimum.

The graph is increasing to the left of 0 and decreasing to the right of 0, so the point $(0, f(0)) = (0, 1)$ is a local maximum.

STEP 5 The graph of f has a horizontal tangent line at the points $(-1, f(-1)) = (-1, 0)$, $(1, f(1)) = (1, 0)$, $(0, f(0)) = (0, 1)$.

The first derivative is never unbounded, so there is no vertical tangent.

STEP 6 Since f is a polynomial function, its end behavior is that of $y = x^4$. Polynomial functions have no asymptotes.

STEP 7 To identify the inflection point, if any, we find $f''(x)$:

$$f''(x) = 12x^2 - 4$$

$f''(x) = 0$ when $12x^2 - 4 = 0$

$$12x^2 = 4$$

$$x^2 = \frac{1}{3} \quad \text{or} \quad x = \pm \frac{1}{\sqrt{3}} \approx \pm 0.577$$

We use the numbers ± 0.577 to separate the number line into three parts

$$-\infty < x < -0.577 \qquad -0.577 < x < 0.577 \qquad 0.577 < x < \infty$$

and choose a test number from each part.

For $x = -1$: $\qquad f''(-1) = 8$

For $x = 0$: $\qquad f''(0) = -4$

For $x = 1$: $\qquad f''(1) = 8$

We conclude that the graph of f is concave up on the intervals $(-\infty, -0.577)$ and $(0.577, \infty)$ and is concave down on the interval $(-0.577, 0.577)$. Since the concavity changes at the points $(-0.577, 0.445)$ and $(0.577, 0.445)$ they are inflection points.

33. $f(x) = x^5 - 10x^4$

STEP 1 Since f is a polynomial, the domain of f is all real numbers.

STEP 2 Let $x = 0$. Then $y = f(0) = 0$. The y-intercept is $(0, 0)$.
Now let $y = 0$. Then $x^5 - 10x^4 = 0$
$$x^4(x - 10) = 0$$
$$x = 0 \qquad \text{or} \qquad x - 10 = 0$$
$$x = 10$$
The x-intercepts are $(0, 0)$ and $(10, 0)$.

STEP 3 To find where the graph is increasing or decreasing, we find $f'(x)$:
$$f'(x) = 5x^4 - 40x^3$$
$f'(x) = 0$ when $5x^4 - 40x^3 = 0$
$$5x^3(x - 8) = 0$$
$$5x^3 = 0 \quad \text{or} \quad x - 8 = 0$$
$$x = 0 \quad \text{or} \quad x = 8$$
We use the numbers to separate the number line into three parts:
$$-\infty < x < 0 \qquad 0 < x < 8 \qquad 8 < x < \infty$$
and choose a test number from each part.

For $x = -1$: $\qquad f'(-1) = 5(-1)^4 - 40(-1)^3 = 45$

For $x = 1$: $\qquad f'(1) = 5(1)^4 - 40(1)^3 = -35$

For $x = 10$: $\qquad f'(10) = 5(10)^4 - 40(10)^3 = 10,000$

We conclude that the graph of f is increasing on the intervals $(-\infty, 0)$ and $(8, \infty)$; f is decreasing on the interval $(0, 8)$.

STEP 4 The graph is increasing to the left of 0 and decreasing to the right of 0, so the point $(0, f(0)) = (0, 0)$ is a local maximum.
 The graph is decreasing to the left of 8 and increasing to the right of 8, so the point $(8, f(8)) = (8, -8192)$ is a local minimum.

STEP 5 The graph of f has a horizontal tangent line at the points $(0, 0)$ and $(8, -8192)$.
 The first derivative is never unbounded, so there is no vertical tangent.

STEP 6 Since f is a polynomial function, its end behavior is that of $y = x^5$. Polynomial functions have no asymptotes.

STEP 7 To identify the inflection point, if any, we find $f''(x)$:

$$f''(x) = 20x^3 - 120x^2$$

$f''(x) = 0$ when $20x^3 - 120x^2 = 0$

$$20x^2(x-6) = 0$$

$$20x^2 = 0 \quad \text{or} \quad x - 6 = 0$$
$$x = 0 \quad \text{or} \quad x = 6$$

We use the numbers to separate the number line into three parts

$$-\infty < x < 0 \qquad 0 < x < 6 \qquad 6 < x < \infty$$

and choose a test number from each part.

For $x = -1$: $f''(-1) = 20(-1)^3 - 120(-1)^2 = -140$

For $x = 1$: $f''(1) = 20(1)^3 - 120(1)^2 = -100$

For $x = 7$: $f''(7) = 20(7)^3 - 120(7)^2 = 980$

We conclude that the graph of f is concave down on the intervals $(-\infty, 0)$ and $(0, 6)$ and is concave up on the interval $(6, \infty)$. Since the concavity changes at the point $(6, -5184)$, it is an inflection point.

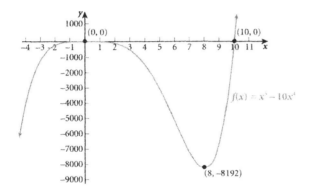

35. $f(x) = x^6 - 3x^5$

STEP 1 Since f is a polynomial, the domain of f is all real numbers.

STEP 2 Let $x = 0$. Then $y = f(0) = 0$. The y-intercept is $(0, 0)$.
Now let $y = 0$. Then $x^6 - 3x^5 = 0$

$$x^5(x-3) = 0$$
$$x^5 = 0 \quad \text{or} \quad x - 3 = 0$$
$$x = 0 \quad \text{or} \quad x = 3$$

The x-intercepts are $(0, 0)$ and $(3, 0)$.

STEP 3 To find where the graph is increasing or decreasing, we find $f'(x)$:

$$f'(x) = 6x^5 - 15x^4$$

$f'(x) = 0$ when $6x^5 - 15x^4 = 0$

$$3x^4(2x - 5) = 0$$

$$3x^4 = 0 \quad \text{or} \quad 2x - 5 = 0$$

$$x = 0 \quad \text{or} \quad x = \frac{5}{2}$$

We use the numbers to separate the number line into three parts:

$$-\infty < x < 0 \qquad 0 < x < 2.5 \qquad 2.5 < x < \infty$$

and choose a test number from each part.

For $x = -1$: $\quad f'(-1) = 6(-1)^5 - 15(-1)^4 = -21$

For $x = 1$: $\quad f'(1) = 6(1)^5 - 15(1)^4 = -9$

For $x = 3$: $\quad f'(3) = 6(3)^5 - 15(3)^4 = 243$

We conclude that the graph of f is increasing on the interval $(2.5, \infty)$; f is decreasing on the intervals $(-\infty, 0)$ and $(0, 2.5)$.

STEP 4 The graph is decreasing to the left of 2.5 and increasing to the right of 2.5, so the point $(2.5, f(2.5)) = (2.5, -48.83)$ is a local minimum.
            There is no local maximum.

STEP 5 The graph of f has a horizontal tangent line at the point $(2.5, -48.83)$ and $(0, 0)$. The first derivative is never unbounded, so there is no vertical tangent.

STEP 6 Since f is a polynomial function, its end behavior is that of $y = x^6$. Polynomial functions have no asymptotes.

STEP 7 To identify the inflection point, if any, we find $f''(x)$:

$$f''(x) = 30x^4 - 60x^3$$

$f''(x) = 0$ when $30x^4 - 60x^3 = 0$

$$30x^3(x - 2) = 0$$

$$30x^3 = 0 \quad \text{or} \quad x - 2 = 0$$

$$x = 0 \quad \text{or} \quad x = 2$$

We use the numbers to separate the number line into three parts

$$-\infty < x < 0 \qquad 0 < x < 2 \qquad 2 < x < \infty$$

and choose a test number from each part.

For $x = -1$: $\quad f''(-1) = 30(-1)^4 - 60(-1)^3 = 90$

For $x = 1$: $\quad f''(1) = 30(1)^4 - 60(1)^3 = -30$

For $x = 3$: $\quad f''(3) = 30(3)^4 - 60(3)^3 = 810$

We conclude that the graph of f is concave up on the intervals $(-\infty, 0)$ and $(2, \infty)$ and is concave down on the interval $(0, 2)$. Since the concavity changes at the points $(0, 0)$ and $(2, -32)$, they are inflection points.

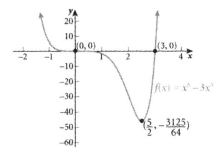

37. $f(x) = 3x^4 - 12x^3$

STEP 1 Since f is a polynomial, the domain of f is all real numbers.

STEP 2 Let $x = 0$. Then $y = f(0) = 0$. The y-intercept is $(0, 0)$.
Now let $y = 0$. Then $3x^4 - 12x^3 = 0$

$$3x^3(x-4) = 0$$

$$3x^3 = 0 \qquad \text{or} \qquad x - 4 = 0$$
$$x = 0 \qquad \text{or} \qquad x = 4$$

The x-intercepts are $(0, 0)$ and $(4, 0)$.

STEP 3 To find where the graph is increasing or decreasing, we find $f'(x)$:

$$f'(x) = 12x^3 - 36x^2$$

$f'(x) = 0$ when $12x^3 - 36x^2 = 0$

$$12x^2(x-3) = 0$$

$$12x^2 = 0 \quad \text{or} \quad x - 3 = 0$$
$$x = 0 \quad \text{or} \quad x = 3$$

We use the numbers to separate the number line into three parts:

$$-\infty < x < 0 \qquad 0 < x < 3 \qquad 3 < x < \infty$$

and choose a test number from each part.

For $x = -1$: $\qquad f'(-1) = 12(-1)^3 - 36(-1)^2 = -48$

For $x = 1$: $\qquad f'(1) = 12(1)^3 - 36(1)^2 = -24$

For $x = 4$: $\qquad f'(4) = 12(4)^3 - 36(4)^2 = 192$

We conclude that the graph of f is increasing on the interval $(3, \infty)$; f is decreasing on the intervals $(-\infty, 0)$ and $(0, 3)$.

STEP 4 The graph is decreasing to the left of 3 and increasing to the right of 3, so the point $(3, f(3)) = (3, -81)$ is a local minimum.
There is no local maximum.

STEP 5 The graph of f has a horizontal tangent line at the points $(0, 0)$ and $(3, -81)$. The first derivative is never unbounded, so there is no vertical tangent.

STEP 6 Since f is a polynomial function, its end behavior is that of $y = 3x^4$. Polynomial functions have no asymptotes.

STEP 7 To identify the inflection point, if any, we find $f''(x)$:
$$f''(x) = 36x^2 - 72x = 36x(x-2)$$
$f''(x) = 0$ when $36x(x-2) = 0$

$$36x = 0 \quad \text{or} \quad x - 2 = 0$$
$$x = 0 \quad \text{or} \quad x = 2$$

We use the numbers 0 and 2 to separate the number line into three parts
$$-\infty < x < 0 \qquad 0 < x < 2 \qquad 2 < x < \infty$$
and choose a test number from each part.

For $x = -1$: $\qquad f''(-1) = 36(-1)(-1-2) = 108$

For $x = 1$: $\qquad f''(1) = 36(1)(1-2) = -36$

For $x = 3$: $\qquad f''(3) = 36(3)(3-2) = 108$

We conclude that the graph of f is concave up on the intervals $(-\infty, 0)$ and $(2, \infty)$ and is concave down on the interval $(0, 2)$. Since the concavity changes at the points $(0, 0)$ and $(2, -48)$, they are inflection points.

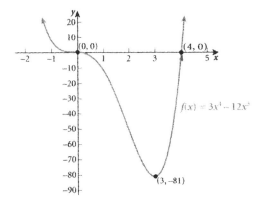

39. $\quad f(x) = x^5 - 10x^2 + 4$

STEP 1 Since f is a polynomial, the domain of f is all real numbers.

STEP 2 Let $x = 0$. Then $y = f(0) = 4$. The y-intercept is $(0, 4)$.
The x-intercept is hard to find, so we skip it.

STEP 3 To find where the graph is increasing or decreasing, we find $f'(x)$:
$$f'(x) = 5x^4 - 20x$$
$f'(x) = 0$ when $5x^4 - 20x = 0$

$$5x(x^3 - 4) = 0$$
$$5x = 0 \quad \text{or} \quad x^3 - 4 = 0$$
$$x = 0 \quad \text{or} \quad x^3 = 4$$
$$x = \sqrt[3]{4} \approx 1.587$$

We use the numbers to separate the number line into three parts:
$$-\infty < x < 0 \qquad 0 < x < 1.587 \qquad 1.587 < x < \infty$$

and choose a test number from each part.

For $x = -1$: $f'(-1) = 5(-1)^4 - 20(-1) = 25$

For $x = 1$: $f'(1) = 5(1)^4 - 20(1) = -15$

For $x = 2$: $f'(2) = 5(2)^4 - 20(2) = 40$

We conclude that the graph of f is increasing on the intervals $(-\infty, 0)$ and $(1.587, \infty)$; f is decreasing on the interval $(0, 1.587)$.

STEP 4 The graph is increasing to the left of 0 and decreasing to the right of 0, so the point $(0, f(0)) = (0, 4)$ is a local maximum.

The graph is decreasing to the left of 1.587 and increasing to the right of 1.587, so the point $(1.587, f(1.587)) = (1.587, -11.119)$ is a local minimum.

STEP 5 The graph of f has a horizontal tangent line at the points $(0, 4)$ and $(1.587, -11.119)$. The first derivative is never unbounded, so there is no vertical tangent.

STEP 6 Since f is a polynomial function, its end behavior is that of $y = x^5$. Polynomial functions have no asymptotes.

STEP 7 To identify the inflection point, if any, we find $f''(x)$:

$$f''(x) = 20x^3 - 20 = 20(x^3 - 1)$$

$f''(x) = 0$ when $20(x^3 - 1) = 0$

$$x^3 - 1 = 0 \quad \text{or } x = 1$$

We use the number 1 to separate the number line into two parts:

$$-\infty < x < 1 \qquad 1 < x < \infty$$

and choose a test number from each part.

For $x = 0$: $f''(0) = 20[0 - 1] = -20$

For $x = 2$: $f''(2) = 20[2^3 - 1] = 140$

We conclude that the graph of f is concave up on the interval $(1, \infty)$ and is concave down on the interval $(-\infty, 1)$. Since the concavity changes at the point $(1, -5)$, it is an inflection point.

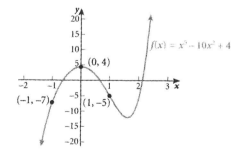

41. $f(x) = x^{2/3}(x-10)$

STEP 1 The domain of f is all real numbers.

STEP 2 Let $x = 0$. Then $y = f(0) = 0$. The y-intercept is $(0, 0)$.
Now let $y = 0$. Then $x^{2/3}(x-10) = 0$

$$x^{2/3} = 0 \quad \text{or} \quad x - 10 = 0$$
$$x = 0 \quad \text{or} \quad x = 10$$

The x-intercepts are $(0, 0)$ and $(10, 0)$.

STEP 3 To find where the graph is increasing or decreasing, we find $f'(x)$:

$$f'(x) = x^{2/3} + \frac{2}{3}x^{-1/3}(x-10) = \frac{3x^{2/3} \cdot x^{1/3}}{3x^{1/3}} + \frac{2x-20}{3x^{1/3}} = \frac{5x-20}{3x^{1/3}}$$

$f'(x)$ is unbounded when $x = 0$. $f'(x) = 0$ when $5x - 20 = 0$ or when $x = 4$.
We use the critical numbers to separate the number line into three parts:

$$-\infty < x < 0 \qquad 0 < x < 4 \qquad 4 < x < \infty$$

and choose a test number from each part.

For $x = -1$: $f'(-1) = \dfrac{5(-1)-20}{3(-1)^{1/3}} = 8.333$

For $x = 1$: $f'(1) = \dfrac{5(1)-20}{3(1)^{1/3}} = -5$

For $x = 5$: $f'(5) = \dfrac{5(5)-20}{3(5)^{1/3}} = 0.975$

We conclude that the graph of f is increasing on the intervals $(-\infty, 0)$ and $(4, \infty)$; f is decreasing on the interval $(0, 4)$.

STEP 4 The graph is increasing to the left of 0 and decreasing to the right of 0, so the point $(0, f(0)) = (0, 0)$ is a local maximum.
  The graph is decreasing to the left of 4 and increasing to the right of 4, so the point $(4, f(4)) = (4, -15.12)$ is a local minimum.

STEP 5 The graph of f has a horizontal tangent line at the point $(4, -15.12)$.
  The first derivative is unbounded at $x = 0$, so there is a vertical tangent line at the point $(0, f(0)) = (0, 0)$.

STEP 6 For the end behavior of f, we look at the two limits at infinitiy:
  As x approaches $-\infty$, f becomes unbounded in the negative direction; as x approaches ∞, f becomes unbounded in the positive direction.

STEP 7 To identify the inflection point, if any, we find $f''(x)$:

$$f''(x) = \frac{10}{9}x^{-1/3} + \frac{20}{9}x^{-4/3} = \frac{10x+20}{9x^{4/3}}$$

$f''(x)$ is unbounded at $x = 0$.

$f''(x) = 0$ when $10x + 20 = 0$, that is when $x = -2$.

We use the numbers 0 and -2 to separate the number line into three parts:

$$-\infty < x < -2 \qquad -2 < x < 0 \qquad 0 < x < \infty$$

and choose a test number from each part.

For $x = -3$: $\qquad f''(-3) = \dfrac{10(-3) + 20}{9(-3)^{4/3}} = -0.257$

For $x = -1$: $\qquad f''(-1) = \dfrac{10(-1) + 20}{9(-1)^{4/3}} = 1.111$

For $x = 1$: $\qquad f''(1) = \dfrac{10(1) + 20}{9(1)^{4/3}} = 3.333$

We conclude that the graph of f is concave up on the intervals $(-2, 0)$ and $(0, \infty)$ and is concave down on the interval $(-\infty, -2)$. Since the concavity changes at the point $(-2, -19.049)$, it is an inflection point.

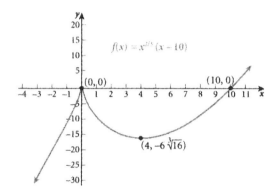

43. $f(x) = x^{2/3}(x^2 - 16)$

STEP 1 The domain of f is all real numbers.

STEP 2 Let $x = 0$. Then $y = f(0) = 0$. The y-intercept is $(0, 0)$.

Now let $y = 0$. Then $\quad x^{2/3}(x^2 - 16) = 0$

$$x^{2/3}(x - 4)(x + 4) = 0$$

$$x^{2/3} = 0 \quad \text{or} \quad x - 4 = 0 \quad \text{or} \quad x + 4 = 0$$

$$x = 0 \quad \text{or} \quad x = 4 \quad \text{or} \quad x = -4$$

The x-intercepts $(0, 0)$, $(-4, 0)$, and $(4, 0)$.

STEP 3 To find where the graph is increasing or decreasing, we find $f'(x)$:

$$f'(x) = x^{2/3} \cdot 2x + \frac{2}{3}x^{-1/3}(x^2 - 16) = \frac{2x^{5/2} \cdot 3x^{1/3}}{3x^{1/3}} + \frac{2(x^2 - 16)}{3x^{1/3}}$$

$$= \frac{6x^2 + 2x^2 - 32}{3x^{1/3}} = \frac{8x^2 - 32}{3x^{1/3}}$$

$f'(x)$ is not defined when $x = 0$. $f'(x) = 0$ when $8x^2 - 32 = 0$.

$$8x^2 = 32$$
$$x^2 = 4$$
$$x = \pm 2$$

We use the numbers to separate the number line into four parts:

$$-\infty < x < -2 \qquad -2 < x < 0 \qquad 0 < x < 2 \qquad 2 < x < \infty$$

and choose a test number from each part.

For $x = -3$: $\qquad f'(-3) = \dfrac{8(-3)^2 - 32}{3(-3)^{1/3}} = -9.245$

For $x = -1$: $\qquad f'(-1) = \dfrac{8(-1)^2 - 32}{3(-1)^{1/3}} = 8$

For $x = 1$: $\qquad f'(1) = \dfrac{8(1)^2 - 32}{3(1)^{1/3}} = -8$

For $x = 3$: $\qquad f'(3) = \dfrac{8(3)^2 - 32}{3(3)^{1/3}} = 9.245$

We conclude that the graph of f is increasing on the intervals $(-2, 0)$ and $(2, \infty)$; f is decreasing on the intervals $(-\infty, -2)$ and $(0, 2)$.

STEP 4 The graph is decreasing to the left of -2 and increasing to the right of -2, so the point $(-2, f(-2)) = (-2, -19.05)$ is a local minimum. The graph is also decreasing to the left of 2 and increasing to the right of 2, so the point $(2, f(2)) = (2, -19.05)$ is a local minimum.

   The graph is increasing to the left of 0 and decreasing to the right of 0, so the point $(0, f(0)) = (0, 0)$ is a local maximum.

STEP 5 The graph of f has a horizontal tangent line at the points $(-2, -19.05)$ and $(2, -19.05)$.

   The first derivative is unbounded at $x = 0$, so there is a vertical tangent line at the point $(0, 0)$.

STEP 6 For the end behavior we look at the limits at infinity. Since f is an even function, we need only to consider the limit as $x \to \infty$. As x approaches ∞, f becomes unbounded in the positive direction.

STEP 7 To identify the inflection point, if any, we find $f''(x)$

$$f''(x) = \frac{40}{9}x^{2/3} + \frac{32}{9}x^{-4/3} = \frac{40x^2 + 32}{9x^{4/3}}$$

$f''(x)$ is unbounded at $x = 0$.

$f''(x)$ is never equal to 0.

We use the number 0 to separate the number line into two parts

$$-\infty < x < 0 \qquad\qquad 0 < x < \infty$$

and choose a test number from each part.

For $x = -1$: $\qquad f''(-1) = \dfrac{40(-1)^2 + 32}{9(-1)^{4/3}} = 8$

For $x = 1$: $\qquad f''(1) = \dfrac{40(1)^2 + 32}{9(1)^{4/3}} = 8$

We conclude that the graph of f is always concave up, and has no inflection point.

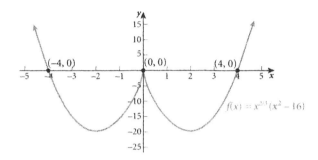

45. $f(x) = xe^x$

STEP 1 The domain of f is all real numbers.

STEP 2 Let $x = 0$. Then $y = f(0) = 0$. The y-intercept is $(0, 0)$.
Now let $y = 0$. Then $xe^x = 0$ or $x = 0$. The x-intercept is $(0, 0)$.

STEP 3 To find where the graph is increasing or decreasing, we find $f'(x)$:
$$f'(x) = xe^x + e^x = e^x(x+1)$$
$f'(x) = 0$ when $e^x(x+1) = 0$
$$x + 1 = 0 \quad \text{or} \quad x = -1$$

We use the numbers to separate the number line into two parts:

$$-\infty < x < -1 \qquad\qquad -1 < x < \infty$$

and choose a test number from each part.

For $x = -2$: $\qquad f'(-2) = e^{-2}(-2+1) = -0.135$

For $x = 0$: $\qquad f'(0) = e^0(0+1) = 1$

We conclude that the graph of f is increasing on the interval $(-1, \infty)$; f is decreasing on the interval $(-\infty, -1)$.

STEP 4 The graph is decreasing to the left of -1 and increasing to the right of -1, so the point $(-1, f(-1)) = (-1, -0.368)$ is a local minimum.
There is no local maximum.

STEP 5 The graph of f has a horizontal tangent line at the point $(-1, -0.368)$. There is no vertical tangent.

STEP 6 For the end behavior of f, we look at the two limits at infinitiy:
As x approaches $-\infty$, the graph of f approaches the x-axis. The line $y = 0$ is a horizontal asymptote as x becomes unbounded in the negative direction.
As x approaches ∞, y becomes unbounded in the positive direction. As x becomes unbounded the function f behaves like $y = e^x$.

STEP 7 To identify the inflection point, if any, we find $f''(x)$
$$f''(x) = xe^x + e^x + e^x = e^x(x+2)$$
$f''(x) = 0$ when $e^x(x+2) = 0$, that is when $x = -2$
We use the number -2 to separate the number line into two parts
$$-\infty < x < -2 \qquad\qquad -2 < x < \infty$$
and choose a test number from each part.
For $x = -3$: $\qquad f''(-3) = e^{-3}(-3+2) = -e^{-3} \approx -0.050$
For $x = 0$: $\qquad f''(0) = e^0(2) = 2$

We conclude that the graph of f is concave up on the interval $(-2, \infty)$ and is concave down on the interval $(-\infty, -2)$. Since the concavity changes at the point $(-2, -0.271)$, it is an inflection point.

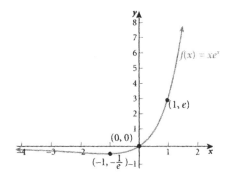

47. First locate the numbers for which $f'(x) = 0$.
$$f'(x) = 3x^2 - 3$$
$$3x^2 - 3 = 0$$
$$3(x^2 - 1) = 0$$
$$3(x-1)(x+1) = 0$$
$$x - 1 = 0 \quad \text{or} \quad x + 1 = 0$$
$$x = 1 \quad \text{or} \quad x = -1$$

Now evaluate $f''(x)$ at these numbers.
$$f''(x) = 6x$$

$f''(-1) = 6(-1) = -6 < 0$. By the Second Derivative Test, f has a local maximum at $(-1, f(-1)) = (-1, 4)$.

$f''(1) = 6(1) = 6 > 0$. By the Second Derivative Test, f has a local minimum at $(1, f(1)) = (1, 0)$.

49. First locate the numbers for which $f'(x) = 0$.
$$f'(x) = 12x^3 + 12x^2$$
$$12x^3 + 12x^2 = 0$$
$$12x^2(x+1) = 0$$
$$x = 0 \quad \text{or} \quad x = -1$$
Now evaluate $f''(x)$ at these numbers.
$$f''(x) = 36x^2 + 24x$$

$f''(0) = 36(0)^2 + 24(0) = 0$. The Second Derivative Test is inconclusive. To determine if a local maximum or minimum exists, we must use the First Derivative Test.
$$f'\left(-\frac{1}{2}\right) = 12\left(-\frac{1}{2}\right)^3 + 12\left(-\frac{1}{2}\right)^2 = -\frac{12}{8} + \frac{12}{4} = \frac{-6+12}{4} = \frac{3}{2} > 0$$
$$f'(1) = 12(1)^3 + 12(1)^2 = 24 > 0$$
Since the first derivative does not change signs, there is no local extreme point at $(0, f(0)) = (0, -3)$.

$f''(-1) = 36(-1)^2 + 24(-1) = 12 > 0$. By the Second Derivative Test, f has a local minimum at $(-1, f(-1)) = (-1, -4)$.

51. First locate the numbers for which $f'(x) = 0$.
$$f'(x) = 5x^4 - 20x^3$$
$$5x^4 - 20x^3 = 0$$
$$5x^3(x-4) = 0$$
$$x = 0 \quad \text{or} \quad x = 4$$

Now evaluate $f''(x)$ at these numbers.
$$f''(x) = 20x^3 - 60x^2$$

$f''(4) = 20(4)^3 - 60(4)^2 = 320$. By the Second Derivative Test, f has a local minimum at $(4, f(4)) = (4, -254)$.

$f''(0) = 20(0)^3 - 60(0)^2 = 0$. The Second Derivative Test is inconclusive. To determine if a local maximum or minimum exists at $x = 0$, we must use the First Derivative Test.
$$f'(-1) = 5(-1)^4 - 20(-1)^3 = 25 > 0$$
$$f'(1) = 5(1)^4 - 20(1)^3 = -15 < 0$$
Since the first derivative is positive to the left of 0 and negative to the right of 0, we conclude that there is a local maximum at $(0, f(0)) = (0, 2)$.

53.
$$f(x) = x + \frac{1}{x} = x + x^{-1} \qquad x \neq 0$$

First locate the numbers for which $f'(x) = 0$.

$$f'(x) = 1 - x^{-2} = 1 - \frac{1}{x^2} = \frac{x^2 - 1}{x^2} \qquad x \neq 0$$

$$\frac{x^2 - 1}{x^2} = 0$$

$$x^2 - 1 = 0$$

$$x = \pm 1$$

Now evaluate $f''(x)$ at these numbers.

$$f''(x) = 2x^{-3} = \frac{2}{x^3} \qquad x \neq 0$$

$f''(-1) = \dfrac{2}{(-1)^3} = -2 < 0$. By the Second Derivative Test, f has a local maximum at $(-1, -2)$.

$f''(1) = \dfrac{2}{(1)^3} = 2 > 0$. By the Second Derivative Test, f has a local minimum at $(1, f(1)) = (1, 2)$.

55. Answers will vary.

57. Answers will vary.

59. If the point $(1, 6)$ is an inflection point of the function f, then $f(1) = 6$ and $f''(1) = 0$. We first find $f''(x)$.

$$f(x) = ax^3 + bx^2 \qquad\qquad f(1) = a + b = 6 \qquad (1)$$
$$f'(x) = 3ax^2 + 2bx$$
$$f''(x) = 6ax + 2b \qquad\qquad f''(1) = 6a + 2b = 0 \qquad (2)$$

We solve the system of equations (1) and (2).

$$a = 6 - b \qquad (1)$$
$$6(6 - b) + 2b = 0 \qquad (2)$$
$$36 - 6b + 2b = 0$$
$$36 - 4b = 0$$
$$b = 9 \qquad (2)$$

Back-substituting $b = 9$ into equation (1) gives

$$a + 9 = 6 \qquad (1)$$
$$a = -3$$

The point $(1, 6)$ is an inflection point of the function $f(x) = -3x^3 + 9x^2$.

61.

(a) The average cost function is given by $\overline{C}(x) = \dfrac{C(x)}{x}$.

$$\overline{C}(x) = \frac{2x^2 + 50}{x} = 2x + \frac{50}{x}$$

(b) We locate points where $\overline{C}'(x) = 0$.

$$\overline{C}'(x) = 2 - 50x^{-2} = 2 - \frac{50}{x^2} = \frac{2x^2 - 50}{x^2}$$

$$\frac{2x^2 - 50}{x^2} = 0$$

$$2x^2 - 50 = 0$$

$$2(x^2 - 25) = 0$$

$$(x - 5)(x + 5) = 0$$

$$x = 5 \quad \text{or} \quad x = -5$$

We only consider $x = 5$, since $x = -5$ is not part of the domain of the function. We evaluate $\overline{C}''(5)$.

$$\overline{C}''(x) = 100x^{-3} = \frac{100}{x^3}$$

$$\overline{C}''(5) = \frac{100}{5^3} = \frac{100}{125} = \frac{4}{5} > 0$$

By the Second Derivative Test $x = 5$ is a local minimum and the minimum average cost will be $\overline{C}(5) = 2(5) + \dfrac{50}{5} = 10 + 10 = \$20$.

(c) The marginal cost function is given by the derivative $C'(x) = 4x$.

(d)

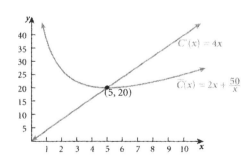

(e) Answers will vary.

63.

(a) The average cost function is given by $\overline{C}(x) = \dfrac{C(x)}{x}$.

$$\overline{C}(x) = \frac{500 + 10x + \frac{x^2}{500}}{x} = \frac{500}{x} + 10 + \frac{x}{500}$$

(b) We locate the points where $\overline{C}'(x) = 0$.

$$\overline{C}'(x) = -500x^{-2} + \frac{1}{500} = -\frac{500}{x^2} + \frac{1}{500} = \frac{x^2 - 500^2}{500x^2}$$

$$\frac{x^2 - 500^2}{500x^2} = 0$$

$$x^2 - 500^2 = 0$$

$$x^2 = 500^2$$

$$x = 500$$

We now evaluate $\overline{C}''(500)$.

$$\overline{C}''(x) = 1000x^{-3} = \frac{1000}{x^3}$$

$$\overline{C}''(500) = \frac{1000}{500^3} = 0.000008 > 0$$

By the Second Derivative Test $x = 500$ is a local minimum and the minimum average

cost will be $\overline{C}(500) = 500 + 10(500) + \frac{500^2}{500} = \$6000$.

(c) The marginal cost function is given by the derivative $C'(x) = 10 + \frac{x}{250}$.

(d)

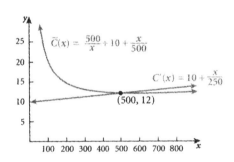

(e) Answers will vary.

65. (a) The domain of the function N is $\{t \mid t \geq 0\}$.

(b) To find the y-intercept we let $t = 0$.

$$N(0) = \frac{50,000}{1 + 49,999e^0} = \frac{50,000}{1 + 49,999} = 1$$

The y-intercept is $(0, 1)$.
There is no t-intercept because $N(t)$ never equals zero.

(c) $N(t) = \dfrac{50,000}{1+49,999e^{-t}} = 50,000\left(1+49,999e^{-t}\right)^{-1}$

$N'(t) = (-1) \cdot 50,000\left(1+49,999e^{-t}\right)^{-2} \cdot (-1) \cdot 49,999e^{-t}$

$N'(t) = \dfrac{50,000 \cdot 49,999e^{-t}}{\left(1+49,999e^{-t}\right)^2} > 0$

since $e^{-t} > 0$. The function is always increasing.

(d)

$N''(t) = \dfrac{-1 \cdot 50,000 \cdot 49,999e^{-t}\left(1+49,999e^{-t}\right)^2 - \left(50,000 \cdot 49,999e^{-t}\right) \cdot 2\left(1+49,999e^{-t}\right)(-1) \, 49,999e^{-t}}{\left(1+49,999e^{-t}\right)^4}$

$= \dfrac{\left[-50,000 \cdot 49,999e^{-t}\left(1+49,999e^{-t}\right)\right]\left[\left(1+49,999e^{-t}\right)-2 \cdot 49,999e^{-t}\right]}{\left(1+49,999e^{-t}\right)^4}$

$= \dfrac{-50,000 \cdot 49,999e^{-t}\left(1+49,999e^{-t}\right)\left(1-49,999e^{-t}\right)}{\left(1+49,999e^{-t}\right)^4}$

$= \dfrac{-50,000 \cdot 49,999e^{-t}\left(1-49,999e^{-t}\right)}{\left(1+49,999e^{-t}\right)^3} = \dfrac{50,000 \cdot 49,999e^{-t}\left(49,999e^{-t}-1\right)}{\left(1+49,999e^{-t}\right)^3}$

The sign of N'' is controlled by $49,999e^{-t}-1$ since the rest of the expression is always positive.

$49,999e^{-t} - 1 = 0$

$49,999e^{-t} = 1$

$e^{-t} = \dfrac{1}{49,999}$ or when $e^{t} = 49,999$

This occurs when $t = \ln 49,999 = 10.820$. If $t < 10.82$, $49,999e^{-t} - 1 > 0$ and $N''(t) > 0$.
We conclude that the function N is concave up on the interval $(0, 10.82)$ and concave down on the interval $(10.82, \infty)$.

(e) The inflection point is $(\ln 49,999, N(\ln 49,999)) \approx (10.82, 25,000)$.

(f)

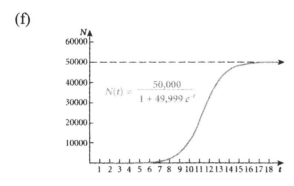

(g) Since $N''(t)$ is positive to the left of 10.82 and negative to the right of 10.82, by the First Derivative Test we conclude that the rumor is spreading at the greatest rate when $t = 10.82$ days.

67. (a) The growth rate of the population is given by $P'(t)$.

$$P(t) = \frac{800e^t}{1+0.1(e^t-1)} = \frac{800e^t}{0.9+0.1e^t} = \frac{8000e^t}{9+e^t}$$

$$P'(t) = \frac{8000e^t(9+e^t)-8000e^t(e^t)}{(9+e^t)^2}$$

$$= \frac{72{,}000e^t+8000e^{2t}-8000e^{2t}}{(9+e^t)^2} = \frac{72{,}000e^t}{(9+e^t)^2}$$

(b) To determine where the growth rate is maximum we look at $P''(t)$ and evaluate where it equal zero.

$$P''(t) = \frac{72{,}000e^t(9+e^t)^2 - 72{,}000e^t\left[2(9+e^t)(e^t)\right]}{(9+e^t)^4}$$

$$= \frac{72{,}000e^t(9+e^t)\left[(9+e^t)-2e^t\right]}{(9+e^t)^4}$$

$$= \frac{72{,}000e^t(9-e^t)}{(9+e^t)^3}$$

$P''(t) = 0$ when $9-e^t = 0$ since the denominator and $72{,}000e^t$ are always positive.

$9-e^t = 0$ when $e^t = 9$ or when $t = \ln 9 \approx 2.197$. Using the First Derivative Test we find that $(\ln 9, P(\ln 9)) \approx (2.197, 4000)$ is a local maximum. So the population is growing the fastest after about 2.197 days.

(c) $\lim\limits_{x \to \infty}\left(\dfrac{800e^t}{1+0.10(e^t-1)}\right) = \lim\limits_{x \to \infty}\left(\dfrac{8000e^t}{9+e^t}\right) = 8000 \cdot \lim\limits_{x \to \infty}\left(\dfrac{e^t}{9+e^t}\right) = 8000 \cdot \lim\limits_{x \to \infty}\left(\dfrac{e^t}{e^t}\right)$

$$= 8000 \cdot \lim\limits_{x \to \infty}(1) = 8000$$

(d)

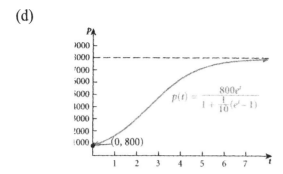

$p(t) = \dfrac{800 e^t}{1 + \frac{1}{10}(e^t - 1)}$

(0, 800)

69. (a) The sales rate of the car model is given by $f'(x)$.

$$f(x) = \frac{20,000}{1 + 50 e^{-x}} = 20,000\left(1 + 50\, e^{-x}\right)^{-1}$$

$$f'(x) = -20,000\left(1 + 50\, e^{-x}\right)^{-2} \cdot \left(-50\, e^{-x}\right) = \frac{100,000\, e^{-x}}{\left(1 + 50\, e^{-x}\right)^{2}}$$

The sales rate is maximum where $f''(x) = 0$.

$$f''(x) = \frac{-100,000\, e^{-x}\left(1 + 50\, e^{-x}\right)^{2} - \left(100,000\, e^{-x}\right)\left[2\left(1 + 50\, e^{-x}\right)\left(-50\, e^{-x}\right)\right]}{\left(1 + 50\, e^{-x}\right)^{4}}$$

$$= \frac{-100,000\, e^{-x}\left(1 + 50\, e^{-x}\right)^{2} + 100 \cdot 100,000\, e^{-2x}\left(1 + 50\, e^{-x}\right)}{\left(1 + 50\, e^{-x}\right)^{4}}$$

$$= \frac{-100,000\, e^{-x}\left(1 + 50\, e^{-x}\right) + 100 \cdot 100,000\, e^{-x}}{\left(1 + 50\, e^{-x}\right)^{3}}$$

$$= \frac{-100,000\, e^{-x} - 50 \cdot 100,000\, e^{-2x} + 100 \cdot 100,000\, e^{-2x}}{\left(1 + 50\, e^{-x}\right)^{3}}$$

$$= \frac{-100,000\, e^{-x} + 50 \cdot 100,000\, e^{-2x}}{\left(1 + 50\, e^{-x}\right)^{3}} = \frac{100,000\, e^{-x}\left(50\, e^{-x} - 1\right)}{\left(1 + 50\, e^{-x}\right)^{3}}$$

$f''(x) = 0$ when $50\, e^{-x} - 1 = 0$ since the denominator and $100,000\, e^{-x}$ are always positive. $50\, e^{-x} - 1 = 0$ when $50\, e^{-x} = 1$ or $e^{x} = 50$ or when $\ln 50 = x$. From the First Derivative Test we see that $(\ln 50, f(\ln 50)) \approx (3.91, 10,000)$ is a relative maximum. The sales rate is a maximum after about 3.91 months of sales.

(b)

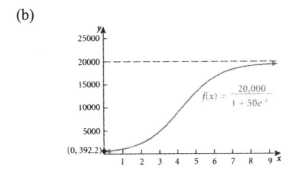

$$f(x) = \frac{20,000}{1 + 50e^{-x}}$$

$(0, 392.2)$

5.4 Optimization

1. $f(x) = x^2 + 2x$ $\qquad$ $f'(x) = 2x + 2$

   $f'(x) = 0$ when $x = -1$

   The critical number -1 is in the interval $[-3, 3]$, so we evaluate f at each of the three points.

   $$f(-3) = (-3)^2 + 2(-3) = 3$$
   $$f(-1) = (-1)^2 + 2(-1) = -1$$
   $$f(3) = (3)^2 + 2(3) = 15$$

   The absolute maximum of f on $[-3, 3]$ is 15 and the absolute minimum is -1.

3. $f(x) = 1 - 6x - x^2$ $\qquad$ $f'(x) = -6 - 2x$

   $f'(x) = 0$ when $x = -3$.

   The critical number -3 is not in the interval $[0, 4]$, so we evaluate f only at the endpoints.

   $$f(0) = 1 - 6(0) - 0^2 = 1$$
   $$f(4) = 1 - 6(4) - 4^2 = -39$$

   The absolute maximum of f on $[0, 4]$ is 1 and the absolute minimum is -39.

5. $f(x) = x^3 - 3x^2$ $\qquad$ $f'(x) = 3x^2 - 6x$

   $f'(x) = 0$ when $3x^2 - 6x = 0$
   $$3x(x - 2) = 0$$
   $$x = 0 \quad \text{or} \quad x = 2$$

   The critical number 0 is not in the interval $[1, 4]$, so we ignore it and evaluate f at each of the other three numbers.

   $$f(1) = 1^3 - 3 \cdot 1^2 = -2$$
   $$f(2) = 2^3 - 3 \cdot 2^2 = -4$$
   $$f(4) = 4^3 - 3 \cdot 4^2 = 16$$

   The absolute maximum of f on $[1, 4]$ is 16 and the absolute minimum is -4.

7. $f(x) = x^4 - 2x^2 + 1$ $f'(x) = 4x^3 - 4x$

$f'(x) = 0$ when $4x^3 - 4x = 0$

$$4x(x^2 - 1) = 0$$

$$4x(x-1)(x+1) = 0$$

$$x = 0 \quad \text{or} \quad x = 1 \quad \text{or} \quad x = -1$$

The critical number -1 is not in the interval $[0, 1]$, so we ignore it. The other two critical numbers are the endpoints of the closed interval, so we evaluate f at each of those two numbers.

$$f(0) = 0^4 - 2(0)^2 + 1 = 1$$

$$f(1) = 1^4 - 2(1)^2 + 1 = 0$$

The absolute maximum of f on $[0, 1]$ is 1 and the absolute minimum is 0.

9.
$f(x) = x^{2/3}$ $f'(x) = \dfrac{2}{3}x^{-1/3} = \dfrac{2}{3x^{1/3}}$

$f'(x)$ is not defined at $x = 0$. It is never equal to zero. So the only critical number is 0.

The critical number 0 is in the interval $[-1, 1]$, so we evaluate f at each of the three numbers.

$$f(-1) = (-1)^{2/3} = 1$$

$$f(0) = 0^{2/3} = 0$$

$$f(1) = 1^{2/3} = 1$$

The absolute maximum of f on $[-1, 1]$ is 1 and the absolute minimum is 0.

11.
$f(x) = 2\sqrt{x} = 2x^{1/2}$ $f'(x) = \dfrac{1}{2} \cdot 2x^{-1/2} = x^{-1/2} = \dfrac{1}{\sqrt{x}}$

$f'(x)$ is not defined at $x = 0$, and it never equals zero. The critical number 0 is not in the interval $[1, 4]$, so we evaluate f only at the endpoints.

$$f(1) = 2\sqrt{1} = 2$$

$$f(4) = 2\sqrt{4} = 4$$

The absolute maximum of f on $[1, 4]$ is 4 and the absolute minimum is 2.

13. $f(x) = x\sqrt{1 - x^2} = x\left(1 - x^2\right)^{1/2}$

$$f'(x) = x \cdot \frac{1}{2}\left(1 - x^2\right)^{-1/2} \cdot (-2x) + 1 \cdot \left(1 - x^2\right)^{1/2}$$

$$= \frac{-x^2}{\left(1 - x^2\right)^{1/2}} + \left(1 - x^2\right)^{1/2} \cdot \frac{\left(1 - x^2\right)^{1/2}}{\left(1 - x^2\right)^{1/2}}$$

$$= \frac{-x^2 + 1 - x^2}{\left(1 - x^2\right)^{1/2}} = \frac{1 - 2x^2}{\left(1 - x^2\right)^{1/2}}$$

$f'(x)$ is not defined at $x = -1$ and $x = 1$; $f'(x) = 0$ when

$$1 - 2x^2 = 0$$

$$x^2 = \frac{1}{2}$$

$$x = \pm \sqrt{\frac{1}{2}}$$

The critical numbers $\sqrt{\dfrac{1}{2}}$ and $-\sqrt{\dfrac{1}{2}}$ are in the interval $[-1, 1]$ and the critical numbers -1 and 1 are the endpoints of the interval $[-1, 1]$, so we evaluate f at each of these numbers.

$$f(-1) = -1\sqrt{1 - (-1)^2} = 0$$

$$f\left(-\sqrt{\frac{1}{2}}\right) = -\sqrt{\frac{1}{2}} \cdot \sqrt{1 - \left(-\sqrt{\frac{1}{2}}\right)^2} = -\sqrt{\frac{1}{2}} \cdot \sqrt{1 - \frac{1}{2}} = -\sqrt{\frac{1}{2}} \cdot \sqrt{\frac{1}{2}} = -\frac{1}{2}$$

$$f\left(\sqrt{\frac{1}{2}}\right) = \sqrt{\frac{1}{2}} \cdot \sqrt{1 - \left(\sqrt{\frac{1}{2}}\right)^2} = \sqrt{\frac{1}{2}} \cdot \sqrt{1 - \frac{1}{2}} = \sqrt{\frac{1}{2}} \cdot \sqrt{\frac{1}{2}} = \frac{1}{2}$$

$$f(1) = 1\sqrt{1 - 1^2} = 0$$

The absolute maximum of f on $[-1, 1]$ is $\dfrac{1}{2}$ and the absolute minimum is $-\dfrac{1}{2}$.

15.

$$f(x) = \frac{x^2}{x - 1}$$

$$f'(x) = \frac{2x(x-1) - x^2(1)}{(x-1)^2} = \frac{2x^2 - 2x - x^2}{(x-1)^2} = \frac{x^2 - 2x}{(x-1)^2}$$

$f'(x)$ is not defined at $x = 1$; $f'(x) = 0$ when $x^2 - 2x = 0$.

$$x^2 - 2x = 0$$

$$x(x-2) = 0$$

$$x = 0 \quad \text{or} \quad x = 2$$

The critical number 0 is in the interval $\left[-1, \dfrac{1}{2}\right]$, the critical numbers 1 and 2 are not in the domain of f. So we evaluate f at 0 and the endpoints.

when $x = -1$: $f(-1) = \dfrac{(-1)^2}{(-1)-1} = -\dfrac{1}{2}$

when $x = 0$: $f(0) = \dfrac{0^2}{0-1} = 0$

when $x = \dfrac{1}{2}$: $f\left(\dfrac{1}{2}\right) = \dfrac{\left(\dfrac{1}{2}\right)^2}{\dfrac{1}{2}-1} = -\dfrac{1}{2}$

The absolute maximum of f on $\left[-1, \dfrac{1}{2}\right]$ is 0, and the absolute minimum is $-\dfrac{1}{2}$.

17. $f(x) = (x+2)^2 (x-1)^{2/3}$

$f'(x) = (x+2)^2 \cdot \dfrac{2}{3}(x-1)^{-1/3} + 2(x+2)(x-1)^{2/3}$

$\quad = \dfrac{2(x+2)^2}{3(x-1)^{1/3}} + (2x+4)(x-1)^{2/3}$

$\quad = \dfrac{2(x+2)^2}{3(x-1)^{1/3}} + (2x+4)(x-1)^{2/3} \cdot \dfrac{3(x-1)^{1/3}}{3(x-1)^{1/3}}$

$\quad = \dfrac{2(x^2+4x+4)}{3(x-1)^{1/3}} + \dfrac{3(2x+4)(x-1)}{3(x-1)^{1/3}}$

$\quad = \dfrac{2x^2+8x+8+3(2x^2+2x-4)}{3(x-1)^{1/3}} = \dfrac{2x^2+8x+8+6x^2+6x-12}{3(x-1)^{1/3}}$

$\quad = \dfrac{8x^2+14x-4}{3(x-1)^{1/3}}$

$f'(x)$ is not defined at $x = 1$; $f'(x) = 0$ when $8x^2+14x-4 = 0$.

$$8x^2+14x-4 = 2(4x^2+7x-2) = 0$$
$$4x^2+7x-2 = 0$$
$$(4x-1)(x+2) = 0$$
$$x = \dfrac{1}{4} \quad \text{or} \quad x = -2$$

The critical numbers -2, $\dfrac{1}{4}$, and 1 are all in the interval $[-4, 5]$, so we evaluate f at each of the three numbers and at the endpoints.

when $x = -4$: $\quad f(-4) = (-4+2)^2(-4-1)^{2/3} \approx 11.696$

when $x = -2$: $\quad f(-2) = (-2+2)^2(-2-1)^{2/3} = 0$

when $x = \dfrac{1}{4}$: $\quad f\left(\dfrac{1}{4}\right) = \left(\dfrac{1}{4}+2\right)^2\left(\dfrac{1}{4}-1\right)^{2/3} \approx 4.179$

when $x = 1$: $\quad f(1) = (1+2)^2(1-1)^{2/3} = 0$

when $x = 5$: $\quad f(5) = (5+2)^2(5-1)^{2/3} \approx 123.472$

The absolute maximum of f on $[-4, 5]$ is 123.472 and the absolute minimum is 0.

19.

$$f(x) = \frac{(x-4)^{1/3}}{x-1}$$

$$f'(x) = \frac{\dfrac{1}{3}(x-4)^{-2/3}(x-1) - (x-4)^{1/3}(1)}{(x-1)^2}$$

$$= \frac{\dfrac{1}{3(x-4)^{2/3}} \cdot (x-1) - (x-4)^{1/3}}{(x-1)^2} = \frac{\dfrac{1}{3(x-4)^{2/3}} \cdot (x-1) - (x-4)^{1/3} \cdot \dfrac{3(x-4)^{2/3}}{3(x-4)^{2/3}}}{(x-1)^2}$$

$$= \frac{(x-1) - 3(x-4)}{3(x-4)^{2/3}(x-1)^2} = \frac{x-1-3x+12}{3(x-4)^{2/3}(x-1)^2}$$

$$= \frac{-2x+11}{3(x-4)^{2/3}(x-1)^2}$$

$f'(x)$ is not defined at $x = 1$ and $x = 4$; $f'(x) = 0$ when $-2x + 11 = 0$, or $x = \dfrac{11}{2} = 5.5$.

The critical numbers 4 and 5.5 are in the interval $[2, 12]$, so we evaluate f at each of these numbers and at the endpoints.

when $x = 2$: $\qquad f(2) = \dfrac{(2-4)^{1/3}}{2-1} \approx -1.260$

when $x = 4$: $\qquad f(4) = \dfrac{(4-4)^{1/3}}{4-1} = 0$

when $x = 5.5$: $\qquad f(5.5) = \dfrac{(5.5-4)^{1/3}}{5.5-1} \approx 0.254$

when $x = 12$: $\qquad f(12) = \dfrac{(12-4)^{1/3}}{12-1} \approx 0.182$

The absolute maximum of f on $[2, 12]$ is 0.254 and the absolute minimum is -1.260.

21. $f(x) = xe^x$

$f'(x) = xe^x + e^x = e^x(x+1)$

$f'(x) = 0$ when $e^x(x+1) = 0$ or when $x = -1$.

The critical number -1 is in the interval $[-10, 10]$, so we evaluate f at it and at the endpoints.

when $x = -10$: $f(-10) = -10e^{-10} \approx -0.0005$

when $x = -1$: $f(-1) = -e^{-1} \approx -0.368$

when $x = 10$: $f(10) = 10e^{10} \approx 220,265$

The absolute maximum of f on $[-10, 10]$ is 220,265 and the absolute minimum is -0.368.

23. $f(x) = \dfrac{\ln x}{x}$

$f'(x) = \dfrac{\dfrac{1}{x} \cdot x - (\ln x) \cdot (1)}{x^2} = \dfrac{1 - \ln x}{x^2}$

$f'(x) = 0$ when $1 - \ln x = 0$ or when $x = e$.

The critical number $e \approx 2.718$ is in the interval $[1, 3]$, so we evaluate f at e and the two endpoints.

when $x = 1$: $f(1) = \dfrac{\ln 1}{1} = 0$

when $x = e$: $f(e) = \dfrac{\ln e}{e} = \dfrac{1}{e} \approx 0.368$

when $x = 3$: $f(3) = \dfrac{\ln 3}{3} \approx 0.366$

The absolute maximum of f on $[1, 3]$ is 0.368 and the absolute minimum is 0.

25. **STEP 1** The quantity to be maximized is volume. Denote it by V.

STEP 2 Denote the dimensions of the side of the small square to be cut out by x. Let the y be the dimension one side of the bottom of box after removing square x.

STEP 3 Then $y = 12 - 2x$

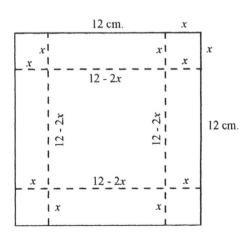

STEP 4 The height of the box is x, while the area of the base of the box is y^2. So,
$$V = xy^2$$
$$V(x) = x(12 - 2x)^2$$
Although the domain of V is the set of real numbers, only values of x between 0 and 6 make sense physically. We need the absolute maximum of V on $[0, 6]$.

STEP 5 Differentiate V and identify the critical numbers.
$$V'(x) = x \cdot 2(12 - 2x) \cdot (-2) + 1 \cdot (12 - 2x)^2$$
$$= -4x(12 - 2x) + (12 - 2x)^2$$
$$= (12 - 2x)[(12 - 2x) - 4x]$$
$$= (12 - 2x)(12 - 6x)$$
If $V'(x) = 0$, then
$$(12 - 2x)(12 - 6x) = 0$$
$$12 - 2x = 0 \quad \text{or} \quad 12 - 6x = 0$$
$$x = 6 \quad \text{or} \quad x = 2$$
We calculate the values of V at the critical number $x = 2$ and at the endpoints 0 and 6.
$$V(0) = 0 \qquad V(2) = 2(12 - 2(2))^2 = 2(8)^2 = 128 \qquad V(6) = 0$$

The maximum volume is 128 cubic centimeters and the dimensions of the box are $x = 2$ centimeters deep by $y = 12 - 2(2) = 8$ centimeters on each side.

27. **STEP 1** We want to minimize the amount of material A which is used to make the box.

    STEP 2 Let x denote dimension of a side of the square bottom of the box, and let y denote the height of the box.

    STEP 3 $\quad V = x^2 y$
    $$8000 = x^2 y$$
    $$y = \frac{8000}{x^2}$$

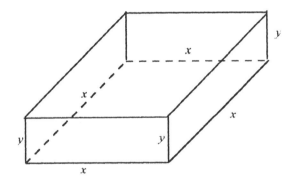

STEP 4 The amount of material used is the area of the base of the box plus the area of the four sides.
$$A = x^2 + 4xy$$
$$A(x) = x^2 + 4x\left(\frac{8000}{x^2}\right)$$
$$= x^2 + \frac{32,000}{x} = x^2 + 32,000x^{-1}$$
The domain of A is all real numbers other than $x = 0$, but only positive values of x make physical sense.

STEP 5 Differentiate A and find the critical numbers.

$$A'(x) = 2x - 32,000x^{-2} = 2x - \frac{32,000}{x^2} = \frac{2x^3 - 32,000}{x^2} \qquad x \neq 0$$

$x = 0$ is a critical number. We solve $A'(x) = 0$ to find other critical numbers, if they exist.

$$\frac{2x^3 - 32,000}{x^2} = 0$$

$$2x^3 - 32,000 = 0$$

$$x^3 = 16,000$$

$$x = 20\sqrt[3]{2}$$

We use the second derivative test to see if $x = 20\sqrt[3]{2}$ locates a maximum or minimum.

$$A''(x) = 2 + 64,000x^{-3}$$

$$A''\left(20\sqrt[3]{2}\right) = 2 + \frac{64,000}{\left(20\sqrt[3]{2}\right)^3} = 2 + \frac{64,000}{16,000} = 2 + 4 = 6 > 0$$

So we conclude that the least amount of material is used when the base measures $20\sqrt[3]{2} \approx 25.20$ centimeters and the height measures $\dfrac{8000}{\left(20\sqrt[3]{2}\right)^2} \approx 12.60$ centimeters.

29. **STEP 1** We want to minimize the cost of the material A which is used to make the cylinder.

STEP 2 Let x denote the radius of the top and bottom of the cylinder, and let y denote the height of the cylinder.

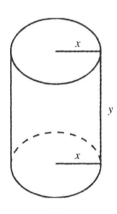

STEP 3
$$V = \pi x^2 y$$
$$4000 = \pi x^2 y$$
$$y = \frac{4000}{\pi x^2}$$

STEP 4 The cost C of producing the can is $0.50 times the area of the top and bottom of the cylinder plus $0.40 times the area of the side of the cylinder.

$$C = 2(0.50)\left(\pi x^2\right) + (0.40)(2\pi xy)$$

$$C(x) = 1.00\left(\pi x^2\right) + 0.80\,\pi\,x\left(\frac{4000}{\pi x^2}\right)$$

$$= \pi x^2 + \frac{3200}{x}$$

The domain of C is $\{x \mid x > 0\}$.

STEP 5 Differentiate C and find the critical numbers.

$$C'(x) = 2\pi x - \frac{3200}{x^2} = \frac{2\pi x^3 - 3200}{x^2}$$

We solve $C'(x) = 0$ to find the critical numbers.

$$2\pi x^3 - 3200 = 0$$

$$x^3 = \frac{3200}{2\pi}$$

$$x = \sqrt[3]{\frac{1600}{\pi}} \approx 7.986$$

We use the second derivative test to see if $x = 7.986$ locates a maximum or minimum.

$$C''(x) = 2\pi - (-2)3200x^{-3} = 2\pi + \frac{6400}{x^3}$$

$$C''(7.986) = 2\pi + \frac{6400}{\left(\dfrac{1600}{\pi}\right)} = 2\pi + 4\pi = 6\pi > 0$$

We conclude that the cost is minimized when the can has a radius of 7.986 cm and a height of $\dfrac{4000}{\pi(7.986)^2} \approx 19.965$ centimeters.

31. **STEP 1** Minimize the cost of installing the telephone line.

STEP 2 Using the hint, we let x denote the distance between the box and the connection.

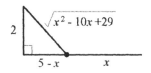

STEP 3 Using the Pythagorean theorem, we calculate the distance the line runs off the road.

$$c^2 = a^2 + b^2$$
$$c^2 = (5-x)^2 + 2^2$$
$$= 25 - 10x + x^2 + 4 = x^2 - 10x + 29$$
$$c = \sqrt{x^2 - 10x + 29}$$

STEP 4 The cost C of laying the line is expressed as

$$C(x) = 50x + 60(x^2 - 10x + 29)^{1/2}$$

The domain of C is $\{x \mid 0 \le x \le 5\}$.

STEP 5 The first derivative gives the critical numbers, if they exist.

$$C'(x) = 50 + 60\left[\frac{1}{2}(x^2 - 10x + 29)^{-1/2}(2x - 10)\right]$$

$$= 50 + 60\left[\frac{x-5}{(x^2 - 10x + 29)^{1/2}}\right] = \frac{50(x^2 - 10x + 29)^{1/2} + 60x - 300}{(x^2 - 10x + 29)^{1/2}}$$

We solve $C'(x) = 0$.

$$50\left(x^2 - 10x + 29\right)^{1/2} + 60x - 300 = 0$$

$$\left(x^2 - 10x + 29\right)^{1/2} = \frac{300 - 60x}{50} = 6 - 1.2x$$

$$x^2 - 10x + 29 = \left(6 - 1.2x\right)^2 = 36 - 14.4x + 1.44x^2$$

$$0.44x^2 - 4.4x + 7 = 0$$

$$x = \frac{4.4 \pm \sqrt{4.4^2 - 4(0.44)(7)}}{2(0.44)} = \frac{4.4 \pm \sqrt{7.04}}{0.88}$$

$$x = \frac{4.4 + \sqrt{7.04}}{0.88} \approx 8.015 \quad \text{or} \quad x = \frac{4.4 - \sqrt{7.04}}{0.88} \approx 1..985$$

$x = 8.015$ is not in the domain of the function, so we test only the endpoints and the critical number 1.985.

when $x = 0$: $\qquad C(0) = 50(0) + 60\left(0^2 - 10(0) + 29\right)^{1/2} = 60\sqrt{29} \approx 323.11$

when $x = 1.985$: $\quad C(1.985) = 50(1.985) + 60\left(1.985^2 - 10(1.985) + 29\right)^{1/2} \approx 316.33$

when $x = 5$: $\qquad C(5) = 50(5) + 60\left(5^2 - 10(5) + 29\right)^{1/2} = 370$

The minimum cost is obtained when the telephone line leaves the road approximately 1.985 kilometers from the box.

33. We want to minimize the cost of operating the truck over the interval $[10, 75]$. We find the derivative of C and locate any critical numbers in the open interval $(10, 75)$.

$$C'(x) = 1.60\left(-\frac{1600}{x^2} + 1\right) = 1.60\left(\frac{x^2 - 1600}{x^2}\right)$$

The critical numbers are those for which $C'(x) = 0$.

$$1.60\left(\frac{x^2 - 1600}{x^2}\right) = 0$$

$$x^2 - 1600 = 0$$

$$x^2 = 1600$$

$$x = \pm 40$$

Only the critical number $x = 40$ is in the open interval, so we evaluate C at 40 and at the endpoints.

when $x = 10$: $\qquad C(10) = 1.60\left(\frac{1600}{10} + 10\right) = 272$

when $x = 40$: $\qquad C(40) = 1.60\left(\frac{1600}{40} + 40\right) = 128$

when $x = 75$: $\qquad C(75) = 1.60\left(\frac{1600}{75} + 75\right) = 154.13$

The cost of operating the truck is minimized when it is driven at 40 miles per hour.

35. **STEP 1** Minimize the size S of the page.

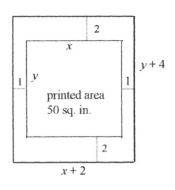

STEP 2 Let x denote the width of the print and y denote the length of the print.

STEP 3 The area of the printed matter is 50 square inches.
$$A = lw$$
$$50 = xy$$
$$y = \frac{50}{x} \text{ inches}$$

STEP 4 The size of the page is expressed by the function
$$S = (y+4)(x+2)$$
$$S(x) = \left(\frac{50}{x}+4\right)(x+2) = 50 + \frac{100}{x} + 4x + 8 = 58 + \frac{100}{x} + 4x$$
The domain of S is $\{x \mid x > 0\}$.

STEP 5 The derivative $S'(x)$ will give critical numbers if they exist.
$$S'(x) = -\frac{100}{x^2} + 4 = \frac{4x^2 - 100}{x^2}$$
We solve the equation $S'(x) = 0$.
$$\frac{4x^2 - 100}{x^2} = 0$$
$$4x^2 - 100 = 0$$
$$x^2 = \frac{100}{4} = 25 \text{ or } x = \pm 5$$

Since $x = -5$ is not in the domain of S we disregard it. We use the second derivative test to see if $x = 5$ locates a maximum or a minimum.
$$S''(x) = 200x^{-3} = \frac{200}{x^3} \qquad\qquad S''(5) = \frac{200}{5^3} = \frac{200}{125} > 0$$

We note that $x = 5$ minimizes S, and we conclude that the most economical page size is one with length $\frac{50}{5} + 4 = 14$ inches and width $5 + 2 = 7$ inches.

37. The relation is justified because an increased tax rate results in a higher price and lower demand. When tax rate is 0, the quantity demanded is 2.45. On the other hand, there is no demand when the tax rate is $t = 18$.

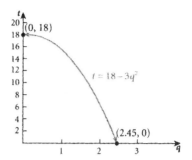

The optimal tax rate maximizes tax revenue. Tax revenue is given by

$$R = qt = q(18 - 3q^2) = 18q - 3q^3$$

The domain of R is $\{q \mid 0 \le q \le \sqrt{6}\}$.

We differentiate R and then set it equal to zero to find the critical numbers.

$$R'(q) = 18 - 9q^2$$
$$18 - 9q^2 = 0$$
$$q^2 = 2$$
$$q = \pm\sqrt{2}$$

The only critical number in the interval $(0, \sqrt{6})$ is $\sqrt{2}$, so we evaluate R at 0, $\sqrt{6}$, and $\sqrt{2}$.

when $q = 0$: $\qquad\qquad R(0) = 18(0) - 3(0)^3 = 0$

when $q = \sqrt{2} \approx 1.414$: $\qquad R(\sqrt{2}) = 18(\sqrt{2}) - 3(\sqrt{2})^3 = 12\sqrt{2} \approx 16.971$

when $q = \sqrt{6}$: $\qquad\qquad R(\sqrt{6}) = 18(\sqrt{6}) - 3(\sqrt{6})^3 = 0$

The revenue is maximized when $q = \sqrt{2}$. The tax rate corresponding to maximum revenue is

$$t = 18 - 3(\sqrt{2})^2 = 12$$

This means that a tax rate of 12% generates a maximum revenue of 16.971 monetary units.

39. The volume of the cylinder is fixed, so

$$V = \pi r^2 h$$

$$h = \frac{V}{\pi r^2}$$

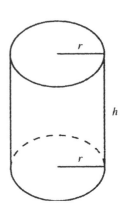

The surface area S of the cylinder is

$$S = (\text{area of side}) + 2(\text{area of the top})$$

$$S = 2\pi r h + 2\pi r^2$$

$$S(r) = 2\pi r \left(\frac{V}{\pi r^2} \right) + 2\pi r$$

$$= \frac{2V}{r} + 2\pi r^2$$

Differentiating S and finding the critical numbers, if there are any gives:

$$S'(r) = -\frac{2V}{r^2} + 4\pi r = \frac{4\pi r^3 - 2V}{r^2} \qquad r \neq 0$$

Solving $S'(r) = 0$,

$$\frac{4\pi r^3 - 2V}{r^2} = 0$$

$$4\pi r^3 - 2V = 0$$

$$r^3 = \frac{2V}{4\pi} = \frac{V}{2\pi}$$

$$r = \sqrt[3]{\frac{V}{2\pi}}$$

Using the second derivative test, we find that

$$S''(r) = \frac{4V}{r^3} + 4\pi \quad \text{and} \quad S''\left(\sqrt[3]{\frac{V}{2\pi}} \right) = \frac{4V}{\left(\sqrt[3]{\frac{V}{2\pi}} \right)^3} + 4\pi = 12\pi > 0$$

We conclude that $r = \sqrt[3]{\frac{V}{2\pi}}$ will minimize the surface area of a cylinder with volume V.

The height of this cylinder should be

$$h = \frac{V}{\pi r^2} = \frac{V}{\pi \left[\sqrt[3]{\frac{V}{2\pi}} \right]^2} = \frac{V}{\pi \sqrt[3]{\frac{V^2}{4\pi^2}}} = \frac{V}{\pi} \cdot \sqrt[3]{\frac{4\pi^2}{V^2}}$$

$$= \sqrt[3]{\frac{V^3}{\pi^3}} \cdot \sqrt[3]{\frac{4\pi^2}{V^2}} = \sqrt[3]{\frac{4\pi^2 V^3}{V^2 \pi^3}} = \sqrt[3]{\frac{4V}{\pi}} = \sqrt[3]{\frac{8V}{2\pi}} = 2\sqrt[3]{\frac{V}{2\pi}} = 2r$$

41.   We use the derivative $C'(t)$ to find when the concentration of the drug is the greatest.

$$C'(t) = \frac{2(16+t^3) - 2t \cdot 3t^2}{(16+t^3)^2} = \frac{32 + 2t^3 - 6t^3}{(16+t^3)^2} = \frac{32 - 4t^3}{(16+t^3)^2}$$

$$\frac{32 - 4t^3}{(16+t^3)^2} = 0$$

$$32 - 4t^3 = 0$$

$$t^3 = \frac{32}{4} = 8$$

$$t = 2$$

Using the first derivative test we see that for $t < 2$, $C'(t) > 0$, and for $t > 2$, $C'(t) < 0$. So we conclude that the concentration of the drug is greatest 2 hours after it is admininstered.

5.5 Elasticity of Demand

1.   (a) $x = f(p) = 4000 - 100p$

(b) $f'(p) = -100$

$$E(p) = \frac{pf'(p)}{f(p)} = \frac{p(-100)}{4000 - 100p} = \frac{-100p}{4000 - 100p} = \frac{p}{p - 40}$$

(c) When $p = \$5$, $E(5) = \dfrac{5}{5 - 40} = -0.143$

Increasing the price by 10% to $5.50, will result in a decrease of approximately 1.43% in quantity demanded.

(d) When $p = \$15$, $E(15) = \dfrac{15}{15 - 40} = -0.6$

Increasing the price by 10% to $16.50, will result in a decrease of approximately 6 % in quantity demanded.

(e) When $p = \$20$, $E(20) = \dfrac{20}{20 - 40} = -1$

Increasing the price by 10% to $22.00, will result in a decrease of approximately 10 % in quantity demanded.

3.   (a) $x = 200(50 - p) = 10,000 - 200p$

(b) $f'(p) = -200$

$$E(p) = \frac{pf'(p)}{f(p)} = \frac{p(-200)}{10,000 - 200p} = \frac{-200p}{10,000 - 200p} = \frac{p}{p - 50}$$

(c) When $p = \$10$, $E(10) = \frac{10}{10 - 50} = -0.25$

Increasing the price by 5% will result in a decrease of approximately $(0.25)(5\%) = 0.0125 = 1.25\%$ in quantity demanded.

(d) When $p = \$25$, $E(25) = \frac{25}{25 - 50} = -1$

Increasing the price by 5% will result in a decrease of approximately 5% in quantity demanded.

(e) When $p = \$35$, $E(35) = \frac{35}{35 - 50} = -2.333$

Increasing the price by 5% will result in a decrease of approximately $(2.333)(5\%) = 0.11665 = 11.665\%$ in quantity demanded.

5. $x = f(p) = 600 - 3p$ $\qquad$ $f'(p) = -3$

$$E(p) = \frac{pf'(p)}{f(p)} = \frac{p(-3)}{600 - 3p} = \frac{-3p}{600 - 3p} = \frac{p}{p - 200}$$

$$E(50) = \frac{50}{50 - 200} = \frac{50}{-150} = -0.333$$

At $p = \$50$ the demand is inelastic.

7. $x = f(p) = \frac{600}{p + 4}$ $\qquad$ $f'(p) = \frac{-600}{(p + 4)^2}$

$$E(p) = \frac{pf'(p)}{f(p)} = \frac{p \left[\dfrac{-600}{(p + 4)^2} \right]}{\dfrac{600}{p + 4}} = \frac{-p}{p + 4}$$

$$E(10) = \frac{-10}{10 + 4} = -0.714$$

At $p = \$10$ the demand is inelastic.

9. $x = f(p) = 10,000 - 10p^2$ $\qquad$ $f'(p) = -20p$

$$E(p) = \frac{pf'(p)}{f(p)} = \frac{p(-20p)}{10,000 - 10p^2} = \frac{-20p^2}{10,000 - 10p^2} = \frac{-2p^2}{1000 - p^2}$$

$$E(10) = \frac{-2(10^2)}{1000 - 10^2} = -0.222$$

At $p = \$10$ the demand is inelastic.

11. $x = f(p) = \sqrt{100 - p}$ $f'(p) = \frac{1}{2}(100 - p)^{-1/2} \cdot (-1) = \frac{-1}{2\sqrt{100 - p}}$

$$E(p) = \frac{pf'(p)}{f(p)} = \frac{p\left[\dfrac{-1}{2\sqrt{100 - p}}\right]}{\sqrt{100 - p}} = \frac{-p}{2(100 - p)}$$

$$E(10) = \frac{-10}{2(100 - 10)} = -0.056$$

At $p = \$10$ the demand is inelastic.

13. $x = f(p) = 40(4 - p)^3$ $f'(p) = 40 \cdot 3(4 - p)^2 \cdot (-1) = -120(4 - p)^2$

$$E(p) = \frac{pf'(p)}{f(p)} = \frac{p\left[-120(4 - p)^2\right]}{40(4 - p)^3} = -\frac{3p}{4 - p}$$

$$E(2) = -\frac{(3)(2)}{4 - 2} = -3$$

At $p = \$2$ the demand is elastic.

15. $x = f(p) = 20 - 3\sqrt{p}$ $f'(p) = \frac{1}{2} \cdot (-3)p^{-1/2} = -\frac{3}{2\sqrt{p}}$

$$E(p) = \frac{pf'(p)}{f(p)} = \frac{p\left[-\dfrac{3}{2\sqrt{p}}\right]}{20 - 3\sqrt{p}} = -\frac{3p}{2\sqrt{p}(20 - 3\sqrt{p})} = -\frac{3p}{40\sqrt{p} - 6p}$$

$$E(4) = -\frac{3(4)}{40\sqrt{4} - 6(4)} = -\frac{12}{80 - 24} = -0.214$$

At $p = \$4$ the demand is inelastic.

17. First we differentiate.

$$\frac{1}{2}x^{-1/2}\frac{dx}{dp} + 2p\frac{dx}{dp} + 2x + 2p = 0$$

$$\frac{dx}{dp} = \frac{-(2x+2p)}{\frac{1}{2}x^{-1/2}+2p} = -\frac{4(x+p)\sqrt{x}}{1+4p\sqrt{x}}$$

$$E(p) = \frac{pf'(p)}{f(p)} = \frac{p}{x}\cdot\left[-\frac{4(x+p)\sqrt{x}}{1+4p\sqrt{x}}\right]$$

When $x = 16$ and $p = 4$,

$$E(4) = \frac{4}{16}\cdot\left[-\frac{4(16+4)\sqrt{16}}{1+4(4)\sqrt{16}}\right] = -\frac{1}{4}\cdot\left[\frac{320}{65}\right] = -\frac{80}{65} = -1.231$$

19. First we differentiate.

$$4x\frac{dx}{dp}+3p\frac{dx}{dp}+3x+20p = 0$$

$$\frac{dx}{dp} = \frac{-(3x+20p)}{4x+3p}$$

$$E(p) = \frac{p}{x}\cdot\left[-\frac{3x+20p}{4x+3p}\right]$$

When $x = 10$ and $p = 5$,

$$E(5) = \frac{5}{10}\cdot\left[-\frac{3\cdot10+20\cdot5}{4\cdot10+3\cdot5}\right] = -\frac{1}{2}\cdot\frac{130}{55} = -1.182$$

21.
$$p = F(x) = 10-\frac{1}{20}x \qquad\qquad p' = F'(x) = -\frac{1}{20}$$

$$E(x) = \frac{F(x)}{xF'(x)} = \frac{10-\frac{1}{20}x}{x\left[-\frac{1}{20}\right]} = -\frac{200-x}{x}$$

$$E(5) = -\frac{200-5}{5} = -39$$

23.
$$p = F(x) = 10-2x^2 \qquad\qquad p' = F'(x) = -4x$$

$$E(x) = \frac{F(x)}{xF'(x)} = \frac{10-2x^2}{x(-4x)} = -\frac{10-2x^2}{4x^2} = -\frac{5-x^2}{2x^2}$$

$$E(2) = -\frac{5-2^2}{2(2^2)} = -\frac{1}{8} = -0.125$$

25.
$$p = F(x) = 50-2\sqrt{x} = 50-2x^{1/2} \qquad\qquad p' = F'(x) = -x^{-1/2} = -\frac{1}{\sqrt{x}}$$

$$E(x) = \frac{F(x)}{xF'(x)} = \frac{50 - 2\sqrt{x}}{x\left(-\dfrac{1}{\sqrt{x}}\right)} = -\frac{50 - 2\sqrt{x}}{\sqrt{x}}$$

$$E(100) = -\frac{50 - 2\sqrt{100}}{\sqrt{100}} = -\frac{50 - 20}{10} = -3$$

27.

$$x = f(p) = \frac{6000}{p} - 500 \qquad\qquad f'(p) = -\frac{6000}{p^2}$$

$$E(p) = \frac{p \cdot f'(p)}{x} = \frac{p\left(-\dfrac{6000}{p^2}\right)}{\dfrac{6000 - 500p}{p}} = -\frac{6000}{6000 - 500p}$$

When $p = 4$,

$$E(4) = -\frac{6000}{6000 - 500(4)} = -1.5$$

(a) The demand is elastic since $|E(4)| > 1$.
(b) If the price is increased, the revenue will decrease.

29.

$$x = f(p) = \sqrt{300 - 6p} = (300 - 6p)^{1/2} \quad f'(p) = \frac{1}{2}(300 - 6p)^{-1/2} \cdot (-6) = -\frac{3}{\sqrt{300 - 6p}}$$

$$E(p) = \frac{p \cdot f'(p)}{x} = \frac{p\left[-\dfrac{3}{\sqrt{300 - 6p}}\right]}{\sqrt{300 - 6p}} = -\frac{3p}{300 - 6p} = -\frac{p}{100 - 2p}$$

When $p = 10$,

$$E(10) = -\frac{10}{100 - 2(10)} = -0.125$$

(a) When the price is \$10, the demand is inelastic.
(b) The revenue will decrease if the price is lowered slightly.

31. When $p = 15$, $x = 2000$ and when $p = 18$ then $x = 1800$.

(a) $m = \dfrac{x_2 - x_1}{p_2 - p_1} = \dfrac{1800 - 2000}{18 - 15} = -\dfrac{200}{3} = -66.67$

$$x - 2000 = -\frac{200}{3}(p - 15)$$

$$x = -\frac{200}{3}p + 1000 + 2000 = -\frac{200}{3}p + 3000$$

(b) $f'(p) = -\dfrac{200}{3}$

$$E(p) = \frac{p \cdot f'(p)}{x} = \frac{p\left(-\dfrac{200}{3}\right)}{-\dfrac{200}{3}p + 3000} = \frac{-200p}{-200p + 9000} = -\frac{p}{45 - p}$$

$$E(18) = -\frac{18}{45 - 18} = -0.667$$

(c) If the price is increased by 5%, the demand will decrease by approximately $(0.667)(0.05) = 0.0333 = 3.33\%$.

(d) Since $|E(18)| < 1$ the demand is inelastic, and the increase in price will cause an increase in revenue.

5.6 Related Rates

1. $x^2 + y^2 = 13$

$2x\dfrac{dx}{dt} + 2y\dfrac{dy}{dt} = 0$

$\dfrac{dx}{dt} = \dfrac{-2y\dfrac{dy}{dt}}{2x}$

When $x = 2$, $y = 3$, and $\dfrac{dy}{dt} = 2$,

$\dfrac{dx}{dt} = \dfrac{-2\cdot3\cdot2}{2\cdot2} = -3$

3. $x^3 y^2 = 72$

$x^3 \cdot 2y\dfrac{dy}{dt} + 3x^2\dfrac{dx}{dt}\cdot y^2 = 0$

$3x^2 y^2\dfrac{dx}{dt} = -2x^3 y\dfrac{dy}{dt}$

$\dfrac{dx}{dt} = \dfrac{-2x^3 y\dfrac{dy}{dt}}{3x^2 y^2}$

When $x = 2$, $y = 3$, and $\dfrac{dy}{dt} = 2$,

$\dfrac{dx}{dt} = \dfrac{-2\cdot2^3\cdot3\cdot2}{3\cdot2^2\cdot3^2} = -\dfrac{8}{9}$

5. $V = 80h^2$ $\dfrac{dV}{dt} = 160h\dfrac{dh}{dt}$

If $h = 3$ and $\dfrac{dh}{dt} = \dfrac{1}{12}$, then $\dfrac{dV}{dt} = 160\cdot3\cdot\dfrac{1}{12} = 40$

7. $V = \dfrac{1}{12}\pi h^3$ $\dfrac{dV}{dt} = \dfrac{3}{12}\pi h^2\dfrac{dh}{dt} = \dfrac{1}{4}\pi h^2\dfrac{dh}{dt}$

If $h = 8$ and $\dfrac{dh}{dt} = \dfrac{5}{16}\pi$, then $\dfrac{dV}{dt} = \dfrac{1}{4}\pi\cdot8^2\cdot\dfrac{5}{16}\pi = 5\pi^2$

9. **STEP 2** The variables are:

    s = length (in centimeters) of a side of the cube

    V = volume (in cubic centimeters) of the cube

    t = time (in seconds).

STEP 1

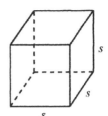

STEP 3 The rates of change are $\dfrac{ds}{dt}$ and $\dfrac{dV}{dt}$.

STEP 4 We know $\dfrac{ds}{dt} = 3$ cm/sec, and we want $\dfrac{dV}{dt}$ when $s = 10$ cm. The volume of a cube is $V = s^3$.

STEP 5 $\dfrac{dV}{dt} = 3s^2\dfrac{ds}{dt}$

STEP 6 When $s = 10$ and $\dfrac{ds}{dt} = 3$, $\dfrac{dV}{dt} = 3 \cdot 10^2 \cdot 3 = 900$ cubic centimeters per second.

11. **STEP 2** The variables are:

    x = length (in centimeters) of one leg of the triangle

    y = length (in centimeters) of a the other leg

    t = time (in minutes).

STEP 1

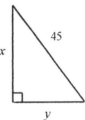

STEP 3 The rates of change are $\dfrac{dx}{dt}$ and $\dfrac{dy}{dt}$.

STEP 4 We know $\dfrac{dx}{dt} = 2$ cm/min, and we want $\dfrac{dy}{dt}$ when $x = 4$ cm. We use the Pythagorean Theorem $x^2 + y^2 = 45^2$, and note that when $x = 4$, $y = \sqrt{45^2 - 4^2} = \sqrt{2009} \approx 44.822$ cm.

STEP 5 $2x\dfrac{dx}{dt} + 2y\dfrac{dy}{dt} = 0$

$$\frac{dy}{dt} = -\frac{2x\dfrac{dx}{dt}}{2y} = -\frac{x}{y}\frac{dx}{dt}$$

STEP 6 When $x = 4$, $y = 44.822$ and $\dfrac{dx}{dt} = 2$, $\dfrac{dy}{dt} = -\dfrac{4}{\sqrt{2009}} \cdot 2 \approx -0.178$ centimeters per minute.

13. **STEP 2** The variables are:

    r = radius (in meters) of the balloon

    V = volume (in cubic meters) of the balloon

    S = surface area (in square meters) of the balloon

    t = time (in minutes).

STEP 1

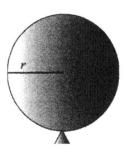

STEP 3 The rates of change are $\dfrac{dr}{dt}$, $\dfrac{dV}{dt}$, and $\dfrac{dS}{dt}$.

STEP 4 We know $\dfrac{dV}{dt} = -1.5$ m^3 per minute

(negative because the air is leaking out). We want $\dfrac{dS}{dt}$

when $r = 4$ m. The volume of a sphere is $V = \dfrac{4}{3}\pi r^3$ and the surface area is $S = 4\pi r^2$.

STEP 5 $\dfrac{dV}{dt} = 4\pi r^2 \dfrac{dr}{dt}$ and $\dfrac{dS}{dt} = 8\pi r \dfrac{dr}{dt}$

We use $\dfrac{dV}{dt}$ to solve for $\dfrac{dr}{dt}$ and then substitute it into $\dfrac{dS}{dt}$.

$$\dfrac{dr}{dt} = \dfrac{1}{4\pi r^2}\dfrac{dV}{dt}$$

$$\dfrac{dS}{dt} = 8\pi r \cdot \dfrac{1}{4\pi r^2}\dfrac{dV}{dt} = \dfrac{2}{r}\dfrac{dV}{dt}$$

STEP 6 When $r = 4$ and $\dfrac{dV}{dt} = -1.5$, $\dfrac{dS}{dt} = \dfrac{2}{4}\cdot(-1.5) = -0.75$ square meters per

minute.

15. **STEP 2** The variables are:

    h = water level (in meters) in the deep end.
    V = volume (in cubic meters) of the pool water
    t = time (in minutess).

STEP 1

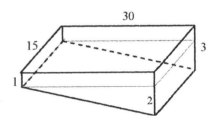

STEP 3 The rates of change are $\dfrac{dh}{dt}$ and $\dfrac{dV}{dt}$.

STEP 4 We know $\dfrac{dV}{dt} = 15$ m^3/min, and we want

$\dfrac{dh}{dt}$ when $h = 2$ m. Once the water hits the 2 meter mark the volume of the new water is
that of a rectangular prism. So $V = lwh$.

STEP 5 $\dfrac{dV}{dt} = lw\dfrac{dh}{dt}$ or $\dfrac{dh}{dt} = \dfrac{1}{lw}\cdot\dfrac{dV}{dt}$

STEP 6 When $l = 30$, $w = 15$ and $\dfrac{dV}{dt} = 15$, $\dfrac{dh}{dt} = \dfrac{1}{30\cdot15}\cdot15 = \dfrac{1}{30} \approx 0.033$ meters per

minute.

17. **STEP 2** The variables are:
  r = radius (in feet) of the oil spill
  A = area (in square feet) of the oil spill
  t = time (in minutes).

STEP 1

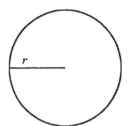

STEP 3 The rates of change are $\dfrac{dr}{dt}$ and $\dfrac{dA}{dt}$.

STEP 4 We know $\dfrac{dr}{dt} = 0.42$ ft/min, and we want $\dfrac{dA}{dt}$

when $r = 120$ feet. The area of a circle is $A = \pi r^2$.

STEP 5 $\dfrac{dA}{dt} = 2\pi r \dfrac{dr}{dt}$

STEP 6 When $r = 120$ and $\dfrac{dr}{dt} = 0.42$, $\dfrac{dA}{dt} = 2\pi \cdot 120 \cdot 0.42 = 100.8\pi \approx 316.673$ square

feet per minute.

19. **STEP 2** x = units produced per day
    C = cost of producing x units
    R = revenue from selling x units
    P = profit from selling x units
    t = time (in days)

STEP 3 The rates of change are $\dfrac{dx}{dt}$, $\dfrac{dC}{dt}$, $\dfrac{dR}{dt}$, and $\dfrac{dP}{dt}$.

STEP 4 We know $\dfrac{dx}{dt} = 100$ units per day and $x = 1000$ units

(a) **STEP 5** $\dfrac{dC}{dt} = 5\dfrac{dx}{dt}$

STEP 6 Substituting $\dfrac{dx}{dt} = 100$, we find $\dfrac{dC}{dt} = 500$. That is cost is increasing at a rate of

$500 per day.

(b) **STEP 5** $\dfrac{dR}{dt} = 15\dfrac{dx}{dt} - \dfrac{x}{5000}\dfrac{dx}{dt}$

STEP 6 When $x = 1000$ and $\dfrac{dx}{dt} = 100$, we find $\dfrac{dR}{dt} = 15 \cdot 100 - \dfrac{1000}{5000} \cdot 100 = 1480$. That

is, revenue is increasing at a rate of $1480 per day.

(c) The revenue is increasing when production is 1000 units per day.

(d) Profit is the difference between revenue and cost.
    $$P(x) = R(x) - C(x)$$

$$P(x) = \left(15x - \frac{x^2}{10,000}\right) - (5x + 5000) = 10x - \frac{x^2}{10,000} - 5000$$

(e) When $x = 1000$, $\dfrac{dP}{dt} = \dfrac{dR}{dt} - \dfrac{dC}{dt} = 1480 - 500 = 980$. That is the profit is increasing at a rate of \$980 per day.

21.   **STEP 2** q = demand (in thousands) of plasma televisions
   p = price of a plasma television
   R = revenue derived from selling laptops
   t = time (in years)

STEP 3 The rates of change are $\dfrac{dp}{dt}$ and $\dfrac{dR}{dt}$.

STEP 4 We know $\dfrac{dp}{dt} = -100$ dollars per year, and we want $\dfrac{dR}{dt}$ when $p = \$7000$. The quantity demanded is $q = 10,000 - 0.90p$ thousand television, and the revenue (in thousands of dollars) is $R = pq = p(10,000 - 0.90p) = 10,000p - 0.90p^2$.

STEP 5 $\dfrac{dR}{dt} = 10,000\dfrac{dp}{dt} - 1.8p\dfrac{dp}{dt}$

STEP 6 When $p = 7000$ and $\dfrac{dp}{dt} = -100$,

$$\frac{dR}{dt} = 10,000 \cdot (-100) - 1.8 \cdot 7000 \cdot (-100) = 260,000.$$

That is, the revenue is increasing at a rate of \$260,000,000 per year.

5.7 The Differential; Linear Approximations

1. $y = x^3 - 2x + 1$

$f'(x) = \dfrac{dy}{dx} = 3x^2 - 2$

$dy = f'(x)dx = \left(3x^2 - 2\right)dx$

3. $y = \dfrac{x-1}{x^2 + 2x - 8}$

$f'(x) = \dfrac{dy}{dx} = \dfrac{1 \cdot \left(x^2 + 2x - 8\right) - (x-1)(2x+2)}{\left(x^2 + 2x - 8\right)^2} = \dfrac{x^2 + 2x - 8 - 2x^2 - 2x + 2x + 2}{\left(x^2 + 2x - 8\right)^2}$

$\qquad = \dfrac{-x^2 + 2x - 6}{\left(x^2 + 2x - 8\right)^2}$

$$dy = f'(x)\,dx = \frac{-x^2 + 2x - 6}{\left(x^2 + 2x - 8\right)^2}\,dx$$

5. Take the differential of each side.

$$d(xy) = d(6)$$

$$x\,dy + y\,dx = 0$$

$$dy = -\frac{y}{x}\,dx \qquad\qquad dx = -\frac{x}{y}\,dy$$

$$\frac{dy}{dx} = -\frac{y}{x} \qquad\qquad \frac{dx}{dy} = -\frac{x}{y}$$

7. Take the differential of each side.

$$d(x^2 + y^2) = d(16)$$

$$2x\,dx + 2y\,dy = 0$$

$$2x\,dx = -2y\,dy$$

$$dy = \frac{2x}{-2y}\,dx \qquad\qquad dx = \frac{-2y}{2x}\,dy$$

$$\frac{dy}{dx} = \frac{2x}{-2y} = -\frac{x}{y} \qquad\qquad \frac{dx}{dy} = \frac{-2y}{2x} = -\frac{y}{x}$$

9. Take the differential of each side.

$$d(x^3 + y^3) = d(3x^2 y)$$

$$3x^2\,dx + 3y^2\,dy = 3x^2 \cdot dy + 6x\,dx \cdot y$$

$$(3y^2 - 3x^2)\,dy = (6xy - 3x^2)\,dx$$

$$dy = \frac{6xy - 3x^2}{3y^2 - 3x^2}\,dx \qquad\qquad dx = \frac{3y^2 - 3x^2}{6xy - 3x^2}\,dy$$

$$\frac{dy}{dx} = \frac{6xy - 3x^2}{3y^2 - 3x^2} = \frac{2xy - x^2}{y^2 - x^2} \qquad\qquad \frac{dx}{dy} = \frac{y^2 - x^2}{2xy - x^2}$$

11. $$d\left(\sqrt{x-2}\right) = d\left[(x-2)^{1/2}\right] = \frac{1}{2}(x-2)^{-1/2} \cdot dx = \frac{dx}{2(x-2)^{1/2}} = \frac{dx}{2\sqrt{x-2}}$$

13. $$d(x^3 - x - 4) = (3x^2 - 1)\,dx$$

15. $f(x) = x^2 - 2x + 1$ and $f(2) = 2^2 - 2(2) + 1 = 1$

$f'(x) = 2x - 2$ and $f'(2) = 2(2) - 2 = 2$

The linear approximation to f near $x = 2$ is
$$f(x) \approx f(2) + f'(2)(x - 2)$$
$$f(x) \approx 1 + 2(x - 2) = 2x - 3$$

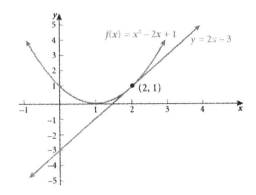

17. $f(x) = \sqrt{x} = x^{1/2}$ and $f(4) = \sqrt{4} = 2$

$f'(x) = \dfrac{1}{2}x^{-1/2} = \dfrac{1}{2\sqrt{x}}$ and $f'(4) = \dfrac{1}{2\sqrt{4}} = \dfrac{1}{4}$

The linear approximation to f near $x = 4$ is
$$f(x) \approx f(4) + f'(4)(x - 4)$$
$$f(x) = 2 + \dfrac{1}{4}(x - 4) = \dfrac{1}{4}x + 1$$

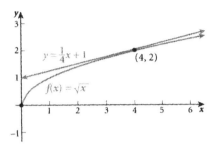

19. $f(x) = e^x$ and $f(0) = e^0 = 1$

$f'(x) = e^x$ and $f'(0) = e^0 = 1$

The linear approximation to f near $x = 0$ is
$$f(x) \approx f(0) + f'(0)(x - 0)$$
$$f(x) = 1 + 1(x - 0) = x + 1$$

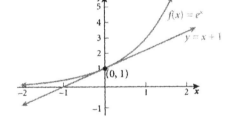

21.   (a) $\Delta y \approx f'(x_0)(x - x_0)$

$\approx 2(3)(3.001 - 3)$

$= 0.006$

(b) $\Delta y \approx f'(x_0)(x - x_0)$

$f'(x) = -\dfrac{1}{(x + 2)^2}$

$\Delta y \approx -\dfrac{1}{(2 + 2)^2}(1.98 - 2) = 0.00125$

23. $A(r) = \pi r^2$ $A'(r) = 2\pi r$

$\Delta A \approx dA = A'(r)\,dr = 2\pi r\,dr$

When $r = 10$ centimeters and increases to 10.1 centimeters,

$\Delta A \approx 2\pi \cdot 10 \cdot (0.1) = 2\pi \approx 6.28$

The area increases by approximately 6.28 square centimeters.

25.

$$V(r) = \frac{4}{3}\pi r^3 \qquad\qquad V'(r) = 4\pi r^2$$

$$\Delta V \approx dV = V'(r)dr = 4\pi r^2 \, dr$$

When $r = 3$ meters and increases to 3.1 meters,

$$\Delta V \approx 4\pi \cdot 3^2 \cdot (0.1) = 3.6\pi \approx 11.310$$

The volume of the balloon increases by approximately 11.310 cubic meters.

27.

$V = s^3$, $V'(s) = 3s^2$ and $\dfrac{\Delta s}{s} = 0.02$. The relative error in the volume,

$$\frac{dV}{V} \approx \frac{dV}{V} = \frac{V'(s)ds}{V} = \frac{3s^2\,ds}{s^3} = 3\frac{ds}{s} = 3\frac{\Delta s}{s} = 3\cdot(0.02) = 0.06$$

The percentage area in volume is 6%.

29. If the diameter d is 4 cm., then the radius r is 2 cm.
Similarly if $d = 3.9$ cm, then $r = 1.95$ cm. We know
that the radius is one-fourth the height, so $h = 4r$.
The volume of a right circular cone is

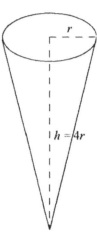

$$V = \frac{1}{3}\pi r^2 h = \frac{4}{3}\pi r^3$$

Taking the differential of V,

$$dV = 4\pi r^2 \cdot dr$$

dr is approximately the change in radius or 0.05 cm.
So, $dV = 4\pi(2^2)\cdot(-0.05) = -2.513$

The cup holds approximately 2.513 cubic
centimeters less than intended.

31. If we let h denote the height of the building and x denote the length of the shadow cast by
the pole, we can use similar triangles to estimate the height of the building.

$$\frac{\text{height}}{\text{base}} = \frac{h}{9+x} = \frac{3}{x}$$

$$h = \frac{27+3x}{x} \text{ meters}$$

If the measured shadow of the pole is 1 meter with an error of 1%, then the percentage

error in the height of the building will be $\dfrac{\Delta h}{h} \approx \dfrac{dh}{h}$.

$$dh = f'(x)dx = \frac{3\cdot x - (3x+27)\cdot 1}{x^2}\cdot dx = -\frac{27}{x^2}dx$$

$$\frac{dh}{h} = \frac{f'(x)dx}{h} = \frac{-\dfrac{27}{x^2}dx}{\dfrac{3x+27}{x}} = -\frac{27}{3x+27}\cdot\frac{dx}{x}$$

When $x = 1$ and $\dfrac{dx}{x} = 0.01$, $\dfrac{\Delta h}{h} \approx -\dfrac{27}{30} \cdot (0.01) = -0.009$

The percentage error in the measured height of the building is approximately 0.9%.

33. If the pendulum is originally 1 meter = 100 centimeters long and increases to 110 centimeters in length, then $\Delta l = 10$ centimeters. From Problem 32, we see that

$\dfrac{\Delta T}{T} \approx \dfrac{1}{2} \cdot \dfrac{dl}{l}$. So, $\dfrac{\Delta T}{T} = \dfrac{1}{2} \cdot \dfrac{10}{100} = \dfrac{1}{20} = 0.05$. The percentage area in the period is

approximately 5%.

5% of a day is 0.05(60 min/hour)(24 hours/day) = 72 minutes.

Chapter 5 Review

TRUE-FALSE ITEMS

1. False   3. False   5. True

FILL IN THE BLANKS

1. decreasing   3. concave up   5. concavity

7. linear approximation

REVIEW EXERCISES

1. $f'(x) = 3x^2 - 2x + 1$

   Horizontal tangent lines occur where $f'(x) = 0$. Since the discriminant $b^2 - 4ac = (-2)^2 - 4(3)(1) = -8 < 0$, f has no horizontal tangent line.

   There is no vertical tangent line since $f'(x)$ is never unbounded.

3. The domain of f is $\{x \mid x \neq -4\}$.

$$f'(x) = \frac{\dfrac{1}{3}x^{-2/3}(x+4) - x^{1/3}(1)}{(x+4)^2} = \frac{\dfrac{x+4}{3x^{2/3}} - x^{1/3}}{(x+4)^2} = \frac{x+4-3x}{3x^{2/3}(x+4)^2} = \frac{4-2x}{3x^{2/3}(x+4)^2}$$

   Horizontal tangent lines occur where $f'(x) = 0$, that is when $4 - 2x = 0$ or when $x = 2$.

   When $x = 2$ $f(2) = \dfrac{2^{1/3}}{2+4} = \dfrac{2^{1/3}}{6}$. So f has a horizontal tangent line at $\left(2, \dfrac{2^{1/3}}{6}\right)$.

   Vertical tangent lines occur where $f'(x)$ is unbounded. $f'(x)$ is unbounded at 0 and at -4. Since $x = -4$ is not in the domain of f we disregard it.

When $x = 0$, $f(0) = \dfrac{0^{1/3}}{0+4} = 0$. So f has a vertical tangent line at $(0, 0)$.

5.   (a) $f(x) = 3x^{1/5}$ $\qquad\qquad$ $f(0) = 3 \cdot 0^{1/5} = 0$

The one-sided limits of f near 0 are
$$\lim_{x \to 0^-} f(x) = \lim_{x \to 0^-} 3x^{1/5} = 0 \qquad\qquad \lim_{x \to 0^+} f(x) = \lim_{x \to 0^+} 3x^{1/5} = 0$$

Since $\lim_{x \to 0} f(x) = f(0)$ the function is continuous at 0.

(b)   The derivative of f at 0 is $f'(0) = \lim_{x \to 0} \dfrac{f(x) - f(0)}{x - 0} = \lim_{x \to 0} \dfrac{3x^{1/5} - 0}{x - 0} = \lim_{x \to 0} \dfrac{3x^{1/5}}{x}$.

$f'(x)$ is unbounded at $x = 0$ and does not exist.

(c)   Since the function is continuous at $x = 0$, and the derivative is unbounded at $x = 0$, so there is a vertical tangent line at $(0, 0)$.

7. $f(x) = \begin{cases} 3x + 1 & x < 3 \\ x^2 + 1 & x \geq 3 \end{cases}$

(a)   When $x = 3, f(3) = 3^2 + 1 = 10$

The one-sided limits are
$$\lim_{x \to 3^-} f(x) = \lim_{x \to 3^-} (3x + 1) = 10 \qquad\qquad \lim_{x \to 3^+} f(x) = \lim_{x \to 3^+} (x^2 + 1) = 10$$

Since $\lim_{x \to 3} f(x) = f(x) = 10$, the function f is continuous at 3.

(b)   The derivative of f at 3 is $f'(3) = \lim_{x \to 3} \dfrac{f(x) - f(3)}{x - 3} = \lim_{x \to 3} \dfrac{f(x) - 10}{x - 3}$.

Looking at one-sided limits,
$$\lim_{x \to 3^-} \dfrac{f(x) - 10}{x - 3} = \lim_{x \to 3^-} \dfrac{(3x+1) - 10}{x - 3} = \lim_{x \to 3^-} \dfrac{3x - 9}{x - 3} = \lim_{x \to 3^-} \dfrac{3(x - 3)}{x - 3} = \lim_{x \to 3^-} 3 = 3$$

$$\lim_{x \to 3^+} \dfrac{f(x) - 10}{x - 3} = \lim_{x \to 3^+} \dfrac{(x^2+1) - 10}{x - 3} = \lim_{x \to 3^+} \dfrac{x^2 - 9}{x - 3} = \lim_{x \to 3^+} \dfrac{(x+3)(x - 3)}{x - 3} = \lim_{x \to 3^+} (x + 3) = 6$$

Since the one-sided limits are not equal we concluded that $f'(3)$ does not exist.

(c)   The one-sided limits in (b) are unequal, so there is no tangent line at the point $(3, 10)$.

9. $f(x) = \dfrac{1}{5}x^5 - x^3 - 4x$

(a)   **STEP 1** $f'(x) = x^4 - 3x^2 - 4$

STEP 2   Solve $f'(x) = 0$.
$$x^4 - 3x^2 - 4 = 0$$

$$\left(x^2-4\right)\left(x^2+1\right)=0$$
$$\left(x-2\right)\left(x+2\right)\left(x^2+1\right)=0$$
$$x-2=0 \quad \text{or} \quad x+2=0 \quad \text{or} \quad x^2+1=0$$
$$x=2 \quad \text{or} \qquad x=-2$$

These numbers separate the number line into three parts: $-\infty<x<-2, -2<x<2$, and $2<x<\infty$, and we use $-3, 0$, and 3 as test numbers.

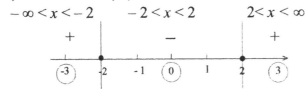

$$f'(-3)=(-3)^4-3(-3)^2-4=50 \quad f'(0)=-4 \quad f'(3)=3^4-3\left(3^2\right)-4=50$$

We conclude that the function is increasing on the intervals $(-\infty,-2)$ and $(2,\infty)$ and is decreasing on the interval $(-2,2)$.

(b) f is increasing for $-\infty<x<-2$ and decreasing for $-2<x<2$. When $x=-2$, $y=f(-2)=9.6$. So by the First Derivative Test, f has a local maximum at the point $(-2, 9.6)$.

f is decreasing for $-2<x<2$ and increasing for $2<x<\infty$. When $x=2, y=f(2)=-9.6$, and by the First Derivative Test, f has a local minimum at $(2,-9.6)$.

11. $$f(x)=\frac{x^2}{x^2-8} \qquad x\neq\pm\sqrt{8}$$

(a) **STEP 1** $$f'(x)=\frac{2x\cdot\left(x^2-8\right)-x^2\cdot\left(2x\right)}{\left(x^2-8\right)^2}=\frac{2x^3-16x-2x^3}{\left(x^2-8\right)^2}=\frac{-16x}{\left(x^2-8\right)^2}$$

STEP 2 $f'(x)$ is not defined at $x=\pm 2\sqrt{2}$
$\qquad$ Solve $f'(x)=0$
$$-16x=0 \text{ or } x=0$$

These 3 numbers separate the number line into four parts, we use the test numbers $-3, -1$, 1, and 3 to see if $f'(x)$ is positive or negative in each part.

$$-\infty<x<-2\sqrt{2} \qquad -2\sqrt{2}<x<0 \qquad 0<x<2\sqrt{2} \qquad 2\sqrt{2}<x<\infty$$

$$f'(-3)=48 \qquad f'(-1)=0.327 \qquad f'(1)=-0.327 \qquad f'(3)=-48$$

We conclude that f is increasing on the intervals $\left(-\infty, -2\sqrt{2}\right)$ and $\left(-2\sqrt{2}, 0\right)$, and that f is decreasing on the intervals $\left(0, 2\sqrt{2}\right)$ and $\left(2\sqrt{2}, \infty\right)$.

(b) f is increasing for $x < 0$ and decreasing for $x > 0$. When $x = 0$, $y = f(0) = 0$. So by the First Derivative Test, there is a local maximum at $(0, 0)$.

13. $f(x) = 1 + 3e^{-x}$

(a) **STEP 1** $f'(x) = -3e^{-x}$

STEP 2 The domain of f' is all real numbers. $f'(x) \neq 0$.

We select $x = 0$ as a test point $f'(0) = -3e^{0} = -3$. We conclude that f is always decreasing.

(b) From the First Derivative Test we conclude that f has no local maximum nor local minimum points.

15. $f(x) = x^{3} - 3x^{2} + 3x - 1$

STEP 1 f is a polynomial, so the domain is all real numbers.

STEP 2 The y-intercept occurs when $x = 0$; $y = f(0) = -1$. The y-intercept is $(0, -1)$. The x-intercept(s) occur when $y = 0$. So we solve $x^{3} - 3x^{2} + 3x - 1 = 0$. This is hard to do so we skip it.

STEP 3 To find where the graph is increasing or decreasing, we find $f'(x)$:
$$f'(x) = 3x^{2} - 6x + 3$$
The solutions to $f'(x) = 0$ are $3x^{2} - 6x - 3 = 0$

$$x = \frac{-b \pm \sqrt{b^{2} - 4ac}}{2a} = \frac{6 \pm \sqrt{36 - 4(3)(-3)}}{2(3)} = \frac{6 \pm \sqrt{72}}{6} = \frac{6 \pm 6\sqrt{2}}{6} = 1 \pm \sqrt{2}$$

$$x \approx -0.414 \quad \text{or} \quad x \approx 2.414$$

We use these two numbers to separate the number line into three parts:
$$-\infty < x < -0.414 \qquad -0.414 < x < 2.414 \qquad 2.414 < x < \infty$$
and choose a test number from each part.

For $x = -1$: $f'(-1) = 12$

For $x = 0$: $f'(0) = 3$

For $x = 3$ $f'(3) = 12$

We conclude that the graph of f is always increasing.

STEP 4 Since the graph of f is always increasing, there is no relative minimum point or relative maximum point.

STEP 5 $f'(x) = 0$ for $x = -0.414$ and for $x = 2.414$. The graph of f has a horizontal tangent line at the points $(-0.414, f(-0.414)) = (-0.414, -2.828)$ and $(2.414, f(2.414)) = (2.414, 2.828)$

The first derivative is never unbounded, so there is no vertical tangent.

STEP 6 Since f is a polynomial function, its end behavior is that of $y = x^3$. Polynomial functions have no asymptotes.

STEP 7 To identify the inflection point, if any, we find $f''(x)$:

$$f''(x) = 6x - 6$$

$f''(x) = 0$ when $6x - 6 = 0$ or when $x = 1$.

We use the number 1 to separate the number line into two parts:

$$-\infty < x < 1 \quad \text{and} \quad 1 < x < \infty$$

and choose a test number from each part.

For $x = 0$: $f''(0) = -6$

For $x = 2$: $f''(2) = 6$

We conclude that the graph of f is concave up on the interval $(1, \infty)$ and is concave down on the interval $(-\infty, 1)$. Since the concavity changes at the point $x = 1$, the point $(1, f(1)) = (1, 0)$ is an inflection point.

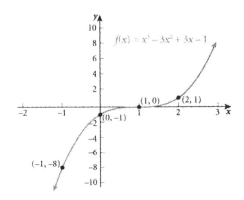

17. $f(x) = x^5 - 5x$

STEP 1 f is a polynomial, so the domain is all real numbers.

STEP 2 The y-intercept occurs when $x = 0$; $y = f(0) = 0$. The y-intercept is $(0, 0)$. The x-intercept(s) occur when $y = 0$. So we solve $x^5 - 5x = 0$.

$$x\left(x^4 - 5\right) = 0$$
$$x = 0 \quad \text{or} \quad x^4 - 5 = 0$$
$$x = \pm \sqrt[4]{5} \approx \pm 1.495$$

The x-intercepts are $(0, 0)$, $(-1.495, 0)$, and $(1.495, 0)$.

STEP 3 To find where the graph is increasing or decreasing, we find $f'(x)$:

$$f'(x) = 5x^4 - 5$$

The solutions to $f'(x) = 0$ are $5x^4 - 5 = 0$

$$x^4 = 1$$
$$x = \pm 1$$

We use the 2 numbers to separate the number line into three parts:

$$-\infty < x < -1 \qquad -1 < x < 1 \qquad 1 < x < \infty$$

and choose a test number from each part.

For $x = -2$: $\qquad f'(x) = 5 \cdot (-2)^4 - 5 = 75$

For $x = 0$: $\qquad f'(0) = 5 \cdot 0 - 5 = -5$

For $x = 2$: $\qquad f'(2) = 5 \cdot 2^4 - 5 = 75$

We conclude that the graph of f is increasing on the intervals $(-\infty, -1)$ and $(1, \infty)$, and f is decreasing on the interval $(-1, 1)$.

STEP 4 The graph is increasing to the left of -1 and decreasing to the right of -1, so the point $(-1, f(-1)) = (-1, 4)$ is a local maximum.

The graph is decreasing to the left of 1 and increasing to the right of 1, so the point $(1, f(1)) = (1, -4)$ is a local minimum.

STEP 5 $f'(x) = 0$ for $x = -1$ and $x = 1$. The graph of f has a horizontal tangent lines at the points $(-1, f(-1)) = (-1, 4)$ and $(1, f(1)) = (1, -4)$.

The first derivative is never unbounded, so there is no vertical tangent.

STEP 6 Since f is a polynomial function, its end behavior is that of $y = x^5$. Polynomial functions have no asymptotes.

STEP 7 To identify the inflection point, if any, we find $f''(x)$:

$$f''(x) = 20x^3$$
$$f''(x) = 0 \text{ when } 20x^3 = 0 \text{ or when } x = 0.$$

We use the numbers 0 to separate the number line into two parts:

$$-\infty < x < 0 \qquad \text{and} \qquad 0 < x < \infty$$

and choose a test number from each part.

For $x = -1$: $\qquad f''(-1) = -20$

For $x = 1$: $\qquad f''(1) = 20$

We conclude that the graph of f is concave up on the interval $(0, \infty)$ and is concave down on the interval $(-\infty, 0)$. Since the concavity changes at $x = 0$, the point $(0, 0)$ is an inflection point.

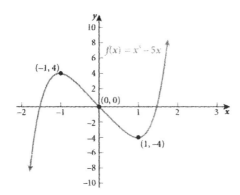

19. $f(x) = x^{4/3} + 4x^{1/3}$

STEP 1   The domain of f is all real numbers.

STEP 2   The y-intercept occurs when $x = 0$; $y = f(0) = 0$. The y-intercept is $(0, 0)$. The x-intercept(s) occur when $y = 0$. So we solve $x^{4/3} + 4x^{1/3} = 0$.

$$x^{1/3}(x + 4) = 0$$

$$x^{1/3} = 0 \quad \text{or} \quad x + 4 = 0$$
$$x = 0 \quad \text{or} \quad x = -4$$

The x-intercepts are $(0, 0)$ and $(-4, 0)$.

STEP 3   To find where the graph is increasing or decreasing, we find $f'(x)$:

$$f'(x) = \frac{4}{3}x^{1/3} + \frac{4}{3}x^{-2/3} = \frac{4(x+1)}{3x^{2/3}}$$

$f'(x)$ is unbounded when $x = 0$.
The solution to $f'(x) = 0$ is $4(x + 1) = 0$ or $x = -1$.

We use these two numbers to separate the number line into three parts:
$$-\infty < x < -1 \qquad -1 < x < 0 \qquad 0 < x < \infty$$
and choose a test number from each part.

For $x = -8$: $\qquad f'(-8) = \dfrac{4(-8+1)}{3(-8)^{2/3}} = \dfrac{-28}{12} \approx -2.333$

For $x = -\dfrac{1}{8}$: $\qquad f'\left(-\dfrac{1}{8}\right) = \dfrac{4\left(-\dfrac{1}{8}+1\right)}{3\left(-\dfrac{1}{8}\right)^{2/3}} = \dfrac{\dfrac{7}{2}}{\dfrac{3}{4}} = \dfrac{14}{3} \approx 4.667$

For $x = 1$: $\qquad f'(1) = \dfrac{4(1+1)}{3(1)^{2/3}} = \dfrac{8}{3} \approx 2.667$

We conclude that the graph of f is increasing on the intervals $(-1, 0)$ and $(0, \infty)$, and f is decreasing on the interval $(-\infty, -1)$.

STEP 4   The graph is decreasing to the left of -1 and increasing to the right of -1, so the point $(-1, f(-1)) = (-1, -3)$ is a local minimum.

There is no local maximum.

STEP 5 $f'(x) = 0$ for $x = -1$. The graph of f has a horizontal tangent line at the point $(-1, f(-1)) = (-1, -3)$.

The first derivative is unbounded at $x = 0$, so there is a vertical tangent line at the point $(0, f(0)) = (0, 0)$.

STEP 6 For the end behavior of f, we look at the two limits at infinity:
$$\lim_{x \to \infty} f(x) = \lim_{x \to \infty} \left[x^{4/3} + 4x^{1/3} \right] = \infty \qquad \lim_{x \to -\infty} f(x) = \lim_{x \to -\infty} \left[x^{4/3} + 4x^{1/3} \right] = \infty$$

STEP 7 To identify the inflection points, if any, we find $f''(x)$:
$$f''(x) = \frac{4}{9} x^{-2/3} - \frac{8}{9} x^{-5/3} = \frac{4}{9x^{2/3}} - \frac{8}{9x^{5/3}} = \frac{4x - 8}{9x^{5/3}}$$
$f''(x)$ is unbounded at $x = 0$, and $f''(x) = 0$ when $4x - 8 = 0$ or when $x = 2$.
We use the numbers 0 and 2 to separate the number line into three parts:
$$-\infty < x < 0 \qquad 0 < x < 2 \qquad 2 < x < \infty$$
and choose a test number from each part.

For $x = -1$: $f''(-1) = \dfrac{-12}{-9} = \dfrac{4}{3}$

For $x = 1$: $f''(1) = \dfrac{-4}{9}$

For $x = 8$: $f''(8) = \dfrac{24}{288} = \dfrac{1}{12}$

We conclude that the graph of f is concave up on intervals $(-\infty, 0)$ and $(2, \infty)$ and is concave down on interval $(0, 2)$. Since the concavity changes at $x = 0$ and $x = 2$, the points $(0, 0)$ and $(2, 7.560)$ are inflection points.

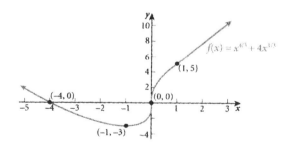

21.
$$f(x) = \frac{2x}{x^2 + 1}$$
STEP 1 f is a rational function. Since the denominator never equals 0, the domain is all real numbers.

STEP 2 The y-intercept occurs when $x = 0$; $y = f(0) = 0$. The y-intercept is $(0, 0)$. The x-intercept(s) occur when $y = 0$. So we solve $2x = 0$ and find $x = 0$. The x-intercept is $(0, 0)$.

STEP 3   To find where the graph is increasing or decreasing, we find $f'(x)$:

$$f'(x) = \frac{2 \cdot (x^2 + 1) - 2x \cdot 2x}{(x^2 + 1)^2} = \frac{2x^2 + 2 - 4x^2}{(x^2 + 1)^2} = \frac{2 - 2x^2}{(x^2 + 1)^2}$$

The solutions to $f'(x) = 0$ are $2 - 2x^2 = 0$

$$2(1 - x)(1 + x) = 0$$
$$1 - x = 0 \quad \text{or} \quad 1 + x = 0$$
$$x = 1 \quad \text{or} \quad x = -1$$

We use the numbers to separate the number line into three parts:

$$-\infty < x < -1 \qquad -1 < x < 1 \qquad 1 < x < \infty$$

and choose a test number from each part.

For $x = -2$: $f'(-2) = \dfrac{2 - 2(-2)^2}{\left[(-2)^2 + 1\right]^2} = -0.24$

For $x = 0$: $f'(0) = \dfrac{2 - 2(0)^2}{\left[(0)^2 + 1\right]^2} = 2$

For $x = 2$: $f'(2) = \dfrac{2 - 2(2)^2}{\left[2^2 + 1\right]^2} = -0.24$

We conclude that the graph of f is increasing on the interval $(-1, 1)$, and that the graph of f is decreasing on the intervals $(-\infty, -1)$ and $(1, \infty)$.

STEP 4   The graph is decreasing to the left of -1 and increasing to the right of -1, so the point $(-1, f(-1)) = (-1, -1)$ is a local minimum.

The graph is increasing to the left of 1 and decreasing to the right of 1, so the point $(1, f(1)) = (1, 1)$ is a local maximum.

STEP 5 $f'(x) = 0$ for $x = 1$ and $x = -1$. The graph of f has a horizontal tangent line at the points $(1, f(1)) = (1, 1)$ and at $(-1, f(-1)) = (-1, -1)$.

The first derivative is never unbounded, so there is no vertical tangent.

STEP 6   For the end behavior we look at the limits at infinity. Since f is an odd function, we need only to consider the limit as $x \to \infty$, the limit as $x \to -\infty$ will be the opposite.

$$\lim_{x \to \infty} f(x) = \lim_{x \to \infty} \frac{2x}{x^2 + 1} = \lim_{x \to \infty} \frac{2}{x} = 0$$

As x becomes unbounded in both the positive and negative directions, $y = 0$ is a horizontal asymptote.

STEP 7   To identify the inflection point, if any, we find $f''(x)$:

$$f''(x) = \frac{-4x(x^2+1)^2 - (2-2x^2) \cdot 2(x^2+1)(2x)}{(x^2+1)^4} = \frac{-4x(x^2+1) - 4x(2-2x^2)}{(x^2+1)^3}$$

$$= \frac{-4x(-x^2+3)}{(x^2+1)^3} = \frac{4x(x^2-3)}{(x^2+1)^3}$$

$f''(x) = 0$ when $4x(x^2-3) = 0$

$$4x = 0 \quad \text{or} \quad x^2 - 3 = 0$$

$$x = 0 \quad \text{or} \quad x = \pm\sqrt{3}$$

We use the numbers 0, $\sqrt{3}$, and $-\sqrt{3}$ to separate the number line into four parts:

$$-\infty < x < -\sqrt{3} \qquad -\sqrt{3} < x < 0 \qquad 0 < x < \sqrt{3} \qquad \sqrt{3} < x < \infty$$

and choose a test number from each part.

For $x = -2$: $\quad f''(-2) = -0.064$

For $x = -1$: $\quad f''(-1) = 1$

For $x = 1$: $\quad f''(1) = -1$

For $x = 2$: $\quad f''(2) = 0.064$

We conclude that the graph of f is concave up on the intervals $(-\sqrt{3}, 0)$ and $(\sqrt{3}, \infty)$, and is concave down on the intervals $(-\infty, -\sqrt{3})$ and $(0, \sqrt{3})$. Since the concavity changes at the points $x = 0$, $x = \sqrt{3}$, and $x = -\sqrt{3}$, the points $(0, 0)$, $(\sqrt{3}, 0.866)$, and $(-\sqrt{3}, -0.866)$ are inflection points.

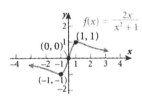

23. First locate the numbers for which $f'(x) = 0$.

$$f'(x) = 12x^2 - 3$$

$$12x^2 - 3 = 0$$

$$3(4x^2 - 1) = 0$$

$$3(2x-1)(2x+1) = 0$$

$$2x - 1 = 0 \quad \text{or} \quad 2x + 1 = 0$$

$$x = \frac{1}{2} \quad \text{or} \quad x = -\frac{1}{2}$$

Now evaluate $f''(x)$ at these numbers.

$$f''(x) = 24x$$

$f''\left(\dfrac{1}{2}\right) = 24\left(\dfrac{1}{2}\right) = 12 > 0$. By the Second Derivative Test, f has a local minimum

at $\left(\dfrac{1}{2}, f\left(\dfrac{1}{2}\right)\right) = \left(\dfrac{1}{2}, -1\right)$

$f''\left(-\dfrac{1}{2}\right) = 24\left(-\dfrac{1}{2}\right) = -12$. By the Second Derivative Test, f has a local maximum

at $\left(-\dfrac{1}{2}, f\left(-\dfrac{1}{2}\right)\right) = \left(-\dfrac{1}{2}, 1\right)$.

25. First locate the numbers for which $f'(x) = 0$.
$$f'(x) = 4x^3 - 4x$$
$$4x^3 - 4x = 0$$
$$4x\left(x^2 - 1\right) = 0$$
$$4x = 0 \quad \text{or} \quad x^2 - 1 = 0$$
$$x = 0 \quad \text{or} \quad x = \pm\sqrt{1} = \pm 1$$

Now evaluate $f''(x)$ at these numbers.
$$f''(x) = 12x^2 - 4$$
$f''(0) = 12\left(0^2\right) - 4 = -4 < 0$. By the Second Derivative Test, f has a local maximum
at $(0, f(0)) = (0, 0)$.

$f''(-1) = 12(-1)^2 - 4 = 12 - 4 = 8 > 0$. By the Second Derivative Test, f has a local
minimum at $(-1, f(-1)) = (-1, -1)$.

$f''(1) = 12(1)^2 - 4 = 12 - 4 = 8 > 0$. By the Second Derivative Test, f has a local minimum
at $(1, f(1)) = (1, -1)$.

27. First locate the numbers for which $f'(x) = 0$.
$$f'(x) = xe^x + e^x$$
$$xe^x + e^x = 0$$
$$e^x(x + 1) = 0$$
$$e^x = 0 \quad \text{or} \quad x + 1 = 0$$
$$x = -1$$

Now evaluate $f''(x)$ at this number.
$$f''(x) = xe^x + e^x + e^x = xe^x + 2e^x$$
$f''(-1) = (-1)e^x + 2e^x = e^x > 0$. By the Second Derivative Test, f has a local minimum

at $(-1, f(-1)) = \left(-1, -\dfrac{1}{e}\right) \approx (-1, -0.368)$.

29. $f(x) = x^3 - 3x^2 + 3x - 1$ $\qquad$ $f'(x) = 3x^2 - 6x + 3$

$f'(x) = 0$ when $3x^2 - 6x + 3 = 0$.
$$x^2 - 2x + 1 = 0$$
$$(x-1)^2 = 0$$
$$x = 1$$

The critical number 1 is in the interval $(0, 3)$, so we evaluate f at the critical number and the two endpoints.
$$f(0) = 0^3 - 3(0)^2 + 3(0) - 1 = -1$$
$$f(1) = 1^3 - 3(1)^2 + 3(1) - 1 = 0$$
$$f(3) = 3^3 - 3(3)^2 + 3(3) - 1 = 8$$

The absolute maximum of f on $[0, 3]$ is 8 and the absolute minimum is -1.

31. $f(x) = x^4 - 4x^3 + 4x^2$ $\qquad$ $f'(x) = 4x^3 - 12x^2 + 8x$

$f'(x) = 0$ when $4x^3 - 12x^2 + 8x = 0$.
$$4x(x^2 - 3x + 2) = 0$$
$$4x(x-2)(x-1) = 0$$

$$4x = 0 \quad \text{or} \quad x - 2 = 0 \quad \text{or} \quad x - 1 = 0$$
$$x = 0 \quad \text{or} \quad x = 2 \quad \text{or} \quad x = 1$$

The critical number 2 is in the interval $(1, 3)$, so we evaluate f at the critical number and the two endpoints.
$$f(1) = 1^4 - 4(1)^3 + 4(1)^2 = 1$$
$$f(2) = 2^4 - 4(2)^3 + 4(2)^2 = 0$$
$$f(3) = 3^4 - 4(3)^3 + 4(3)^2 = 9$$

The absolute maximum of f on $[1, 3]$ is 9 and the absolute minimum is 0.

33.
$$f(x) = x^{4/3} - 4x^{1/3} \qquad f'(x) = \frac{4}{3}x^{1/3} - \frac{4}{3}x^{-2/3} = \frac{4x^{1/3}}{3} - \frac{4}{3x^{2/3}} = \frac{4x-4}{3x^{2/3}}$$

$f'(x)$ is unbounded when $x = 0$.
$f'(x) = 0$ when $4x - 4 = 0$ or $x = 1$.

The critical numbers 0 and 1 are in the interval $(-1, 8)$, so we evaluate f at the critical numbers and the two endpoints.
$$f(-1) = (-1)^{4/3} - 4(-1)^{1/3} = 5$$
$$f(0) = (0)^{4/3} - 4(0)^{1/3} = 0$$
$$f(1) = (1)^{4/3} - 4(1)^{1/3} = -3$$

$$f(8) = (8)^{4/3} - 4(8)^{1/3} = 8$$

The absolute maximum of f on $[-1, 8]$ is 8 and the absolute minimum is -3.

35. $x = f(p) = 1000 - 2p^2$ $\qquad$ $f'(p) = -4p$

$$E(p) = \frac{pf'(p)}{f(p)} = \frac{p(-4p)}{1000 - 2p^2} = -\frac{4p^2}{1000 - 2p^2} = -\frac{2p^2}{500 - p^2}$$

$$E(20) = -\frac{2(20)^2}{500 - 20^2} = -8$$

At $p = \$20$ the demand is elastic.

37. $x = f(p) = \sqrt{500 - p^2} = \left(500 - p^2\right)^{1/2}$

$$f'(p) = \frac{1}{2}\left(500 - p^2\right)^{-1/2} \cdot (-2p) = \frac{-2p}{2\left(500 - p^2\right)^{1/2}} = -\frac{p}{\sqrt{500 - p^2}}$$

$$E(p) = \frac{pf'(p)}{f(p)} = \frac{p\left[-\dfrac{p}{\sqrt{500 - p^2}}\right]}{\sqrt{500 - p^2}} = -\frac{p^2}{500 - p^2}$$

$$E(10) = -\frac{10^2}{500 - 10^2} = -\frac{100}{400} = -\frac{1}{4} = -0.25$$

At $p = \$10$ the demand is inelastic.

39. $x = f(p) = 40 - 2\sqrt{p} = 40 - 2p^{1/2}$ $\qquad$ $f'(p) = -\frac{1}{2} \cdot 2p^{-1/2} = -\frac{1}{\sqrt{p}}$

$$E(p) = \frac{pf'(p)}{f(p)} = \frac{p\left[-\dfrac{1}{\sqrt{p}}\right]}{40 - 2\sqrt{p}} = -\frac{\sqrt{p}}{40 - 2\sqrt{p}}$$

(a) $E(300) = -\dfrac{\sqrt{300}}{40 - 2\sqrt{300}} = -3.232$

At $p = \$300$ the demand is elastic.

(b)   Since the demand is elastic, increasing the price to \$310 will cause a decrease in revenue.

41. $y = 3x^4 - 2x^3 + x$ $\qquad$ $f'(x) = \dfrac{dy}{dx} = 12x^3 - 6x^2 + 1$

$$dy = f'(x)dx = \left(12x^3 - 6x^2 + 1\right)dx$$

43.
$$y = \frac{3 - 2x}{1 + x}$$

$$f'(x) = \frac{dy}{dx} = \frac{(-2)(1+x) - (3-2x)(1)}{(1+x)^2} = \frac{-2 - 2x - 3 + 2x}{(1+x)^2} = \frac{-5}{(1+x)^2}$$

$$dy = f'(x)dx = -\frac{5}{(1+x)^2}\, dx$$

45.
$$f(x) = x^2 - 9 \qquad\qquad f'(x) = 2x$$
$$f(3) = 3^2 - 9 = 0 \qquad\qquad f'(3) = 2 \cdot 3 = 6$$
$$f(x) = f(3) + f'(3)(x-3) = 0 + 6(x-3)$$
$$f(x) = 6x - 18$$

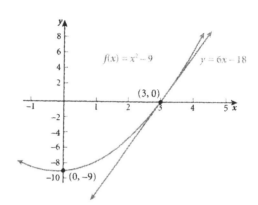

47.
$$x^2 + y^2 = 8$$

$$2x\frac{dx}{dt} + 2y\frac{dy}{dt} = 0$$

$$2x\frac{dx}{dt} = -2y\frac{dy}{dt}$$

$$\frac{dx}{dt} = -\frac{2y}{2x}\frac{dy}{dt} = -\frac{y}{x}\frac{dy}{dt}$$

When $x = 2$, $y = 2$, and $\dfrac{dy}{dt} = 3$,

$$\frac{dx}{dt} = -\frac{2}{2}\cdot 3 = -3$$

49.
$$xy + 6x + y^3 = -2$$

$$x\frac{dy}{dt} + y\frac{dx}{dt} + 6\frac{dx}{dt} + 3y\frac{dy}{dt} = 0$$

$$(x + 3y)\frac{dy}{dt} + (y + 6)\frac{dx}{dt} = 0$$

$$(x + 3y)\frac{dy}{dt} = -(y + 6)\frac{dx}{dt}$$

$$\frac{dy}{dt} = -\frac{y+6}{x+3y}\frac{dx}{dt}$$

When $x = 2$, $y = -3$, and $\frac{dx}{dt} = 3$, $\frac{dy}{dt} = -\frac{(-3)+6}{2+3(-3)} \cdot 3 = \frac{9}{7}$.

51.   **STEP 2** The variables are: **STEP 1**

r = radius (in meters) of the balloon
V = volume (in cubic meters) of the balloon
S = surface area (in square meters) of the balloon
t = time (in minutes).

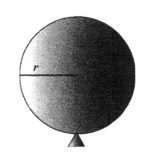

STEP 3 The rates of change are $\frac{dr}{dt}$, $\frac{dV}{dt}$, and $\frac{dS}{dt}$.

STEP 4 We know $\frac{dV}{dt} = 10 \text{ m}^3$ per minute. We want

$\frac{dS}{dt}$ when $r = 3$ m. The volume of a sphere is $V = \frac{4}{3}\pi r^3$ and the surface area is

$S = 4\pi r^2$.

STEP 5 $\dfrac{dV}{dt} = 4\pi r^2 \dfrac{dr}{dt}$ and $\dfrac{dS}{dt} = 8\pi r\dfrac{dr}{dt}$

We use $\dfrac{dV}{dt}$ to solve for $\dfrac{dr}{dt}$ and then substitute it into $\dfrac{dS}{dt}$.

$$\frac{dr}{dt} = \frac{1}{4\pi r^2}\frac{dV}{dt}$$

$$\frac{dS}{dt} = 8\pi r \cdot \frac{1}{4\pi r^2}\frac{dV}{dt} = \frac{2}{r}\frac{dV}{dt}$$

STEP 6 When $r = 3$ and $\dfrac{dV}{dt} = 10$, $\dfrac{dS}{dt} = \dfrac{2}{3}\cdot 10 = \dfrac{20}{3} \approx 6.667$ square meters per minute.

53. $C(x) = 5x^2 + 1125$

(a) $\overline{C}(x) = \dfrac{C(x)}{x} = \dfrac{5x^2 + 1125}{x} = 5x + \dfrac{1125}{x}$

(b) To find the minimum average cost, we check the critical numbers.

$$\overline{C}'(x) = 5 - \frac{1125}{x^2} = \frac{5x^2 - 1125}{x^2}$$

$\overline{C}'(x)$ is unbounded at $x = 0$.

$\overline{C}'(x) = 0$ when $5x^2 - 1125 = 0$

$$x^2 - 225 = 0$$

$$x = \pm\sqrt{225} = \pm 15$$

Since x denotes the number of items produced, we disregard $x = -15$.

$\overline{C}''(x) = 2250x^{-3}$ and $\overline{C}''(15) = \dfrac{2250}{3375} = \dfrac{2}{3}$. By the Second Derivative Test, $x = 15$ locates a

local minimum. The minimum average cost is $\overline{C}(15) = 5(15) + \dfrac{1125}{15} = 150$.

(c) The marginal cost function is the derivative of $C(x)$.
   $C'(x) = 10x$

(d)

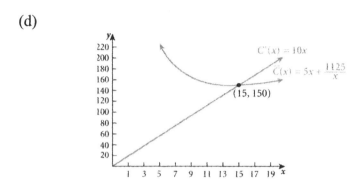

55. Profit = Revenue – Cost, $P(x) = R(x) - C(x)$, where $R(x) = xp$.
$$P(x) = x\left(62,402.50 - 0.5x^2\right) - \left(48,002.50x + 1500\right)$$
$$= 62,402.50x - 0.5x^3 - 48,002.50x - 1500$$
$$= -0.5x^3 + 14,400x - 1500$$

$$P'(x) = -1.5x^2 + 14,400$$
$P'(x) = 0$ when $-1.5x^2 + 14,400 = 0$
$$x^2 = 9600$$
$$x = \pm\sqrt{9600} = \pm 97.98$$

Since x denotes the units to be sold, we disregard $x = -98$, and use $x = 98$ in the second derivative.

$\quad P''(x) = -3x \qquad\qquad P''(98) = -3(98) = -294 < 0$

From the Second Derivative Test $x = 98$ locates a local maximum. 98 units must be sold to maximize profit.

57. **STEP 1** We want to minimize the cost of the material A which is used to make the can.

   STEP 2 Let r denote the radius of the top and bottom of the can, and
   let h denote the height of the can.

STEP 3 $V = \pi r^2 h$

$500 = \pi r^2 h$

$h = \dfrac{500}{\pi r^2}$

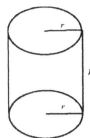

STEP 4 Since the cost of top, bottom and sides are the same, we need to minimize the surface area of the can.

$$S = 2\left(\pi r^2\right) + 2\pi r h$$

$$S(r) = 2\pi r^2 + 2\pi r \left(\dfrac{500}{\pi r^2}\right)$$

$$= 2\pi r^2 + \dfrac{1000}{r}$$

The domain of S is $\{r \mid r > 0\}$.

STEP 5 Differentiate S and find the critical numbers.

$$S'(r) = 4\pi r - \dfrac{1000}{r^2} = \dfrac{4\pi r^3 - 1000}{r^2}$$

We solve $S'(x) = 0$ to find the critical numbers.

$$4\pi r^3 - 1000 = 0$$

$$r^3 = \dfrac{1000}{4\pi}$$

$$r = \sqrt[3]{\dfrac{250}{\pi}} \approx 4.301$$

We use the Second Derivative Test to see if $r = 4.301$ locates a maximum or a minimum.

$$S''(r) = 4\pi - (-2)1000 r^{-3} = 4\pi + \dfrac{2000}{r^3}$$

$$S''\left(\sqrt[3]{\dfrac{250}{\pi}}\right) = 4\pi + \dfrac{2000}{\left(\sqrt[3]{\dfrac{250}{\pi}}\right)^3} = 4\pi + 8\pi = 12\pi > 0$$

We conclude that the cost is minimized when the can has a radius of 4.301 centimeters and a height of $\dfrac{500}{\pi r^2} \approx \dfrac{500}{\pi(4.301)^2} = 8.604$ centimeters.

59. $A = \pi r^2$ $dA = 2\pi r\, dr$

If $r = 10$ and changes to 8, then $dr = \Delta r = -2$, and

$$dA = 2\pi(10)(-2) = -40\pi \approx -125.66.$$

The area of the burn decreases by approximately 125.66 square centimeters.

61. $D(x) = -4x^3 - 3x^2 + 2000$ $\qquad dD = -\left(12x^2 + 6x\right)dx$

(a) If $x = 1.50$ and changes to 2.00, then $dx = \Delta x = 0.50$, and

$$dD = -\left(12(1.5)^2 + 6(1.5)\right)(0.50) = -18$$

The demand for peanuts will decrease by approximately 1800 pounds.

(b) If $x = 2.50$ and changes to 3.50, then $dx = \Delta x = 1.00$, and

$$dD = -\left(12(2.5)^2 + 6(2.5)\right)(1.00) = -90$$

The demand for peanuts will decrease by approximately 9000 pounds.

63. $c(x) = \dfrac{3x}{4 + 2x^2}$

$$c'(x) = \frac{3\left(4 + 2x^2\right) - 3x(4x)}{\left(4 + 2x^2\right)^2} = \frac{12 + 6x^2 - 12x^2}{\left(4 + 2x^2\right)^2} = \frac{12 - 6x^2}{\left(4 + 2x^2\right)^2}$$

$$dc = \frac{12 - 6x^2}{\left(4 + 2x^2\right)^2}\,dx$$

(a) If the time is $x = 1.2$ and changes to 1.3, then $dx = \Delta x = 0.1$, and

$$dc = \frac{12 - 6(1.2)^2}{\left(4 - 2(1.2)^2\right)^2}(0.1) = 0.0071$$

The concentration of the drug increases by approximately 0.0071.

(b) If the time is $x = 2$ and changes to 2.25, then $dx = \Delta x = 0.25$, and

$$dc = \frac{12 - 6(2)^2}{\left(4 - 2(2)^2\right)^2}(0.25) = -0.0208$$

The concentration of the drug decreases by approximately 0.0208.

65. (b)

CHAPTER 5 PROJECT

(a) Since the demand remains constant throughout the year, on average there are $\dfrac{x}{2}$ vacuum cleaners in the store at any time t. So the average holding costs will be $H \cdot \dfrac{x}{2} = \dfrac{Hx}{2}$.

(b) If the total demand is denoted by D and x vacuum cleaners are shipped per order, x times the number of orders must equal the yearly demand D, or the number of orders must equal $\dfrac{D}{x}$. So the yearly reorder costs will be R times the number of orders placed or $R \cdot \dfrac{D}{x} = \dfrac{RD}{x}$.

(c) Total cost is the sum of ordering cost and holding cost.

$$C(x) = \dfrac{RD}{x} + \dfrac{Hx}{2}$$

The domain of C is $\{x \mid 0 < x \le 500\}$.

(d) First we find the derivative $C'(x)$ and the critical numbers.

$$C'(x) = -\dfrac{RD}{x^2} + \dfrac{H}{2} = \dfrac{-2RD + Hx^2}{2x^2}$$

$C'(x)$ is unbounded at $x = 0$, but 0 is not in the domain of C.

$C'(x) = 0$ when $-2RD + Hx^2 = 0$.

$$Hx^2 = 2RD$$

$$x^2 = \dfrac{2RD}{H}$$

$$x = \sqrt{\dfrac{2RD}{H}}$$

$$C''(x) = \dfrac{2RD}{x^3}$$

$$C''\left(\sqrt{\dfrac{2RD}{H}}\right) = \dfrac{2RD}{\left(\sqrt{\dfrac{2RD}{H}}\right)^3} = 2RD \cdot \dfrac{H^{3/2}}{(2RD)^{3/2}} = \dfrac{H^{3/2}}{\sqrt{2RD}} > 0$$

By the Second Derivative Test, we see that cost C is minimized when $x = \sqrt{\dfrac{2RD}{H}}$.

(e) Assuming $D = 500$ vacuum cleaners, $H = \$10$/vacuum cleaner, and $R = \$40$/order, the lot size that will minimize cost is

$$x = \sqrt{\dfrac{2RD}{H}} = \sqrt{\dfrac{2 \cdot 40 \cdot 500}{10}} = \sqrt{4000} \approx 63.24$$

Ordering 63 vacuum cleaners at a time will minimize cost.

If 63 vacuum cleaners are ordered at a time, there needs to be 8 orders placed per year.

(f) If $H = \$3$/vacuum cleaner, the lot size that will minimize cost is

$$x = \sqrt{\dfrac{2 \cdot 40 \cdot 500}{3}} = \sqrt{\dfrac{40,000}{3}} \approx 115.47$$

115 vacuum cleaners should be ordered at a time to minimize cost.

If 115 vacuum cleaners are ordered, there needs to be 5 orders placed per year.

(g) If the cost C is revised to reflect shipping costs, then

$$C(x) = \frac{Hx}{2} + \frac{D(R + Sx)}{x} = \frac{Hx}{2} + \frac{DR}{x} + \frac{DSx}{x} = \frac{Hx}{2} + \frac{DR}{x} + DS$$

$$C'(x) = \frac{H}{2} - \frac{DR}{x^2} = \frac{Hx^2 - 2DR}{2x^2}$$

$C'(x) = 0$ when $Hx^2 - 2DR = 0$ or when $x = \sqrt{\dfrac{2RD}{H}}$.

The lot size that will minimize the total cost is still $\sqrt{\dfrac{2RD}{H}}$.

MATHEMATICAL QUESTIONS FROM THE PROFESSIONAL EXAMS

1. **(b)** The marginal cost function is the derivative of C.
$$C'(X) = 6X^2 + 8X + 3$$

3. **(a)**

5. **(a)**

7. **(d)** $D = E^2(100 - I)$

$$D'(t) = -E^2 \frac{dI}{dt} + 2E(100 - I)\frac{dE}{dt} = -(95)^2(3) + 2(95)(100 - 6)(2) = 8645$$

9.

(c) $f(x) = \dfrac{1}{6}x^3 - 2x$ $\qquad$ $f'(x) = \dfrac{1}{2}x^2 - 2$ $\qquad$ $f''(x) = x$

f is decreasing when $f'(x) < 0$ or when $-2 < x < 2$.

f is concave up when $f''(x) > 0$ or when $x > 0$.

So f is both decreasing and concave up on the interval $(0, 2)$.

11.

(c) $f'(x) = \dfrac{(3xe^{3x} + e^{3x})(1 + x) - (xe^{3x})(1)}{(1 + x)^2}$

$$f'(1) = \frac{(3e^3 + e^3)(1 + 1) - (e^3)}{(1 + 1)^2} = \frac{7e^3}{4}$$

13. (a) Using similar triangles, we find

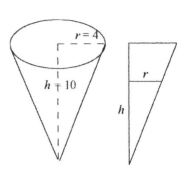

$$\frac{h}{r} = \frac{10}{4}$$

$$r = \frac{2}{5}h$$

$$V = \frac{1}{3}\pi r^2 \cdot h = \frac{1}{3}\pi \left(\frac{2}{5}h\right)^2 \cdot h = \frac{4}{75}\pi h^3$$

$$\frac{dV}{dt} = \frac{4}{25}\pi h^2 \frac{dh}{dt}$$

$$\frac{dh}{dt} = \frac{25}{4\pi h^2}\frac{dV}{dt}$$

When $h = 5$, $\dfrac{dh}{dt} = \dfrac{25}{4\pi (5)^2} \cdot 2 = \dfrac{1}{2\pi}$

15.

(d) $r = \dfrac{\Delta s}{\Delta t} = \dfrac{ds}{dt}$

We are told that $s_2 = (s_1)^2$, $r_2 = \dfrac{ds_1}{dt} = \dfrac{d}{dx}\left(s_1^{\,2}\right) = 2s_1 \cdot \dfrac{ds_1}{dt} = 2 \cdot 9 \cdot 3 = 54$

Chapter 6
The Integral of a Function and Applications

6.1 Antiderivatives; the Indefinite Integral; Marginal Analysis

1. $F(x) = \dfrac{x^4}{4} + K$

3. $F(x) = \dfrac{2x^2}{2} + 3x + K = x^2 + 3x + K$

5. $F(x) = 4\ln|x| + K$

7. $f(x) = \sqrt[3]{x} = x^{1/3}$ $\qquad$ $F(x) = \dfrac{x^{4/3}}{\frac{4}{3}} + K = \dfrac{3x^{4/3}}{4} + K$

9. $\displaystyle\int 3\,dx = 3x + K$

11. $\displaystyle\int x\,dx = \dfrac{x^2}{2} + K$

13. $\displaystyle\int x^{1/3}\,dx = \dfrac{x^{4/3}}{\frac{4}{3}} + K = \dfrac{3x^{4/3}}{4} + K$

15. $\displaystyle\int x^{-2}\,dx = \dfrac{x^{-1}}{-1} + K = -\dfrac{1}{x} + K$

17. $\displaystyle\int x^{-1/2}\,dx = \dfrac{x^{1/2}}{\frac{1}{2}} + K = 2x^{1/2} + K$

19. $\displaystyle\int (2x^3 + 5x)\,dx = \dfrac{2x^4}{4} + \dfrac{5x^2}{2} + K = \dfrac{x^4}{2} + \dfrac{5x^2}{2} + K$

21. $\displaystyle\int (x^2 + 2e^x)\,dx = \dfrac{x^3}{3} + 2e^x + K$

23. $\displaystyle\int (x^3 - 2x^2 + x - 1)\,dx = \dfrac{x^4}{4} - \dfrac{2x^3}{3} + \dfrac{x^2}{2} - x + K$

25. $\displaystyle\int \left(\dfrac{x-1}{x}\right)dx = \int\left(1 - \dfrac{1}{x}\right)dx = x - \ln|x| + K$

27.
$$\int \left(2e^x - \frac{3}{x}\right) dx = 2e^x - 3\ln|x| + K$$

29.
$$\int \left(\frac{3\sqrt{x} + 1}{\sqrt{x}}\right) dx = \int (3 + x^{-1/2}) dx = 3x + \frac{x^{1/2}}{\frac{1}{2}} + K = 3x + 2\sqrt{x} + K$$

31.
$$\int \frac{x^2 - 4}{x + 2} dx = \int \frac{(x-2)(x+2)}{x+2} dx = \int (x-2) dx = \frac{x^2}{2} - 2x + K$$

33.
$$\int x(x-1) dx = \int (x^2 - x) dx = \frac{x^3}{3} - \frac{x^2}{2} + K$$

35.
$$\int \left(\frac{3x^5 + 2}{x}\right) dx = \int \left(3x^4 + \frac{2}{x}\right) dx = \frac{3x^5}{5} + 2\ln|x| + K$$

37.
$$\int \frac{4e^x + e^{2x}}{e^x} dx = \int (4 + e^x) dx = 4x + e^x + K$$

39. $R'(x) = 600$ $R(x) = 600x + K$
If $R = 0$ when $x = 0$, then $K = 0$, and $R(x) = 600x$.

41. $R'(x) = 20x + 5$ $R(x) = 10x^2 + 5x + K$
If $R = 0$ when $x = 0$, then $K = 0$, and $R(x) = 10x^2 + 5x$.

43.
$$C(x) = \int C'(x) dx = \int (14x - 2800) dx = \frac{14x^2}{2} - 2800x + K$$

Fixed cost $= K = \$4300$ so $C(x) = 7x^2 - 2800x + 4300$.
Cost is minimum when $C'(x) = 0$.
$$14x - 2800 = 0$$
$$14x = 2800$$
$$x = 200$$
Cost is minimum when 200 units are produced.

45.
$$C(x) = \int (20x - 8000) dx = \frac{20x^2}{2} - 8000x + K$$
Fixed cost $= K = \$500$ so $C(x) = 10x^2 - 8000x + 500$.

Cost is minimum when $C'(x) = 0$.
$$20x - 8000 = 0$$
$$20x = 8000$$
$$x = 400$$
Cost is minimum when 400 units are produced.

47.

(a) $C(x) = \int (1000 - 20x + x^2)\,dx = 1000x - \dfrac{20x^2}{2} + \dfrac{x^3}{3} + K$

Fixed cost $= K = \$9000$, so $C(x) = 1000x - 10x^2 + \dfrac{x^3}{3} + 9000$

(b) Revenue is the product of the price and the number of items sold, $R = px$.
$$R = R(x) = 3400x$$

(c) Profit is the difference between revenue and cost, $P = R - C$.
$$P = P(x) = 3400x - \left(1000x - 10x^2 + \dfrac{x^3}{3} + 9000\right)$$
$$= 3400x - 1000x + 10x^2 - \dfrac{x^3}{3} - 9000 = -\dfrac{x^3}{3} + 10x^2 + 2400x - 9000$$

(d) The maximum profit occurs at a critical number.
$$P'(x) = -x^2 + 20x + 2400$$
$P'(x) = 0$ when $-x^2 + 20x + 2400 = 0$
$$x^2 - 20x - 2400 = 0$$
$$(x - 60)(x + 40) = 0$$
$$x - 60 = 0 \quad \text{or} \quad x + 40 = 0$$
$$x = 60 \quad \text{or} \quad x = -40$$
Since x denotes the number of units produced we disregard -40.

$P''(x) = -2x + 20$ $\qquad\qquad$ $P''(60) = -2(60) + 20 = -100 < 0$
By the Second Derivative Test, selling $x = 60$ units yields maximum profit.

(e) $P(60) = -\dfrac{60^3}{3} + 10 \cdot 60^2 + 2400 \cdot 60 - 9000 = 99{,}000$

The maximum profit is $\$99{,}000$.

(f)

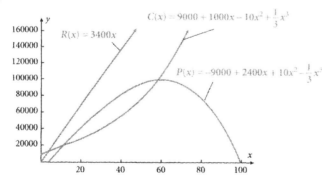

49. Let P denote the prison population. Then
$$P = P(t) = \int (7000t + 20{,}000)\,dt = \frac{7000t^2}{2} + 20{,}000t + K$$

We let t denote time (in years), and letting $t = 0$ represent 1998, we find K.
$$P(0) = 3500(0)^2 + 20{,}000(0) + K = 592{,}462$$
$$K = 592{,}462$$
$$P(t) = 3500t^2 + 20{,}000t + 592{,}462$$

In 2008 $t = 10$, and
$$P(10) = 3500\,(10)^2 + 20{,}000\,(10) + 592{,}462 = 1{,}142{,}462$$

According to this model in 2008 there will be 1,142,462 inmates in the United States.

51. Let P denote the population of the town. Then
$$P'(t) = 2 + t^{4/5} \qquad P(t) = \int P'(t)\,dt = \int (2 + t^{4/5})\,dt = 2t + \frac{t^{9/5}}{\frac{9}{5}} + K$$

$$= 2t + \frac{5t^{9/5}}{9} + K$$

We let t denote time (in months), and let $t = 0$ represent the time when $P = 20{,}000$.
Since $P(0) = 20{,}000$, $K = 20{,}000$. In 10 months $t = 10$, and the population will be
$$P(10) = 2(10) + \frac{5 \cdot 10^{9/5}}{9} + 20{,}000 = 20{,}055 \text{ people.}$$

53. Let P denote the voting population (in thousands). Then
$$P = P(t) = \int (2.2t - 0.8t^2)\,dt = \frac{2.2t^2}{2} - \frac{0.8t^3}{3} + K$$

If t denotes time (in years), and $P(0) = 20$, then we find K
$$P(0) = 1.1 \cdot 0^2 - \frac{0.8}{3} \cdot 0^3 + K = 20$$
$$K = 20$$

In 3 years, $t = 3$, and
$$P(3) = 1.1 \cdot 3^2 - \frac{0.8}{3} \cdot 3^3 + 20 = 22.7$$

In three years the voting population will be 22,700 citizens.

55. Let f denote the amount of end product present in the reaction. Then
$$f = \int \left(\frac{\sqrt{t} - 1}{t} \right) dt = \int \left(\frac{1}{\sqrt{t}} - \frac{1}{t} \right) dt = \int t^{-1/2}\,dt - \int \frac{1}{t}\,dt = \frac{t^{1/2}}{\frac{1}{2}} - \ln|t| + K = 2t^{1/2} - \ln|t| + K$$

Since the reaction started at $t = 1$, there is no end product at that time. So we find K
$$f(1) = 2 \cdot 1^{1/2} - \ln 1 + K = 0$$

$$2 - 0 + K = 0$$
$$K = -2$$

After 4 minutes, $t = 4$, and

$$f(4) = 2 \cdot 4^{1/2} - \ln 4 - 2 = 4 - 2 \ln 2 - 2 = 2 - 2 \ln 2$$

There will be $2 - 2 \ln 2$ milligrams of end product after 4 minutes of reaction.

57. Let V denote the amount of water in the reservoir at time t. Then

$$V = V(t) = \int \left(15,000 - \frac{5}{2}t \right) dt = \int 15,000 \, dt - \frac{5}{2} \int t \, dt = 15,000t - \frac{5t^2}{4} + K$$

If initially (at $t = 0$) there are 100,000 gallons in the reservoir, we find K

$$V(0) = 15,000 \cdot 0 - \frac{5}{4} \cdot 0^2 + K = 100,000$$

$$K = 100,000$$

$$V(t) = 15,000 \, t - \frac{5}{4}t^2 + 100,000$$

The reservoir will be empty when $V(t) = 0$.

$$15,000 \, t - \frac{5}{4} \, t^2 + 100,000 = 0$$

$$-5t^2 + 60,000 \, t + 400,000 = 0$$

$$t^2 - 12,000 \, t - 80,000 = 0$$

$$t = \frac{12,000 \pm \sqrt{12,000^2 - 4(-80,000)}}{2}$$

$$t \approx 12,006.66$$

(We disregard the negative value of t.)

The reservoir will be empty in approximately 12,007 hours (500.28 days).

6.2 Integration Using Substitution

1.
Let $u = 2x + 1$. Then $du = 2 \, dx$ so $dx = \frac{1}{2} \, du$

$$\int (2x+1)^5 \, dx = \frac{1}{2} \int u^5 \, du = \frac{1}{2} \frac{u^6}{6} + K = \frac{(2x+1)^6}{12} + K$$

3.
Let $u = 2x - 3$. Then $du = 2 \, dx$, so $dx = \frac{1}{2} \, du$.

$$\int e^{2x-3} \, dx = \frac{1}{2} \int e^u \, du = \frac{1}{2} \, e^u + K = \frac{e^{2x-3}}{2} + K$$

5. Let $u = -2x + 3$. Then $du = -2\,dx$, so $dx = -\dfrac{1}{2}\,du$.

$$\int (-2x+3)^{-2}\,dx = -\frac{1}{2}\int u^{-2}\,du = -\frac{1}{2}\frac{u^{-1}}{-1} + K$$

$$= \frac{1}{2}(-2x+3)^{-1} + K = \frac{1}{2(-2x+3)} + K = \frac{1}{6-4x} + K$$

7. Let $u = x^2 + 4$. Then $du = 2x\,dx$, so $x\,dx = \dfrac{1}{2}\,du$.

$$\int x(x^2+4)^2\,dx = \frac{1}{2}\int u^2\,du = \frac{1}{2}\cdot\frac{u^3}{3} + K = \frac{1}{2}\cdot\frac{(x^2+4)^3}{3} + K = \frac{(x^2+4)^3}{6} + K$$

9. Let $u = x^3 + 1$. Then $du = 3x^2\,dx$, so $x^2\,dx = \dfrac{1}{3}du$.

$$\int e^{x^3+1}x^2\,dx = \frac{1}{3}\int e^u\,du = \frac{1}{3}\cdot e^u + K = \frac{1}{3}\cdot e^{x^3+1} + K = \frac{e^{x^3+1}}{3} + K$$

11. Let $u = x^2$. Then $du = 2x\,dx$, so $x\,dx = \dfrac{1}{2}\,du$.

$$\int\left(e^{x^2}+e^{-x^2}\right)x\,dx = \int e^{x^2}x\,dx + \int e^{-x^2}x\,dx = \frac{1}{2}\int e^u\,du + \frac{1}{2}\int e^{-u}\,du = \frac{1}{2}\cdot e^u - \frac{1}{2}\cdot e^{-u} + K$$

$$= \frac{e^{x^2}}{2} - \frac{e^{-x^2}}{2} + K = \frac{e^{x^2}-e^{-x^2}}{2} + K$$

13. Let $u = x^3 + 2$. Then $du = 3x^2\,dx$, so $x^2\,dx = \dfrac{1}{3}du$.

$$\int x^2(x^3+2)^6\,dx = \frac{1}{3}\int u^6\,du = \frac{1}{3}\cdot\frac{u^7}{7} + K = \frac{(x^3+2)^7}{21} + K$$

15. Let $u = 1 + x^2$. Then $du = 2x\,dx$, so $x\,dx = \dfrac{1}{2}\,du$.

$$\int \frac{x}{\sqrt[3]{1+x^2}}\,dx = \int x(1+x^2)^{-1/3}\,dx = \frac{1}{2}\int u^{-1/3}\,du$$

$$= \frac{1}{2}\cdot\frac{u^{2/3}}{\frac{2}{3}} + K = \frac{3u^{2/3}}{4} + K$$

$$= \frac{3(1+x^2)^{2/3}}{4} + K$$

17. Let $u = x + 3$. Then $du = dx$ and $x = u - 3$.

$$\int x\sqrt{x+3}\, dx = \int (u-3)\sqrt{u}\, du = \int \left(u^{3/2} - 3u^{1/2}\right) du = \frac{u^{5/2}}{\frac{5}{2}} - \frac{3u^{3/2}}{\frac{3}{2}} + K$$

$$= \frac{2(x+3)^{5/2}}{5} - \frac{6(x+3)^{3/2}}{3} + K = \frac{2(x+3)^{5/2}}{5} - 2(x+3)^{3/2} + K$$

19. Let $u = e^x + 1$. Then $du = e^x\, dx$.

$$\int \frac{e^x}{e^x + 1}\, dx = \int \frac{1}{u}\, du = \ln|u| + K = \ln\left(e^x + 1\right) + K$$

21. Let $u = \sqrt{x} = x^{1/2}$. Then $du = \frac{1}{2}x^{-1/2}\, dx = \frac{1}{2\sqrt{x}}\, dx$, so $2\, du = \frac{1}{\sqrt{x}}\, dx$.

$$\int \frac{e^{\sqrt{x}}}{\sqrt{x}}\, dx = 2 \int e^u\, du = 2 \cdot e^u + K = 2e^{\sqrt{x}} + K$$

23. Let $u = x^{1/3} - 1$. Then $du = \frac{1}{3}x^{-2/3}\, dx = \frac{1}{3x^{2/3}}\, dx$, so $3\, du = \frac{1}{x^{2/3}}\, dx$.

$$\int \frac{\left(x^{1/3} - 1\right)^6}{x^{2/3}}\, dx = 3 \int u^6\, du = 3 \cdot \frac{u^7}{7} + K = \frac{3\left(x^{1/3} - 1\right)^7}{7} + K$$

25. Let $u = x^2 + 2x + 3$. Then $du = (2x + 2)\, dx$, so $\frac{1}{2}\, du = (x + 1)\, dx$

$$\int \frac{(x+1)\, dx}{\left(x^2 + 2x + 3\right)^2} = \frac{1}{2} \int \frac{1}{u^2}\, du = \frac{1}{2} \int u^{-2}\, du = \frac{1}{2}\frac{u^{-1}}{(-1)} + K = -\frac{1}{2u} + K$$

$$= -\frac{1}{2\left(x^2 + 2x + 3\right)} + K = -\frac{1}{2x^2 + 4x + 6} + K$$

27. Let $u = 1 + \sqrt{x}$. Then $du = \frac{1}{2\sqrt{x}}\, dx$, so $2\, du = \frac{1}{\sqrt{x}}\, dx$.

$$\int \frac{dx}{\sqrt{x}\left(1 + \sqrt{x}\right)^4} = 2 \int \frac{1}{u^4}\, du = 2 \int u^{-4}\, du = 2 \cdot \frac{u^{-3}}{(-3)} + K = -\frac{2}{3u^3} + K$$

$$= -\frac{2}{3\left(1 + \sqrt{x}\right)^3} + K$$

29. Let $u = 2x + 3$. Then $du = 2\, dx$, so $\frac{1}{2}\, du = dx$.

$$\int \frac{dx}{2x + 3} = \frac{1}{2} \int \frac{1}{u}\, du = \frac{1}{2} \cdot \ln|u| = \frac{1}{2} \ln|2x + 3| + K$$

31.
Let $u = 4x^2 + 1$. Then $du = 8x\ dx$, so $\dfrac{1}{8}\ du = x\ dx$.

$$\int \frac{x\ dx}{4x^2+1} = \frac{1}{8}\int \frac{du}{u} = \frac{1}{8}\cdot \ln|u| + K = \frac{1}{8}\ln\left(4x^2+1\right) + K$$

33.
Let $u = x^2 + 2x + 2$. Then $du = (2x+2)dx = 2(x+1)\ dx$, so $\dfrac{1}{2}\ du = (x+1)\ dx$

$$\int \frac{x+1}{x^2+2x+x}\ dx = \frac{1}{2}\int \frac{1}{u}\ du = \frac{1}{2}\ln|u| + K = \frac{1}{2}\ln\left(x^2+2x+x\right) + K$$

35. The value V of the car is the antiderivative of the depreciation rate with $V(0) = \$\,27{,}000$.

$$V = \int V'(t)\,dt = \int -6000e^{-0.5t}\ dt = -6000\int e^{-0.5t}\ dt$$

$$= -6000\left(\frac{1}{-0.5}e^{-0.5t}\right) + K = 12{,}000\ e^{-0.5t} + K$$

We use $V(0) = 27{,}000$ to determine K.

$$V(0) = 12{,}000\ e^{-0.5(0)} + K = 27{,}000$$
$$K = 27{,}000 - 12{,}000 = 15{,}000$$

$$V(t) = 12{,}000\ e^{-0.5t} + 15{,}000$$

After 2 years $t = 2$, and the car is worth
$$V(2) = 12{,}000\ e^{-0.5(2)} + 15{,}000 = 12{,}000\ e^{-1} + 15{,}000 = \$19{,}414.55$$
After 4 years $t = 4$, and the car is worth
$$V(4) = 12{,}000\ e^{-0.5(4)} + 15{,}000 = 12{,}000\ e^{-2} + 15{,}000 = \$16{,}624.02$$

37. (a) The budget B is the antiderivative of the growth rate with $B(0) = 68.6$ billion dollars.

$$B = \int B'(t)\,dt = \int 1.715\ e^{0.025t}\ dt = 1.715\int e^{0.025t}\ dt = 1.715 \cdot \frac{1}{0.025}e^{0.025t} + K$$

We use $B(0) = 68.6$ to determine K.

$$B(0) = 68.6\ e^{0.025\,(0)} + K = 68.6$$
$$68.6 + K = 68.6$$
$$K = 0$$
$$B(t) = 68.6\ e^{0.025t}$$

(b) We need to find t so that $B(t) > 100$

$$68.6\ e^{0.025t} > 100$$

$$e^{0.025t} > \frac{100}{68.6} \approx 1.4577$$

Changing the exponential to logarithmic form we find

$$\ln\left(\frac{100}{68.6}\right) = 0.025t$$

$$t = \frac{\ln\left(\frac{100}{68.6}\right)}{0.025} \approx 15.075$$

The budget will exceed $ 100 billion in just over 15 years, that is in 2016.

39. (a) The number of employees N is given by the antiderivative of $N'(t)$ with $N(0) = 400$.

$$N = \int N'(t)\, dt = \int 20e^{0.01t}\, dt = 20\int e^{0.01t}\, dt = 20 \cdot \frac{1}{0.01} e^{0.01t} + K = 2000e^{0.01t} + K$$

We use $N(0) = 400$ to determine K.

$$N(0) = 2000\, e^0 + K = 400$$
$$K = 400 - 2000 = -1600$$
$$N(t) = 2000\, e^{0.01t} - 1600$$

(b) We need to find t so that $N(t) = 800$.

$$N(t) = 2000\, e^{0.01t} - 1600 = 800$$
$$2000\, e^{0.01t} = 2400$$
$$e^{0.01t} = \frac{2400}{2000} = 1.2$$
$$\ln 1.2 = 0.01\, t$$
$$t = \frac{\ln 1.2}{0.01} = 18.232$$

It will take about 18.232 years for the number of employees to reach 800.

41. Let $u = ax + b$. Then $du = a\, dx$, so $\dfrac{1}{a}\, du = dx$.

$$\int (ax+b)^n\, dx = \frac{1}{a}\int u^n\, du = \frac{1}{a} \cdot \frac{u^{n+1}}{n+1} + K = \frac{u^{n+1}}{a(n+1)} + K = \frac{(ax+b)^{n+1}}{a(n+1)} + K$$

6.3 Integration by Parts

1.
$$\int u\, dv = uv - \int v\, du$$

If $u = x$, then $du = dx$; and if $dv = e^{4x}\, dx$, then $v = \dfrac{1}{4} e^{4x}$.

$$\int x\, e^{4x}\, dx = \frac{1}{4} x e^{4x} - \int \frac{1}{4} e^{4x}\, dx = \frac{1}{4} x e^{4x} - \frac{1}{4} \cdot \frac{1}{4} e^{4x} + K$$
$$= \frac{1}{4} x e^{4x} - \frac{1}{16} e^{4x} + K$$

3. Choose $\quad u = x \quad$ and $\quad dv = e^{2x}\,dx$

Then $\quad du = dx \quad$ and $\quad v = \dfrac{1}{2}e^{2x}$

$$\int u\,dv = uv - \int v\,du$$

$$\int xe^{2x}\,dx = \dfrac{1}{2}xe^{2x} - \dfrac{1}{2}\int e^{2x}\,dx = \dfrac{1}{2}xe^{2x} - \dfrac{1}{4}e^{2x} + K$$

5. Choose $\quad u = x^2 \quad$ and $\quad dv = e^{-x}\,dx$

Then $\quad du = 2x\,dx \quad$ and $\quad v = -e^{-x}$

$$\int u\,dv = uv - \int v\,du$$

$$\int x^2 e^{-x}\,dx = x^2\left(-e^{-x}\right) - \int -e^{-x}\cdot 2x\,dx$$

$$= -x^2 e^{-x} + 2\int xe^{-x}\,dx$$

We use integration by parts once more. This time
choose $\quad u = x \quad$ and $\quad dv = e^{-x}\,dx$

Then $\quad du = dx \quad$ and $\quad v = -e^{-x}$

$$\int x^2 e^{-x}\,dx = -x^2 e^{-x} + 2\left[x\cdot\left(-e^{-x}\right) - \int -e^{-x}\cdot dx\right]$$

$$= -x^2 e^{-x} - 2x\,e^{-x} + 2\int e^{-x}\,dx$$

$$= -x^2 e^{-x} - 2x\,e^{-x} - 2e^{-x} + K$$

$$= -e^{-x}\left(x^2 + 2x + 2\right) + K$$

7. Choose $\quad u = \ln x \quad$ and $\quad dv = \sqrt{x}\,dx$

$\quad du = \dfrac{1}{x}\,dx \quad$ and $\quad v = \dfrac{2}{3}x^{3/2}$

$$\int u\,dv = uv - \int v\,du$$

$$\int \sqrt{x}\,\ln x\,dx = \ln x\cdot\dfrac{2}{3}x^{3/2} - \int \dfrac{2}{3}x^{3/2}\cdot\dfrac{1}{x}\,dx$$

$$= \dfrac{2}{3}x^{3/2}\ln x - \dfrac{2}{3}\int x^{1/2}\,dx$$

$$= \dfrac{2}{3}x^{3/2}\ln x - \dfrac{2}{3}\cdot\dfrac{x^{3/2}}{\dfrac{3}{2}} + K = \dfrac{2}{3}x^{3/2}\ln x - \dfrac{4}{9}x^{3/2} + K$$

9. Choose $\quad u = (\ln x)^2 \quad$ and $\quad dv = dx$

$\quad du = 2\ln x\cdot\dfrac{1}{x}\,dx \quad$ and $\quad v = x$

$$= \frac{2}{x} \ln x \, dx$$

$$\int u \, dv = uv - \int v \, du$$

$$\int (\ln x)^2 \, dx = (\ln x)^2 \cdot x - \int \left(x \cdot \frac{2}{x} \ln x \right) dx$$

$$= x (\ln x)^2 - 2 \int \ln x \, dx$$

We use integration by parts once more.
This time choose $u = \ln x$ and $dv = dx$

$$du = \frac{1}{x} \, dx \qquad \text{and} \qquad v = x$$

$$\int (\ln x)^2 \, dx = x (\ln x)^2 - 2 \left[x \ln x - \int x \cdot \frac{1}{x} \, dx \right]$$

$$= x (\ln x)^2 - 2x \ln x + 2 \int dx$$

$$= x (\ln x)^2 - 2x \ln x + 2x + K$$

11.   Choose $u = \ln 3x$ and $dv = x^2 \, dx$

$$du = \frac{1}{x} \, dx \qquad \text{and} \qquad v = \frac{x^3}{3}$$

$$\int u \, dv = uv - \int v \, du$$

$$\int x^2 \ln 3x \, dx = \ln 3x \cdot \frac{x^3}{3} - \int \frac{x^3}{3} \cdot \frac{1}{x} \, dx$$

$$= \frac{x^3 \ln 3x}{3} - \frac{1}{3} \int x^2 \, dx$$

$$= \frac{x^3 \ln 3x}{3} - \frac{1}{3} \cdot \frac{x^3}{3} + K$$

$$= \frac{x^3 \ln 3x}{3} - \frac{x^3}{9} + K = \frac{x^3}{9} \left[3 \ln 3x - 1 \right] + K$$

13.   Choose $u = (\ln x)^2$ and $dv = x^2 \, dx$

$$du = 2 \ln x \cdot \frac{1}{x} \, dx \qquad \text{and} \qquad v = \frac{x^3}{3}$$

$$= \frac{2 \ln x}{x} \, dx$$

$$\int u \, dv = uv - \int v \, du$$

$$\int x^2 (\ln x)^2 \, dx = (\ln x)^2 \cdot \frac{x^3}{3} - \int \frac{x^3}{3} \cdot \frac{2 \ln x}{x} \, dx$$

$$= \frac{x^3 (\ln x)^2}{3} - \frac{2}{3} \int x^2 \ln x \, dx$$

We use integration by parts a second time.

This time choose $\qquad u = \ln x \qquad$ and $\qquad dv = x^2 \, dx$

Then $\qquad\qquad du = \frac{1}{x} \, dx \qquad$ and $\qquad v = \frac{x^3}{3}$

$$\int x^2 (\ln x)^2 \, dx = \frac{x^3 (\ln x)^2}{3} - \frac{2}{3} \left[\ln x \cdot \frac{x^3}{3} - \int \left(\frac{x^3}{3} \cdot \frac{1}{x} \right) dx \right]$$

$$= \frac{x^3 (\ln x)^2}{3} - \frac{2}{3} \left[\frac{x^3 \ln x}{3} - \frac{1}{3} \int x^2 \, dx \right]$$

$$= \frac{x^3 (\ln x)^2}{3} - \frac{2}{3} \left[\frac{x^3 \ln x}{3} - \frac{1}{3} \cdot \frac{x^3}{3} + K \right]$$

$$= \frac{x^3 (\ln x)^2}{3} - \frac{2x^3 \ln x}{9} + \frac{2x^3}{27} + K$$

15. Choose $\qquad u = \ln x \qquad$ and $\qquad dv = x^{-3} \, dx$

$$du = \frac{1}{x} \, dx \qquad \text{and} \qquad v = \frac{x^{-2}}{-2}$$

$$\int u \, dv = uv - \int v \, du$$

$$\int \frac{\ln x}{x^3} \, dx = \ln x \cdot \left(-\frac{x^{-2}}{2} \right) - \int -\frac{x^{-2}}{2} \cdot \frac{1}{x} \, dx = -\frac{x^{-2} \ln x}{2} + \frac{1}{2} \int x^{-3} \, dx$$

$$= -\frac{x^{-2} \ln x}{2} + \frac{1}{2} \cdot \frac{x^{-2}}{-2} + K = -\frac{x^{-2} \ln x}{2} - \frac{x^{-2}}{4} + K$$

$$= -\frac{2 \ln x + 1}{4x^2} + K$$

17. The function P is the antiderivative of $P'(t)$ with $P(0) = 5000$.

$$P = \int P'(t) \, dt = \int \left(90 \sqrt{t} - 100 t e^{-t} \right) dt = 90 \int \sqrt{t} \, dt - 100 \int \left(t e^{-t} \right) dt$$

We use integration by parts to integrate the second integral.

Choose $\qquad\qquad u = t \qquad\qquad\qquad dv = e^{-t} \, dt$

$\qquad\qquad\qquad du = dt \qquad\qquad\qquad v = -e^{-t}$

$$P = 90 \frac{t^{3/2}}{\frac{3}{2}} - 100 \left[-t e^{-t} - \int -e^{-t} \, dt \right] = 60 t^{3/2} + 100 t e^{-t} + 100 e^{-t} + K$$

We use $P(0) = 5000$ to determine K.

$$P(0) = 60(0)^{3/2} + 100(0)e^0 + 100e^0 + K = 5000$$
$$100 + K = 5000$$
$$K = 4900$$
$$P(t) = 60t^{3/2} + 100te^{-t} + 100e^{-t} + 4900$$

In 4 days $t = 4$, and $P(4) = 60(4)^{3/2} + 100(4)e^{-4} + 100e^{-4} + 4900 = 5389$ ants.

In one week $t = 7$, and $P(7) = 60(7)^{3/2} + 100(7)e^{-7} + 100e^{-7} + 4900 = 6012$ ants.

6.4 The Definite; Learning Curves; Total Sales Over Time

1.
$$\int_1^2 (3x-1)\,dx = \left(\frac{3x^2}{2} - x\right)\Bigg|_1^2 = \left(\frac{12}{2} - 2\right) - \left(\frac{3}{2} - 1\right) = 6 - 2 - \frac{3}{2} + 1 = \frac{7}{2}$$

3.
$$\int_0^1 (3x^2 + e^x)\,dx = \frac{3x^3}{3} + e^x\Bigg|_0^1 = (1^3 + e^1) - (0^3 + e^0) = 1 + e - 0 - 1 = e$$

5.

$$\int_0^1 \sqrt{u}\ du = \int_0^1 u^{1/2}\ du = \frac{u^{3/2}}{\frac{3}{2}}\Bigg|_0^1 = \frac{2u^{3/2}}{3}\Bigg|_0^1 = \frac{2(1)^{3/2}}{3} - \frac{2(0)^{2/3}}{3} = \frac{2}{3}$$

7.

$$\int_0^1 \left(t^2 - t^{3/2}\right) dt = \left(\frac{t^3}{3} - \frac{t^{5/2}}{\frac{5}{2}}\right)\Bigg|_0^1 = \left(\frac{t^3}{3} - \frac{2t^{5/2}}{5}\right)\Bigg|_0^1 = \left(\frac{1^3}{3} - \frac{2(1)^{5/2}}{5}\right) - \left(\frac{0^3}{3} - \frac{2(0)^{5/2}}{5}\right)$$

$$= \frac{1}{3} - \frac{2}{5} - 0 = \frac{5-6}{15} = -\frac{1}{15}$$

9.

$$\int_{-2}^3 (x-1)(x+3)\ dx = \int_{-2}^3 \left(x^2 + 2x - 3\right) dx = \left(\frac{x^3}{3} + \frac{2x^2}{2} - 3x\right)\Bigg|_{-2}^3$$

$$= \left[\frac{3^3}{3} + 3^2 - 3(3)\right] - \left[\frac{(-2)^3}{3} + (-2)^2 - 3(-2)\right] = 9 + 9 - 9 + \frac{8}{3} - 4 - 6$$

$$= -1 + \frac{8}{3} = \frac{5}{3}$$

11.

$$\int_1^2 \frac{x^2 - 1}{x^4}\ dx = \int_1^2 \left(x^{-2} - x^{-4}\right) dx = \left(\frac{x^{-1}}{-1} - \frac{x^{-3}}{-3}\right)\Bigg|_1^2 = \left(-\frac{1}{x} + \frac{1}{3x^3}\right)\Bigg|_1^2$$

$$= \left[-\frac{1}{2} + \frac{1}{3(2)^3}\right] - \left[-\frac{1}{1} + \frac{1}{3(1)^3}\right] = -\frac{1}{2} + \frac{1}{24} + 1 - \frac{1}{3} = \frac{5}{24}$$

13.

$$\int_1^8 \left(\sqrt[3]{t^2} + \frac{1}{t}\right) dt = \int_1^8 \left(t^{2/3} + \frac{1}{t}\right) dt = \left(\frac{t^{5/3}}{\frac{5}{3}} + \ln|t|\right)\Bigg|_1^8 = \left(\frac{3t^{5/3}}{5} + \ln|t|\right)\Bigg|_1^8$$

$$= \left[\frac{3(8)^{5/3}}{5} + \ln 8\right] - \left[\frac{3(1)^{5/3}}{5} + \ln 1\right]$$

$$= \frac{3(32)}{5} + \ln 8 - \frac{3}{5} - 0 = \frac{96-3}{5} + \ln 8 = \frac{93}{5} + \ln 8 = 3\ln 2 + \frac{93}{5}$$

15.

$$\int_1^4 \frac{x+1}{\sqrt{x}}\ dx = \int_1^4 \left(\frac{x}{\sqrt{x}} + \frac{1}{\sqrt{x}}\right) dx = \int_1^4 \left(x^{1/2} + x^{-1/2}\right) dx = \left(\frac{x^{3/2}}{\frac{3}{2}} + \frac{x^{1/2}}{\frac{1}{2}}\right)\Bigg|_1^4 = \left(\frac{2x^{3/2}}{3} + 2x^{1/2}\right)\Bigg|_1^4$$

$$= \left[\frac{2(4)^{3/2}}{3} + 2(4)^{1/2}\right] - \left[\frac{2(1)^{3/2}}{3} + 2(1)^{1/2}\right] = \frac{16}{3} + 4 - \frac{2}{3} - 2 = \frac{14+6}{3} = \frac{20}{3}$$

17. $\displaystyle\int_3^3 (5x^4 + 1)^{3/2}\, dx = 0$ \qquad Property 4 $\displaystyle\int_a^a f(x)\, dx = 0$

19.

$$\int_{-1}^1 (x+1)^2\, dx = \int_{-1}^1 (x^2 + 2x + 1)\, dx = \left(\frac{x^3}{3} + \frac{2x^2}{2} + x\right)\Bigg|_{-1}^1$$

$$= \left[\frac{1^3}{3} + 1^2 + 1\right] - \left[\frac{(-1)^3}{3} + (-1)^2 + (-1)\right] = \frac{1}{3} + 2 + \frac{1}{3} - 1 + 1 = \frac{8}{3}$$

21.

$$\int_1^e \left(x - \frac{1}{x}\right) dx = \frac{x^2}{2} - \ln|x|\Bigg|_1^e = \left[\frac{e^2}{2} - \ln e\right] - \left[\frac{1}{2} - \ln 1\right] = \frac{e^2}{2} - 1 - \frac{1}{2} + 0 = \frac{e^2}{2} - \frac{3}{2} = \frac{e^2 - 3}{2}$$

23.

$$\int_0^1 e^{-x}\, dx = \frac{e^{-x}}{-1}\Bigg|_0^1 = \left[-e^{-1}\right] - \left[-e^0\right] = -e^{-1} + 1 = 1 - \frac{1}{e}$$

25. We use the method of substitution to evaluate the integral.
Let $u = x + 1$. Then $du = dx$.

We adjust the limits of integration.
When $x = 1$, $u = 1 + 1 = 2$, and when $x = 3$, $u = 3 + 1 = 4$. So

$$\int_1^3 \frac{dx}{x+1} = \int_2^4 \frac{du}{u} = \ln|u|\Big|_2^4 = \ln 4 - \ln 2 = \ln 2$$

27. We use the method of substitution to evaluate the integral.
Let $u = x^{3/2} + 1$. Then $du = \frac{3}{2} x^{1/2}\, dx$. So $\frac{2}{3}\, du = x^{1/2}\, dx = \sqrt{x}\, dx$

We adjust the limits of integration.
When $x = 0$, $u = 0^{3/2} + 1 = 1$, and when $x = 1$, $u = 1^{3/2} + 1 = 2$. So

$$\int_0^1 \frac{\sqrt{x}}{x^{3/2} + 1}\, dx = \frac{2}{3}\int_1^2 \frac{du}{u} = \frac{2}{3}\ln|u|\Big|_1^2 = \frac{2}{3}(\ln 2 - \ln 1) = \frac{2}{3}\ln 2$$

29. We integrate by parts. We choose $u = x$ and $dv = e^{2x}\, dx$. Then $du = dx$ and $v = \frac{1}{2}e^{2x}$.

$$\int_1^3 x \cdot e^{2x}\, dx = \left[x \cdot \frac{1}{2}e^{2x}\right]_1^3 - \int_1^3 \frac{1}{2}e^{2x}\, dx = \frac{xe^{2x}}{2}\Bigg|_1^3 - \frac{1}{4}e^{2x}\Bigg|_1^3$$

$$= \left[\frac{3e^6}{2} - \frac{e^2}{2}\right] - \left[\frac{e^6}{4} - \frac{e^2}{4}\right] = \frac{6e^6 - 2e^2 - e^6 + e^2}{4} = \frac{5e^5 - e^2}{4}$$

31.

We integrate by parts. Choose $u = x$ and $dv = e^{-3x}dx$. Then $du = dx$, and $v = -\frac{1}{3}e^{-3x}$.

$$\int_1^2 xe^{-3x}dx = x\left(-\frac{1}{3}e^{-3x}\right)\Big|_1^2 - \int_1^2 -\frac{1}{3}e^{-3x}\,dx = -\frac{xe^{-3x}}{3}\Big|_1^2 + \frac{1}{3}\int_1^2 e^{-3x}\,dx$$

$$= \left[-\frac{2e^{-6}}{3} + \frac{e^{-3}}{3}\right] + \frac{1}{3}\left[-\frac{1}{3}e^{-3x}\Big|_1^2\right] = -\frac{2e^{-6}}{3} + \frac{e^{-3}}{3} + \frac{1}{3}\left[-\frac{e^{-6}}{3} + \frac{e^{-3}}{3}\right]$$

$$= -\frac{2e^{-6}}{3} + \frac{e^{-3}}{3} - \frac{e^{-6}}{9} + \frac{e^{-3}}{9} = -\frac{6e^{-6}}{9} + \frac{3e^{-3}}{9} - \frac{e^{-6}}{9} + \frac{e^{-3}}{9}$$

$$= -\frac{7e^{-6}}{9} + \frac{4e^{-3}}{9} = -\frac{7}{9e^6} + \frac{4}{9e^3}$$

33.

Integrating by parts, we choose $u = \ln x$ and $dv = dx$. Then $du = \frac{1}{x}dx$ and $v = x$.

$$\int_1^5 \ln x\, dx = \ln x \cdot x\Big|_1^5 - \int_1^5 x \cdot \frac{1}{x}\,dx = x\ln x\Big|_1^5 - \int_1^5 dx = x\ln x\Big|_1^5 - x\Big|_1^5$$

$$= [5\ln 5 - 1\ln 1] - [5 - 1] = 5\ln 5 - 0 - 4 = 5\ln 5 - 4 \approx 4.047$$

35. $\displaystyle\int_2^2 e^{x^2}\,dx = 0$ \qquad Property 4 \quad $\displaystyle\int_a^a f(x)\,dx = 0$

37. $\displaystyle\int_0^1 e^{-x^2}\,dx + \int_1^0 e^{-x^2}\,dx = \int_0^1 e^{-x^2}\,dx - \int_0^1 e^{-x^2}\,dx = 0$ \quad Property 3 \quad $\displaystyle\int_a^b f(x)\,dx = -\int_b^a f(x)\,dx$

39. $\displaystyle\int_1^3 [f(x) + g(x)]\,dx = \int_1^3 f(x)\,dx + \int_1^3 g(x)\,dx = 4 + (-2) = 2$

↑

(Property 7: $\displaystyle\int_a^b [f(x) \pm g(x)]\,dx = \int_a^b f(x)\,dx \pm \int_a^b g(x)\,dx$)

41. $\displaystyle\int_3^6 8f(x)\,dx = 8\int_3^6 f(x)\,dx = 8 \cdot 8 = 64$

↑

(Property 6: $\displaystyle\int_a^b c\,f(x)\,dx = c\int_a^b f(x)\,dx$)

43. $\int_3^6 \left[3f(x)+4g(x)\right] dx = \int_3^6 3f(x)\,dx + \int_3^6 4g(x)\,dx$

↑

(Property 7: $\int_a^b \left[f(x) \pm g(x)\right] dx = \int_a^b f(x)\ dx \pm \int_a^b g(x)\ dx$)

$= 3\int_3^6 f(x)\,dx + 4\int_3^6 g(x)\,dx = 3 \cdot 8 + 4 \cdot 3 = 36$

↑

(Property 6: $\int_a^b c\,f(x)\,dx = c\int_a^b f(x)\ dx$)

45. $\int_1^6 f(x)\,dx = \int_1^3 f(x)\,dx + \int_3^6 f(x)\,dx = 4 + 8 = 12$

↑

(Property 5: $\int_a^b f(x)\,dx = \int_a^c f(x)\ dx + \int_c^b f(x)\ dx$)

47. $C'(x) = 6x^2 - 100x + 1000$

The increase in cost in raising production from 100 units to 110 units is

$C(110) - C(100) = \int_{100}^{110} C'(x)\,dx$

$= \int_{100}^{110} \left(6x^2 - 100x + 1000\right) dx$

$= \left(\dfrac{6x^3}{3} - \dfrac{100x^2}{2} + 1000x\right)\Bigg|_{100}^{110}$

$= \left[2(110)^3 - 50(110)^2 + 1000(110)\right] - \left[2(100)^3 - 50(100)^2 + 1000(100)\right]$

$= \$\,567{,}000$

49. (a) $f(x) = 1272 x^{-0.35}$

$\int_{30}^{80} 1272 x^{-0.35}\,dx = \dfrac{1272 x^{0.65}}{0.65}\Bigg|_{30}^{80} = 1956.92\left(80^{0.65} - 30^{0.65}\right) = 15{,}921.34$

The total labor hours needed is 15,921.

(b) $f(x) = 1272 x^{-0.15}$

$\int_{30}^{80} 1272 x^{-0.15}\,dx = \dfrac{1272 x^{0.85}}{0.85}\Bigg|_{30}^{80} = 1496.47\left(80^{0.85} - 30^{0.85}\right) = 35{,}089.39$

The total labor hours needed is 35,089.

(c) Answers will vary.

51. $D(x) = -8.93x + 70$

$$\int_0^{12} (-8.93x + 70)\,dx = \left(-\frac{8.93x^2}{2} + 70x\right)\Bigg|_0^{12} = \left[-\frac{8.93(12)^2}{2} + 70(12)\right] - [0] = 197.04$$

The total budget deficit for 2002-2003 is projected to be 197.04 billion dollars.

53. The total sales during the first year is

$$\int_0^{12} (1200 - 950e^{-x})\,dx = (1200x + 950e^{-x})\Big|_0^{12}$$
$$= \left[1200(12) + 950e^{-12}\right] - \left[0 + 950e^0\right]$$
$$= \$13,450.01$$

55. $f(x) = 1000x^{-0.5}$

To produce an additional 25 units

$$\int_{35}^{60} 1000x^{-0.05}\,dx = \frac{1000\,x^{0.5}}{0.5}\Bigg|_{35}^{60} = 2000\left[60^{0.5} - 35^{0.5}\right] = 3659.77$$

The job will take 3660 labor hours.

57. (a) $f(x) = x^2$ is an even function because $f(-x) = (-x)^2 = x^2 = f(x)$.

$$\int_{-1}^{1} x^2\,dx = \frac{x^3}{3}\Bigg|_{-1}^{1} = \frac{1^3}{3} - \frac{(-1)^3}{3} = \frac{1}{3} + \frac{1}{3} = \frac{2}{3}$$

$$2\int_0^1 x^3\,dx = 2\left[\frac{x^3}{3}\Bigg|_0^1\right] = 2\left[\frac{1^3}{3} - 0\right] = \frac{2}{3}$$

So $\int_{-1}^{1} x^2\,dx = 2\int_0^1 x^3\,dx$.

(b) $f(x) = x^4 + x^2$ is an even function because

$$f(-x) = (-x)^4 + (-x)^2 = x^4 + x^2 = f(x).$$

$$\int_{-1}^{1} (x^4 + x^2)\,dx = \left(\frac{x^5}{5} + \frac{x^3}{3}\right)\Bigg|_{-1}^{1} = \left[\frac{1^5}{5} + \frac{1^3}{3}\right] - \left[\frac{(-1)^5}{5} + \frac{(-1)^3}{3}\right]$$
$$= \frac{1}{5} + \frac{1}{3} + \frac{1}{5} + \frac{1}{3} = \frac{2}{5} + \frac{2}{3} = \frac{6+10}{15} = \frac{16}{15}$$

$$2\int_0^1 (x^4 + x^2)\,dx = 2\left[\left(\frac{x^5}{5} + \frac{x^3}{3}\right)\Bigg|_0^1\right] = 2\left[\left(\frac{1}{5} + \frac{1}{3}\right) - 0\right] = \frac{16}{15}$$

So $\int_{-1}^{1} (x^4 + x^2)\,dx = 2\int_0^1 (x^4 + x^2)\,dx$.

6.5 Finding Areas; Consumer's Surplus; Producer's Surplus; Maximizing Profit over Time

1.

$$A = \int_2^6 (3x+2)\, dx = \left(\frac{3x^2}{2} + 2x\right)\Big|_2^6$$

$$= \left[\frac{3\cdot 6^2}{2} + 2(6)\right] - \left[\frac{3\cdot 2^2}{2} + 2(2)\right]$$

$$= 54 + 12 - 6 - 4 = 56$$

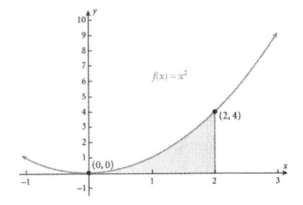

3.

$$A = \int_0^2 x^2\, dx = \frac{x^3}{3}\Big|_0^2 = \frac{2^3}{3} - 0 = \frac{8}{3}$$

5.

$$A = \int_{-2}^{-1} (x^2 - 1)\, dx - \int_{-1}^{1} (x^2 - 1)\, dx = \left(\frac{x^3}{3} - x\right)\Big|_{-2}^{-1} - \left(\frac{x^3}{3} - x\right)\Big|_{-1}^{1}$$

$$= \left\{\left[\frac{(-1)^3}{3} - (-1)\right] - \left[\frac{(-2)^3}{3} - (-2)\right]\right\} - \left\{\left[\frac{1^3}{3} - 1\right] - \left[\frac{(-1)^3}{3} - (-1)\right]\right\}$$

$$= -\frac{1}{3} + 1 + \frac{8}{3} - 2 - \frac{1}{3} + 1 - \frac{1}{3} + 1 = \frac{8}{3}$$

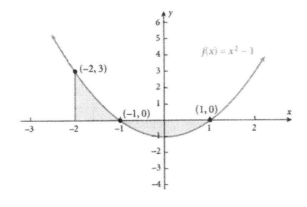

7. $f(x) = \sqrt[3]{x} = x^{1/3}$

$$A = -\int_{-1}^{0} x^{1/3}\, dx + \int_{0}^{8} x^{1/3}\, dx$$

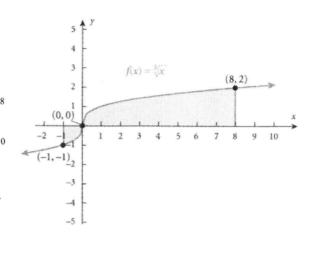

$$= -\left.\frac{x^{4/3}}{\frac{4}{3}}\right|_{-1}^{0} + \left.\frac{x^{4/3}}{\frac{4}{3}}\right|_{0}^{8} = -\left.\frac{3x^{4/3}}{4}\right|_{-1}^{0} + \left.\frac{3x^{4/3}}{4}\right|_{0}^{8}$$

$$= \left\{[-0] - \left[-\frac{3(-1)^{4/3}}{4}\right]\right\} + \left\{\left[\frac{3(8)^{4/3}}{4}\right] - [0]\right\}$$

$$= \frac{3}{4} + 12 = \frac{51}{4}$$

9. $f(x) = e^{x}$

$$A = \int_{0}^{1} e^{x}\, dx = \left. e^{x}\right|_{0}^{1} = e^{1} - e^{0}$$

$$= e - 1 \approx 1.718$$

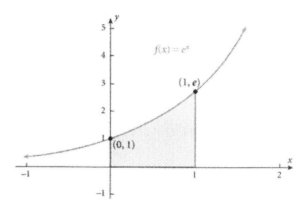

11.

$$A = \int_{0}^{1} g(x)\, dx - \int_{0}^{1} f(x)\, dx$$

$$= \int_{0}^{1} 2x\, dx - \int_{0}^{1} x\, dx$$

$$= \left.\frac{2x^{2}}{2}\right|_{0}^{1} - \left.\frac{x^{2}}{2}\right|_{0}^{1}$$

$$= \left[1^{2} - 0\right] - \left[\frac{1^{2}}{2} - 0\right]$$

$$= 1 - \frac{1}{2} = \frac{1}{2}$$

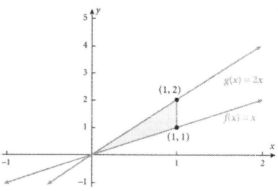

13. First we find where the graphs intersect.

$$f(x) = g(x)$$
$$x^2 = x$$
$$x^2 - x = 0$$
$$x(x-1) = 0$$
$$x = 0 \quad \text{or} \quad x = 1$$

$$f(0) = 0^2 = 0 \qquad\qquad f(1) = 1^2 = 1$$
$$g(0) = 0 \qquad\qquad\qquad g(1) = 1$$

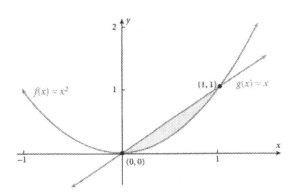

$$A = \int_0^1 g(x)\,dx - \int_0^1 f(x)\,dx = \int_0^1 x\,dx - \int_0^1 x^2\,dx$$

$$= \frac{x^2}{2}\bigg|_0^1 - \frac{x^3}{3}\bigg|_0^1 = \left[\frac{1}{2} - 0\right] - \left[\frac{1}{3} - 0\right]$$

$$= \frac{1}{2} - \frac{1}{3} = \frac{3-2}{6} = \frac{1}{6}$$

15. We graph $f(x) = x^2 + 1$ and $g(x) = x + 1$.
Then we find where the graphs intersect by
solving the equation $f(x) = g(x)$.

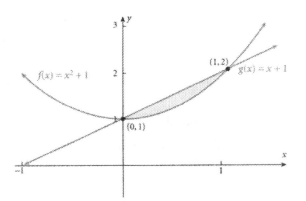

$$x^2 + 1 = x + 1$$
$$x^2 - x = 0$$
$$x(x-1) = 0$$
$$x = 0 \quad \text{or} \quad x - 1 = 0$$
$$\qquad\qquad\qquad x = 1$$

$$A = \int_0^1 \left(g(x) - f(x)\right) dx = \int_0^1 \left[(x+1) - (x^2 + 1)\right] dx$$

$$= \int_0^1 \left(x - x^2\right) dx = \left(\frac{x^2}{2} - \frac{x^3}{3}\right)\bigg|_0^1$$

$$= \left[\frac{1^2}{2} - \frac{1^3}{3}\right] - [0]$$

$$= \frac{1}{2} - \frac{1}{3} = \frac{1}{6}$$

17. We graph $f(x) = \sqrt{x} = x^{1/2}$ and $g(x) = x^3$.

Then we find where the graphs intersect by solving the equation $f(x) = g(x)$.

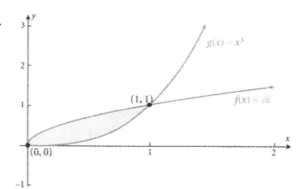

$$\sqrt{x} = x^3$$
$$x = x^6$$
$$x\left(x^5 - 1\right) = 0$$
$$x = 0 \quad \text{or} \quad x^5 - 1 = 0$$
$$x = 1$$

$$A = \int_0^1 \left(f(x) - g(x)\right) dx = \int_0^1 \left(x^{1/2} - x^3\right) dx$$

$$= \left(\frac{x^{3/2}}{\frac{3}{2}} - \frac{x^4}{4}\right)\Bigg|_0^1$$

$$= \left[\frac{2}{3} \cdot 1^{3/2} - \frac{1^4}{4}\right] - [0]$$

$$= \frac{2}{3} - \frac{1}{4} = \frac{8-3}{12} = \frac{5}{12}$$

19. We graph $f(x) = x^2$ and $g(x) = x^4$.

Then we find where the graphs intersect by solving the equation $f(x) = g(x)$.

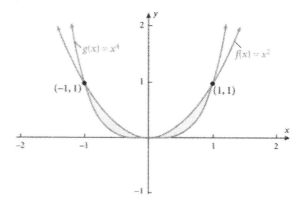

$$x^2 = x^4$$
$$x^4 - x^2 = 0$$
$$x^2(x^2 - 1) = 0$$
$$x^2(x - 1)(x + 1) = 0$$
$$x^2 = 0 \quad \text{or} \quad x - 1 = 0 \quad \text{or} \quad x + 1 = 0$$
$$x = 0 \quad \text{or} \quad x = 1 \quad \text{or} \quad x = -1$$

$$A = \int_{-1}^0 \left(f(x) - g(x)\right) dx + \int_0^1 \left(f(x) - g(x)\right) dx = \int_{-1}^1 \left(f(x) - g(x)\right) dx = \int_{-1}^1 \left(x^2 - x^4\right) dx$$

$$A = \left(\frac{x^3}{3} - \frac{x^5}{5}\right)\Bigg|_{-1}^1$$

$$= \left[\frac{1}{3} - \frac{1}{5}\right] - \left[\frac{(-1)^3}{3} - \frac{(-1)^5}{5}\right]$$

$$= \frac{1}{3} - \frac{1}{5} + \frac{1}{3} - \frac{1}{5} = \frac{5}{15} - \frac{3}{15} + \frac{5}{15} - \frac{3}{15} = \frac{4}{15}$$

21. We graph $f(x) = x^2 - 4x$ and $g(x) = -x^2$.

Then we find where the graphs intersect by solving the equation $f(x) = g(x)$.

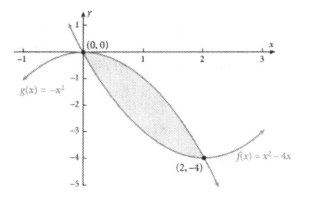

$$x^2 - 4x = -x^2$$
$$2x^2 - 4x = 0$$
$$2x(x-2) = 0$$
$$2x = 0 \quad \text{or} \quad x - 2 = 0$$
$$x = 0 \quad \text{or} \quad x = 2$$

$$A = \int_0^2 (g(x) - f(x))\, dx = \int_0^2 \left[-x^2 - (x^2 - 4x) \right] dx = \int_0^2 \left(-2x^2 + 4x \right) dx$$

$$= \left(-\frac{2x^3}{3} + \frac{4x^2}{2} \right) \Bigg|_0^2$$

$$= \left[-\frac{2 \cdot 2^3}{3} + 2 \cdot 2^2 \right] - [0]$$

$$= -\frac{16}{3} + 8 = \frac{8}{3}$$

23. We graph $f(x) = 4 - x^2$ and $g(x) = x + 2$.

Then we find where the graphs intersect by solving the equation $f(x) = g(x)$.

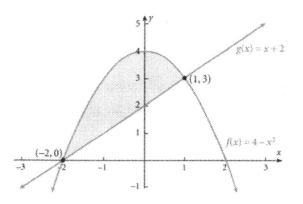

$$4 - x^2 = x + 2$$
$$x^2 + x - 2 = 0$$
$$(x+2)(x-1) = 0$$
$$x + 2 = 0 \quad \text{or} \quad x - 1 = 0$$
$$x = -2 \quad \text{or} \quad x = 1$$

$$A = \int_{-2}^1 [f(x) - g(x)]\, dx = \int_{-2}^1 \left[(4 - x^2) - (x+2) \right] dx = \int_{-2}^1 \left[(2 - x^2 - x) \right] dx$$

$$= \left(2x - \frac{x^3}{3} - \frac{x^2}{2} \right) \Bigg|_{-2}^1$$

$$= \left[2 \cdot 1 - \frac{1^3}{3} - \frac{1^2}{2} \right] - \left[2(-2) - \frac{(-2)^3}{3} - \frac{(-2)^2}{2} \right]$$

$$= 2 - \frac{1}{3} - \frac{1}{2} + 4 - \frac{8}{3} + 2 = \frac{9}{2}$$

25. We graph $f(x) = x^3$ and $g(x) = 4x$.

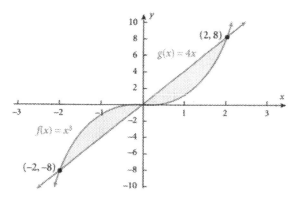

Then we find where the graphs intersect by solving the equation $f(x) = g(x)$.

$$x^3 = 4x$$
$$x^3 - 4x = 0$$
$$x(x^2 - 4) = 0$$
$$x(x-2)(x+2) = 0$$
$$x = 0 \quad \text{or} \quad x - 2 = 0 \quad \text{or} \quad x + 2 = 0$$
$$x = 2 \quad \text{or} \quad x = -2$$

$$A = \int_{-2}^{0} \left(f(x) - g(x) \right) dx + \int_{0}^{2} \left(g(x) - f(x) \right) dx$$

$$= \int_{-2}^{0} \left(x^3 - 4x \right) dx + \int_{0}^{2} \left(4x - x^3 \right) dx$$

$$= \left(\frac{x^4}{4} - \frac{4x^2}{2} \right) \Bigg|_{-2}^{0} + \left(\frac{4x^2}{2} - \frac{x^4}{4} \right) \Bigg|_{0}^{2}$$

$$= \left\{ [0] - \left[\frac{(-2)^4}{4} - 2(-2)^2 \right] \right\} + \left\{ \left[2(2)^2 - \frac{2^4}{4} \right] - [0] \right\}$$

$$0 - [4 - 8] + [8 - 4] + 0 = 4 + 4 = 8$$

27. We graph $y = x^2$, $y = x$, and $y = -x$.

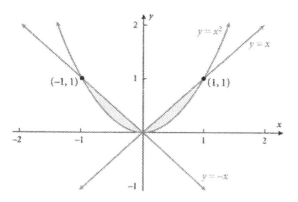

Then we find where the graphs intersect by solving the equations $x^2 = x$ and $x^2 = -x$.

$$x^2 = x \qquad\qquad x^2 = -x$$
$$x^2 - x = 0 \qquad\qquad x^2 + x = 0$$
$$x(x-1) = 0 \qquad\qquad x(x+1) = 0$$
$$x = 0 \quad \text{or} \quad x - 1 = 0 \qquad x = 0 \quad \text{or} \quad x + 1 = 0$$
$$x = 1 \qquad\qquad\qquad\qquad x = -1$$

$$A = \int_{-1}^{0} \left(-x - x^2 \right) dx + \int_{0}^{1} \left(x - x^2 \right) dx$$

$$= \left(-\frac{x^2}{2} - \frac{x^3}{3} \right) \Bigg|_{-1}^{0} + \left(\frac{x^2}{2} - \frac{x^3}{3} \right) \Bigg|_{0}^{1}$$

$$= \left\{ [0] - \left[-\frac{(-1)^2}{2} - \frac{(-1)^3}{3} \right] \right\} + \left\{ \left[\frac{1}{2} - \frac{1}{3} \right] - [0] \right\}$$

$$= -\left[\frac{-3+2}{6} \right] + \left[\frac{3-2}{6} \right] = \frac{1}{6} + \frac{1}{6} = \frac{2}{6} = \frac{1}{3}$$

29. (a) $\int_0^4 (3x+1)\, dx$ represents the area under the graph of $f(x) = 3x+1$ from $x = 0$ to $x = 4$.

(b)
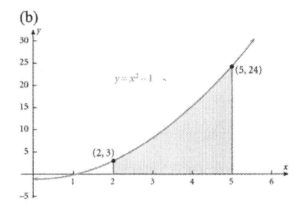

(c) $\int_0^4 (3x+1)\, dx = \left(\frac{3x^2}{2} + x \right) \Big|_0^4$

$$= \left[\frac{3 \cdot 4^2}{2} + 4 \right] - [0]$$

$$= 28$$

31. (a) $\int_2^5 (x^2 - 1)\, dx$ represents the area under the graph of $f(x) = x^2 - 1$ from $x = 2$ to $x = 5$.

(b)
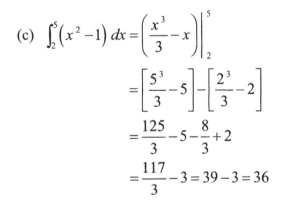

(c) $\int_2^5 (x^2 - 1)\, dx = \left(\frac{x^3}{3} - x \right) \Big|_2^5$

$$= \left[\frac{5^3}{3} - 5 \right] - \left[\frac{2^3}{3} - 2 \right]$$

$$= \frac{125}{3} - 5 - \frac{8}{3} + 2$$

$$= \frac{117}{3} - 3 = 39 - 3 = 36$$

33. (a) $\int_0^2 e^x\, dx$ represents the area under the graph of $f(x) = e^x$ from $x = 0$ to $x = 2$.

(b)
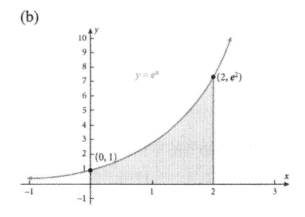

(c) $\int_0^2 e^x\, dx = e^x \Big|_0^2$

$$= e^2 - e^0$$

$$= e^2 - 1$$

35. We first find the equilibrium point (x^*, p^*), by solving the equation $D(x^*) = S(x^*)$.

$$-5x^* + 20 = 4x^* + 8$$
$$-9x^* = -12$$
$$x^* = \frac{12}{9} = \frac{4}{3}$$

and $p^* = D(x^*) = S(x^*)$

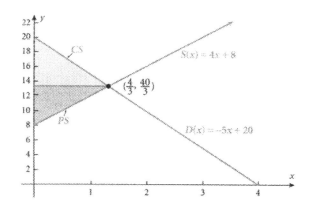

$$= -5\left(\frac{4}{3}\right) + 20 = -\frac{20}{3} + \frac{60}{3} = \frac{40}{3} \approx 13.33$$

The consumer's surplus, $CS = \int_0^{x^*} D(x)\, dx - p^* x^*$

$$CS = \int_0^{4/3}(-5x + 20)\, dx - \left(\frac{4}{3}\right)\left(\frac{40}{3}\right)$$

$$= \left(\frac{-5x^2}{2} + 20x\right)\Bigg|_0^{4/3} - \frac{160}{9}$$

$$= \left[-\frac{5}{2}\cdot\left(\frac{4}{3}\right)^2 + 20\left(\frac{4}{3}\right)\right] - [0] - \frac{160}{9}$$

$$= -\frac{80}{18} + \frac{80}{3} - \frac{160}{9} = \frac{-40 + 240 - 160}{9} = \frac{40}{9} = \$4.44$$

The producer's surplus, $PS = p^* x^* - \int_0^{x^*} S(x)\, dx$

$$PS = \frac{160}{9} - \int_0^{4/3}(4x + 8)\, dx$$

$$= \frac{160}{9} - \left(\frac{4x^2}{2} + 8x\right)\Bigg|_0^{4/3}$$

$$= \frac{160}{9} - \left\{\left[2\left(\frac{4}{3}\right)^2 + 8\left(\frac{4}{3}\right)\right] - [0]\right\}$$

$$\frac{160}{9} - \left[\frac{32}{9} + \frac{32}{3}\right] = \frac{160}{9} - \frac{32}{9} - \frac{96}{9} = \frac{32}{9} = \$3.56$$

37. The best time to terminate operations is when $R'(t) = C'(t)$.

$$19 - t^{1/2} = 3 + 3t^{1/2}$$
$$-4t^{1/2} = -16$$
$$t^{1/2} = 4 \text{ or } t = 16$$

Operations should be terminated after 16 years.

At $t = 16$, $R'(16) = 19 - 16^{1/2} = 19 - 4 = 15$ million dollars.

$$C'(16) = 3 + 3 \cdot 16^{1/2} = 3 + 3 \cdot 4 = 15 \text{ million dollars.}$$

The profit at $t = 16$ years is

$$P(16) = \int_0^{16} \left[\left(19 - t^{1/2} \right) - \left(3 + 3t^{1/2} \right) \right] dt = \int_0^{16} \left[16 - 4t^{1/2} \right] dt$$

$$= \left(16t - \frac{4t^{3/2}}{\frac{3}{2}} \right) \Bigg|_0^{16} = \left[16 \cdot 16 - \frac{8}{3} \cdot 16^{3/2} \right] - [0]$$

$$= 256 - \frac{512}{3} = \frac{768}{3} - \frac{512}{3} = \frac{256}{3} = 85.33 \text{ million dollars.}$$

39. (a) $f(x) = x^2$; $a = 0$ $b = 1$

$$\int_0^1 x^2 \, dx = \frac{x^3}{3} \Bigg|_0^1 = \frac{1}{3}$$

So, $f(c)(b-a) = \frac{1}{3}$

$$c^2(1-0) = \frac{1}{3}$$

$$c^2 = \frac{1}{3} \quad \text{or} \quad c = \frac{1}{\sqrt{3}} = \frac{\sqrt{3}}{3}$$

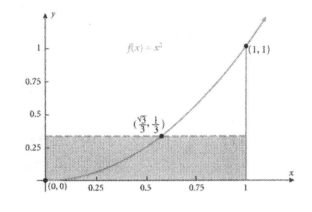

(b) $f(x) = \frac{1}{x^2} = x^{-2}$; $a = 1$ $b = 4$

$$\int_1^4 x^{-2} \, dx = \frac{x^{-1}}{-1} \Bigg|_1^4 = -\frac{1}{4} + 1 = \frac{3}{4}$$

So, $f(c)(b-a) = \frac{3}{4}$

$$\frac{1}{c^2}(4-1) = \frac{3}{4}$$

$$\frac{3}{c^2} = \frac{3}{4}$$

$$c^2 = 4 \quad \text{or} \quad c = 2$$

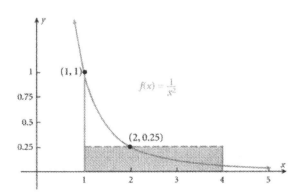

41. $\dfrac{d}{dx}\displaystyle\int_a^x f(t)\,dt = f(x)$

(a) $\dfrac{d}{dx}\displaystyle\int_1^x t^2\,dt = f(x) = x^2$ (b) $\dfrac{d}{dx}\displaystyle\int_2^x \sqrt{t^2 - 2}\,dt = f(x) = \sqrt{x^2 - 2}$

(c) $\dfrac{d}{dx}\displaystyle\int_5^x \sqrt{t^t + 2t}\,dt = f(x) = \sqrt{x^x + 2x}$

6.6 Approximating Definite Integrals

1. **STEP 1** $[1, 3]$ has been divided into two subintervals of equal length
  $[1, 2]$ and $[2, 3]$
STEP 2 $f(1) = 1$ $f(2) = 2$
STEP 3 $\displaystyle\int_1^3 f(x)\,dx \approx f(1) \cdot 1 + f(2) \cdot 1 = 1 + 2 = 3$

3. **STEP 1** $[0, 8]$ has been divided into 4 intervals, each of width 2.
STEP 2 $f(0) = 10$ $f(2) = 6$ $f(4) = 7$ $f(6) = 5$
STEP 3 $\displaystyle\int_0^8 f(x)\,dx \approx f(0) \cdot 2 + f(2) \cdot 2 + f(4) \cdot 2 + f(6) \cdot 2$
  $= 10 \cdot 2 + 6 \cdot 2 + 7 \cdot 2 + 5 \cdot 2 = 56$

5. (a)

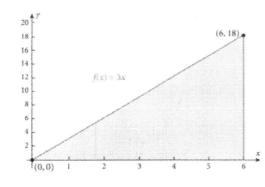

(b) **STEP 1** When $[0, 6]$ is separated into three subintervals each will have width 2.
  $[0, 2]$ $[2, 4]$ $[4, 6]$

STEP 2 $f(0) = 3 \cdot 0 = 0$
  $f(2) = 3 \cdot 2 = 6$
  $f(4) = 3 \cdot 4 = 12$

STEP 3 $A \approx f(0) \cdot 2 + f(2) \cdot 2 + f(4) \cdot 2$
  $= 2(0 + 6 + 12) = 36$

(c) We use STEP 1 from part (b).
STEP 2 $f(2) = 3 \cdot 2 = 6$ $f(4) = 3 \cdot 4 = 12$ $f(6) = 3 \cdot 6 = 18$
STEP 3 $A \approx f(2) \cdot 2 + f(4) \cdot 2 + f(6) \cdot 2 = 2(6 + 12 + 18) = 72$

(d) **STEP 1** When $[0, 6]$ is separated into six subintervals each will have width 1.
  $[0, 1]$ $[1, 2]$ $[2, 3]$ $[3. 4]$ $[4, 5]$ $[5, 6]$

STEP 2 $f(0) = 0$ $f(1) = 3 \cdot 1 = 3$ $f(2) = 6$ $f(3) = 3 \cdot 3 = 9$

       $f(4) = 12$ $f(5) = 3 \cdot 5 = 15$

STEP 3 $A \approx f(0) \cdot 1 + f(1) \cdot 1 + f(2) \cdot 1 + f(3) \cdot 1 + f(4) \cdot 1 + f(5) \cdot 1$

       $= 1(0 + 3 + 6 + 9 + 12 + 15) = 45$

(e) We use STEP 1 from part (d).

STEP 2 $f(1) = 3 \cdot 1 = 3$ $f(2) = 6$ $f(3) = 3 \cdot 3 = 9$

       $f(4) = 12$ $f(5) = 3 \cdot 5 = 15$ $f(6) = 3 \cdot 6 = 18$

STEP 3 $A \approx f(1) \cdot 1 + f(2) \cdot 1 + f(3) \cdot 1 + f(4) \cdot 1 + f(5) \cdot 1 + f(6) \cdot 1$

       $= 1(3 + 6 + 9 + 12 + 15 + 18) = 63$

(f) Since A is a triangle, with base 6 and altitude 18, the actual area is

$$A = \frac{1}{2}bh = \frac{1}{2} \cdot 6 \cdot 18 = 54$$

7. (a)

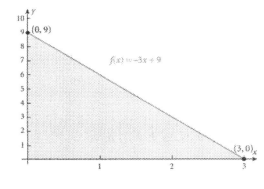

(b) **STEP 1** When [0, 3] is separated into three subintervals each will have width 1.

       [0, 1]   [1, 2]   [2, 3]

STEP 2 $f(0) = -3(0) + 9 = 9$

           $f(1) = -3(1) + 9 = 6$

           $f(2) = -3(2) + 9 = 3$

STEP 3 $A \approx f(0) \cdot 1 + f(1) \cdot 1 + f(2) \cdot 1$

       $= 1(9 + 6 + 3) = 18$

(c) We use STEP 1 from part (b).

STEP 2 $f(3) = -3(3) + 9 = 0$

STEP 3 $A \approx f(1) \cdot 1 + f(2) \cdot 1 + f(3) \cdot 1 = 1(6 + 3 + 0) = 9$

(d) **STEP 1** When [0, 3] is separated into six subintervals each will have width 0.5.

       [0, 0.5]   [0.5, 1]   [1, 1.5]   [1.5, 2]   [2, 2.5]   [2.5, 3]

STEP 2 $f(0) = 9$ $f(0.5) = -3(0.5) + 9 = 7.5$ $f(1) = 6$

       $f(1.5) = -3(1.5) + 9 = 4.5$ $f(2) = 3$ $f(2.5) = -3(2.5) + 9 = 1.5$

STEP 3

$$A \approx f(0) \cdot 0.5 + f(0.5) \cdot 0.5 + f(1) \cdot 0.5 + f(1.5) \cdot 0.5 + f(2) \cdot 0.5 + f(2.5) \cdot 0.5$$
$$= 0.5(9 + 7.5 + 6 + 4.5 + 3 + 1.5) = 0.5(31.5) = 15.75$$

(e) We use STEP 1 from part (d).

STEP 3

$$A \approx f(0.5) \cdot 0.5 + f(1) \cdot 0.5 + f(1.5) \cdot 0.5 + f(2) \cdot 0.5 + f(2.5) \cdot 0.5 + f(3) \cdot 0.5$$
$$= 0.5(7.5 + 6 + 4.5 + 3 + 1.5 + 0) = 0.5(22.5) = 11.25$$

(f) Since A is a triangle, with base 3 and altitude 9, the actual area is
$$A = \frac{1}{2}bh = \frac{1}{2} \cdot 3 \cdot 9 = \frac{27}{2} = 13.5$$

9. (a)

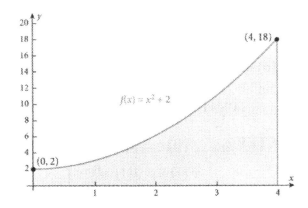

(b) **STEP 1** When [0, 4] is separated into four subintervals each will have width 1.

    [0, 1]   [1, 2]   [2, 3]   [3, 4]

STEP 2 $f(0) = 0^2 + 2 = 2$ $f(1) = 1^2 + 2 = 3$

               $f(2) = 2^2 + 2 = 6$ $f(3) = 3^2 + 2 = 11$

STEP 3 $A \approx \left[f(0) \cdot 1\right] + \left[f(1) \cdot 1\right] + \left[f(2) \cdot 1\right] + \left[f(3) \cdot 1\right] = 1(2 + 3 + 6 + 11) = 22$

(c) **STEP 1** When [0, 4] is separated into eight subintervals each will have width 0.5.

    [0, 0.5]   [0.5, 1]   [1, 1.5]   [1.5, 2]   [2, 2.5]   [2.5, 3]   [3, 3.5]   [3.5, 4]

STEP 2 $f(0.5) = 0.5^2 + 2 = 2.25$ $f(1.5) = 1.5^2 + 2 = 4.25$

               $f(2.5) = 2.5^2 + 2 = 8.25$ $f(3.5) = 3.5^2 + 2 = 14.25$

STEP 3 $A \approx \left[f(0) \cdot 0.5\right] + \left[f(0.5) \cdot 0.5\right] + \left[f(1) \cdot 0.5\right] + \left[f(1.5) \cdot 0.5\right]$

$\qquad + \left[f(2) \cdot 0.5\right] + \left[f(2.5) \cdot 0.5\right] + \left[f(3) \cdot 0.5\right] + \left[f(3.5) \cdot 0.5\right]$

$\qquad = 0.5(2 + 2.25 + 3 + 4.25 + 6 + 8.25 + 11 + 14.25) = 0.5(51)$

$\qquad A \approx 25.5$

(d) $A = \int_0^4 \left(x^2 + 2\right) dx$

(e) $\int_0^4 \left(x^2 + 2\right) dx = \left(\dfrac{x^3}{3} + 2x\right)\Bigg|_0^4 = \left[\dfrac{4^3}{3} + 2(4)\right] - [0] = \dfrac{88}{3} \approx 29.333$

11. (a)

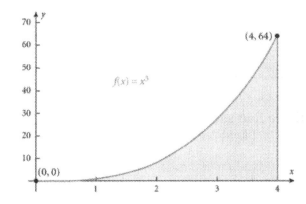

(b) **STEP 1** When [0, 4] is separated into four subintervals each will have width 1.

$\qquad$ [0, 1] $\qquad$ [1, 2] $\qquad$ [2, 3] $\qquad$ [3, 4]

STEP 2 $\qquad f(0) = 0^3 = 0 \qquad\qquad f(1) = 1^3 = 1$

$\qquad\qquad\quad f(2) = 2^3 = 8 \qquad\qquad f(3) = 3^3 = 27$

STEP 3 $A \approx \left[f(0) \cdot 1\right] + \left[f(1) \cdot 1\right] + \left[f(2) \cdot 1\right] + \left[f(3) \cdot 1\right] = 1(0 + 1 + 8 + 27) = 36$

(c) **STEP 1** When [0, 4] is separated into eight subintervals each will have width 0.5.

$\qquad$ [0, 0.5] $\quad$ [0.5, 1] $\quad$ [1, 1.5] $\quad$ [1.5, 2] $\quad$ [2, 2.5] $\quad$ [2.5, 3] $\quad$ [3, 3.5] $\quad$ [3.5, 4]

STEP 2 $\quad f(0.5) = 0.5^3 = 0.125 \qquad\qquad f(1.5) = 1.5^3 = 3.375$

$\qquad\qquad f(2.5) = 2.5^3 = 15.625 \qquad\qquad f(3.5) = 3.5^3 = 42.875$

STEP 3 $\quad A \approx \left[f(0) \cdot 0.5\right] + \left[f(0.5) \cdot 0.5\right] + \left[f(1) \cdot 0.5\right] + \left[f(1.5) \cdot 0.5\right]$

$\qquad\qquad + \left[f(2) \cdot 0.5\right] + \left[f(2.5) \cdot 0.5\right] + \left[f(3) \cdot 0.5\right] + \left[f(3.5) \cdot 0.5\right]$

$$= 0.5(0 + 0.125 + 1 + 3.375 + 8 + 15.625 + 27 + 42.875) = 0.5(98)$$
$$A \approx 49$$

(d) $A = \int_0^4 x^3 \, dx$

(e) $\int_0^4 x^3 \, dx = \left(\dfrac{x^4}{4} \right) \Bigg|_0^4 = \dfrac{4^4}{4} - 0 = 64$

13. (a)

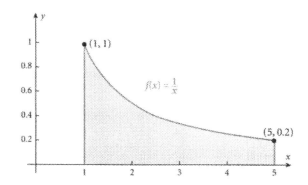

(b) **STEP 1** When $[1, 5]$ is separated into four subintervals each will have width 1.

$$[1, 2] \qquad [2, 3] \qquad [3, 4] \qquad [4, 5]$$

STEP 2 $f(1) = \dfrac{1}{1} = 1 \qquad\qquad f(2) = \dfrac{1}{2}$

$\qquad\qquad\quad f(3) = \dfrac{1}{3} \qquad\qquad\quad f(4) = \dfrac{1}{4}$

STEP 3 $A \approx \left[f(1) \cdot 1 \right] + \left[f(2) \cdot 1 \right] + \left[f(3) \cdot 1 \right] + \left[f(4) \cdot 1 \right]$

$\qquad\qquad = 1\left(1 + \dfrac{1}{2} + \dfrac{1}{3} + \dfrac{1}{4} \right) = \dfrac{12 + 6 + 4 + 3}{12} = \dfrac{25}{12} \approx 2.083$

(c) **STEP 1** When $[1, 5]$ is separated into eight subintervals each will have width 0.5.

$$[1, 1.5] \quad [1.5, 2] \quad [2, 2.5] \quad [2.5, 3] \quad [3, 3.5] \quad [3.5, 4] \quad [4, 4.5] \quad [4.5, 5]$$

STEP 2 $f\left(\dfrac{3}{2} \right) = \dfrac{1}{\frac{3}{2}} = \dfrac{2}{3} \qquad\qquad f\left(\dfrac{5}{2} \right) = \dfrac{1}{\frac{5}{2}} = \dfrac{2}{5}$

$\qquad\qquad\quad f\left(\dfrac{7}{2} \right) = \dfrac{1}{\frac{7}{2}} = \dfrac{2}{7} \qquad\qquad f\left(\dfrac{9}{2} \right) = \dfrac{1}{\frac{9}{2}} = \dfrac{2}{9}$

STEP 3 $A \approx \left[f(1) \cdot \dfrac{1}{2} \right] + \left[f\left(\dfrac{3}{2}\right) \cdot \dfrac{1}{2} \right] + \left[f(2) \cdot \dfrac{1}{2} \right] + \left[f\left(\dfrac{5}{2}\right) \cdot \dfrac{1}{2} \right]$

$\qquad\qquad + \left[f(3) \cdot \dfrac{1}{2} \right] + \left[f\left(\dfrac{7}{2}\right) \cdot \dfrac{1}{2} \right] + \left[f(4) \cdot \dfrac{1}{2} \right] + \left[f\left(\dfrac{9}{2}\right) \cdot \dfrac{1}{2} \right]$

$\qquad\qquad = \dfrac{1}{2}\left(1 + \dfrac{2}{3} + \dfrac{1}{2} + \dfrac{2}{5} + \dfrac{1}{3} + \dfrac{2}{7} + \dfrac{1}{4} + \dfrac{2}{9} \right) = \dfrac{4609}{2520} \approx 1.829$

(d) $A = \displaystyle\int_{1}^{5} \dfrac{1}{x}\, dx$ 　　　　　　(e) $\displaystyle\int_{1}^{5} \dfrac{1}{x}\, dx = \ln x \Big|_{1}^{5} = \ln 5 - \ln 1 = \ln 5 \approx 1.609$

15. (a)

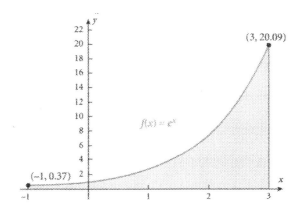

(b) **STEP 1** When $[-1, 3]$ is separated into four subintervals each will have width 1.
　　　　$[-1, 0]$　　$[0, 1]$　　　$[1, 2]$　　　$[2, 3]$

STEP 2　$f(-1) = e^{-1}$　　　　　$f(0) = e^0 = 1$
　　　　　$f(1) = e^1$　　　　　　$f(2) = e^2$

STEP 3
$A \approx \left[f(-1) \cdot 1 \right] + \left[f(0) \cdot 1 \right] + \left[f(1) \cdot 1 \right] + \left[f(2) \cdot 1 \right] = 1\left(e^{-1} + 1 + e + e^2 \right) \approx 11.475$

(c) **STEP 1** When $[0, 4]$ is separated into eight subintervals each will have width 0.5.
　　$[-1, -0.5]$　$[-0.5, 0]$　$[0, 0.5]$　$[0.5, 1]$　　$[1, 1.5]$　　$[1.5, 2]$　　$[2, 2.5]$　　$[2.5, 3]$

STEP 2　$f(-0.5) = e^{-0.5}$　　　　$f(0.5) = e^{0.5}$
　　　　　$f(1.5) = e^{1.5}$　　　　　$f(2.5) = e^{2.5}$

STEP 3　$A \approx \left[f(-1) \cdot 0.5 \right] + \left[f(-0.5) \cdot 0.5 \right] + \left[f(0) \cdot 0.5 \right] + \left[f(0.5) \cdot 0.5 \right]$

$\qquad\qquad + \left[f(1) \cdot 0.5 \right] + \left[f(1.5) \cdot 0.5 \right] + \left[f(2) \cdot 0.5 \right] + \left[f(2.5) \cdot 0.5 \right]$

$$= 0.5\left(e^{-1} + e^{-0.5} + 1 + e^{0.5} + e + e^{1.5} + e^2 + e^{2.5}\right) \approx 0.5(30.39465) \approx 15.197$$

(d) $A = \int_{-1}^{3} e^x \, dx$

(e) $\int_{-1}^{3} e^x \, dx = e^x \Big|_{-1}^{3} = e^3 - e^{-1} = \dfrac{e^4 - 1}{e} \approx 19.718$

17.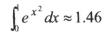

$$\int_{0}^{1} e^{x^2} \, dx \approx 1.46$$

19.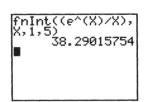

$$\int_{1}^{5} \dfrac{e^x}{x} \, dx \approx 38.29$$

21. (a)

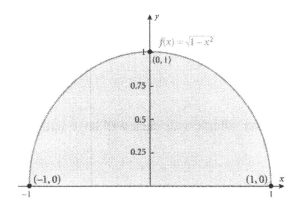

(b) **STEP 1** When $[-1, 1]$ is separated into five subintervals each will have width 0.4.
$[-1, -0.6]$ $[-0.6, -0.2]$ $[-0.2, 0.2]$ $[0.2, 0.6]$ $[0.6, 1]$

STEP 2 $f(-1) = \sqrt{1 - (-1)^2} = 0$ $f(-0.6) = \sqrt{1 - (-0.6)^2} = 0.8$

$f(-0.2) = \sqrt{1 - (-0.2)^2} = \sqrt{0.96}$ $f(0.2) = \sqrt{1 - (0.2)^2} = \sqrt{0.96}$

$f(0.6) = \sqrt{1 - (0.6)^2} = 0.8$

STEP 3 $A \approx \left[f(-1) \cdot 0.4\right] + \left[f(-0.6) \cdot 0.4\right] + \left[f(-0.2) \cdot 0.4\right]$
$\qquad\qquad + \left[f(0.2) \cdot 0.4\right] + \left[f(0.6) \cdot 0.4\right]$
$\qquad = 0.4\left(0 + 0.8 + \sqrt{0.96} + \sqrt{0.96} + 0.8\right) \approx 0.4(3.55959) \approx 1.424$

(c) **STEP 1** When $[-1, 1]$ is separated into ten subintervals each will have width 0.2.

$\quad$ $[-1, -0.8]$ $\quad$ $[-0.8, -0.6]$ $\quad$ $[-0.6, -0.4]$ $\quad$ $[-0.4, -0.2]$ $\quad$ $[-0.2, 0]$

$\quad$ $[0, 0.2]$ $\quad$ $[0.2, 0.4]$ $\quad$ $[0.4, .06]$ $\quad$ $[0.6, 0.8]$ $\quad$ $[0.8, 1]$

STEP 2 $\quad f(-0.8) = \sqrt{1-(-0.8)^2} = 0.6$ $\qquad f(-0.4) = \sqrt{1-(-0.4)^2} = \sqrt{0.84}$

$\qquad\qquad f(0) = \sqrt{1-0^2} = 1$ $\qquad\qquad\qquad f(0.4) = \sqrt{1-(0.4)^2} = \sqrt{0.84}$

$\qquad\qquad f(0.8) = \sqrt{1-(0.8)^2} = 0.6$

STEP 3 $\quad A \approx \left[f(-1) \cdot 0.2\right] + \left[f(-0.8) \cdot 0.2\right] + \left[f(-0.6) \cdot 0.4\right] + \left[f(-0.4) \cdot 0.2\right]$

$\qquad\qquad\quad + \left[f(-0.2) \cdot 0.2\right] + \left[f(0) \cdot 0.2\right] + \left[f(0.2) \cdot 0.2\right] + \left[f(0.4) \cdot 0.2\right]$

$\qquad\qquad\quad + \left[f(0.6) \cdot 0.2\right] + \left[f(0.8) \cdot 0.2\right]$

$\qquad\qquad = 0.2\left(0 + 0.6 + 0.8 + \sqrt{0.84} + \sqrt{0.96} + 1 + \sqrt{0.96} + \sqrt{0.84} + 0.8 + 0.6\right)$

$\qquad\qquad = 0.2(7.59262) = 1.519$

(d) $\quad A = \int_{-1}^{1} \sqrt{1-x^2}\, dx = \int_{-1}^{1} \left(1-x^2\right)^{1/2}\, dx$

(e)

```
fnInt(√(1-X²),X,
-1,1)
        1.570796729
```

(f) Since the graph is a semi-circle with radius 1, the area is

$\quad A = \frac{1}{2}\left(\pi r^2\right) = \frac{1}{2}\pi$

6.7 Differential Equations

1.

The general solution to the differential equation $\dfrac{dy}{dx} = x^2 - 1$ is $y = \dfrac{x^3}{3} - x + K$.

We use $x = 0$ and $y = 0$ to find K.

$\qquad 0 = \dfrac{0}{3} - 0 + K$

$\qquad K = 0$

The particular solution to the differential equation is

$\qquad y = \dfrac{x^3}{3} - x$

3.

The general solution to the differential equation $\dfrac{dy}{dx} = x^2 - x$ is $y = \dfrac{x^3}{3} - \dfrac{x^2}{2} + K$

We use $x = 3$ and $y = 3$ to find K.

$$3 = \frac{3^3}{3} - \frac{3^2}{2} + K = 9 - \frac{9}{2} + K$$

$$K = 3 - \frac{9}{2} = -\frac{3}{2}$$

The particular solution to the differential equation is

$$y = \frac{x^3}{3} - \frac{x^2}{2} - \frac{3}{2}$$

5.

The general solution to the equation $\dfrac{dy}{dx} = x^3 - x + 2$ is $y = \dfrac{x^4}{4} - \dfrac{x^2}{2} + 2x + K$.

We use $x = -2$ and $y = 1$ to find K.

$$1 = \frac{(-2)^4}{4} - \frac{(-2)^2}{2} + 2(-2) + K$$

$$K = 1 - 4 + 2 + 4 = 3$$

The particular solution to the differential equation is

$$y = \frac{x^4}{4} - \frac{x^2}{2} + 2x + 3$$

7.

The general solution to the differential equation $\dfrac{dy}{dx} = e^x$ is $y = e^x + K$.

We use $x = 0$ and $y = 4$ to find K.

$$4 = e^0 + K$$

$$K = 4 - 1 = 3$$

The particular solution to the differential equation is

$$y = e^x + 3$$

9.

The general solution to the differential equation $\dfrac{dy}{dx} = \dfrac{x^2 + x + 1}{x} = x + 1 + \dfrac{1}{x}$

is $y = \dfrac{x^2}{2} + x + \ln|x| + K$.

We use $x = 1$ and $y = 0$ to find K.

$$0 = \frac{1^2}{2} + 1 + \ln|1| + K$$

$$K = -\frac{3}{2}$$

The particular solution to the differential equation is

$$y = \frac{x^2}{2} + x + \ln|x| - \frac{3}{2}$$

11.
The differential equation describing the population growth is $\dfrac{dN}{dt} = kN$, where N is the population size and t is the time in minutes.

The general solution of the equation is $N(t) = N_0 e^{kt}$ where N_0 is the population at $t = 0$ and k is the constant of proportionality. Using $N(5) = 150$ and $N_0 = 100$, we solve for k.

$$150 = 100 e^{5k}$$
$$e^{5k} = 1.5$$
$$5k = \ln 1.5$$
$$k = \frac{\ln 1.5}{5}$$

After 1 hour $t = 60$ and there will be 12,975 bacteria.

$$N(60) = 100\ e^{(60)\left[\frac{\ln 1.5}{5}\right]} = 100 e^{(12)\ln 1.5} = 12,974.63$$

After 90 minutes, there will be 147,789 bacteria.

$$N(90) = 100\ e^{(90)\left[\frac{\ln 1.5}{5}\right]} = 100 e^{(18)\ln 1.5} = 147,789.188$$

There will be 1,000,000 bacteria after $t = 113.58$ minutes.

$$1,000,000 = 100\ e^{\left[\frac{\ln 1.5}{5}\right]t}$$
$$10,000 = e^{\left[\frac{\ln 1.5}{5}\right]t}$$
$$\ln 10,000 = \left[\frac{\ln 1.5}{5}\right]t$$
$$t = \frac{5 \cdot \ln 10,000}{\ln 1.5} \approx 113.58$$

13.
The differential equation describing the radioactive decay is $\dfrac{dA}{dt} = kA$, where A is the amount of radium present and t is the time in years.

The general solution of the equation is $A(t) = A_0 e^{kt}$ where A_0 is the amount at $t = 0$ and k is the constant of proportionality. Using $A_0 = 8$ grams and $A(1690) = 4$ grams, we solve for k.

$$4 = 8\ e^{1690k}$$
$$\frac{1}{2} = e^{1690k}$$
$$\ln \frac{1}{2} = 1690k$$
$$k = \frac{\ln 0.5}{1690} \approx -0.0041015$$

In 100 years there will be 7.679 grams of radium present.

$$N(100) = 8e^{\left[\frac{\ln 0.5}{1690}\right] \cdot 100} = 7.679$$

15. If we begin with 100 grams of carbon and $t = 5600$ years is the half life, then we use the half-life to find k.

$$N(5600) = 50 = 100 e^{5600k}$$

$$\frac{1}{2} = e^{5600k}$$

$$\ln 0.5 = 5600 k$$

$$k = \frac{\ln 0.5}{5600} \approx 0.000124$$

We now find t so that $N(t) = 30$.

$$30 = 100 e^{\left[\frac{\ln 0.50}{5600}\right]t}$$

$$0.3 = e^{\left[\frac{\ln 0.50}{5600}\right]t}$$

$$\ln 0.30 = \frac{\ln 0.50}{5600} t$$

$$t = \frac{5600 \ln(0.30)}{\ln 0.5} \approx 9727$$

The tree is 9727 years old.

17. Since the population obeys the law of uninhibited growth and $N(0) = 1500$, we use $N(24) = 2500$ to find the value of k.

$$N(t) = 1500\, e^{kt}$$

$$2500 = 1500\, e^{24k}$$

$$\ln\left(\frac{25}{15}\right) = 24k$$

$$k = \frac{1}{24} \cdot \ln\left(\frac{5}{3}\right)$$

After 3 days $t = 72$ hours, and the population is 6944 mosquitos.

$$N(72) = 1500 e^{72\left[\frac{1}{24}\ln\frac{5}{3}\right]} = 6944.44$$

19. The differential equation describing the population growth is

$$\frac{dN}{dt} = 3000 e^{2t/5}$$

We solve the equation to find the function that describes the population.

$$\int dN = \int 3000\, e^{2t/5}\, dt$$

$$N = \frac{3000\, e^{2t/5}}{\dfrac{2}{5}} + K$$

$$N(t) = 7500\, e^{2t/5} + K$$

To find K we use $N(0) = 7500$.

$$7500 = 7500\, e^{0} + K$$

$$K = 0$$

So $N(t) = 7500\, e^{2t/5}$, and when $t = 5$

$$N(5) = 7500\, e^{2 \cdot 5/5} = 7500\, e^{2} = 55{,}417.9$$

There are 55,418 bacteria present.

21. (a) We use $N(0) = 10{,}000$ and $N(t_1) = 20{,}000$ to find the constant of proportionality k.

$$20{,}000 = 10{,}000\, e^{kt_1}$$

$$2 = e^{kt_1}$$

$$\ln 2 = k\, t_1$$

$$k = \frac{\ln 2}{t_1}$$

To find $N(t)$, we use $N(t_1 + 10) = 100{,}000$.

$$100{,}000 = 10{,}000\, e^{\frac{\ln 2}{t_1}(t_1 + 10)}$$

$$10 = e^{\frac{\ln 2}{t_1}(t_1 + 10)}$$

$$\ln 10 = \frac{\ln 2}{t_1}(t_1 + 10) = \ln 2 + \frac{10 \ln 2}{t_1}$$

$$\ln 10 - \ln 2 = \frac{10 \ln 2}{t_1}$$

$$\ln 5 = \frac{10 \ln 2}{t_1}$$

$$t_1 = \frac{10 \ln 2}{\ln 5} \approx 4.30677$$

and $k = \dfrac{\ln 2}{\dfrac{10 \ln 2}{\ln 5}} = \dfrac{\ln 5}{10}$.

So $N(t) = 10{,}000\, e^{\left(\frac{\ln 5}{10}\right) t} = 10{,}000(5^{\,t/10})$

(b) $N(20) = 10{,}000\, e^{\left(\frac{\ln 5}{10}\right)(20)} = 250{,}000$ bacteria

(c) The value of t_1 is 4.30667. So after 4.3 minutes there were 20,000 bacteria present.

23. $\dfrac{dA}{dt} = -\alpha A$ then $A(t) = A_0 e^{-\alpha t}$.

$$\frac{1}{2} = e^{-\alpha t}$$

$$\ln \frac{1}{2} = -\alpha t$$

$$t = -\frac{\ln 0.5}{1.5 \times 10^{-7}} = 4,620,981.2 \text{ years.}$$

25.

(a) The differential equation is $\dfrac{dp}{dx} = k\,p$.

We solve the equation.

$$\frac{dp}{p} = k\,dx$$

$$\ln p = kx + K$$

To find K we use $p(0) = 300$, and find $\ln 300 = 0 + K$ or $K = \ln 300$, and we then find

$$\ln p = kx + \ln 300$$
$$\ln p - \ln 300 = kx$$
$$\ln \frac{p}{300} = kx$$
$$\frac{p}{300} = e^{kx}$$
$$p = 300 e^{kx}$$

To find k we use the second boundary condition $p(200) = 150$.

$$150 = 300 e^{200\,k}$$

$$\frac{1}{2} = e^{200k}$$

$$\ln\left(\frac{1}{2}\right) = 200\,k$$

$$k = \frac{\ln 0.5}{200} \approx -0.0034657$$

So the price-demand equation is $p = 300 e^{\left(\frac{\ln 0.5}{200}\right)x}$.

(b) To sell 300 units, the price p should be

$$p = 300 e^{300\left(\frac{\ln 0.5}{200}\right)} = \$106.07$$

(c) To sell 350 units, the price p should be

$$p = 300e^{350\left(\frac{\ln 0.5}{200}\right)} = \$89.19$$

Chapter 6 Review

TRUE-FALSE ITEMS

1.	True	**3.**	False	**5.**	False
7.	True	**9.**	False		

FILL-IN-THE-BLANKS

1. $F'(x) = f(x)$ **3.** integration by parts   **5.** 0

7. $\displaystyle\int_0^2 \sqrt{x^2 + 1}\, dx$

REVIEW EXERCISES

1. $F(x) = \dfrac{6x^6}{6} + K = x^6 + K$

3. $F(x) = \dfrac{x^4}{4} + \dfrac{x^2}{2} + K$

5. $f(x) = \dfrac{1}{\sqrt{x}} = x^{-1/2}$

$F(x) = \dfrac{x^{1/2}}{\dfrac{1}{2}} + K = 2x^{1/2} + K = 2\sqrt{x} + K$

7. $\displaystyle\int 7\, dx = 7 \int dx = 7x + K$

9. $\displaystyle\int \left(5x^3 + 2\right) dx = \dfrac{5x^4}{4} + 2x + K$

11. $\displaystyle\int \left(x^4 - 3x^2 + 6\right) dx = \dfrac{x^5}{5} - \dfrac{3x^3}{3} + 6x + K = \dfrac{x^5}{5} - x^3 + 6x + K$

13. $\displaystyle\int \dfrac{3}{x}\, dx = 3 \int \dfrac{1}{x}\, dx = 3 \ln |x| + K$

15. We use the method of substitution to evaluate the integral. Let $u = x^2 - 1$, then

$du = 2x\, dx$.

$$\int \frac{2x}{x^2-1}\, dx = \int \frac{1}{u}\, du = \ln|u| + k = \ln\left|x^2 - 1\right| + K$$

17.
$$\int e^{3x}\, dx = \frac{1}{3} e^{3x} + K$$

19. We use the method of substitution to evaluate the integral.
Let $u = x^3 + 3x$. Then $du = \left(3x^2 + 3\right) dx$
$$du = 3\left(x^2 + 1\right) dx$$
$$\frac{1}{3}\, du = \left(x^2 + 1\right) dx$$
$$\int \left(x^3 + 3x\right)^5 \left(x^2 + 1\right) dx = \frac{1}{3}\int u^5\, du = \frac{1}{3} \cdot \frac{u^6}{6} + K = \frac{\left(x^3 + 3x\right)^6}{18} + K$$

21.
$$\int 2x(x-3)\, dx = \int \left(2x^2 - 6x\right) dx = \frac{2x^3}{3} - \frac{6x^2}{2} + K = \frac{2x^3}{3} - 3x^2 + K$$

23. We use the method of substitution to evaluate the integral.
Let $u = 3x^2 + x$. Then $du = (6x+1)\, dx$.
$$\int e^{3x^2 + x}(6x+1)\, dx = \int e^u\, du = e^u + K = e^{3x^2 + x} + K$$

25. We use the method of substitution to evaluate the integral.
Let $u = x - 5$. Then $x = u + 5$ and $du = dx$.
$$\int x\sqrt{x-5}\, dx = \int (u+5)\sqrt{u}\, du = \int \left(u^{3/2} + 5u^{1/2}\right) du = \frac{u^{5/2}}{\frac{5}{2}} + \frac{5u^{3/2}}{\frac{3}{2}} + K$$
$$= \frac{2(x-5)^{5/2}}{5} + \frac{10(x-5)^{3/2}}{3} + K$$

27. We use the method of integration by parts to evaluate the integral.
Choose $u = x$ $dv = e^{4x}\, dx$.

Then $du = dx$ $v = \frac{1}{4} e^{4x}$
$$\int x e^{4x}\, dx = \int u \cdot dv = uv - \int v\, du$$
$$= x \cdot \frac{1}{4} e^{4x} - \int \frac{1}{4} e^{4x}\, dx$$

$$= \frac{xe^{4x}}{4} - \frac{1}{4} \cdot \frac{1}{4} e^{4x} + K = \frac{xe^{4x}}{4} - \frac{e^{4x}}{16} + K$$

29. We use the method of integration by parts to evaluate the integral.

Choose $\quad u = \ln 2x \qquad dv = x^{-2} dx$

Then $\qquad du = \frac{1}{x} dx \qquad v = -x^{-1} = -\frac{1}{x}$

$$\int x^{-2} \ln 2x \, dx = \int u \cdot dv = uv - \int v \, du$$

$$= \ln 2x \cdot \left(-\frac{1}{x}\right) - \int \left(-\frac{1}{x}\right) \cdot \frac{1}{x} dx$$

$$= -\frac{\ln 2x}{x} + \int x^{-2} \, dx$$

$$= -\frac{\ln 2x}{x} - \frac{1}{x} + K$$

$$= -\frac{1}{x}(1 + \ln 2x) + K$$

31.
$$R(x) = \int R'(x) \, dx = \int (5x + 2) \, dx = \frac{5x^2}{2} + 2x + K$$

Since $R(0) = 0$, we have $K = 0$, and the revenue function R is

$$R(x) = \frac{5x^2}{2} + 2x$$

33.
$$C(x) = \int C'(x) \, dx = \int (5x + 120,000) \, dx = \frac{5x^2}{2} + 120,000x + K$$

Since fixed cost is $7500, $C(0) = 7500$ and $K = 7500$. The cost function C is

$$C(x) = \frac{5x^2}{2} + 120,000x + 7500$$

The minimum cost occurs either at the vertex (since C is a quadratic function) or

at $x = 0$. At the vertex $x = -\frac{b}{2a} = -\frac{120,000}{2\left(\frac{5}{2}\right)} = -24,000$ which is negative and so not in

the domain. At $x = 0$, $C(0) = 7500$ and is minimum.

35.

(a) $R(x) = \int R'(x) \, dx = \int (500 - 0.01x) \, dx = 500x - \frac{0.01x^2}{2} + K$

Using the fact that $R(0) = 0$ we solve for K.

$$R(0) = 500(0) - 0.005(0) + K = 0$$

So $K = 0$, and $R(x) = 500x - 0.005x^2$

(b) Since $R(x)$ is a quadratic function, and since $a = 0.005$ is negative, the maximum

value occurs at the x-value of the vertex, if it is in the domain.
$$x = -\frac{b}{2a} = -\frac{500}{2(-0.005)} = 50,000$$
Revenue is maximized when 50,000 televisions are sold.

(c) The maximum revenue that can be obtained is
$$R(50,000) = 500(50,000) - 0.005(50,000^2) = \$\ 12,500,000$$

(d) If sales increase from 35,000 to 40,000 televisions, the revenue will increase by
$$R(40,000) - R(35,000) = 12,000,000 - 11,375,000 = \$\ 625,000$$

37.
$$\int_{-2}^{1}\left(x^2 + 3x - 1\right)dx = \left(\frac{x^3}{3} + \frac{3x^2}{2} - x\right)\Bigg|_{-2}^{1} = \left(\frac{1}{3} + \frac{3}{2} - 1\right) - \left(\frac{(-2)^3}{3} + \frac{3(-2)^2}{2} - (-2)\right)$$
$$= \frac{1}{3} + \frac{3}{2} - 1 + \frac{8}{3} - 6 - 2 = -\frac{9}{2}$$

39.
$$\int_{4}^{9} 8\sqrt{x}\ dx = 8 \int_{4}^{9} x^{1/2}\ dx = 8\left[\left(\frac{x^{3/2}}{\frac{3}{2}}\right)\Bigg|_{4}^{9}\right] = 8\left[\frac{2(9^{3/2})}{3} - 2\left(\frac{4^{3/2}}{3}\right)\right] = \frac{16}{3}[27 - 8] = \frac{304}{3}$$

41.
$$\int_{0}^{1}\left(e^x - e^{-x}\right)dx = \left(e^x + e^{-x}\right)\Big|_{0}^{1} = \left(e^1 + e^{-1}\right) - \left(e^0 + e^0\right) = e + e^{-1} - 2 = e + \frac{1}{e} - 2$$

43. We use the method of substitution to evaluate this integral.

Let $u = 3x + 2$. Then $du = 3dx$ and $\frac{1}{3}du = dx$.

When $x = 0$, $u = 3(0) + 2 = 2$, and when $x = 4$, $u = 3(4) + 2 = 14$.

$$\int_{0}^{4}\frac{dx}{(3x+2)^2} = \frac{1}{3}\int_{2}^{14}\frac{1}{u^2}\ du = \frac{1}{3}\int_{2}^{14}u^{-2}\ du = \frac{1}{3}\left[\frac{u^{-1}}{-1}\Bigg|_{2}^{14}\right]$$
$$= -\frac{1}{3}\left[\frac{1}{14} - \frac{1}{2}\right] = -\frac{1}{3}\left[\frac{1-7}{14}\right] = \frac{2}{14} = \frac{1}{7}$$

45.
$$\int_{-2}^{2}e^{3x}\ dx = \frac{1}{3}e^{3x}\Big|_{-2}^{2} = \frac{1}{3}\left[e^6 - e^{-6}\right] \approx 134.475$$

47. We use integration by parts to evaluate this integral by choosing
$$u = x + 2 \qquad\qquad dv = e^{-x}\ dx$$
$$du = dx \qquad\qquad v = -e^{-x}$$

$$\int_0^1 u\ dv = \int_0^1 (x+2)e^{-x}dx = (x+2)\left(-e^{-x}\right)\Big|_0^1 - \int_0^1 \left(-e^{-x}\right)\ dx$$

$$= -(x+2)\ e^{-x}\Big|_0^1 + \int_0^1 e^{-x}\ dx$$

$$= -(x+2)\ e^{-x}\Big|_0^1 - e^{-x}\Big|_0^1$$

$$= \left[-3e^{-1}+2e^0\right]-\left[e^{-1}-e^0\right]$$

$$= -3e^{-1}+2-e^{-1}+1 = 3-4e^{-1} = 3 - \frac{4}{e}$$

49. $$\int_0^9 f(x)\ dx = \int_0^5 f(x)\ dx + \int_5^9 f(x)\ dx$$

$$= 3 + (-2) = 1 \qquad\qquad \text{Property 5} \quad \int_a^b f(x)\ dx = \int_a^c f(x)\ dx + \int_c^b f(x)\ dx$$

51. $$\int_9^5 g(x)\ dx = -\int_5^9 g(x)\ dx = -10 \qquad\qquad \text{Property 3} \quad \int_a^b f(x)\ dx = -\int_b^a f(x)\ dx$$

53. $$A = \int_{-1}^2 (x^2+4)\ dx = \left(\frac{x^3}{3}+4x\right)\Bigg|_{-1}^2$$

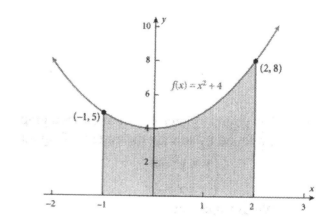

$$= \left[\frac{2^3}{3}+4\cdot 2\right]-\left[\frac{(-1)^3}{3}+4\cdot(-1)\right]$$

$$= \frac{8}{3}+8+\frac{1}{3}+4$$

$$= 15$$

55. $$A = \int_0^1 (e^x+x)\ dx = \left(e^x+\frac{x^2}{2}\right)\Bigg|_0^1$$

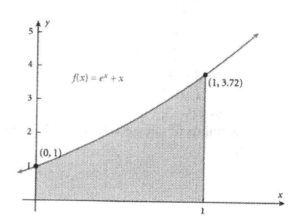

$$= \left[e^1+\frac{1^2}{2}\right]-\left[e^0+0\right]$$

$$= e+\frac{1}{2}-1$$

$$= e-\frac{1}{2}$$

57.

$$A = -\int_0^2 \left(x^2 - x - 2\right)dx + \int_2^3 \left(x^2 - x - 2\right)dx = -\left(\frac{x^3}{3} - \frac{x^2}{2} - 2x\right)\Bigg|_0^2 + \left(\frac{x^3}{3} - \frac{x^2}{2} - 2x\right)\Bigg|_2^3$$

$$= -\left\{\left[\frac{2^3}{3} - \frac{2^2}{2} - 2(2)\right] - [0]\right\} + \left\{\left[\frac{3^3}{3} - \frac{3^2}{2} - 2(3)\right] - \left[\frac{2^3}{3} - \frac{2^2}{2} - 2(2)\right]\right\}$$

$$= \left[9 - \frac{9}{2} - 6\right] - 2\left[\frac{8}{3} - \frac{4}{2} - 4\right] = \left[3 - \frac{9}{2}\right] - 2\left[\frac{8}{3} - 6\right]$$

$$= 3 - \frac{9}{2} - \frac{16}{3} + 12 = \frac{31}{6}$$

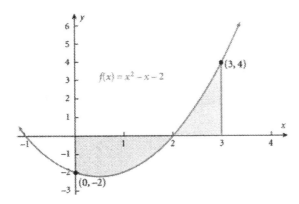

59. The points of intersection of the two graphs are found by solving the system of equations

$$\begin{cases} y = x^2 - 4 \\ x + y = 2 \end{cases}$$

We use substitution.

$$x + \left(x^2 - 4\right) = 2$$

$$x^2 + x - 6 = 0$$

$$(x+3)(x-2) = 0$$

$$x + 3 = 0 \quad \text{or} \quad x - 2 = 0$$

$$x = -3 \quad \text{or} \qquad x = 2$$

When $x = -3$, $y = 5$, and when $x = 2$, $y = 0$. So the points of intersection are $(-3, 5)$ and $(2, 0)$.

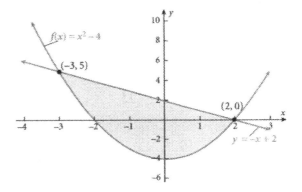

$$A = \int_{-3}^{2}\left[(2-x)-(x^2-4)\right]dx = \int_{-3}^{2}\left(-x^2-x+6\right)dx = \left(-\frac{x^3}{3}-\frac{x^2}{2}+6x\right)\Bigg|_{-3}^{2}$$

$$= \left[-\frac{2^3}{3}-\frac{2^2}{2}+6(2)\right]-\left[-\frac{(-3)^3}{3}-\frac{(-3)^2}{2}+6(-3)\right]$$

$$= -\frac{8}{3}-2+12-9+\frac{9}{2}+18 = \frac{125}{6}$$

61. The points of intersection of the two graphs are found by solving the system of equations

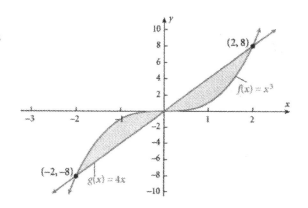

$$\begin{cases} f(x)=x^3 \\ g(x)=4x \end{cases}$$

$$x^3 = 4x$$

$$x^3 - 4x = 0$$

$$x\left(x^2-4\right)=0$$

$$x = 0 \quad \text{or} \quad x^2 = 4$$

$$x = \pm 2$$

The points of intersection are $(0, 0)$, $(-2, -8)$, and $(2, 8)$.

$$A = \int_{-2}^{0}\left[f(x)-g(x)\right]dx + \int_{0}^{2}\left[g(x)-f(x)\right]dx = \int_{-2}^{0}\left[x^3-4x\right]dx + \int_{0}^{2}\left[4x-x^3\right]dx$$

$$= \left(\frac{x^4}{4}-\frac{4x^2}{2}\right)\Bigg|_{-2}^{0} + \left(\frac{4x^2}{2}-\frac{x^4}{4}\right)\Bigg|_{0}^{2}$$

$$= \left[0-\left(\frac{(-2)^4}{4}-2(-2)^2\right)\right]+\left[\left(2(-2)^2-\frac{(-2)^4}{4}\right)-0\right]$$

$$= -4+8+8-4 = 8$$

63.

$$\int_{2000}^{2500} P'(x)\,dx = \int_{2000}^{2500}(9-0.004x)\,dx = \left(9x-\frac{0.004x^2}{2}\right)\Bigg|_{2000}^{2500}$$

$$= \left[9(2500)-0.002\left(2500^2\right)\right]-\left[9(2000)-0.002(2000)^2\right]=0$$

There is no change in monthly profit obtained by increasing production from 2000 to 2500 pairs of jeans.

65. (a) The time t_{max} of optimal termination is found when

$$R'(x)=C'(x)$$

$$-10t = 2t-12$$

$$-12t = -12$$

$$t_{max} = 1$$

To maximize profit the owner should keep the machine for 1 time unit.

(b) The total profit that the machine will generate in the unit of time is

$$P(t_{max}) = \int_0^1 \left[R'(t) - C'(t) \right] dt = \int_0^1 \left[-10t - (2t - 12) \right] dt$$

$$= \left(\frac{-12t^2}{2} + 12t \right) \Big|_0^1 = [-6 + 12] - [0] = 6 \text{ monetary units.}$$

67. (a)

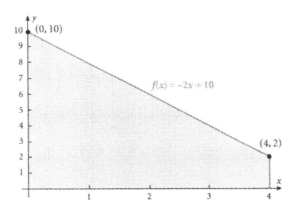

Divide the interval into 4 subintervals of length $\Delta x = 1$.

$$[0, 1] \quad [1, 2] \quad [2, 3] \quad [3, 4]$$

Evaluate f at the endpoints.

$$f(0) = 10$$

$$f(1) = -2 + 10 = 8$$

$$f(2) = -4 + 10 = 6$$

$$f(3) = -6 + 10 = 4$$

$$f(4) = -8 + 10 = 2$$

(b) Using the left endpoints, we have

$$A \approx f(0) \cdot 1 + f(1) \cdot 1 + f(2) \cdot 1 + f(3) \cdot 1$$

$$\approx (10 + 8 + 6 + 4) \cdot 1 = 28$$

(c) Using right endpoints, we have

$$A \approx f(1) \cdot 1 + f(2) \cdot 1 + f(3) \cdot 1 + f(4) \cdot 1$$

$$\approx (8 + 6 + 4 + 2) \cdot 1 = 20$$

(d) Divide the interval into 8 subintervals of length $\Delta x = \frac{1}{2}$.

$$\left[0, \frac{1}{2}\right] \quad \left[\frac{1}{2}, 1\right] \quad \left[1, \frac{3}{2}\right] \quad \left[\frac{3}{2}, 2\right] \quad \left[2, \frac{5}{2}\right] \quad \left[\frac{5}{2}, 3\right] \quad \left[3, \frac{7}{2}\right] \quad \left[\frac{7}{2}, 4\right]$$

Evaluate f at the new endpoints.

$$f\left(\frac{1}{2}\right) = -2\left(\frac{1}{2}\right) + 10 = 9 \qquad f\left(\frac{3}{2}\right) = -2\left(\frac{3}{2}\right) + 10 = 7$$

$$f\left(\frac{5}{2}\right) = -2\left(\frac{5}{2}\right) + 10 = 5 \qquad f\left(\frac{7}{2}\right) = -2\left(\frac{7}{2}\right) + 10 = 3$$

Using the left endpoints, we have

$$A \approx f(0) \cdot \frac{1}{2} + f\left(\frac{1}{2}\right) \cdot \frac{1}{2} + f(1) \cdot \frac{1}{2} + f\left(\frac{3}{2}\right) \cdot \frac{1}{2} + f(2) \cdot \frac{1}{2} + f\left(\frac{5}{2}\right) \cdot \frac{1}{2} + f(3) \cdot \frac{1}{2} + f\left(\frac{7}{2}\right) \cdot \frac{1}{2}$$

$$\approx (10+9+8+7+6+5+4+3) \cdot \frac{1}{2} = 26$$

(e) Using the right endpoints, we have

$$A \approx f\left(\frac{1}{2}\right) \cdot \frac{1}{2} + f(1) \cdot \frac{1}{2} + f\left(\frac{3}{2}\right) \cdot \frac{1}{2} + f(2) \cdot \frac{1}{2} + f\left(\frac{5}{2}\right) \cdot \frac{1}{2} + f(3) \cdot \frac{1}{2} + f\left(\frac{7}{2}\right) \cdot \frac{1}{2} + f(4) \cdot \frac{1}{2}$$

$$\approx (9+8+7+6+5+4+3+2) \cdot \frac{1}{2} = 22$$

(f) $A = \int_0^4 f(x)\, dx = \int_0^4 (-2x+10)\, dx$

(g) $\int_0^4 (-2x+10)\, dx = \left(\frac{-2x^2}{2} + 10x \right) \Big|_0^4 = \left[-4^2 + 10(4) \right] - [0] = 24$

69. $\int_1^{10} x \ln xx\, dx \approx 90.38$

```
fnInt(Xln(X),X,1
,10)
        90.37925465
```

71. The general solution to the differential equation $\dfrac{dy}{dx} = x^2 + 5x - 10$ is

$$y = \int (x^2 + 5x - 10)\, dx$$

$$y = \frac{x^3}{3} + \frac{5x^2}{2} - 10x + K$$

We use $x = 0$ and $y = 1$ to find K.

$$1 = 0 + 0 - 0 + K$$
$$1 = K$$

The particular solution to the differential equation is

$$y = \frac{x^3}{3} + \frac{5x^2}{2} - 10x + 1$$

73. The general solution to the differential equation $\dfrac{dy}{dx} = e^{2x} - x$ is

$$y = \int (e^{2x} - x)\, dx$$

$$y = \frac{1}{2} e^{2x} - \frac{x^2}{2} + K$$

We use $x = 0$ and $y = 3$ to find K.

$$3 = \frac{1}{2}e^0 - 0 + K$$

$$\frac{5}{2} = K$$

The particular solution to the differential equation is

$$y = \frac{1}{2}e^{2x} - \frac{x^2}{2} + \frac{5}{2}$$

75. The general solution to the differential equation $\frac{dy}{dx} = 10y$ is

$$\frac{dy}{y} = 10\,dx$$

$$\int \frac{dy}{y} = \int 10\,dx$$

$$\ln|y| = 10x + K$$

$$y = e^{10x} + K$$

We use $x = 0$ and $y = 1$ to find K.

$$\ln 1 = 0 + K$$

$$0 = K$$

The particular solution to the differential equation is

$$y = e^{10x}$$

77. $N(0) = 2000$ $\qquad$ $N(2) = 3 \cdot 2000 = 6000$

$$N(t) = N(0)\, e^{kt} = 2000\, e^{kt}$$

We use $N(2) = 6000$ to find k.

$$6000 = 2000\, e^{kt}$$

$$3 = e^{k \cdot 2}$$

$$\ln 3 = 2k$$

$$k = \frac{\ln 3}{2}$$

So in $4\frac{1}{2}$ hours $t = \frac{9}{2}$, and there will be

$$N\!\left(\frac{9}{2}\right) = 2000\, e^{\left(\frac{\ln 3}{2}\right)\left(\frac{9}{2}\right)} = 23{,}689 \ \text{bacteria.}$$

79.
If $A(0) = 1$, then $A(5600) = \frac{1}{2}(1) = 0.5$, and $A(t) = 0.4(1) = 0.4$.

$$A(t) = A(0)e^{kt} = e^{kt}$$

We use $A(5600)$ to find k.

$$0.5 = e^{5600\,k}$$
$$\ln\ 0.5 = 5600\ k$$
$$\frac{\ln 0.5}{5600} = k$$

We now solve for t, the age of the bones.

$$0.4 = e^{\left(\frac{\ln 0.5}{5600}\right)t}$$
$$\ln 0.4 = \frac{\ln 0.5}{5600}\,t$$
$$t = \frac{5600\ \ln 0.4}{\ln 0.5} \approx 7402.8$$

The bones are 7403 years old.

81. $E'(t) = 0.02t^2 + t \qquad E(0) = 5$

$$E(t) = \int E'(t)\ dt = \int \left(0.02t^2 + t\right) dt$$
$$= \frac{0.02t^3}{3} + \frac{t^2}{2} + K = \ = \frac{t^3}{150} + \frac{t^2}{2} + 5$$

Since $E(0) = 5$, then $K = 5$, and

$$E(5) = \frac{0.02(5)^3}{3} + \frac{(5)^2}{2} + 5 = 18.33\ \text{million dollars.}$$

83.

$$\int_8^{18} f(x)\ dx = \int_8^{18} 700x^{-0.1}\ dx = \left(700\ \cdot\ \frac{x^{0.9}}{0.9}\right)\bigg|_8^{18}$$
$$= \frac{700}{0.9}\left[18^{0.9} - 8^{0.9}\right] = 5431.77$$

Margo should allow an additional 5432 labor hours to produce the 500 additional tennis ball servers.

85. (a) At market equilibrium $D(x) = S(x)$.

$$12 - \frac{x^*}{50} = \frac{x^*}{20} + 5$$
$$1200 - 2x^* = 5x^* + 500$$
$$700 = 7x^*$$
$$x^* = 100$$

The price p^* at $x^* = 100$ is $D(100) = 12 - \dfrac{100}{50} = 12 - 2 = 10$.

So at market equilibrium 100 units will be sold at $10 each.

(b)   The consumer's surplus $CS = \int_0^{x^*} D(x)\,dx - p^* x^*$

$$CS = \int_0^{100} \left(12 - \frac{x}{50}\right) dx - 10 \cdot 100 = \left(12x - \frac{x^2}{100}\right)\Big|_0^{100} - 1000$$

$$= 12(100) - \frac{100^2}{100} - 1000 = 1200 - 100 - 1000 = 100$$

The supplier's surplus $PS = p^* s^* - \int_0^{x^*} S(x)\,dx$.

$$PS = 10 \cdot 100 - \int_0^{x^*} \left(\frac{x}{20} + 5\right) dx = 1000 - \left[\left(\frac{x^2}{40} + 5x\right)\Big|_0^{100}\right]$$

$$= 1000 - \left[\frac{100^2}{40} + 5(100)\right]$$

$$= 1000 - [250 + 500] = \$250$$

(c)

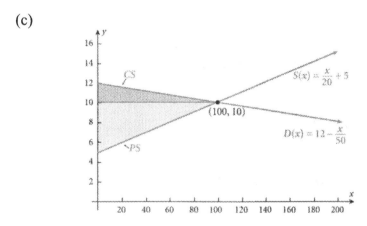

87.
   (a) $$\frac{dp}{dx} = k\,p$$

$$\frac{dp}{p} = k\,dx$$

$$\ln p = k\,x + k$$

We find k by using the facts that $x = 0$ when $p = 800$.

$$\ln 800 = k \cdot 0 + k$$

So, $\ln p = kx + \ln 800$

$$\ln p - \ln 800 = kx$$

$$\ln \frac{p}{800} = kx$$

$$\frac{p}{800} = e^{kx}$$

$$p = 800 e^{kx}$$

We use the fact that $x = 80$ when $p = 600$ to find k.

$$600 = 800\, e^{80k}$$
$$0.75 = e^{80\,k}$$
$$\ln 0.75 = 80\,k$$
$$k = \frac{\ln 0.75}{80} \approx -0.003596$$

The demand equation is $p(x) = 800\, e^{\left(\frac{\ln 0.75}{80}\right)x}$

(b) If the manufacturer produces $x = 200$ units, the price will be
$$p(200) = 800\, e^{200\left(\frac{\ln 0.75}{80}\right)} = \$389.71$$

(c) If the manufacturer produces $x = 250$ units, the price will be
$$p(250) = 800\, e^{250\left(\frac{\ln 0.75}{80}\right)} = \$325.58$$

CHAPTER 6 PROJECT

1.
$$A = \int_0^1 x\, dx = \left(\frac{x^2}{2}\right)\Bigg|_0^1 = \frac{1}{2}$$

3.
$$G = \frac{\text{area of the top piece}}{\text{area of the triangle}} = \frac{\text{area of the top piece}}{\dfrac{1}{2}}$$

$$= \frac{\text{area of the triangle - area under Lorenz Curve}}{\dfrac{1}{2}}$$

$$= \frac{\dfrac{1}{2} - \int_0^1 L(x)\, dx}{\dfrac{1}{2}}$$

$$= 1 - 2\int_0^1 L(x)\, dx$$

5. Answers will vary.

7. The Gini coefficient for the US economy in 1993 is
$$G = 1 - 2\int_0^1 \left(0.442x^2 + 5.8x^3 - 23.71x^4 + 31.036x^5 + 5.71x^6 - 38.842x^7 + 20.564x^8\right) dx$$

$$= 1 - 2\left(\frac{0.442x^3}{3} + \frac{5.8x^4}{4} - \frac{23.71x^5}{5} + \frac{31.036x^6}{6} + \frac{5.71x^7}{7} - \frac{38.842x^8}{8} + \frac{20.564x^9}{9}\right)\Big|_0^1$$

$$= 1 - 2\left[\left(\frac{0.442}{3} + \frac{5.8}{4} - \frac{23.71}{5} + \frac{31.036}{6} + \frac{5.71}{7} - \frac{38.842}{8} + \frac{20.564}{9}\right) - 0\right]$$

$$= 0.453$$

MATHEMATICAL QUESTIONS FROM PROFESSIONAL EXAMS

1. (b)

Evaluate the integral using integration by parts. Choose

$$u = \ln x \qquad dv = \frac{1}{x}\,dx$$

$$dv = \frac{1}{x}\,dx \qquad v = \ln x$$

$$\int_1^e \frac{1}{x}\ln x\,dx = (\ln x)^2\Big|_1^e - \int_1^e \frac{1}{x}\ln x\,dx$$

$$2\int_1^e \frac{1}{x}\ln x\,dx = (\ln x)^2\Big|_1^e$$

$$\int_1^e \frac{1}{x}\ln x\,dx = \frac{1}{2}\left[(\ln x)^2\Big|_1^e\right] = \frac{1}{2}\left[1^2 - 0\right] = \frac{1}{2}$$

3. (c)

If $\int_1^b f(x)\,dx = b^2 e^b - e$, then $\int f(x)\,dx = x^2 e^x + K$, and the derivative of $x^2 e^x$ is $f(x)$.

$$\frac{d}{dx}\left(x^2 e^x\right) = x^2 e^x + 2x e^x = f(x).$$

5. (c)

Let $t = 0$ denote 4 years ago. Then $P(0) = 25,000$ and $P(4) = 36,000$.

$$P(t) = P(0)e^{kt}$$

$$P(t) = 25,000 e^{kt}$$

We use $P(4)$ to find k.

$$36,000 = 25,000 e^{4k}$$

$$1.44 = e^{4k}$$

$$\ln 1.44 = 4k$$

$$k = \frac{\ln 1.44}{4}$$

Sis years from now $t = 10$, and

$$P(10) = 25,000 e^{10\left(\frac{\ln 1.44}{4}\right)} = 62,208$$

Chapter 7
Other Applications and Extensions of the Integral

7.1 Improper Integrals

1.

$\int_0^{\infty} x^2 \, dx$ is an improper integral because the upper limit of integration is not finite.

3.

$\int_0^1 \frac{1}{x} \, dx$ is an improper integral because the function is not continuous at $x = 0$.

5.

$\int_1^2 \frac{dx}{x-1}$ is an improper integral because the function is not continuous at $x = 1$.

7.

$$\int_1^{\infty} e^{-4x} \, dx = \lim_{b \to \infty} \int_1^b e^{-4x} \, dx = \lim_{b \to \infty} \left[-\frac{1}{4} e^{-4x} \Big|_1^b \right]$$

$$= -\frac{1}{4} \lim_{b \to \infty} \left(e^{-4b} \right) + \frac{1}{4} \lim_{b \to \infty} \left(e^{-4} \right) = -\frac{1}{4} \lim_{b \to \infty} \left(\frac{1}{e^{4b}} \right) + \frac{1}{4} e^{-4}$$

$$= \frac{1}{4} e^{-4}$$

9.

$$\int_0^{\infty} \sqrt{x} \, dx = \lim_{b \to \infty} \int_0^{\infty} x^{1/2} \, dx = \lim_{b \to \infty} \left[\frac{x^{3/2}}{\frac{3}{2}} \Big|_0^b \right]$$

$$= \lim_{b \to \infty} \left[\frac{2b^{3/2}}{3} - \frac{2(0^{3/2})}{3} \right] = \frac{2}{3} \lim_{b \to \infty} b^{2/3} = \infty$$

Since the limit is infinite, the integral has no value.

11.

$$\int_{-1}^0 \frac{1}{\sqrt[5]{x}} \, dx = \lim_{b \to 0^-} \int_{-1}^b x^{-1/5} \, dx = \lim_{b \to 0^-} \left[\frac{x^{4/5}}{\frac{4}{5}} \Big|_{-1}^b \right]$$

$$= \lim_{b \to 0^-} \left[\frac{5b^{4/5}}{4} - \frac{5(-1)^{4/5}}{4} \right]$$

$$= \lim_{b \to 0^-} \frac{5b^{4/5}}{4} - \lim_{b \to 0^-} \frac{5}{4}$$

$$= 0 - \frac{5}{4} = -\frac{5}{4}$$

13.

$$\int_0^1 \frac{1}{x}\,dx = \lim_{a \to 0^+} \int_a^1 \frac{1}{x}\,dx = \lim_{a \to 0^+} \left[\ln x \Big|_a^1\right]$$

$$= \lim_{a \to 0^+} \left[\ln 1 - \ln a\right] = -\lim_{a \to 0^+} \left[\ln a\right] = \infty$$

Since the limit is infinite, the integral has no value.

15.

$$A = \int_0^1 f(x)\,dx = \int_0^1 \frac{1}{\sqrt{x}}\,dx = \lim_{a \to 0^+} \int_a^1 x^{-1/2}\,dx = \lim_{a \to 0^+} \left[\frac{x^{1/2}}{\frac{1}{2}}\Big|_a^1\right]$$

$$= 2\left[\lim_{a \to 0^+} \left(1 - a^{1/2}\right)\right] = 2(1) = 2$$

17.

$$V(t) = \int_0^\infty 5124 e^{-0.05t}\,dt = \lim_{t \to b} \int_0^b 5124 e^{-0.05t}\,dt = 5124 \lim_{t \to b} \frac{e^{-0.05t}}{-0.05}\Big|_0^b$$

$$= -\frac{5124}{0.05} \lim_{t \to b} \left[e^{-0.05\,b} - e^0\right]$$

$$= -\frac{5124}{0.05} \lim_{t \to b} \left[\frac{1}{e^{0.05\,b}} - 1\right]$$

$$= \frac{5124}{0.05} = \$102,480$$

19. (a) Answers will vary.

(b) Total reaction is

$$\int_0^\infty t e^{-t^2}\,dt = \lim_{b \to \infty} \int_0^b t e^{-t^2}\,dt$$

We use the method of substitution to evaluate the integral.

Let $u = -t^2$. Then $du = -2t\,dt$, and $-\frac{1}{2}\,du = t\,dt$.

When $t = 0$, $u = 0$, and when $t = b$, $u = -b^2$.

$$\int_0^\infty t e^{-t^2}\,dt = \lim_{b \to \infty}\left[-\frac{1}{2}\int_0^{-b^2} e^u\,du\right] = -\frac{1}{2}\lim_{b \to \infty}\left[e^u\Big|_0^{-b^2}\right]$$

$$= -\frac{1}{2}\lim_{b \to \infty}\left[e^{-b^2} - e^0\right] = -\frac{1}{2}[0 - 1] = \frac{1}{2}$$

21. (a) $\displaystyle \int_{-1}^1 \frac{1}{x^2}\,dx = \int_{-1}^0 x^{-2}\,dx + \int_0^1 x^{-2}\,dx = \lim_{c \to 0^-} \int_{-1}^c x^{-2}\,dx + \lim_{c \to 0^+} \int_c^1 x^{-2}\,dx$

$$= \lim_{c \to 0^-} \left[-x^{-1} \Big|_{-1}^{c} \right] + \lim_{c \to 0^+} \left[-x^{-1} \Big|_{c}^{1} \right]$$

$$= \lim_{c \to 0^-} \left[-\frac{1}{c} + 1 \right] + \lim_{c \to 0^+} \left[-1 + \frac{1}{c} \right] = \infty$$

Since the limit is infinite, the integral has no value.

(b) $\displaystyle \int_0^4 \frac{x\,dx}{\sqrt[3]{x^2 - 4}} = \int_0^2 \frac{x\,dx}{\sqrt[3]{x^2 - 4}} + \int_2^4 \frac{x\,dx}{\sqrt[3]{x^2 - 4}} = \lim_{c \to 2^-} \int_0^c \frac{x\,dx}{\sqrt[3]{x^2 - 4}} + \lim_{c \to 2^+} \int_c^4 \frac{x\,dx}{\sqrt[3]{x^2 - 4}}$

We use the method of substitution to evaluate the integrals.

Let $u = x^2 - 4$. Then $du = 2x\,dx$ and $\dfrac{1}{2}\,du = x\,dx$.

When $x = 0$, $u = -4$; when $x = 2$, $u = 0$; and when $x = 4$, $u = 12$.

$$\int_0^4 \frac{x\,dx}{\sqrt[3]{x^2 - 4}} = \lim_{c \to 2^-} \left[\frac{1}{2} \int_{-4}^{c^2 - 4} u^{-1/3}\,du \right] + \lim_{c \to 2^+} \left[\frac{1}{2} \int_{c^2 - 4}^{12} u^{-1/3}\,du \right]$$

$$= \frac{1}{2} \lim_{c \to 2^-} \left[\frac{3u^{2/3}}{2} \Big|_{-4}^{c^2 - 4} \right] + \frac{1}{2} \lim_{c \to 2^+} \left[\frac{3u^{2/3}}{2} \Big|_{c^2 - 4}^{12} \right]$$

$$= \frac{3}{4} \left\{ \lim_{c \to 2^-} \left[\left(c^2 - 4 \right)^{2/3} - \left(-4 \right)^{2/3} \right] + \lim_{c \to 2^+} \left[12^{2/3} - \left(c^2 - 4 \right)^{2/3} \right] \right\}$$

$$= \frac{3}{4} \left\{ \left[0^{2/3} - 4^{2/3} \right] + \left[12^{2/3} - 0 \right] \right\}$$

$$= \frac{3}{4} \left[-4^{2/3} + 12^{2/3} \right] \approx 2.041$$

7.2 Average Value of a Function

1.

$$AV = \frac{1}{1 - 0} \int_0^1 x^2\,dx = \frac{x^3}{3} \Big|_0^1 = \frac{1}{3} - 0 = \frac{1}{3}$$

3.

$$AV = \frac{1}{1 - (-1)} \int_{-1}^1 (1 - x^2)\,dx = \frac{1}{2} \left[\left(x - \frac{x^3}{3} \right) \Big|_{-1}^1 \right]$$

$$= \frac{1}{2} \left[\left(1 - \frac{1}{3} \right) - \left(-1 - \frac{(-1)^3}{3} \right) \right]$$

$$= \frac{1}{2} \left[1 - \frac{1}{3} + 1 - \frac{1}{3} \right] = \frac{1}{2} \left[\frac{4}{3} \right] = \frac{2}{3}$$

5.

$$AV = \frac{1}{5-1} \int_1^5 3x \, dx = \frac{1}{4}\left[\frac{3x^2}{2} \Big|_1^5 \right] = \frac{1}{4}\left[\frac{3(5^2)}{2} - \frac{3(1^2)}{2} \right] = \frac{1}{4}\left[\frac{75}{2} - \frac{3}{2} \right] = \frac{72}{8} = 9$$

7.

$$AV = \frac{1}{2-(-2)} \int_{-2}^2 \left(-5x^4 + 4x - 10 \right) dx = \frac{1}{4}\left[\left(\frac{-5x^5}{5} + \frac{4x^2}{2} - 10x \right) \Big|_{-2}^2 \right]$$

$$= \frac{1}{4}\left[\left(-2^5 + 2^3 - 10 \cdot 2 \right) - \left(-(-2)^5 + 2(-2)^2 - 10(-2) \right) \right]$$

$$= \frac{1}{4}\left[-32 + 8 - 20 - 32 - 8 - 20 \right] = \frac{1}{4}\left[-104 \right] = -26$$

9.

$$AV = \frac{1}{1-0} \int_0^1 e^x \, dx = e^x \Big|_0^1 = e^1 - e^0 = e - 1$$

11. The average value of the population during the next 20 years is

$$AV = \frac{1}{20-0} \int_0^{20} \left(6 \cdot 10^9 \right) e^{0.03t} \, dt = \frac{6 \cdot 10^9}{20} \int_0^{20} e^{0.03t} \, dt = 3 \cdot 10^8 \, \frac{e^{0.03t}}{0.03} \Big|_0^{20}$$

$$= 3 \cdot 10^8 \cdot \frac{\left(e^{0.6} - 1 \right)}{0.03} \approx 8.22 \cdot 10^9$$

13. The average temperature AT of the rod is

$$AT = \frac{1}{3-0} \int_0^3 25x \, dx = \frac{1}{3}\left[\frac{25x^2}{2} \Big|_0^3 \right] = \frac{1}{3}\left[\frac{25 \cdot 3^2}{2} - 0 \right] = \frac{75}{2} = 37.5°\, C.$$

15. If the car is accelerating at a rate of 3 meters per second per second, its velocity is

$$v = \int 3 \, dt = 3t + K$$

Since at time $t = 0$, the car is at rest, we have $v(0) = 0$. We use this condition to solve for K.

$$v(0) = 3(0) + K = 0$$
$$K = 0$$

So the average speed during the first 8 seconds of acceleration is

$$AV = \frac{1}{8-0} \int_0^8 3t \, dt = \frac{1}{8}\left[\frac{3t^2}{2} \Big|_0^8 \right] = \frac{1}{8}\left[\frac{3 \cdot 8^2}{2} - 0 \right] = 12 \text{ meters per}$$

second.

17.

$$AR = \frac{1}{6-1} \int_1^6 \left(-4.43x^3 + 46.17x^2 - 132.5x + 290 \right) dx$$

$$= \frac{1}{5}\left[\left(\frac{-4.43x^4}{4} + \frac{46.17x^3}{3} - \frac{132.5x^2}{2} + 290x \right) \Big|_1^6 \right]$$

$$= \frac{1}{5}\left[\left(\frac{-4.43\left(6^4\right)}{4}+\frac{46.17\left(6^3\right)}{3}-\frac{132.5\left(6^2\right)}{2}+290(6)\right)-\left(\frac{-4.43}{4}+\frac{46.17}{3}-\frac{132.5}{2}+290\right)\right]$$

$$= \frac{1}{5}\left[1243.92-238.0325\right]=201.1775$$

The average annual revenue of Exxon-Mobil Corporation between 1997 and 2002 is 201.1775 billion dollars.

19.

$$AR = \frac{1}{90-0}\int_0^{90}\left(-0.000414x+0.206748\right)dx$$

$$= \frac{1}{90}\left[\left(-0.000414\frac{x^2}{2}+0.206748x\right)\Big|_0^{90}\right]$$

$$= \frac{1}{90}\left[\left(-0.000414\left(\frac{90^2}{2}\right)+0.206748(90)\right)-0\right]$$

$$= -0.000414(45)+0.206748(1)=0.188118$$

There was an average of 0.188 inches of rain per day during the first 90 days of the year.

7.3 Continuous Probability Functions

1. We show the two conditions are satisfied.

Condition 1: $f(x)=\frac{1}{2}>0$ everywhere.

Condition 2: $\int_0^2 f(x)\,dx=1$

$$\int_0^2 \frac{1}{2}\,dx=\frac{1}{2}x\Big|_0^2=\frac{1}{2}(2-0)=1$$

So f is a probability density function.

3. We show the two conditions are satisfied.

Condition 1: $f(x)=2x\geq 0$ when $x\geq 0$. So f is nonnegative on the interval $[0, 1]$.

Condition 2: $\int_0^1 f(x)\,dx=1$

$$\int_0^1 2x\,dx=\frac{2x^2}{2}\Big|_0^1=1^2-0=1$$

So f is a probability density function.

5. We show the two conditions are satisfied.

Condition 1: $f(x) = \dfrac{3}{250}\left(10x - x^2\right)$. We find where $f = 0$, use the points to separate the number line into 3 parts and test a point from each part for its sign.

$$\frac{3}{250}\left(10x - x^2\right) = 0$$
$$10x - x^2 = 0$$
$$x(10 - x) = 0$$
$$x = 0 \quad \text{or} \quad 10 - x = 0$$
$$x = 10$$

When $x = -1, f(-1) = -0.132$
When $x = 1, f(1) = 0.108$
When $x = 11, f(11) = -0.132$
We conclude that f is positive on the interval $(0, 10)$. So f is nonnegative on the interval $[0, 5]$.

Condition 2: $\displaystyle\int_0^5 f(x)\,dx = 1$

$$\int_0^5 \frac{3}{250}\left(10x - x^2\right)dx = \frac{3}{250}\left[\left(\frac{10x^2}{2} - \frac{x^3}{3}\right)\bigg|_0^5\right]$$

$$= \frac{3}{250}\left[\left(\frac{250}{2} - \frac{125}{3}\right) - (0)\right] = \frac{3}{250}\left[\frac{750}{6} - \frac{250}{6}\right]$$

$$= \frac{3}{250}\cdot\frac{500}{6} = 1$$

So f is a probability density function.

7. We show the two conditions are satisfied.

Condition 1: $f(x) = \dfrac{1}{x} > 0$ when $x > 0$. So f is nonnegative on the interval $[1, e]$.

Condition 2: $\displaystyle\int_1^e f(x)\,dx = 1$

$$\int_1^e \frac{1}{x}\,dx = \ln x\,\Big|_1^e = \ln e - \ln 1 = 1 - 0 = 1$$

So f is a probability density function.

9. $\displaystyle\int_0^3 k\,dx = kx\,\Big|_0^3 = 3k - 0 = 1.$ So $k = \dfrac{1}{3}$.

11. $\displaystyle\int_0^2 kx\,dx = k\cdot\frac{x^2}{2}\,\bigg|_0^2 = k(2 - 0) = 1.$ So $k = \dfrac{1}{2}$.

13.

$$\int_0^5 k\left(10x - x^2\right) dx = k \int_0^5 \left(10x - x^2\right) dx = k\left[\left(\frac{10x^2}{2} - \frac{x^3}{3}\right)\Big|_0^5\right] = k\left[125 - \frac{125}{3}\right] = \frac{250}{3} k = 1$$

So $k = \dfrac{3}{250}$.

15.

$$\int_1^2 \frac{k}{x} dx = k \int_1^2 \frac{1}{x} dx = k\left[\ln x \Big|_1^2\right] = k\left[\ln 2 - \ln 1\right] = k \ln 2 = 1. \text{ So } k = \frac{1}{\ln 2}.$$

17.

$$E(x) = \int_0^2 x f(x) dx = \int_0^2 \frac{1}{2} x\, dx = \frac{x^2}{4}\Big|_0^2 = \frac{1}{4}(4 - 0) = 1$$

19.

$$E(x) = \int_0^1 x f(x) dx = \int_0^1 2x^2\, dx = \frac{2x^3}{3}\Big|_0^1 = \frac{2}{3}(1 - 0) = \frac{2}{3}$$

21.

$$E(x) = \int_0^5 x f(x) dx = \int_0^5 \frac{3}{250}\left(10x^2 - x^3\right) dx = \frac{3}{250}\left[\left(\frac{10x^3}{3} - \frac{x^4}{4}\right)\Big|_0^5\right]$$

$$= \frac{3}{250}\left(\frac{10(5^3)}{3} - \frac{5^4}{4}\right) = \frac{3}{250}\left(\frac{1250}{3} - \frac{625}{4}\right)$$

$$= \frac{3}{250}\left(\frac{5000 - 1875}{12}\right) = \frac{1}{250} \cdot \frac{3125}{4} = \frac{25}{8}$$

23.

$$E(x) = \int_1^e x f(x) dx = \int_1^e dx = x \Big|_1^e = e - 1$$

25.

$$P(1 \le X \le 3) = \int_1^3 \frac{1}{5} dx = \frac{1}{5} x \Big|_1^3 = \frac{1}{5}(3 - 1) = \frac{2}{5}$$

There is a 40% probability that the number selected is from the interval [1, 3].

27. Let the random variable X denote the interval between incoming calls.

$$P(X \ge 6) = 1 - P(X < 6) = 1 - \int_0^6 0.5\, e^{-0.5x}\, dx$$

$$= 1 - \left[-e^{-0.5x}\Big|_0^6\right]$$

$$= 1 + e^{-3} - e^0 = e^{-3} \approx 0.0498$$

The probability of waiting at least 6 minutes for the next call is 4.98%.

29.
$$P(X<5)=\int_0^5 0.4e^{-0.4x}\,dx=-e^{-0.4x}\Big|_0^5=-e^{-2}+e^0=1-e^{-2}\approx 0.86466$$
The probability a subject makes a choice in fewer than 5 seconds is 0.865.

31.
Since the average life of the light bulb is 2000 hours, $\lambda=\dfrac{1}{2000}$, and

$$f(x)=\begin{cases} \dfrac{1}{2000}e^{-x/2000} & x\ge 0 \\[2mm] 0 & x<0 \end{cases}$$

(a) If the random variable X denotes the length of the light bulb's life, then
$$P(1800\le X\le 2200)=\int_{1800}^{2200}\frac{1}{2000}e^{-x/2000}\,dx$$
$$=-e^{-x/2000}\Big|_{1800}^{2200}$$
$$=-e^{-22/20}+e^{-18/20}\approx 0.07370$$

The probability a light bulb lasts between 1800 and 2200 hours is 0.074.

(b) $P(X\ge 2500)=1-P(X\le 2500)=1-\int_0^{2500}\frac{1}{2000}e^{-x/2000}\,dx$
$$=1-\left[-e^{-x/2000}\Big|_0^{2500}\right]$$
$$=1-\left[-e^{-25/20}+e^0\right]$$
$$=1+e^{-25/20}-1\approx 0.2865$$

There is a 28.7% probability that a light bulb will last at least 2500 hours.

33.
Since the average life of the light bulb is 2000 hours, $\lambda=\dfrac{1}{2000}$, and

$$f(x)=\begin{cases} \dfrac{1}{2000}e^{-x/2000} & x\ge 0 \\[2mm] 0 & x<0 \end{cases}$$

Let the random variable X denote the length of the light bulb's life.

(a) $P(X<1500)=\int_0^{1500}\frac{1}{2000}e^{-x/2000}\,dx=-e^{-x/2000}\Big|_0^{1500}$
$$=-e^{-15/20}+e^0=1-e^{-3/4}\approx 0.52763$$

The probability the light burns out in under 1500 hours is 0.528.

(b) $P(1750 \leq X \leq 2000) = \int_{1750}^{2000} \dfrac{1}{2000} e^{-x/2000}\, dx = -e^{-x/2000} \Big|_{1750}^{2000}$

$$= -e^{-1} + e^{-175/200} \approx 0.04898$$

The probability the light bulb lasts between 1750 and 2000 hours is 0.049.

(c) $P(X > 1900) = 1 - P(X \leq 1900) = 1 - \int_{0}^{1900} \dfrac{1}{2000} e^{-x/2000}\, dx$

$$= 1 - \left[-e^{-x/2000} \Big|_{0}^{1900} \right]$$

$$= 1 - \left[-e^{-19/20} + e^{0} \right]$$

$$= 1 + e^{-19/20} - 1 = e^{-19/20} \approx 0.38674$$

The probability that a light bulb lasts longer than 1900 hours is 0.387.

(d) Answers may vary.

35.

$$P(X > 10) = 1 - P(X \leq 10) = 1 - \int_{0}^{10} \dfrac{1}{8} e^{-x/8}\, dx = 1 - \left[-e^{-x/8} \Big|_{0}^{10} \right]$$

$$= 1 - \left[-e^{-10/8} + e^{0} \right]$$

$$= 1 + e^{-5/4} - 1 = e^{-5/4} = 0.28650$$

A customer waits longer than 10 minutes for lunch 28.7% of the time.

37.

(a) Since the average wait time is 10 minutes, $\lambda = \dfrac{1}{10}$.

$$f(x) = \begin{cases} \dfrac{1}{10} e^{-x/10} & x \geq 0 \\ 0 & x < 0 \end{cases}$$

(b) $P(7 \leq x \leq 12) = \int_{7}^{12} \dfrac{1}{10} e^{-x/10}\, dx = -e^{-x/10} \Big|_{7}^{12} = -e^{-12/10} + e^{-7/10} \approx 0.19539$

There is a 19.5% probability of waiting in line between 7 and 12 minutes.

(c) $P(X > 15) = 1 - P(X \leq 15) = 1 - \int_{0}^{15} \dfrac{1}{10} e^{-x/10}\, dx$

$$= 1 - \left[-e^{-x/10} \Big|_{0}^{15} \right]$$

$$= 1 + e^{-15/10} - 1 \approx 0.22313$$

The probability a customer waits longer than 15 minutes is 0.223.

(d) The mother will be on time to the bus stop if she waits less than 15 minutes on line.

$$P(X < 15) = e^{-15/10} - 1 = 0.77687$$

The probability of her getting out in less than 15 minutes is 0.777.

39. The expected waiting time between tee-offs is the average time between tee-offs. So it will be 9 minutes.

41.
$$E(x) = \int_0^4 x f(x) \, dx = \int_0^4 \frac{3}{56} \left(5x^2 - x^3\right) dx$$

$$= \frac{3}{56} \left[\left(\frac{5x^3}{3} - \frac{x^4}{4} \right) \Big|_0^4 \right]$$

$$= \frac{3}{56} \left[\frac{320}{3} - \frac{256}{4} \right] = 2.286$$

We can expect the contractor's cost estimate to be off by 2.286%.

43. (a) We want the probability the pregnancy lasts longer than 287 days. We use the fact that f is a probability density function, and the properties of the definite integral. That is, $\int_a^b f(x) \, dx = 1$ and $\int_a^b f(x) \, dx = \int_a^c f(x) \, dx + \int_c^b f(x) \, dx$ to determine the probability.

$$P(X > 287) = 1 - P(X \le 287)$$

$$P(X > 287) = 1 - \int_0^{287} \frac{1}{10\sqrt{2\pi}} e^{-(x-280)^2/200} \, dx$$

$$= 1 - \frac{1}{10\sqrt{2\pi}} \int_0^{287} e^{-(x-280)^2/200} \, dx$$

We evaluate the integral using a graphing utility. (We used a TI-83 Plus.)
In the math subroutine we go to 9: fnInt (and enter the integral in the following way.

fnInt(function to be integrated, the variable, the lower limit, the upper limit)
fnInt($e \wedge (-(x-280)^2 / 200), x, 0, 287)$

Then we calculated $1 - \dfrac{1}{10\sqrt{2\pi}}$ * ans. (See the screen shot below.)

```
fnInt(e^(-(X-280
)²/200),X,0,287)
        19.00115343
1-(1/(10√(2π)))*
Ans
        .2419636522
■
```

The probability a pregnancy lasts more than one week beyond the mean is 0.242.

(b) We want the probability the pregnancy lasts between 273 and 287 days. That is,

$$P(273 < X < 287) = \int_{273}^{287} \frac{1}{10\sqrt{2\pi}} e^{-(x-280)^2/200} \, dx = \frac{1}{10\sqrt{2\pi}} \int_{273}^{287} e^{-(x-280)^2/200} \, dx$$

Again we use a graphing utility. This time the lower limit of integration is 273.

```
fnInt(1/(10√(2π)
)e^(-(X-280)²/20
0),X,273,287)
        .5160726956
```

The probability a baby is born within one week of the mean gestation period is 0.516.

45.   (a) Since the numbers are so large, we will use the cost of the automobiles in thousands of dollars.

$$c = 17 \qquad f(c) = f(17) = \frac{2}{20-10} = \frac{1}{5}$$

We first find $m_1 x + b_1$.

$$m_1 = \frac{y_2 - y_1}{x_2 - x_1} = \frac{\dfrac{1}{5} - 0}{17 - 10} = \frac{1}{35}$$

$$y = m_1 x + b_1$$
$$0 = \frac{1}{35}(10) + b_1$$
$$b_1 = -\frac{2}{7}$$
$$y = \frac{1}{35}x - \frac{2}{7}$$

Then we find $m_2 x + b_2$

$$m_2 = \frac{y_2 - y_1}{x_2 - x_1} = \frac{\dfrac{1}{5} - 0}{17 - 20} = -\frac{1}{15}$$

$$y = m_2 x + b_2$$
$$0 = -\frac{1}{15}(20) + b_2$$
$$b_2 = \frac{4}{3}$$
$$y = -\frac{1}{15}x + \frac{4}{3}$$

So the probability density function is

$$f(x) = \begin{cases} \dfrac{1}{35}x - \dfrac{2}{7} & \text{if} \quad 10 \le x \le 17 \\[2mm] -\dfrac{1}{15}x + \dfrac{4}{3} & \text{if} \quad 17 < x \le 20 \end{cases}$$

(b) The probability the car will cost less than $15,000 is 0.357.

$$P(X < 15) = \int_{10}^{15}\left(\frac{1}{35}x - \frac{2}{7}\right)dx = \left(\frac{x^2}{70} - \frac{2x}{7}\right)\Bigg|_{10}^{15}$$

$$= \left(\frac{225}{70} - \frac{300}{70}\right) - \left(\frac{100}{70} - \frac{200}{70}\right) = \frac{25}{70} = \frac{5}{14} = 0.35714$$

(c)

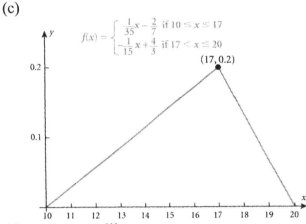

(d) Answers will vary.

(e) $E(x) = \int_{10}^{20} x f(x)\, dx = \int_{10}^{17}\left(\frac{x^2}{35} - \frac{2x}{7}\right)dx + \int_{17}^{20}\left(-\frac{x^2}{15} + \frac{4x}{3}\right)dx$

$$= \left(\frac{x^3}{105} - \frac{x^2}{7}\right)\Bigg|_{10}^{17} + \left(-\frac{x^3}{45} + \frac{2x^2}{3}\right)\Bigg|_{17}^{20}$$

$$=$$

$$\left[\left(\frac{17^3}{105} - \frac{17^2}{7}\right) - \left(\frac{10^3}{105} - \frac{10^2}{7}\right)\right] + \left[\left(-\frac{20^3}{45} + \frac{2(20^2)}{3}\right) - \left(-\frac{17^3}{45} + \frac{2(17^2)}{3}\right)\right]$$

$$= 10.26667 + 5.4 = 15.66667$$

The expected price of the car is $15, 666.67

(f) Answers will vary.

47. (a) Results will vary.

(b) $P(0.6 \le X < 0.9) = \int_{0.6}^{0.9} dx = x\Big|_{0.6}^{0.9} = 0.9 - 0.6 = 0.3$

49. The uniform probability density function is

$$f(x) = \begin{cases} \dfrac{1}{b-a} & \text{if} \quad a \le x \le b \\ 0 & \text{if} \quad a < x \text{ or } x > b \end{cases}$$

$$\sigma^2 = \int_a^b x^2 f(x)\, dx - \left[E(x)\right]^2$$

$$= \int_a^b x^2 \cdot \frac{1}{b-a}\, dx - \left[\int_a^b x \cdot \frac{1}{b-a}\, dx\right]^2$$

$$= \frac{1}{b-a}\left(\frac{x^3}{3}\Big|_a^b\right) - \left[\frac{1}{b-a}\left(\frac{x^2}{2}\Big|_a^b\right)\right]^2$$

$$= \frac{1}{b-a}\left[\frac{b^3-a^3}{3}\right] - \frac{1}{(b-a)^2}\left[\frac{b^2-a^2}{2}\right]^2$$

$$= \frac{b^3-a^3}{3(b-a)} - \frac{\left(b^2-a^2\right)^2}{4(b-a)^2}$$

$$= \frac{4(b-a)\left(b^3-a^3\right) - 3\left(b^2-a^2\right)^2}{12(b-a)^2}$$

$$= \frac{4(b-a)(b-a)\left(b^2+ab+a^2\right) - 3(b-a)^2(b+a)^2}{12(b-a)^2}$$

$$= \frac{4\left(b^2+ab+a^2\right) - 3(b+a)^2}{12}$$

$$= \frac{4b^2+4ab+4a^2 - 3b^2 - 6ab - 3a^2}{12}$$

$$= \frac{b^2+a^2-2ab}{12} = \frac{(b-a)^2}{12}$$

Chapter 7 Review

TRUE-FALSE ITEMS

1.   False   **3.**   False

FILL-IN-THE-BLANKS

1.   average value   **3.**   probability density   **5.** $\displaystyle\lim_{t \to 2^-} \int_0^t f(x)\, dx$

REVIEW EXERCISES

1. $\displaystyle\int_0^\infty 20e^{-20x}\, dx = \lim_{b \to \infty} \int_0^b 20e^{-20x}\, dx = \lim_{b \to \infty}\left(-e^{-20x}\right)\Big|_0^b$

$$= \lim_{b \to \infty} \left(-e^{-20b} + 1 \right)$$

$$= \lim_{b \to \infty} 1 - \lim_{b \to \infty} \frac{1}{e^{20b}} = 1 - 0 = 1$$

3.

$$\int_0^8 \frac{1}{\sqrt[3]{x}} \, dx = \lim_{t \to 0^+} \int_t^8 \frac{1}{\sqrt[3]{x}} \, dx = \lim_{t \to 0^+} \int_t^8 x^{-1/3} \, dx = \lim_{t \to 0^+} \left[\frac{x^{2/3}}{\frac{2}{3}} \Big|_t^8 \right]$$

$$= \lim_{t \to 0^+} \left[\frac{3 \left(8^{2/3} \right)}{2} - \frac{3 t^{2/3}}{2} \right]$$

$$= \lim_{t \to 0^+} \left(\frac{3 \left(8^{2/3} \right)}{2} \right) - \lim_{t \to 0^+} \left(\frac{3 t^{2/3}}{2} \right)$$

$$= 6 - 0 = 6$$

5.

$$\int_0^1 \frac{x+1}{x} \, dx = \lim_{t \to 0^+} \left[\int_t^1 \frac{x+1}{x} \, dx \right] = \lim_{t \to 0^+} \left[\int_t^1 \left(1 + \frac{1}{x} \right) dx \right] = \lim_{t \to 0^+} \left[\left(x + \ln |x| \right) \Big|_t^1 \right]$$

$$= \lim_{t \to 0^+} \left[\left(1 + \ln 1 \right) - \left(t + \ln |t| \right) \right]$$

$$= \lim_{t \to 0^+} 1 - \lim_{t \to 0^+} \ln |t|$$

$$= - \lim_{t \to 0^+} \ln |t| = \infty$$

Since the limit is infinite, the integral has no value.

7. Since the graph of f lies below the x-axis, the area is given by

$$A = - \int_0^\infty f(x) \, dx = \lim_{b \to \infty} \left[- \int_0^b f(x) \, dx \right] = \lim_{b \to \infty} \left[- \int_0^b -e^{-x} \, dx \right] = \lim_{b \to \infty} \left[\int_0^b e^{-x} \, dx \right]$$

$$= \lim_{b \to \infty} \left[\left(-e^{-x} \right) \Big|_0^b \right]$$

$$= \lim_{b \to \infty} \left[\left(-e^{-b} \right) - \left(-e^0 \right) \right]$$

$$= \lim_{b \to \infty} \left[\frac{1}{e^b} + 1 \right] = 1$$

The area is 1 square unit.

9.

$$AV = \frac{1}{3 - (-1)} \int_{-1}^3 x^3 \, dx = \frac{1}{4} \left[\frac{x^4}{4} \Big|_{-1}^3 \right] = \frac{1}{16} \left[3^4 - (-1)^4 \right] = \frac{80}{16} = 5$$

11.

$$AV = \frac{1}{6-2}\int_2^6 \left(x^2 + x\right) dx = \frac{1}{4}\left[\left(\frac{x^3}{3} + \frac{x^2}{2}\right)\Big|_2^6\right]$$

$$= \frac{1}{4}\left[\left(\frac{6^3}{3} + \frac{6^2}{2}\right) - \left(\frac{2^3}{3} + \frac{2^2}{2}\right)\right]$$

$$= \frac{1}{4}\left[72 + 18 - \frac{8}{3} - 2\right] = 22 - \frac{2}{3} = \frac{64}{3}$$

13.

$$AV = \frac{1}{2-(-2)}\int_{-2}^2 3x^2\, dx = \frac{1}{4}\left[\frac{3x^3}{3}\Big|_{-2}^2\right] = \frac{1}{4}\left[2^3 - (-2)^3\right] = \frac{1}{4}[8+8] = 4$$

15. (a) We show the two conditions are satisfied.
Condition 1: $f(x) \geq 0$

$$f(x) = \frac{8}{9}x$$

$\frac{8}{9}x \geq 0$ when $x \geq 0$. So $f(x) \geq 0$ for all x in the interval $\left[0, \frac{3}{2}\right]$.

Condition 2: $\int_0^{3/2} f(x)\, dx = 1$

$$\int_0^{3/2} \frac{8}{9}x\, dx = \frac{8}{9}\left[\frac{x^2}{2}\Big|_0^{3/2}\right] = \frac{8}{9}\left[\frac{\left(\frac{3}{2}\right)^2}{2} - 0\right] = \frac{8}{9}\cdot\frac{1}{2}\cdot\frac{9}{4} = 1$$

So f is a probability density function.

(b) $E(x) = \int_a^b x f(x)\, dx = \int_0^{3/2} \frac{8}{9}x^2\, dx = \frac{8}{9}\left[\frac{x^3}{3}\Big|_0^{3/2}\right] = \frac{8}{27}\left[\left(\frac{3}{2}\right)^3 - 0\right] = \frac{8}{27}\cdot\frac{27}{8} = 1$

The expected value is 1.

17. (a) We show the two conditions are satisfied.
Condition 1: $f(x) \geq 0$

$$f(x) = 12x^3\left(1 - x^2\right)$$

To determine where $f(x) \geq 0$ we solve $12x^3\left(1 - x^2\right) = 0$ and choose test points.

$$12x^3(1-x)(1+x) = 0$$

$12x^3 = 0$ or $1 - x = 0$ or $1 + x = 0$
$x = 0$ or $\quad x = 1$ or $\quad x = -1$

These numbers separate the number line into 4 parts. We are only interested in the interval [0, 1], and we choose $x = \dfrac{1}{2}$ and test it. $f\left(\dfrac{1}{2}\right) = \dfrac{9}{8} = 1.125$. So we conclude $f(x) \geq 0$ on the interval [0, 1].

Condition 2: $\displaystyle\int_0^1 f(x)\,dx = 1$

$$\int_0^1 \left[12x^3\left(1-x^2\right)\right] dx = \int_0^1 \left(12x^3 - 12x^5\right) dx = 12\left[\left(\frac{x^4}{4} - \frac{x^6}{6}\right)\Big|_0^1\right]$$

$$= 12\left[\left(\frac{1^4}{4} - \frac{1^6}{6}\right) - 0\right] = 12\left(\frac{3-2}{12}\right) = 1$$

So f is a probability density function.

(b) $E(x) = \displaystyle\int_a^b x f(x)\,dx = \int_0^1 \left(12x^4 - 12x^6\right) dx = 12\left[\left(\frac{x^5}{5} - \frac{x^7}{7}\right)\Big|_0^1\right]$

$$= 12\left[\left(\frac{1^5}{5} - \frac{1^7}{7}\right) - 0\right] = 12\left(\frac{7-5}{35}\right) = \frac{24}{35} = 0.6857$$

The expected value is $\dfrac{24}{35}$.

19. $AS = \dfrac{1}{10-0}\displaystyle\int_0^{10}\left(1340 - 850\,e^{-t}\right) dt = \dfrac{1}{10}\left[\left(1340t + 850e^{-t}\right)\Big|_0^{10}\right]$

$$= \frac{1}{10}\left[\left(13{,}400 + 850e^{-10}\right) - \left(0 + 850e^0\right)\right]$$

$$= \frac{1}{10}\left(13{,}400 + 850e^{-10} - 850\right) = 1255.0$$

On the average, 1255 units are sold each year.

21. $AP = \dfrac{1}{150-100}\displaystyle\int_{100}^{150} 50e^{-0.01x}\,dx = \dfrac{1}{50}\cdot 50\int_{100}^{150} e^{-0.01x}\,dx = -\dfrac{1}{0.01}e^{-0.01x}\Big|_{100}^{150}$

$$= -100\left(e^{-1.5} - e^{-1}\right) = 14.47$$

The average price of the sandals is \$14.47.

23.

(a) $P(X \leq 1) = \displaystyle\int_{-2}^1 \frac{1}{12}\,dx = \frac{1}{12}x\Big|_{-2}^1 = \frac{1}{12}\left[1 - (-2)\right] = \frac{1}{4}$

(b) $P(X \geq 5) = \displaystyle\int_5^{10} \frac{1}{12}\,dx = \frac{1}{12}x\Big|_5^{10} = \frac{1}{12}\left[10 - 5\right] = \frac{5}{12}$

(c) The expected value of X is

$$E(x) = \int_{-2}^{10} \frac{1}{12} x \, dx = \frac{1}{12} \cdot \frac{x^2}{2} \Big|_{-2}^{10} = \frac{1}{24}\left[10^2 - (-2)^2\right] = \frac{96}{24} = 4$$

25. (a) We show the two conditions are satisfied.
Condition 1: $f(x) \geq 0$

$$\frac{3}{635,840}\left(x^2 - 28x + 196\right) \geq 0 \quad \text{when} \quad x^2 - 28x + 196 \geq 0$$

$$x^2 - 28x + 196 = (x-14)^2 \text{ is always nonnegative. So condition 1 is satisfied.}$$

Condition 2: $\int_{20}^{100} f(x) \, dx = 1$

$$\int_{20}^{100} \frac{3}{635,840}\left(x^2 - 28x + 196\right) dx = \frac{3}{635,840} \int_{20}^{100} \left(x^2 - 28x + 196\right) dx$$

$$= \frac{3}{635,840}\left[\frac{x^3}{3} - 14x^2 + 196x \right]_{20}^{100}$$

$$= \frac{3}{635,840}\left[\left(\frac{100^3}{3} - 14\left(100^2\right) + 196(100)\right) - \left(\frac{20^3}{3} - 14\left(20^2\right) + 196(20)\right)\right]$$

$$= 1$$

Both conditions are satisfied, making f a probability density function.

(b) $P(X \leq 40) = \int_{20}^{40} \frac{3}{635,840}\left(x^2 - 28x + 196\right) dx = \frac{3}{635,840} \int_{20}^{40}\left(x^2 - 28x + 196\right) dx$

$$= \frac{3}{635,840}\left[\frac{x^3}{3} - 14x^2 + 196x \right]_{20}^{40}$$

$$= \frac{3}{635,840}\left[\left(\frac{40^3}{3} - 14\left(40^2\right) + 196(40)\right) - \left(\frac{20^3}{3} - 14\left(20^2\right) + 196(20)\right)\right]$$

$$= = \frac{3}{635,840}[373.33333 - 986.66667] = \frac{217}{7948} = 0.0273$$

The probability a man dies at or before age 40 is 0.032.

(c) $P(X \leq 60) = \int_{20}^{60} \frac{3}{635,840}\left(x^2 - 28x + 196\right) dx = \frac{3}{635,840} \int_{20}^{60}\left(x^2 - 28x + 196\right) dx$

$$= \frac{3}{635,840}\left[\frac{x^3}{3} - 14x^2 + 196x \right]_{20}^{60}$$

$$= \frac{3}{635,840}\left[\left(\frac{60^3}{3} - 14\left(60^2\right) + 196(60)\right) - \left(\frac{20^3}{3} - 14\left(20^2\right) + 196(20)\right)\right]$$

$$= \frac{3}{635,840}[33,360 - 986.66667] = \frac{607}{3974} = 0.15274$$

The probability a man dies at or before age 60 is 0.153.

(d) The expected age of death is

$$E(x) = \int_{20}^{100} x f(x)\, dx = \int_{20}^{100} \frac{3}{635,840}\left(x^3 - 28x^2 + 196x\right) dx$$

$$= \frac{3}{635,840} \int_{20}^{100}\left(x^3 - 28x^2 + 196x\right) dx$$

$$= \frac{3}{635,840}\left[\frac{x^4}{4} - \frac{28x^3}{3} + 98x^2\right]\Bigg|_{20}^{100}$$

$$= \frac{3}{635,840}\left[\left(\frac{100^4}{4} - \frac{28(100^3)}{3} + 98(100^2)\right) - \left(\frac{20^4}{4} - \frac{28(20^3)}{3} + 98(20^2)\right)\right]$$

$$= 78.52 \text{ years.}$$

27. (a) We show the two conditions are satisfied.
Condition 1: $f(x) \ge 0$

$$f(x) = \frac{1}{2}x \ge 0 \quad \text{whenever } x \ge 0$$

So f is nonnegative on the interval [0, 2].

Condition 2: $\int_0^2 f(x)\, dx = 1$

$$\int_0^2 \frac{1}{2}x\, dx = \frac{1}{2}\left[\frac{x^2}{2}\right]\Bigg|_0^2 = \frac{1}{2}[2 - 0] = 1$$

Both conditions are satisfied, making f a probability density function.

(b) The probability X is less than one is

$$P(X < 1) = \int_0^1 \frac{1}{2}x\, dx = \frac{1}{2}\left[\frac{x^2}{2}\right]\Bigg|_0^1 = \frac{1}{2}\left[\frac{1}{2} - 0\right] = \frac{1}{4} = 0.25$$

(c) The probability X is between 1 and 1.5 is

$$P(1 < X < 1.5) = \int_1^{1.5} \frac{1}{2}x\, dx = \frac{1}{2}\left[\frac{x^2}{2}\right]\Bigg|_1^{1.5} = \frac{1}{4}[2.25 - 1] = \frac{1.25}{4} = \frac{5}{16} = 0.3125$$

(d) The probability X is greater than 1.5 is

$$P(X > 1.5) = \int_{1.5}^2 \frac{1}{2}x\, dx = \frac{1}{2}\left[\frac{x^2}{2}\right]\Bigg|_{1.5}^2 = \frac{1}{4}[4 - 2.25] = \frac{1.75}{4} = \frac{7}{16} = 0.4375$$

(e) The expected value of X is

$$E(x) = \int_0^2 x f(x) \, dx = \int_0^2 \frac{1}{2} x^2 \, dx = \frac{1}{2} \left[\frac{x^3}{3} \right]_0^2 = \frac{1}{6} [8 - 0] = \frac{8}{6} = \frac{4}{3}$$

29. The probability density function f is

$$f(x) = \begin{cases} \dfrac{1}{15} & \text{if} \quad 0 \le x \le 15 \\ 0 & \text{if} \quad x < 0 \text{ or } x > 15 \end{cases}$$

(a) The probability the tourist waits fewer than 3 minutes to hear the chimes is

$$P(X < 3) = \int_0^3 \frac{1}{15} \, dx = \frac{1}{15} x \Big|_0^3 = \frac{1}{15} [3 - 0] = \frac{1}{5} = 0.20$$

(b) The probability the tourist must wait more than 10 minutes to hear the clock is

$$P(X > 10) = \int_{10}^{15} \frac{1}{15} \, dx = \frac{1}{15} x \Big|_{10}^{15} = \frac{1}{15} [15 - 10] = \frac{5}{15} = \frac{1}{3}$$

(c) The expected time a person will wait to hear the chimes is

$$E(X) = \int_0^{15} x f(x) \, dx = \int_0^{15} \frac{1}{15} x \, dx = \frac{1}{15} \cdot \frac{x^2}{2} \Big|_0^{15} = \frac{1}{15} \left[\frac{15^2}{2} - 0 \right] = \frac{15}{2} = 7.5$$

minutes.

31. Let the random variable X denote the time one waits for a call. The probability density function is

$$f(x) = \begin{cases} 2.5 e^{-2.5x} & \text{if} \quad x \ge 0 \\ 0 & \text{if} \quad x < 0 \end{cases}$$

The probability the switchboard is idle for more than one minute is

$$P(X \ge 1) = 1 - P(X < 1) = 1 - \int_0^1 2.5 e^{-2.5x} \, dx = 1 + \left[e^{-2.5x} \Big|_0^1 \right]$$

$$= 1 + \left[e^{-2.5} - e^0 \right] = e^{-2.5} \approx 0.0821$$

33. Let the random variable X denote the life of the light bulb. If the bulb has an average life of 1750 hours, then $\lambda = \dfrac{1}{1750}$, and the probability density function is

$$f(x) = \begin{cases} \dfrac{1}{1750} e^{-x/1750} & \text{if} \quad x \ge 0 \\ 0 & \text{if} \quad x < 0 \end{cases}$$

(a) The probability the light bulb lasts between 1500 and 2000 hours is

$$P(1500 \le X \le 2000) = \int_{1500}^{2000} \frac{1}{1750} e^{-x/1750} \, dx = - \left. e^{-x/1750} \right|_{1500}^{2000}$$
$$= - e^{-2000/1750} + e^{-1500/1750} = 0.1055$$

(b) The probability the light bulb burns for more than 2000 hours is

$$P(X > 2000) = 1 - P(X \le 2000) = 1 - \int_{0}^{2000} \frac{1}{1750} e^{-x/1750} \, dx$$
$$= 1 - \left[- \left. e^{-x/1750} \right|_{0}^{2000} \right] = 1 + e^{-2000/1750} - e^{0}$$
$$= e^{-2000/1750} \approx 0.3189$$

35. The probability the toll collector waits more than 1 minute for the first car is

$$P(X > 1) = 1 - P(X \le 1) = 1 - \left[\int_{0}^{1} \frac{2}{3} e^{-2t/3} \, dt \right] = 1 - \left[- \left. e^{-2t/3} \right|_{0}^{1} \right]$$
$$= 1 + e^{-2/3} - 1 = 0.5134$$

CHAPTER 7 PROJECT

1.

Interval	Width	Tally	Frequency	Relative Frequency	Probability
1	0 – 5	ⅣⅣ ⅣⅣ ⅣⅣ ⅣⅣ I	21	0.23596	0.26288
2	5 – 10	ⅣⅣ ⅣⅣ ⅣⅣ III	18	0.20225	0.19377
3	10 – 15	ⅣⅣ ⅣⅣ II	12	0.13483	0.14283
4	15 – 20	ⅣⅣ ⅣⅣ	10	0.11236	0.10529
5	20 – 25	ⅣⅣ ⅣⅣ II	12	0.13483	0.07761
6	25 – 30	II	2	0.02247	0.05721
7	30 – 35	II	2	0.02247	0.04217
8	35 – 40	III	3	0.03371	0.03108
9	40 – 45	II	2	0.02247	0.02291
10	45 – 50	IIII	4	0.04494	0.01689
11	50 – 55	I	1	0.01124	0.01245
12	55 – 60		0	0	0.00918
13	60 – 65	I	1	0.01124	0.00676
14	65 – 70		0	0	0.00499
15	70 – 75		0	0	0.00368
16	75 – 80		0	0	0.00271
17	80 – 85		0	0	0.00200
18	85 – 90	I	1	0.01124	0.00147

3.   Answers will vary.

5.

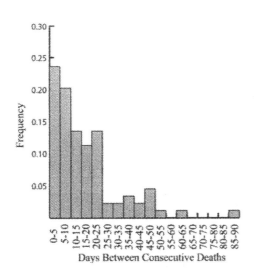

7. See column 6 of the table in Problem 1.

9.

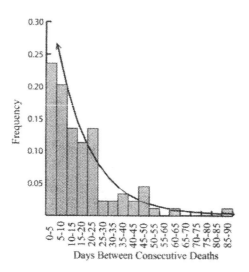

Days Between Consecutive Deaths

Comments will vary.

MATHEMATICAL QUESTIONS FROM PROFESSIONAL EXAMS

1. (d)
$$\int_0^1 x \ln x \, dx = \lim_{t \to 0^+} \left[\int_t^1 x \ln x \, dx \right]$$

We evaluate the integral by using integration by parts. We choose
$$u = \ln x \qquad\qquad dv = x \, dx$$
$$du = \frac{1}{x} \, dx \qquad\qquad v = \frac{x^2}{2}$$

$$\lim_{t \to 0^+} \left[\int_t^1 x \ln x \, dx \right] = \lim_{t \to 0^+} \left[uv \Big|_t^1 - \int_t^1 v \, du \right]$$

$$= \lim_{t \to 0^+} \left[\frac{x^2 \ln x}{2} \Big|_t^1 - \int_t^1 \frac{x}{2} \, dx \right]$$

$$= \lim_{t \to 0^+} \left[\left(\frac{x^2 \ln x}{2} \Big|_t^1 \right) - \left(\frac{x^2}{4} \Big|_t^1 \right) \right]$$

$$= \lim_{t \to 0^+} \left[\left(\frac{1 \cdot \ln 1}{2} - \frac{t \cdot \ln t}{2} \right) - \left(\frac{1}{4} - \frac{t^2}{4} \right) \right]$$

$$= \lim_{t \to 0^+} \left[-\frac{t \cdot \ln t}{2} - \frac{1}{4} + \frac{t^2}{4} \right]$$

$$= -\lim_{t \to 0^+} \left(\frac{t \cdot \ln t}{2} \right) - \lim_{t \to 0^+} \left(\frac{1}{4} \right) + \lim_{t \to 0^+} \left(\frac{t^2}{4} \right)$$

$$= - \lim_{t \to 0^+} \left(\frac{t \cdot \ln t}{2} \right) - \frac{1}{4}$$

To determine $\lim_{t \to 0^+} (t \ln t)$ we use a table.

t	1	0.1	0.01	0.001	0.0001	0.00001
$t \ln t$	0	-0.2303	-0.0461	-0.0069	-0.0009	-0.0001

We conclude $\lim_{t \to 0^+} (t \ln t) = 0$, and $\int_0^1 x \ln x \, dx = -\frac{1}{4}$

3. (b)

$$\int_0^\infty \frac{x+1}{\left(x^2 + 2x + 2 \right)^2} \, dx = \lim_{b \to \infty} \left[\int_0^b \frac{x+1}{\left(x^2 + 2x + 2 \right)^2} \, dx \right]$$

We use the method of substitution to evaluate the integral.

Let $u = x^2 + 2x + 2$. Then $du = 2(x+1) \, dx$, and $\frac{1}{2} du = (x+1) \, dx$.

When $x = 0$, $u = 2$, and when $x = b$, $u = b^2 + 2b + 2$.

$$\lim_{b \to \infty} \left[\int_0^b \frac{x+1}{\left(x^2 + 2x + 2 \right)^2} \, dx \right] = \lim_{b \to \infty} \left[\frac{1}{2} \int_2^{b^2+2b+2} u^{-2} \, du \right]$$

$$= \frac{1}{2} \lim_{b \to \infty} \left[-u^{-1} \Big|_2^{b^2+2b+2} \right]$$

$$= \frac{1}{2} \lim_{b \to \infty} \left[-\frac{1}{b^2 + 2b + 2} + \frac{1}{2} \right] = \frac{1}{4}$$

5. (d)

We are only concerned with the time after 8:30.

The probability X arrives after 8:30 is $\frac{1}{2}$.

The probability Y arrives after 8:30 is 1.

As long as X arrives after 8:30, Y arrives first $\frac{1}{2}$ the time. So the probability Y arrives first if $\frac{1}{2} \cdot \frac{1}{2} = \frac{1}{4}$.

Chapter 8
Calculus of Functions of
Two or More Variables

8.1 Rectangular Coordinates in Space

1.

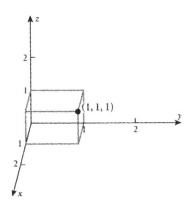

3.

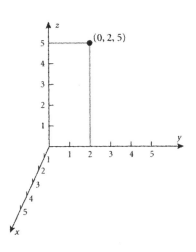

5.

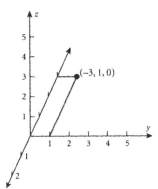

7. Since one vertex is the origin and the edges of the box are parallel to the axes, three of the remaining vertices are also on an axis.

    (2, 0, 0), (0, 1, 0), (0, 0, 3)

    Because the figure is a rectangular box, the remaining three corners will have one coordinate that is 0.

    (2, 1, 0), (2, 0, 3), (0, 1, 3)

9. Draw a rectangular box (prism) and label the two given vertices. The other 6 vertices are determined by changing the value of one coordinate to that of the other given vertex.

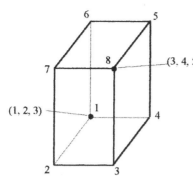
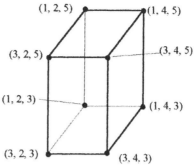

Vertex 2: change $x = 1$ in vertex 1 to $x = 3$ giving (3, 2, 3).
Vertex 3: change $y = 2$ in vertex 2 to $y = 4$ giving (3, 4, 3).
Vertex 4: change $x = 3$ in vertex 3 to $x = 1$ giving (1, 4, 3).
Vertex 5: change $z = 3$ in vertex 4 to $z = 5$ giving (1, 4, 5).
Vertex 6: change $y = 4$ in vertex 5 to $y = 2$ giving (1, 2, 5).
Vertex 7: change $x = 1$ in vertex 6 to $x = 3$ giving (3, 2, 5).
Changing $y = 2$ in vertex 7 to $y = 4$ will give you (3, 4, 5) which is vertex 8.

11. Draw a rectangular box (prism) and label the two given vertices. The other 6 vertices are determined by changing the value of one coordinate to that of the other given vertex.

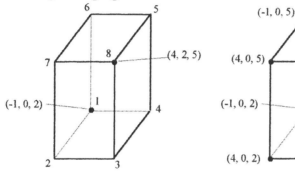

Vertex 2: change $x = -1$ in vertex 1 to $x = 4$ giving (4, 0, 2).
Vertex 3: change $y = 0$ in vertex 2 to $y = 2$ giving (4, 2, 2).
Vertex 4: change $x = 4$ in vertex 3 to $x = -1$ giving $(-1, 2, 2)$.
Vertex 5: change $z = 2$ in vertex 4 to $z = 5$ giving $(-1, 2, 5)$.
Vertex 6: change $y = 2$ in vertex 5 to $y = 0$ giving $(-1, 0, 5)$.
Vertex 7: change $x = -1$ in vertex 6 to $x = 4$ giving (4, 0, 5).
Changing $y = 0$ in vertex 7 to $y = 2$ will give you (4, 2, 5) which vertex 8.

13. $y = 3$ describes a plane parallel to the xz-plane and passing through the point (0, 3, 0).

15. $x = 0$ describes the yz-plane.

17. $z = 5$ describes a plane parallel to the xy-plane and passing through the point (0, 0, 5).

19. We use the distance formula with $\left(x_1, y_1, z_1\right) = (1, 3, 0)$ and $\left(x_2, y_2, z_2\right) = (4, 1, 2)$.

$$d = \sqrt{\left(x_2 - x_1\right)^2 + \left(y_2 - y_1\right)^2 + \left(z_2 - z_1\right)^2}$$
$$d = \sqrt{\left(4-1\right)^2 + \left(1-3\right)^2 + \left(2-0\right)^2}$$
$$d = \sqrt{9+4+4} = \sqrt{17}$$

21. We use the distance formula with $\left(x_1, y_1, z_1\right) = (-1, 2, -3)$ and $\left(x_2, y_2, z_2\right) = (4, -2, 1)$.

$$d = \sqrt{\left(x_2 - x_1\right)^2 + \left(y_2 - y_1\right)^2 + \left(z_2 - z_1\right)^2}$$
$$d = \sqrt{\left(4-(-1)\right)^2 + \left(-2-2\right)^2 + \left(1-(-3)\right)^2}$$
$$d = \sqrt{25+16+16} = \sqrt{57}$$

23. We use the distance formula with $\left(x_1, y_1, z_1\right) = (4, -2, -2)$ and $\left(x_2, y_2, z_2\right) = (3, 2, 1)$.

$$d = \sqrt{\left(x_2 - x_1\right)^2 + \left(y_2 - y_1\right)^2 + \left(z_2 - z_1\right)^2}$$
$$d = \sqrt{\left(3-4\right)^2 + \left(2-(-2)\right)^2 + \left(1-(-2)\right)^2}$$
$$d = \sqrt{1+16+9} = \sqrt{26}$$

25. The equation of the sphere whose center is (3, 1, 1) and whose radius is 1 is
$$\left(x-3\right)^2 + \left(y-1\right)^2 + \left(z-1\right)^2 = 1$$

27. The equation of the sphere whose center is (– 1, 1, 2) and whose radius is 3 is
$$\left(x+1\right)^2 + \left(y-1\right)^2 + \left(z-2\right)^2 = 9$$

29. Complete the squares.
$$x^2 + y^2 + z^2 + 2x - 2y = 2$$
$$\left(x^2 + 2x\right) + \left(y^2 - 2y\right) + \left(z^2\right) = 2$$
$$\left(x^2 + 2x + 1\right) + \left(y^2 - 2y + 1\right) + \left(z^2\right) = 2 + 1 + 1$$
$$\left(x+1\right)^2 + \left(y-1\right)^2 + z^2 = 4$$
The center of the sphere is (– 1, 1, 0) and the radius is 2.

31. Complete the squares.
$$x^2 + y^2 + z^2 + 4x + 4y + 2z = 0$$
$$\left(x^2 + 4x\right) + \left(y^2 + 4y\right) + \left(z^2 + 2z\right) = 0$$
$$\left(x^2 + 4x + 4\right) + \left(y^2 + 4y + 4\right) + \left(z^2 + 2z + 1\right) = 0 + 4 + 4 + 1$$
$$\left(x+2\right)^2 + \left(y+2\right)^2 + \left(z+1\right)^2 = 9$$
The center of the sphere is (– 2, – 2, – 1), and the radius is 3.

33. Complete the squares.
$$2x^2 + 2y^2 + 2z^2 - 8x + 4z = -2$$
$$x^2 + y^2 + z^2 - 4x + 2z = -1$$
$$\left(x^2 - 4x\right) + \left(y^2\right) + \left(z^2 + 2z\right) = -1$$
$$\left(x^2 - 4x + 4\right) + \left(y^2\right) + \left(z^2 + 2z + 1\right) = -1 + 4 + 1$$
$$\left(x-2\right)^2 + y^2 + \left(z+1\right)^2 = 4$$

The center of the sphere is $(2, 0, -1)$, and the radius is 2.

35. To determine the equation of the sphere, we first need to find the center and the radius of the sphere.

The center is the midpoint of the diameter.

$$x_m = \frac{x_1 + x_2}{2} = \frac{-2 + 2}{2} = 0 \qquad\qquad y_m = \frac{y_1 + y_2}{2} = \frac{0 + 6}{2} = 3$$

$$z_m = \frac{z_1 + z_2}{2} = \frac{4 + 8}{2} = 6$$

The center of the sphere is $(0, 3, 6)$.

The radius is the distance from the midpoint to one of endpoints. We use $(2, 6, 8)$.

$$r = \sqrt{\left(2-0\right)^2 + \left(6-3\right)^2 + \left(8-6\right)^2} = \sqrt{4+9+4} = \sqrt{17}$$

The equation of the sphere is
$$x^2 + \left(y-3\right)^2 + \left(z-6\right)^2 = 17$$

8.2 Functions and Their Graphs

1. $f(2,1) = 2^2 + 1 = 5$

3. $f(2,1) = \sqrt{2 \cdot 1} = \sqrt{2}$

5. $f(2,1) = \dfrac{1}{2 \cdot 2 + 1} = \dfrac{1}{5}$

7. $f(2,1) = \dfrac{2^2 - 1}{2 - 1} = 3$

9. $f(2,1) = \sqrt{4 - 2^2 \cdot 1^2} = \sqrt{4-4} = 0$

11. $f(x,y) = 3x + 2y + xy$

(a) $f(1,0) = 3 \cdot 1 + 2 \cdot 0 + 1 \cdot 0 = 3$

(b) $f(0,1) = 3 \cdot 0 + 2 \cdot 1 + 0 \cdot 1 = 2$

(c) $f(2,1) = 3 \cdot 2 + 2 \cdot 1 + 2 \cdot 1 = 10$

(d) $f(x+\Delta x,\, y)=3(x+\Delta x)+2y+(x+\Delta x)y$
$$= 3x+3\Delta x+2y+xy+\Delta xy$$

(e) $f(x,y+\Delta y)=3x+2(y+\Delta y)+x(y+\Delta y)$
$$= 3x+2y+2\Delta y+xy+x\Delta y$$

13. $f(x,\, y)=\sqrt{xy}+x$

(a) $f(0,0)=\sqrt{0\cdot 0}+0=0$ (b) $f(0,1)=\sqrt{0\cdot 1}+0=0$

(c) $f\left(a^2,t^2\right)=\sqrt{a^2\cdot t^2}+a^2=\sqrt{a^2}\cdot\sqrt{t^2}+a^2=at+a^2$

(d) $f(x+\Delta x,\, y)=\sqrt{(x+\Delta x)y}+x+\Delta x$

(e) $f(x,\, y+\Delta y)=\sqrt{x(y+\Delta y)}+x$

15. $f(x,\, y,\, z)=x^2y+y^2z$

(a) $f(1,2,3)=1^2\cdot 2+2^2\cdot 3=2+12=14$

(b) $f(0,1,2)=0^2\cdot 1+1^2\cdot 2=0+2=2$

(c) $f(-1,-2,-3)=(-1)^2\cdot(-2)+(-2)^2\cdot(-3)=-2-12=-14$

17. $z=f(x,\, y)=\sqrt{x}\sqrt{y}$

Since the radicand must be nonnegative, we have $x\ge 0$ and $y\ge 0$. The domain is the set $\{(x,y)\mid x\ge 0 \text{ and } y\ge 0\}$. That is, the domain is the first quadrant and the positive x- and y- axes.

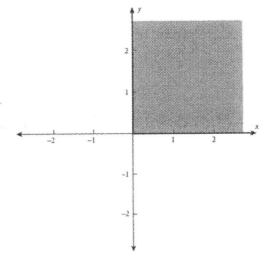

19. $z = f(x, y) = \sqrt{9 - x^2 - y^2}$

Since the radicand must be nonnegative,
$$9 - x^2 - y^2 \geq 0$$
$$x^2 + y^2 \leq 9$$

This inequality describes the domain of f. That is, the domain of f is the points (x, y) that are either on the circle $x^2 + y^2 = 9$ or inside the circle.

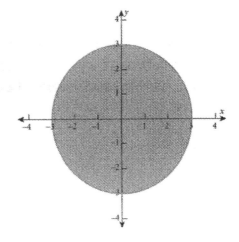

21. $z = f(x, y) = \dfrac{\ln x}{\ln y}$

The logarithmic function is defined only for positive numbers, so x and y must be positive. However, $\ln 1 = 0$, so we must eliminate $y = 1$ from the domain. The domain of f is the set $\{(x, y) \mid x > 0 \text{ and } y > 0, \text{ but } y \neq 1\}$.

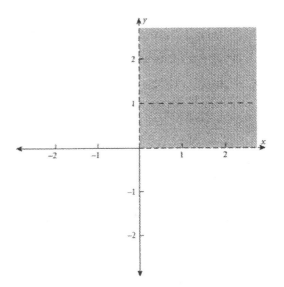

23. $z = f(x, y) = \dfrac{3}{x^2 + y^2 - 4}$

This is a rational function, so its domain is all real numbers except those which make the denominator zero. That is, $x^2 + y^2 - 4 \neq 0$.
$$x^2 + y^2 \neq 4$$

The domain of f is the set of all real numbers excluding the boundary of the circle $x^2 + y^2 = 4$.

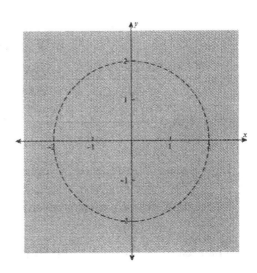

25. $z = f(x, y) = \ln(x^2 + y^2)$

Logarithm functions are defined only for positive numbers. $x^2 + y^2 > 0$ provided both x and y are not both equal to zero. The domain of f is the set $\{(x, y) \mid (x, y) \neq (0, 0)\}$.

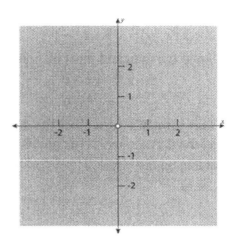

27. $w = f(x, y, z) = \sqrt{x^2 + y^2 + z^2 - 16}$

Since the radicand must be nonnegative, $x^2 + y^2 + z^2 - 16 \geq 0$. That is, $x^2 + y^2 + z^2 \geq 16$. This describes the surface of a sphere of radius 4 and all the space outside it. The domain of f is the set $\{(x, y, z) \mid x^2 + y^2 + z^2 \geq 16\}$.

29. $w = f(x, y, z) = \dfrac{4}{x^2 + y^2 + z^2}$

f is a rational function, so the domain is all real numbers except those that make the denominator zero. That is, $x^2 + y^2 + z^2 \neq 0$. The domain of f is the set $\{(x, y, z) \mid (x, y, z) \neq (0, 0, 0)\}$.

31. $z = f(x, y) = 3x + 4y$

(a) $f(x + \Delta x, y) = 3(x + \Delta x) + 4y$

(b) $f(x + \Delta x, y) - f(x, y) = [3(x + \Delta x) + 4y] - [3x + 4y]$
$$= 3x + 3\Delta x + 4y - 3x - 4y = 3\Delta x$$

(c) $\dfrac{f(x + \Delta x, y) - f(x, y)}{\Delta x} = \dfrac{3\Delta x}{\Delta x}$

(d) $\lim\limits_{\Delta x \to 0} \dfrac{f(x + \Delta x, y) - f(x, y)}{\Delta x} = \lim\limits_{\Delta x \to 0} \dfrac{3\Delta x}{\Delta x} = \lim\limits_{\Delta x \to 0} 3 = 3$

33. Let r denote the radius of the tank, and
  h denote the height of the tank.
The area A_1 of the top (or bottom) of the tank is given by
$$A_1 = \pi r^2$$
The area A_2 of the side of the tank is given by
$$A_2 = 2\pi r \cdot h = 2\pi rh$$

The cost function C is then
$$C = 2\left[300A_1\right] + 500A_2$$
$$C(r,h) = 600\pi r^2 + 1000\pi rh \text{ dollars.}$$

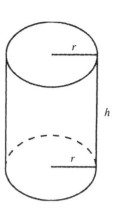

35.
$$A(N, I) = 9\left(\frac{N}{I}\right)$$

(a) $A(3, 4) = 9\left(\dfrac{3}{4}\right) = \dfrac{27}{4} = 6.75$

(b) $A(6, 3) = 9\left(\dfrac{6}{3}\right) = 18.0$

(c) $A(2, 9) = 9\left(\dfrac{2}{9}\right) = 2.0$

(d) $A(3, 18) = 9\left(\dfrac{3}{18}\right) = \dfrac{3}{2} = 1.5$

37.   Here $x = 650 - 500 = 150$, and $y = 1600 - 1500 = 100$. So,
$$B(150, 100) = 79.99 + 0.4(150) + 0.02(100) = 79.99 + 60 + 2 = \$141.99$$

39. $H = -42.379 + 2.04901523t + 10.14333127r - 0.22475541tr - 0.00683783t^2$
$\quad\quad - 0.05481717r^2 + 0.00122874t^2r + 0.00085282tr^2 - 0.00000199t^2r^2$

(a) When $t = 95°$ and $r = 50$, then $H(t, r) = 105.216°$ F.

(b) When $t = 97°$ F, the function $H = H(r)$.

$$H = \left[-42.379 + 2.04901523t - 0.00683783t^2\right] + \left[10.14333127 - 0.22475541t + 0.00122874t^2\right]r$$
$$+ \left[-0.05481717 + 0.00085282t - 0.00000199t^2\right]r^2$$
$$= 0.00918246r^2 - 0.09672884r + 92.03833484$$

We want to find r that will make $H = 105$. That is, we want to solve
$$0.00918246r^2 - 0.09672884r + 92.03833484 = 105$$
$$0.00918246r^2 - 0.09672884r + 92.03833484 - 105 = 0$$
$$0.00918246r^2 - 0.09672884r - 12.96166516 = 0$$
This is a quadratic function in one variable r and can be solved using the quadratic formula or a graphing utility. We used a TI-83 Plus by graphing
$$Y1 = 0.00918246x^2 - 0.09672884x - 12.96166516$$
and using 2nd CALC 2: zero.

 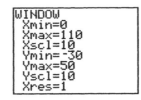 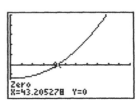

43% relative humidity results in a heat index of 105° F

(c) If $t = 102°$, then
$$H = 0.01146651r^2 + 0.00209041r + 95.47977014$$
We want to find r that will make $H = 130$. That is, we want to solve
$$0.01146651r^2 + 0.00209041r + 95.47977014 = 130$$
$$0.01146651r^2 + 0.00209041r + 95.47977014 - 130 = 0$$
$$0.01146651r^2 + 0.00209041r - 34.52022986 = 0$$
Again using a TI-83. Plus, we graph
$$Y1 = 0.01146651x^2 + 0.00209041x - 34.52022986$$

 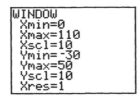 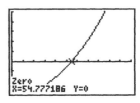

55% relative humidity will result in a heat index of 130° F.

8.3 Partial Derivatives

1. $f_x(x, y) = 3$ $f_y(x, y) = -2 + 9y^2$

   $f_x(2, -1) = 3$ $f_y(-2, 3) = -2 + 9 \cdot 3^2 = 79$

3. $f_x(x, y) = 2(x - y)$ $f_y(x, y) = 2(x - y)(-1) = -2(x - y)$

   $f_x(2, -1) = 2(2 - (-1)) = 2 \cdot 3 = 6$ $f_y(-2, 3) = -2[(-2) - 3] = 10$

5. $f_x(x, y) = \dfrac{1}{2}(x^2 + y^2)^{-1/2} \cdot 2x$ $f_y(x, y) = \dfrac{1}{2}(x^2 + y^2)^{-1/2} \cdot 2y$

   $= \dfrac{x}{\sqrt{x^2 + y^2}}$ $= \dfrac{y}{\sqrt{x^2 + y^2}}$

$$f_x(2,-1) = \frac{2}{\sqrt{2^2+(-1)^2}} = \frac{2}{\sqrt{5}} = \frac{2\sqrt{5}}{5} \qquad f_y(-2,3) = \frac{3}{\sqrt{(-2)^2+3^2}} = \frac{3}{\sqrt{13}} = \frac{3\sqrt{13}}{13}$$

7. $\quad f_x(x,y) = -2y - 24x \qquad\qquad\qquad f_y(x,y) = 3y^2 - 2x + 2y$

$\quad\quad f_{xx}(x,y) = -24 \qquad\qquad\qquad\qquad f_{yy}(x,y) = 6y + 2$

$\quad\quad f_{xy}(x,y) = -2 \qquad\qquad\qquad\qquad\quad f_{yx}(x,y) = -2$

9. $\quad f_x(x,y) = e^y + ye^x + 1 \qquad\qquad\quad f_y(x,y) = xe^y + e^x$

$\quad\quad f_{xx}(x,y) = ye^x \qquad\qquad\qquad\qquad f_{yy}(x,y) = xe^y$

$\quad\quad f_{xy}(x,y) = e^y + e^x \qquad\qquad\qquad f_{yx}(x,y) = e^y + e^x$

11.

$$f(x,y) = \frac{x}{y} = x \cdot y^{-1}$$

$$f_x(x,y) = \frac{1}{y} \qquad\qquad\qquad\qquad f_y(x,y) = -x \cdot y^{-2} = -\frac{x}{y^2}$$

$$f_{xx}(x,y) = 0 \qquad\qquad\qquad\quad\; f_{yy}(x,y) = -x \cdot (-2)y^{-3} = \frac{2x}{y^3}$$

$$f_{xy}(x,y) = -y^{-2} = -\frac{1}{y^2} \qquad\qquad f_{yx}(x,y) = -1 \cdot y^{-2} = -\frac{1}{y^2}$$

13. $\quad f(x,y) = \ln\left(x^2 + y^2\right)$

$$f_x(x,y) = \frac{1}{x^2+y^2} \cdot 2x = \frac{2x}{x^2+y^2} \qquad f_y(x,y) = \frac{1}{x^2+y^2} \cdot 2y = \frac{2y}{x^2+y^2}$$

$$f_{xx}(x,y) = \frac{2\left(x^2+y^2\right) - 2x(2x)}{\left(x^2+y^2\right)^2} = \frac{2x^2+2y^2-4x^2}{\left(x^2+y^2\right)^2} = \frac{2y^2-2x^2}{\left(x^2+y^2\right)^2}$$

$$f_{yy}(x,y) = \frac{2\left(x^2+y^2\right) - 2y(2y)}{\left(x^2+y^2\right)^2} = \frac{2x^2+2y^2-4y^2}{\left(x^2+y^2\right)^2} = \frac{2x^2-2y^2}{\left(x^2+y^2\right)^2}$$

$$f_{xy}(x,y) = \frac{-2x(2y)}{\left(x^2+y^2\right)^2} = \frac{-4xy}{\left(x^2+y^2\right)^2}$$

$$f_{yx}(x,y) = \frac{-2y(2x)}{\left(x^2+y^2\right)^2} = \frac{-4xy}{\left(x^2+y^2\right)^2}$$

15.

$$f(x, y) = \frac{10 - x + 2y}{xy}$$

$$f_x(x, y) = \frac{(-1) \cdot xy - (10 - x + 2y) \cdot y}{(xy)^2} = \frac{-xy - 10y + xy - 2y^2}{x^2 y^2} = \frac{-10\cancel{y} - 2y^{\cancel{2}}}{x^2 y^{\cancel{2}}}$$

$$= \frac{-10 - 2y}{x^2 y}$$

$$f_y(x, y) = \frac{2 \cdot xy - (10 - x + 2y) \cdot x}{(xy)^2} = \frac{2xy - 10x + x^2 - 2xy}{x^2 y^2} = \frac{-10\cancel{x} + x^{\cancel{2}}}{x^{\cancel{2}} y^2} = \frac{-10 + x}{xy^2}$$

$$f_{xx}(x, y) = \frac{0 \cdot x^2 y - (-10 - 2y)(2xy)}{(x^2 y)^2} = \frac{20\cancel{xy} + 4x\cancel{y}^{\cancel{2}}}{x^{\cancel{3}} y^{\cancel{2}}} = \frac{20 + 4y}{x^3 y}$$

$$f_{yy}(x, y) = \frac{-(-10 + x)(2xy)}{(xy^2)^2} = \frac{(20 - 2x)\cancel{xy}}{x^{\cancel{2}} y^{\cancel{3}}} = \frac{20 - 2x}{xy^3}$$

$$f_{xy}(x, y) = \frac{-2(x^2 y) - (-10 - 2y) \cdot x^2}{(x^2 y)^2} = \frac{-2x^2 y + 10x^2 + 2x^2 y}{x^4 y^2} = \frac{10x^2}{x^4 y^2} = \frac{10}{x^2 y^2}$$

$$f_{yx}(x, y) = \frac{1 \cdot (xy^2) - (-10 + x) \cdot y^2}{(x^2 y)^2} = \frac{\cancel{xy^2} + 10y^2 - \cancel{xy^2}}{x^4 y^2} = \frac{10\cancel{x^2}}{x^{\cancel{2}} y^2} = \frac{10}{x^2 y^2}$$

17.

$$f(x, y) = x^3 + y^2$$

$$f_x(x, y) = 3x^2 \qquad\qquad f_y(x, y) = 2y$$

$$f_{xy}(x, y) = 0 \qquad\qquad f_{yx}(x, y) = 0$$

19.

$$f(x, y) = 3x^4 y^2 + 7x^2 y$$

$$f_x(x, y) = 12x^3 y^2 + 14xy \qquad\qquad f_y(x, y) = 6x^4 y + 7x^2$$

$$f_{xy}(x, y) = 24x^3 y + 14x \qquad\qquad f_{yx}(x, y) = 24x^3 y + 14x$$

21.

$$f(x, y) = \frac{y}{x^2} = y \cdot x^{-2}$$

$$f_x(x, y) = -2yx^{-3} = \frac{-2y}{x^3} \qquad\qquad f_y(x, y) = x^{-2} = \frac{1}{x^2}$$

$$f_{xy}(x, y) = -2x^{-3} = \frac{-2}{x^3} \qquad\qquad f_{yx}(x, y) = -2x^{-3} = \frac{-2}{x^3}$$

23. $f_x(x, y, z) = 2xy - 3yz$

$f_y(x, y, z) = x^2 - 3xz$

$f_z(x, y, z) = -3xy + 3z^2$

25. $f_x(x, y, z) = e^y$

$f_y(x, y, z) = xe^y + e^z$

$f_z(x, y, z) = ye^z$

27. $f_x(x, y, z) = \ln(yz) + y \cdot \dfrac{1}{x} = \ln(yz) + \dfrac{y}{x}$

$f_y(x, y, z) = x \cdot \dfrac{1}{y} + \ln(xz) = \dfrac{x}{y} + \ln(xz)$

$f_z(x, y, z) = x \cdot \dfrac{1}{z} + y \cdot \dfrac{1}{z} = \dfrac{x}{z} + \dfrac{y}{z} = \dfrac{x+y}{z}$

29. $f_x(x, y, z) = \dfrac{2x}{x^2 + y^2 + z^2}$

$f_y(x, y, z) = \dfrac{2y}{x^2 + y^2 + z^2}$

$f_z(x, y, z) = \dfrac{2z}{x^2 + y^2 + z^2}$

31. $z = f(x, y) = 5x^2 + 3y^2$

$f_x(x, y) = 10x$

$f_x(2, 3) = 10 \cdot 2 = 20$

The slope of the tangent line to the curve of intersection of the surface z and the plane $y = 3$ is 20.

33. $z = f(x, y) = \sqrt{16 - x^2 - y^2} = \left(16 - x^2 - y^2\right)^{1/2}$

$f_y(x, y) = \dfrac{1}{2}\left(16 - x^2 - y^2\right)^{-1/2} \cdot (-2y) = \dfrac{-y}{\left(16 - x^2 - y^2\right)^{-1/2}} = \dfrac{-y}{\sqrt{16 - x^2 - y^2}}$

$f_y(1, 2) = \dfrac{-2}{\sqrt{16 - 1^2 - 2^2}} = \dfrac{-2}{\sqrt{11}} = -\dfrac{2\sqrt{11}}{11}$

The slope of the tangent line to the curve of intersection of the surface z and the plane $x = 1$ is $-\dfrac{2\sqrt{11}}{11}$.

35. $z = f(x, y) = e^x \ln y$

$$f_y(x, y) = e^x \cdot \frac{1}{y} = \frac{e^x}{y}$$

$$f_y(0, 1) = \frac{e^0}{1} = 1$$

The slope of the tangent line to the curve of intersection of the surface z and the plane $x = 0$ is 1.

37. $z = f(x, y) = 2\ln \sqrt{x^2 + y^2} = 2\ln \left(x^2 + y^2\right)^{1/2} = \ln \left(x^2 + y^2\right)$

$$f_y(x, y) = \frac{2y}{x^2 + y^2}$$

$$f_y(1, 1) = \frac{2 \cdot 1}{1^2 + 1^2} = \frac{2}{2} = 1$$

The slope of the tangent line to the curve of intersection of the surface z and the plane $x = 1$ is 1.

39. $z = x^2 + 4y^2$

$$\frac{\partial z}{\partial x} = 2x \qquad\qquad\qquad\qquad\qquad \frac{\partial z}{\partial y} = 8y$$

$$x\frac{\partial z}{\partial x} + y\frac{\partial z}{\partial y} = x \cdot 2x + y \cdot 8y = 2x^2 + 8y^2 = 2\left(x^2 + 4y^2\right) = 2z$$

41. $z = \ln \sqrt{x^2 + y^2} = \ln \left(x^2 + y^2\right)^{1/2} = \frac{1}{2}\ln \left(x^2 + y^2\right)$

$$\frac{\partial z}{\partial x} = \frac{1}{2} \cdot \frac{2x}{x^2 + y^2} = \frac{x}{x^2 + y^2} \qquad\qquad \frac{\partial z}{\partial y} = \frac{1}{2} \cdot \frac{2y}{x^2 + y^2} = \frac{y}{x^2 + y^2}$$

$$\frac{\partial^2 z}{\partial x^2} = \frac{1 \cdot \left(x^2 + y^2\right) - x \cdot 2x}{\left(x^2 + y^2\right)^2} \qquad\qquad \frac{\partial^2 z}{\partial y^2} = \frac{1 \cdot \left(x^2 + y^2\right) - y \cdot 2y}{\left(x^2 + y^2\right)^2}$$

$$= \frac{y^2 - x^2}{\left(x^2 + y^2\right)^2} \qquad\qquad\qquad\qquad = \frac{x^2 - y^2}{\left(x^2 + y^2\right)^2}$$

$$\frac{\partial^2 z}{\partial x^2} + \frac{\partial^2 z}{\partial y^2} = \frac{y^2 - x^2}{\left(x^2 + y^2\right)^2} + \frac{x^2 - y^2}{\left(x^2 + y^2\right)^2} = \frac{y^2 - x^2 + x^2 - y^2}{\left(x^2 + y^2\right)^2} = 0$$

43. $z = 1000 - 20x - 50y$

(a) $\dfrac{\partial z}{\partial x} = -20 \qquad\qquad\qquad\qquad\qquad \dfrac{\partial z}{\partial y} = -50$

(b)   Answers will vary.

45.
$$A = 9\left(\frac{N}{I}\right)$$

(a) $\dfrac{\partial A}{\partial N} = \dfrac{9}{I}$ $\dfrac{\partial A}{\partial I} = -9\left(\dfrac{N}{I^2}\right)$

(b) If $N = 78$ and $I = 217$, then

$\dfrac{\partial A}{\partial N} = \dfrac{9}{217} = 0.0415$ $\dfrac{\partial A}{\partial I} = -9\left(\dfrac{78}{217^2}\right) = -0.0149$

(c) Answers will vary.

47.

(a) $\dfrac{\partial H}{\partial t} = 2.049015323 - 0.22475541r - 0.01367566t + 0.00245748\,tr + 0.00085282r^2$
$-\,0.00000398tr^2$

(b) Answers will vary.

(c) $\dfrac{\partial H}{\partial r} = 10.14333127 - 0.22475541t - 0.10963434r + 0.00122874t^2 + 0.00170564tr$
$-\,0.00000398t^2r$

(d) Answers will vary.

49. No, you should not believe it. Explanations will vary.

8.4 Local Maxima and Local Minima

1. We find the partial derivatives f_x and f_y, set each equal to zero, and solve the system of equations.

$f_x(x, y) = 4x^3 - 4x = 0$ $f_y(x, y) = 2y = 0$
$\quad\quad 4x(x^2 - 1) = 0$ $\quad\quad\quad y = 0$
$\quad 4x(x-1)(x+1) = 0$
$\quad 4x = 0 \quad$ or $\quad x - 1 = 0 \quad$ or $\quad x + 1 = 0$
$\quad\quad x = 0 \quad$ or $\quad\quad x = 1 \quad$ or $\quad\quad x = -1$
The critical points are $(0, 0)$, $(1, 0)$, and $(-1, 0)$.

3. We find the partial derivatives f_x and f_y, set each equal to zero, and solve the system of equations.

$f_x(x, y) = 4y - 4x^3 = 0$ $f_y(x, y) = 4x - 4y^3 = 0$
$\quad\quad\quad y = x^3$ $\quad\quad\quad x - (x^3)^3 = 0$
$\quad\quad\quad\quad\quad\quad\quad\quad\quad\quad x(1 - x^8) = 0$
$\quad\quad\quad\quad\quad\quad\quad x = 0 \quad$ or $\quad 1 - x^8 = 0$
$\quad\quad\quad\quad\quad\quad\quad\quad\quad\quad\quad\quad\quad\quad x = \pm 1$
The critical points are $(x, x^3) = (0, 0)$, $(1, 1)$, and $(-1, -1)$.

5. We find the partial derivatives f_x and f_y, set each equal to zero, and solve the system of equations.

$$f_x(x, y) = 4x^3 = 0 \qquad\qquad f_y(x, y) = 4y^3 = 0$$
$$x = 0 \qquad\qquad y = 0$$

The critical point is $(0, 0)$.

7. We find the partial derivatives f_x and f_y, set each equal to zero, and solve the system of equations.

$$f_x(x, y) = 6x - 2y = 0 \qquad\qquad f_y(x, y) = -2x + 2y = 0$$

Adding $\qquad 6x - 2y = 0$
and $\qquad -2x + 2y = 0$
we get $\qquad 4x \quad\quad = 0 \quad$ or $\quad x = 0$.

Substituting $x = 0$ into the first equation, we find $-2y = 0$ or $y = 0$.
The critical point is $(0, 0)$.

To determine the character of the critical point we find the two second order partial derivatives and the mixed partial derivative, evaluate them at the critical point, and calculate D.

$$f_{xx}(x, y) = 6 \qquad\qquad f_{yy}(x, y) = 2 \qquad\qquad f_{xy}(x, y) = -2$$
$$f_{xx}(0, 0) = 6 \qquad\qquad f_{yy}(0, 0) = 2 \qquad\qquad f_{xy}(0, 0) = -2$$

$$D = f_{xx}(x_0, y_0) \cdot f_{yy}(x_0, y_0) - \left[f_{xy}(x_0, y_0) \right]^2 = 6 \cdot 2 - (-2)^2 = 8 > 0$$

Since $f_{xx}(0, 0) > 0$ and $D > 0$, f has a local minimum at $(x_0, y_0, z_0) = (0, 0, 0)$.

9. We find the partial derivatives f_x and f_y, set each equal to zero, and solve the system of equations.

$$f_x(x, y) = 2x - 3 = 0 \qquad\qquad f_y(x, y) = 2y = 0$$
$$x = \frac{3}{2} = 1.5 \qquad\qquad y = 0$$

The critical point is $\left(\dfrac{3}{2}, 0 \right)$.

To determine the character of the critical point we find the two second order partial derivatives and the mixed partial derivative, evaluate them at the critical point, and calculate D.

$$f_{xx}(x, y) = 2 \qquad\qquad f_{yy}(x, y) = 2 \qquad\qquad f_{xy}(x, y) = 0$$
$$f_{xx}\left(\frac{3}{2}, 0 \right) = 2 \qquad\qquad f_{yy}\left(\frac{3}{2}, 0 \right) = 2 \qquad\qquad f_{xy}\left(\frac{3}{2}, 0 \right) = 0$$

$$D = f_{xx}(x_0, y_0) \cdot f_{yy}(x_0, y_0) - \left[f_{xy}(x_0, y_0) \right]^2 = 2 \cdot 2 - 0^2 = 4 > 0$$

Since $f_{xx}\left(\dfrac{3}{2}, 0\right) > 0$ and $D > 0$, f has a local minimum at $\left(x_0, y_0, z_0\right) = \left(\dfrac{3}{2}, 0, \dfrac{39}{4}\right)$.

11. We find the partial derivatives f_x and f_y, set each equal to zero, and solve the system of equations.

$$f_x(x, y) = 2x + 4 = 0 \qquad\qquad f_y(x, y) = -2y + 8 = 0$$
$$x = -2 \qquad\qquad\qquad\qquad y = 4$$

The critical point is $(-2, 4)$.

To determine the character of the critical point we find the two second order partial derivatives and the mixed partial derivative, evaluate them at the critical point, and calculate D.

$$f_{xx}(x, y) = 2 \qquad\qquad f_{yy}(x, y) = -2 \qquad\qquad f_{xy}(x, y) = 0$$
$$f_{xx}(-2, 4) = 2 \qquad\qquad f_{yy}(-2, 4) = -2 \qquad\qquad f_{xy}(-2, 4) = 0$$
$$D = f_{xx}(x_0, y_0) \cdot f_{yy}(x_0, y_0) - \left[f_{xy}(x_0, y_0)\right]^2 = 2 \cdot (-2) - 0^2 = -4$$

Since $D < 0$, $\left(x_0, y_0, z_0\right) = (-2, 4, 12)$ is a saddle point.

13. We find the partial derivatives f_x and f_y, set each equal to zero, and solve the system of equations.

$$f_x(x, y) = 2x - 4 = 0 \qquad\qquad f_y(x, y) = 8y + 8 = 0$$
$$x = 2 \qquad\qquad\qquad\qquad y = -1$$

The critical point is $(2, -1)$.

To determine the character of the critical point we find the two second order partial derivatives and the mixed partial derivative, evaluate them at the critical point, and calculate D.

$$f_{xx}(x, y) = 2 \qquad\quad f_{yy}(x, y) = 8 \qquad\qquad f_{xy}(x, y) = 0$$
$$D = f_{xx}(x_0, y_0) \cdot f_{yy}(x_0, y_0) - \left[f_{xy}(x_0, y_0)\right]^2 = 2 \cdot 8 - 0^2 = 16$$

Since $f_{xx}(2, -1) > 0$ and $D > 0$, $\left(x_0, y_0, z_0\right) = (2, -1, -9)$ is a local minimum.

15. We find the partial derivatives f_x and f_y, set each equal to zero, and solve the system of equations.

$$f_x(x, y) = 2x + y - 6 = 0 \qquad\qquad f_y(x, y) = 2y + x = 0$$
$$x = -2y$$
$$2(-2y) + y - 6 = 0$$
$$-3y = 6 \text{ or } y = -2 \qquad\qquad x = -2(-2) = 4$$

The critical point is $(4, -2)$.

To determine the character of the critical point we find the two second order partial derivatives and the mixed partial derivative, evaluate them at the critical point, and calculate D.

$$f_{xx}(x, y) = 2 \qquad\qquad f_{yy}(x, y) = 2 \qquad\qquad f_{xy}(x, y) = 1$$

$$D = f_{xx}(x_0, y_0) \cdot f_{yy}(x_0, y_0) - \left[f_{xy}(x_0, y_0) \right]^2 = 2 \cdot 2 - 1^2 = 3$$

Since $f_{xx}(4, -2) > 0$ and $D > 0$, $(x_0, y_0, z_0) = (4, -2, -6)$ is a local minimum.

17. We find the partial derivatives f_x and f_y, set each equal to zero, and solve the system of equations.

$$f_x(x, y) = 2x + y = 0 \qquad\qquad f_y(x, y) = -2y + x = 0$$

$$y = -2x \qquad\qquad\qquad -2(-2x) + x = 0$$
$$\qquad\qquad\qquad\qquad\qquad 5x = 0$$
$$\qquad\qquad\qquad\qquad\qquad x = 0 \text{ and } y = 0$$

The critical point is $(0, 0)$.

To determine the character of the critical point we find the two second order partial derivatives and the mixed partial derivative, evaluate them at the critical point, and calculate D.

$$f_{xx}(x, y) = 2 \qquad\qquad f_{yy}(x, y) = -2 \qquad\qquad f_{xy}(x, y) = 1$$

$$D = f_{xx}(x_0, y_0) \cdot f_{yy}(x_0, y_0) - \left[f_{xy}(x_0, y_0) \right]^2 = 2(-2) - 1^2 = -5$$

Since $D < 0$, $(x_0, y_0, z_0) = (0, 0, 2)$ is a saddle point.

19. We find the partial derivatives f_x and f_y, set each equal to zero, and solve the system of equations.

$$f_x(x, y) = 3x^2 - 6y = 0 \qquad\qquad f_y(x, y) = -6x + 3y^2 = 0$$

$$y = \frac{1}{2}x^2 \qquad\qquad -6x + 3\left(\frac{x^2}{2} \right)^2 = -2x + \frac{x^4}{4} = 0$$

$$\qquad\qquad\qquad\qquad\qquad x\left(\frac{x^3}{4} - 2 \right) = 0$$

$$\qquad\qquad\qquad\qquad\qquad x = 0 \quad \text{or} \quad \frac{x^3}{4} - 2 = 0$$

$$\qquad\qquad\qquad\qquad\qquad\qquad\qquad x^3 = 8$$

$$\qquad\qquad\qquad\qquad\qquad\qquad\qquad x = 2$$

When $x = 0$, $y = 0$, and when $x = 2$, $y = 2$. So the critical points are $(0, 0)$ and $(2, 2)$.

To determine the character of the critical points we find the two second order partial derivatives and the mixed partial derivative, evaluate them at the critical points, and calculate D.

$$f_{xx}(x, y) = 6x \qquad\qquad f_{yy}(x, y) = 6y \qquad\qquad f_{xy}(x, y) = -6$$

$$f_{xx}(0, 0) = 0 \qquad f_{yy}(0, 0) = 0 \qquad f_{xy}(0, 0) = -6$$

$$D = f_{xx}(0, 0) \cdot f_{yy}(0, 0) - \left[f_{xy}(0, 0)\right]^2 = 0 \cdot 0 - (-6)^2 = -36$$

Since $D < 0$ $(x_0, y_0, z_0) = (0, 0, 0)$ is a saddle point.

$$f_{xx}(2, 2) = 6 \cdot 2 = 12 \qquad f_{yy}(2, 2) = 6 \cdot 2 = 12 \qquad f_{xy}(2, 2) = -6$$

$$D = f_{xx}(2, 2) \cdot f_{yy}(2, 2) - \left[f_{xy}(2, 2)\right]^2 = 12 \cdot 12 - (-6)^2 = 108$$

Since $f_{xx}(2, 2) > 0$ and $D > 0$, $(x_0, y_0, z_0) = (2, 2, -8)$ is a local minimum.

21. We find the partial derivatives f_x and f_y, set each equal to zero, and solve the system of equations.

$$f_x(x, y) = 3x^2 + 2xy = 0 \qquad\qquad f_y(x, y) = x^2 + 2y = 0$$

$$y = -\frac{1}{2}x^2$$

$$3x^2 + 2x\left(-\frac{1}{2}x^2\right) = 0$$

$$3x^2 - x^3 = 0$$

$$x^2(3 - x) = 0$$

$$x^2 = 0 \quad \text{or} \quad 3 - x = 0$$

$$x = 0 \quad \text{or} \qquad x = 3$$

When $x = 0$, then $y = 0$, and when $x = 3$, then $y = -\dfrac{9}{2}$. So the critical points are $(0, 0)$

and $\left(3, -\dfrac{9}{2}\right)$.

To determine the character of the critical points we find the two second order partial derivatives and the mixed partial derivative, evaluate them at the critical points, and calculate D.

$$f_{xx}(x, y) = 6x + 2y \qquad\qquad f_{yy}(x, y) = 2 \qquad\qquad f_{xy}(x, y) = 2x$$

$$f_{xx}(0, 0) = 0 \qquad\qquad f_{yy}(0, 0) = 2 \qquad\qquad f_{xy}(0, 0) = 0$$

$$D = f_{xx}(x_0, y_0) \cdot f_{yy}(x_0, y_0) - \left[f_{xy}(x_0, y_0)\right]^2 = 0 \cdot 2 - 0^2 = 0.$$

Since $D = 0$, no information is given by the test.

$$f_{xx}\left(3, -\frac{9}{2}\right) = 6 \cdot 3 + 2\left(-\frac{9}{2}\right) = 9 \qquad f_{yy}\left(3, -\frac{9}{2}\right) = 2 \qquad f_{xy}\left(3, -\frac{9}{2}\right) = 2 \cdot 3 = 6$$

$$D = f_{xx}\left(3, -\frac{9}{2}\right) \cdot f_{yy}\left(3, -\frac{9}{2}\right) - \left[f_{xy}\left(3, -\frac{9}{2}\right)\right]^2 = 9 \cdot 2 - 6^2 = -18$$

Since $D < 0$, $\left(x_0, y_0, z_0\right) = \left(3, -\dfrac{9}{2}, \dfrac{27}{4}\right)$ is a saddle point.

23. We find the partial derivatives f_x and f_y, set each equal to zero, and solve the system of equations.

$$f_x\left(x,\ y\right) = \frac{0 - y(1)}{\left(x+y\right)^2} = -\frac{y}{\left(x+y\right)^2} = 0 \qquad\qquad f_y\left(x,\ y\right) = \frac{1 \cdot \left(x+y\right) - y \cdot 1}{\left(x+y\right)^2} = \frac{x}{\left(x+y\right)^2} = 0$$

$$y = 0 \qquad\qquad\qquad\qquad\qquad x = 0$$

but the point $(0, 0)$ is not in the domain of f. So f has no local maximum or local minimum point.

25. If we let R denote the revenue function, then
$$R(x) = xp = x(12 - x) \text{ and } R(y) = y(8 - y).$$
The joint revenue function is given by $R(x, y) = R(x) + R(y)$.
$$R\left(x,\ y\right) = 12x - x^2 + 8y - y^2$$
The joint profit function is given by $P(x, y) = R(x, y) - C(x, y)$.
$$P\left(x,\ y\right) = \left(12x - x^2 + 8y - y^2\right) - \left(x^2 + 2xy + 3y^2\right)$$
$$= 12x - 2x^2 + 8y - 4y^2 - 2xy$$
The first-order partial derivatives are
$$P_x\left(x,\ y\right) = 12 - 4x - 2y \qquad\qquad P_y\left(x,\ y\right) = 8 - 8y - 2x$$
The solution to the system is the critical point of P.
$$4x + 2y = 12$$
$$\underline{2x + 8y = \ \ 8}$$
Using the first equation, we find that $y = 6 - 2x$. Substituting for y in the second equation, we have
$$2x + 8(6 - 2x) = 8$$
$$x + 24 - 8x = 4$$
$$20 = 7x \ \text{ or } x = \frac{20}{7}$$

Back-substituting into the first equation we get $y = 6 - \dfrac{40}{7} = \dfrac{2}{7}$. So $\left(\dfrac{20}{7}, \dfrac{2}{7}\right)$ is the critical point.

The second order partial derivatives of P are
$$P_{xx}\left(x,\ y\right) = -4 \qquad\qquad P_{yy}\left(x,\ y\right) = -8 \qquad\qquad P_{xy}\left(x,\ y\right) = -2$$
and $D = (-4)(-8) - (-2)^2 = 32 - 4 = 28$.

Since $P_{xx}\left(\dfrac{20}{7}, \dfrac{2}{7}\right) = -4 < 0$ and $D > 0$, there is a local maximum at $\left(\dfrac{20}{7}, \dfrac{2}{7}\right)$.

When $x = \dfrac{20}{7} \approx 2.857$, $p = 12 - 2.857 = 9.14286$, and when $y = \dfrac{2}{7} \approx 0.286$, $q = 7.71429$

So to maximize profit, the company should sell 2857 units of product x at $9142.86 per unit and 286 units of product y at $7714.29 per unit. The maximum profit is $18,285.71.

27.
$$P(x, y) = 2000x - \frac{1}{5}x^2 + 1150y - 2y^2 + xy + 10,000$$

We find the partial derivatives P_x and P_y, set each equal to zero, and solve the system of equations.

$$P_x(x, y) = 2000 - \frac{2}{5}x + y = 0 \qquad\qquad P_y(x, y) = 1150 - 4y + x = 0$$

$$y = \frac{2}{5}x - 2000 \qquad\qquad 1150 - 4\left(\frac{2}{5}x - 2000\right) + x = 0$$

$$1150 - \frac{8}{5}x + 8000 + x = 0$$

$$9150 - \frac{3}{5}x = 0$$

$$x = \frac{5}{3} \cdot 9150 = 15,250$$

$$y = \frac{2}{5} \cdot 15,250 - 2000 = 4100$$

The critical point is (15250, 4100).

The second order partial derivatives are

$$P_{xx}(x, y) = -\frac{2}{5} \qquad\qquad P_{yy}(x, y) = -4 \qquad\qquad P_{xy}(x, y) = 1$$

$$D = \left(-\frac{2}{5}\right) \cdot (-4) - 1^2 = \frac{8}{2} - 1 = \frac{3}{5}$$

Since $P_{xx}(15250, 4100) = -\frac{2}{5} < 0$ and $D > 0$, there is a local maximum at (15250, 4100).

The maximum profit of $17,617,500 is obtained when 15,250 tons of grade A steel and 4100 tons of grade B steel is produced.

29. (a) If x is constant, then
$$R_y(x, y) = 2x^2 y(a-x)(b-y) + x^2 y^2(a-x)(-1) = 0$$
$$\left[(a-x)x^2 y\right]\left[2(b-y) - y\right] = 0$$
$$\left[(a-x)x^2 y\right] = 0 \quad \text{or} \quad \left[2b - 3y\right] = 0$$
$$y = 0 \quad \text{or} \quad y = \frac{2b}{3}$$

We use the second derivative test to see which value of y locates a relative maximum.
$$R_{yy}(x, y) = \left[(a-x)x^2\right]\left[2b - 3y\right] - \left[3(a-x)x^2 y\right]$$
$$R_{yy}(x, 0) = \left[(a-x)x^2\right]\left[2b\right] > 0 \text{ since } x < a \text{ and } b > 0.$$

$$R_{yy}\left(x, \frac{2b}{3}\right) = \left[(a-x)x^2\right]\left[2b-2b\right] - \left[2(a-x)x^2b\right] = 0 - 2(a-x)x^2b < 0$$

When $y = \dfrac{2b}{3}$, the reaction to the drug is maximized.

(b) If the amount y of the second drug is held constant, then

$$R_x(x, y) = 2xy^2(a-x)(b-y) - x^2y^2(b-y) = 0$$
$$\left[(b-y)xy^2\right]\left[2(a-x) - x\right] = 0$$
$$(b-y)xy^2 = 0 \quad \text{or} \quad 2a - 3x = 0$$
$$x = 0 \quad \text{or} \quad x = \frac{2a}{3}$$

We use the second derivative test to see which value of x locates a relative maximum.

$$R_{xx}(x, y) = -3\left[(by^2 - y^3)x\right] + (by^2 - y^3)(2a - 3x)$$
$$R_{xx}\left(\frac{2a}{3}, y\right) = -2(by^2 - y^3)a + (by^2 - y^3)(2a - 2a) = -2y^2(b-y)a < 0$$
$$R_{xx}(0, y) = -3\left[(by^2 - y^3)0\right] + (by^2 - y^3)(2a - 3(0)) = (by^2 - y^3)(2a) > 0$$

When $x = \dfrac{2a}{3}$ the reaction to the drug is maximized.

(c) If both x and y are variable, then

$$R_{xy}(x, y) = (2a - 3x)(2ybx - 3xy^2) = (2a - 3x)(2b - 3y)xy$$
$$R_{xy}\left(\frac{2a}{3}, \frac{2b}{3}\right) = \left[2a + 3\left(\frac{2a}{3}\right)\right]\left[2b - 3\left(\frac{2b}{3}\right)\right]\left[\left(\frac{2a}{3}\right)\left(\frac{2b}{3}\right)\right] = 0$$

$$R_{xx}\left(\frac{2a}{3}, \frac{2b}{3}\right) = \left[b - \frac{2b}{3}\right]\left[\frac{2b}{3}\right]^2\left[2a - 6\left(\frac{2a}{3}\right)\right]$$
$$= \left[\frac{3b - 2b}{3}\right]\left[\frac{4b^2}{9}\right][-2a] = \frac{b}{3} \cdot \frac{4b^2}{9} \cdot (-2a) = -\frac{8ab^3}{27}$$

$$R_{yy}\left(\frac{2a}{3}, \frac{2b}{3}\right) = \left[a - \frac{2a}{3}\right]\left[\frac{2a}{3}\right]^2\left[2b - 6\left(\frac{2b}{3}\right)\right]$$
$$= \left[\frac{3a - 2a}{3}\right]\left[\frac{4a^2}{9}\right][-2b] = \frac{a}{3} \cdot \frac{4a^2}{9} \cdot (-2b) = -\frac{8a^3b}{27}$$

$$D = \left[-\frac{8ab^3}{27}\right] \cdot \left[-\frac{8a^3b}{27}\right] - 0 = \frac{64a^4b^4}{27^2} > 0$$

Since $R_{xx}\left(\dfrac{2a}{3}, \dfrac{2b}{3}\right) < 0$ and $D > 0$, then the reaction to the drug is a maximum.

31. To find the maximum reaction to the drug, we first find the critical points of y.
$$y = x^2[a-x]t = ax^2t - x^3t$$

$$\frac{\partial y}{\partial x} = 2axt - 3x^2t = 0 \qquad\qquad \frac{\partial y}{\partial t} = ax^2 - x^3 = 0$$
$$x^2(a-x) = 0$$
$$x = 0 \quad \text{or} \quad x = a$$

We find the second order partial derivatives and test to see if the critical point locates a relative maximum point.

$$\frac{\partial^2 y}{\partial x^2} = 2at - 6xt \qquad \frac{\partial^2 y}{\partial t^2} = 0 \qquad \frac{\partial^2 y}{\partial x \partial t} = 2ax - 3x^2$$

$$D = \frac{\partial^2 y}{\partial x^2} \cdot \frac{\partial^2 y}{\partial t^2} - \left[\frac{\partial y}{\partial x \partial t}\right]^2 = 0 \cdot 0 - \left(2a^2 - 3a^2\right)^2 = -a^4$$

Since $D < 0$, there is a saddle point at $(a, 0, 0)$, and there is no maximum reaction to the drug.

33. (a) Let w denote the width, h denote the depth, and l denote the length of the box. There are two restrictions $l \le 108$, and $l + 2w + 2h \le 130$ inches. So $l \le 130 - 2w - 2h$.
$$V = lwh = (130 - 2w - 2h)wh = 130wh - 2w^2h - 2wh^2$$
To find the dimensions of the box that meet regulations while maximizing volume, we find the critical points of V.

$$V_h = 130w - 2w^2 - 4wh = 0 \qquad\qquad V_w = 130h - 4wh - 2h^2 = 0$$
$$2w(65 - w - 2h) = 0 \qquad\qquad\qquad = h[130 - 4w - 2h]$$
$$w = 0 \quad \text{or} \quad 65 - w - 2h = 0$$
$$w = 65 - 2h$$

When $w = 0$, then $130h - 2h^2 = 0$ or $h = 0$ or $h = 65$, but both of these measurements result in $V = 0$ cubic inches, which is a minimum.

When $w = 65 - 2h$, then
$$h[130 - 4(65 - 2h) - 2h] = 0$$
$$h[-130 + 6h] = 0$$
$$h = 0 \quad \text{or} \quad 6h - 130 = 0$$
$$h = \frac{65}{3} \qquad \text{and so } w = 65 - 2\left(\frac{65}{3}\right) = \frac{65}{3} = 21.667$$

To check to see if these values result in the maximum volume, we find the second order partials and evaluate D.

$$V_{hh} = -4w \qquad\qquad V_{ww} = -4h \qquad\qquad V_{hw} = 130 - 4w - 4h$$

$$V_{hh}\left(\frac{65}{3}, \frac{65}{3}\right) = -4\left(\frac{65}{3}\right) = -\frac{260}{3} \qquad\qquad V_{ww}\left(\frac{65}{3}, \frac{65}{3}\right) = -4\left(\frac{65}{3}\right) = -\frac{260}{3}$$

$$V_{lw}\left(\frac{65}{3},\frac{65}{3}\right)=130-4\left(\frac{65}{3}\right)-4\left(\frac{65}{3}\right)=130-8\left(\frac{65}{3}\right)=-\frac{130}{3}$$

$$D=\left(-\frac{260}{3}\right)\left(-\frac{260}{3}\right)-\left(-\frac{130}{3}\right)^2=\frac{4\cdot130^2}{9}-\frac{130^2}{9}=\frac{3\cdot130^2}{9}>0$$

Since $V_{hh}\left(\frac{65}{3},\frac{65}{3}\right)<0$ and $D>0$, the volume is maximum when the width $=21.667$

inches, the depth is 21.667 inches and the length is $130-4(21.667)=43.332$ inches. The maximum volume is 20,342.59 cubic inches.

(b) If we let r denote the radius of the cylinder and h denote the height, then we have the restriction $h+2\pi r\le130$ or $h\le130-2\pi r$. The volume of the cylinder is
$$V=\pi r^2h=\pi r^2\left(130-2\pi r\right)=130\pi r^2-2\pi^2r^3$$

To find the dimensions of the cylinder that meet regulations while maximizing volume, we find the critical points of V.
$$\frac{dV}{dr}=260\pi r-6\pi^2r^2=2\pi r\left(130-3\pi r\right)=0$$

$$r=0 \quad\text{or}\quad 130-3\pi r=0$$

$$r=\frac{130}{3\pi}$$

We use the second derivative test to see if the volume is maximized.
$$\frac{d^2V}{dr^2}=260\pi-12\pi^2r$$

When $r=\frac{130}{3\pi}$, then $\frac{d^2V}{dr^2}=260\pi-12\left(\frac{130}{3\pi}\right)\pi^2=-260\pi<0$. So the volume is

maximized if the cylinder has a radius of $\frac{130}{3\pi}\approx13.79$ inches and a height of 43.33

inches.

35. We find the critical points of W.

$$W_x=\frac{1}{100}\left[\frac{1}{10}x-y-4\right]=0 \qquad\qquad W_y=\frac{1}{100}\left[50y-x\right]=0$$

$$x-10y-40=0 \qquad\qquad\qquad 50y-x=0 \quad\text{or}\quad x=50y$$

$$50y-10y=40$$
$$40y=40$$
$$y=1 \qquad\text{and}\qquad x=50$$

The critical point is (50, 1). We find the second order partial derivatives to test if the critical point locates a minimum value.

$$W_{xx} = \frac{1}{100} \cdot \frac{1}{10} = \frac{1}{1000} \quad W_{yy} = \frac{1}{100}[50] = \frac{1}{2} \quad W_{xy} = \frac{1}{100}[-1] = -\frac{1}{100}$$

$$D = \left[\frac{1}{1000}\right]\left[\frac{1}{2}\right] - \left[-\frac{1}{100}\right]^2 = \frac{1}{2000} - \frac{1}{10,000} = \frac{5-1}{10,000} = \frac{4}{10,000} > 0$$

Since $W_{xx}(50, 1) > 0$ and $D > 0$, waste is minimized when the manufacturer uses 50 tons of steel at the rate of 1 ton per week.

8.5 Lagrange Multipliers

1. **STEP 1** Find the maximum of $z = f(x, y) = 3x + 4y$
    Subject to the constraint $\quad g(x, y) = x^2 + y^2 - 9 = 0$

    STEP 2 Construct the function $F(x, y, \lambda) = 3x + 4y + \lambda\left(x^2 + y^2 - 9\right)$.

    STEP 3 Set up the system of equations

    $$\frac{\partial F}{\partial x} = 3 + 2\lambda x = 0 \qquad (1)$$

    $$\frac{\partial F}{\partial y} = 4 + 2\lambda y = 0 \qquad (2)$$

    $$\frac{\partial F}{\partial \lambda} = x^2 + y^2 - 9 = 0 \qquad (3)$$

    STEP 4 Solve the system of equations for x, y, and λ.

    $$\lambda = -\frac{3}{2x} \qquad (1)$$

    $$4 + 2\left(-\frac{3}{2x}\right)y = 0 \qquad (2)$$

    $$4 - \frac{3y}{x} = 0$$

    $$y = \frac{4x}{3}$$

    $$x^2 + \left(\frac{4x}{3}\right)^2 - 9 = 0 \qquad (3)$$

    $$9x^2 + 16x^2 = 81$$

    $$25x^2 = 81$$

    $$x = \pm\sqrt{\frac{81}{25}} = \pm\frac{9}{5}$$

    $$y = \frac{4}{3} \cdot \left(\pm\frac{9}{5}\right) = \pm\frac{12}{5}, \text{ and } \lambda = -\frac{3}{2} \cdot \left(\pm\frac{5}{9}\right) = \pm\frac{5}{6}$$

STEP 5 Evaluate $z = f(x, y)$.

$$z = f\left(\frac{9}{5}, \frac{12}{5}\right) = 3 \cdot \frac{9}{5} + 4 \cdot \frac{12}{5} = \frac{27 + 48}{5} = \frac{75}{5} = 15$$

$$z = f\left(-\frac{9}{5}, -\frac{12}{5}\right) = 3 \cdot \left(-\frac{9}{5}\right) + 4 \cdot \left(-\frac{12}{5}\right) = \frac{-27 - 48}{5} = \frac{-75}{5} = -15$$

The maximum value of z subject to constraint g, is 15.

3. **STEP 1** Find the minimum of $z = f(x, y) = x^2 + y^2$
    Subject to the constraint $\quad g(x, y) = x + y - 1 = 0$

STEP 2 Construct the function $F(x, y, \lambda) = f(x, y) + \lambda g(x, y)$.
$$F(x, y, \lambda) = x^2 + y^2 + \lambda(x + y - 1) = 0$$

STEP 3 Set up the system of equations
$$\begin{aligned} F_x(x, y, \lambda) &= 2x + \lambda = 0 & \Rightarrow \quad \lambda = -2x & \quad (1) \\ F_y(x, y, \lambda) &= 2y + \lambda = 0 & \Rightarrow \quad \lambda = -2y & \quad (2) \\ F_\lambda(x, y, \lambda) &= x + y - 1 = 0 & & \quad (3) \end{aligned}$$

STEP 4 Solve the system of equations for x, y, and λ.
From (1) and (2) we find that $-2x = -2y$ or $x = y$.

Substituting this result in (3) gives $y + y = 1$ or $2y = 1$ or $y = \frac{1}{2}$ and $x = \frac{1}{2}$.

STEP 5 $\quad z = f\left(\frac{1}{2}, \frac{1}{2}\right) = \left(\frac{1}{2}\right)^2 + \left(\frac{1}{2}\right)^2 = \frac{1}{2}$

We conclude that z attains its minimum value at $\left(\frac{1}{2}, \frac{1}{2}\right)$. The minimum value is $\frac{1}{2}$.

5. **STEP 1** Find the maximum of $z = f(x, y) = 12xy - 3y^2 - x^2$
    Subject to the constraint $\quad g(x, y) = x + y - 16 = 0$
    STEP 2 Construct the function $F(x, y, \lambda) = f(x, y) + \lambda g(x, y)$
    $$F(x, y, \lambda) = 12xy - 3y^2 - x^2 + \lambda(x + y - 16)$$
    STEP 3 Set up the system of equations
    $$\begin{aligned} F_x(x, y, \lambda) &= 12y - 2x + \lambda = 0 & (1) \\ F_y(x, y, \lambda) &= 12x - 6y + \lambda = 0 & (2) \\ F_\lambda(x, y, \lambda) &= x + y - 16 = 0 & (3) \end{aligned}$$
    STEP 4 Solve the system of equations for x and y.
    Subtracting (2) from (1) we get $18y - 14x = 0$
    $$x = \frac{18}{14}y = \frac{9}{7}y$$

Substituting for x in (3), gives $\frac{9}{7}y + y = 16$

$$\frac{16}{7}y = 16 \text{ or } y = 7 \text{ and } x = 9.$$

STEP 5 Evaluate $z = f(x, y)$ at (9, 7).

$$z = 12(9)(7) - 3(7^2) - 9^2 = 528$$

We conclude that z attains its maximum value at (9, 7). The maximum value is 528.

7. **STEP 1** Find the minimum of $z = f(x, y) = 5x^2 + 6y^2 - xy$
   Subject to the constraint $g(x, y) = x + 2y - 24 = 0$
   STEP 2 Construct the function $F(x, y, \lambda) = f(x, y) + \lambda g(x, y)$
   $$F(x, y, \lambda) = 5x^2 + 6y^2 - xy + \lambda(x + 2y - 24)$$

   STEP 3 Set up the system of equations
   $$\begin{array}{ll} F_x(x, y, \lambda) = 10x - y + \lambda = 0 & (1) \\ F_y(x, y, \lambda) = 12y - x + 2\lambda = 0 & (2) \\ F_\lambda(x, y, \lambda) = x + 2y - 24 = 0 & (3) \end{array}$$

   STEP 4 Solve the system of equations for x and y.
   Subtracting twice (1) from (2) gives $-21x + 14y = 0$

   $$y = \frac{21x}{14} = \frac{3}{2}x$$

   Substituting for y in (3) we get $x + 2\left(\frac{3}{2}x\right) = 24$

   $$4x = 24 \text{ or } x = 6 \text{ and } y = \frac{3}{2} \cdot 6 = 9$$

   STEP 5 Evaluate $z = f(x, y)$ at (6, 9).
   $$z = 5(6^2) + 6(9^2) - (6)(9) = 612$$
   We conclude that z attains its minimum value at (6, 9). The minimum value is 612.

9. **STEP 1** Find the maximum of $w = f(x, y, z) = xyz$
   Subject to the constraint $g(x, y, z) = x + 2y + 2z - 120 = 0$
   STEP 2 Construct the function $F(x, y, z, \lambda) = f(x, y, z) + \lambda g(x, y, z)$
   $$= xyz + x(x + 2y + 2z - 120)$$

   STEP 3 Set up the system of equations

$$F_x(x,y,z,\lambda)=yz+\lambda=0 \qquad\qquad \Rightarrow \qquad \lambda=-yz \qquad (1)$$

$$F_y(x,y,z,\lambda)=xz+2\lambda=0 \qquad\qquad \Rightarrow \qquad \lambda=-\frac{xz}{2} \qquad (2)$$

$$F_z(x,y,z,\lambda)=xy+2\lambda=0 \qquad\qquad \Rightarrow \qquad \lambda=-\frac{xy}{2} \qquad (3)$$

$$F_\lambda(x,y,z,\lambda)=x+2y+2z-120=0 \qquad\qquad (4)$$

STEP 4 Solve the system of equations for x, y, and z.

From (1) and (2) we find $-yz=-\dfrac{xz}{2}$ or $y=\dfrac{x}{2}$.

From (1) and (3) we find $-yz=-\dfrac{xy}{2}$ or $z=\dfrac{x}{2}$.

Substituting for y and z in (4) gives $x+2\left(\dfrac{x}{2}\right)+2\left(\dfrac{x}{2}\right)=120$ or $3x=120$ or $x=40$.

Back substituting we find $y=20$ and $z=20$.
STEP 5 Evaluate $w=f(x,y,z)$ at (40, 20, 20).
$$w=(40)(20)(20)=16{,}000$$
We conclude that w attains its maximum value at (40, 20, 20). The maximum is 16,000.

11. **STEP 1** Find the minimum of $w=f(x,y,z)=x^2+y^2+z^2-x-3y-5z$
    Subject to the constraint $g(x,y,z)=x+y+2z-20=0$

    STEP 2 Construct the function $F(x,y,z,\lambda)=f(x,y,z)+\lambda\,g(x,y,z)$
$$F(x,y,z,\lambda)=x^2+y^2+z^2-x-3y-5z+\lambda(x+y+2z-20)$$

    STEP 3 Set up the system of equations
$$F_x(x,y,z,\lambda)=2x-1+\lambda=0 \qquad \Rightarrow \qquad \lambda=1-2x \qquad (1)$$
$$F_y(x,y,z,\lambda)=2y-3+\lambda=0 \qquad \Rightarrow \qquad \lambda=3-2y \qquad (2)$$

$$F_z(x,y,z,\lambda)=2z-5+2\lambda=0 \qquad \Rightarrow \qquad \lambda=\frac{5-2z}{2} \qquad (3)$$

$$F_\lambda(x,y,z,\lambda)=x+y+2z-20=0 \qquad\qquad (4)$$

    STEP 4 Solve the system of equations for x, y, and z.
From (1) and (2) we find $1-2x=3-2y$
$$-2x=2-2y \text{ or } x=y-1$$

From (2) and (3) we find $3-2y=\dfrac{5-2z}{2}$
$$6-4y=5-2z$$
$$1-4y=-2z$$
$$-\frac{1}{2}+2y=z$$

Substituting for x and z in (4) gives
$$y-1+y+2\left(-\frac{1}{2}+2y\right)=20$$

$$y - 1 + y - 1 + 4y = 20$$

$$6y = 22 \quad \text{or} \quad y = \frac{22}{6} = \frac{11}{3}$$

Back substituting we get $x = y - 1 = \dfrac{8}{3}$ and $z = -\dfrac{1}{2} + 2y = -\dfrac{1}{2} + \dfrac{22}{3} = \dfrac{-3 + 44}{6} = \dfrac{41}{6}$.

STEP 5 Evaluate $w = f(x, y, z)$ at $\left(\dfrac{8}{3}, \dfrac{11}{3}, \dfrac{41}{6}\right)$.

$$w = \left(\frac{8}{3}\right)^2 + \left(\frac{11}{3}\right)^2 + \left(\frac{41}{6}\right)^2 - \frac{8}{3} - 3\left(\frac{11}{3}\right) - 5\left(\frac{41}{6}\right) = \frac{233}{12} \approx 19.417$$

We conclude that w attains its minimum at $\left(\dfrac{8}{3}, \dfrac{11}{3}, \dfrac{41}{6}\right)$. The minimum value is $\dfrac{233}{12}$.

13. **STEP 1**   Maximize $z = f(x, y) = xy$
Subject to the constraint $g(x, y) = x + y - 100 = 0$
STEP 2 Construct the function $F(x, y, \lambda) = f(x, y) + \lambda \, g(x, y)$
$$F(x, y, \lambda) = xy + \lambda(x + y - 100)$$

STEP 3 Set up the system of equations

$$
\begin{aligned}
F_x(x, y, \lambda) = y + \lambda = 0 &\quad \Rightarrow \quad \lambda = -y &\quad (1) \\
F_y(x, y, \lambda) = x + \lambda = 0 &\quad \Rightarrow \quad \lambda = -x &\quad (2) \\
F_\lambda(x, y, \lambda) = x + y - 100 = 0 &\quad &\quad (3)
\end{aligned}
$$

STEP 4 Solve the system of equations for x and y.
From (1) and (2) we find $-y = -x$ or $y = x$.
Substituting for y in (3) gives $x + x = 100$
$$2x = 100 \quad \text{or} \quad x = 50$$
Since $x = 50$, $y = 50$.

STEP 5 Evaluate $z = f(x, y)$ at $(50, 50)$.
$$z = (50)(50) = 2500$$
The two numbers whose sum is 100 and whose product is a maximum are 50 and 50.

15. **STEP 1** Find the maximum of $w = f(x, y, z) = x + y + z$
Subject to the constraint $g(x, y) = x^2 + y^2 + z^2 - 25 = 0$
STEP 2 Construct the function $F(x, y, z, \lambda) = f(x, y, z) + \lambda \, g(x, y, z)$
$$F(x, y, z, \lambda) = x + y + z + \lambda\left(x^2 + y^2 + z^2 - 25\right)$$
STEP 3 Set up the system of equations

$$F_x(x, y, z, \lambda) = 1 + 2x\lambda = 0 \qquad \Rightarrow \qquad \lambda = -\frac{1}{2x} \qquad (1)$$

$$F_y(x, y, z, \lambda) = 1 + 2y\lambda = 0 \qquad \Rightarrow \qquad \lambda = -\frac{1}{2y} \qquad (2)$$

$$F_z(x, y, z, \lambda) = 1 + 2z\lambda = 0 \qquad \Rightarrow \qquad \lambda = -\frac{1}{2z} \qquad (3)$$

$$F_\lambda(x, y, z, \lambda) = x^2 + y^2 + z^2 - 25 = 0 \qquad (4)$$

STEP 4 Solve the system of equations for x, y, and z.

From (1) and (2) we find $-\dfrac{1}{2x} = -\dfrac{1}{2y}$

$$2y = 2x \qquad \text{or} \quad y = x$$

From (1) and (3) we find $-\dfrac{1}{2x} = -\dfrac{1}{2z}$

$$2z = 2x \qquad \text{or} \quad z = x$$

Substituting for y and z in (4) gives

$$x^2 + x^2 + x^2 = 25$$
$$3x^2 = 25$$
$$x^2 = \frac{25}{3} \qquad \text{or} \qquad x = \pm\frac{5}{\sqrt{3}} = \pm\frac{5\sqrt{3}}{3}$$

When $x = \dfrac{5\sqrt{3}}{3}$, $y = x = \dfrac{5\sqrt{3}}{3}$ and $z = x = \dfrac{5\sqrt{3}}{3}$.

When $x = -\dfrac{5\sqrt{3}}{3}$, $y = x = -\dfrac{5\sqrt{3}}{3}$ and $z = x = -\dfrac{5\sqrt{3}}{3}$.

STEP 5 Evaluate $w = f(x, y, z)$ at each solution (x_0, y_0, z_0).

At $\left(\dfrac{5\sqrt{3}}{3}, \dfrac{5\sqrt{3}}{3}, \dfrac{5\sqrt{3}}{3} \right)$, $w = \dfrac{5\sqrt{3}}{3} + \dfrac{5\sqrt{3}}{3} + \dfrac{5\sqrt{3}}{3} = 5\sqrt{3}$.

At $\left(-\dfrac{5\sqrt{3}}{3}, -\dfrac{5\sqrt{3}}{3}, -\dfrac{5\sqrt{3}}{3} \right)$, $w = \left(-\dfrac{5\sqrt{3}}{3} \right) + \left(-\dfrac{5\sqrt{3}}{3} \right) + \left(-\dfrac{5\sqrt{3}}{3} \right) = -5\sqrt{3}$

We conclude the sum of the three numbers that satisfy the constraint is maximum when each of the numbers equals $\dfrac{5\sqrt{3}}{3}$.

17. We solve the problem using the method of Lagrange multipliers.
    STEP 1 Find the minimum cost $C = C(x, y) = 18x^2 + 9y^2$
    Subject to the constraint $g(x, y) = x + y - 54 = 0$
    STEP 2 Construct the function $F(x, y, \lambda) = C(x, y) + \lambda g(x, y)$
    $$F(x, y, \lambda) = 18x^2 + 9y^2 + \lambda(x + y - 54)$$

STEP 3 Set up the system of equations

$$F_x(x,y,\lambda)=36x+\lambda=0 \qquad \Rightarrow \qquad \lambda=-36x \qquad (1)$$
$$F_y(x,y,\lambda)=18y+\lambda=0 \qquad \Rightarrow \qquad \lambda=-18y \qquad (2)$$
$$F_\lambda(x,y,\lambda)=x+y-54=0 \qquad\qquad\qquad\qquad (3)$$

STEP 4 Solve the system of equations for x and y.

From (1) and (2) we find $-36x=-18y$ or $y=2x$.

Substituting for y in (3) gives $x+2x=54$

$$3x=54 \quad \text{or} \quad x=18 \quad \text{and} \quad y=2x=2(18)=36$$

STEP 5 Evaluate $C=f(x,y)$ at $(18,36)$.

$$C=18\cdot18^2+9\cdot36^2=17{,}496$$

Cost is minimized when 18 units of product x and 36 units of product y are produced. The minimum cost is \$17,496.

19. We solve the problem using the method of Lagrange multipliers.

STEP 1 Find the maximum of $w=V(l,w,h)=lwh$

Subject to the constraint $g(l,w,h)=l+w+h-62=0$

STEP 2 Construct the function $F(l,w,h,\lambda)=f(l,w,h)+\lambda g(l,w,h)$

$$F(l,w,h,\lambda)=lwh+\lambda(l+w+h-62)$$

STEP 3 Set up the system of equations

$$F_l(l,w,h,\lambda)=wh+\lambda=0 \qquad \Rightarrow \qquad \lambda=-wh \qquad (1)$$
$$F_w(l,w,h,\lambda)=lh+\lambda=0 \qquad \Rightarrow \qquad \lambda=-lh \qquad (2)$$
$$F_h(l,w,h,\lambda)=wl+\lambda=0 \qquad \Rightarrow \qquad \lambda=-wl \qquad (3)$$
$$F_\lambda(l,w,h,\lambda)=l+w+h-62=0 \qquad\qquad\qquad\qquad (4)$$

STEP 4 Solve the system of equations for x, y, and z.

From (1) and (2) we find $wh=lh$ or $w=l$.

From (1) and (3) we find $wh=wl$ or $h=l$.

Substituting for w and h in (4) gives

$$l+l+l-62=0$$
$$3l=62$$
$$l=\frac{62}{3}=20.667$$

Then $w=l=20.667$ and $h=l=20.667$.

STEP 5 Evaluate $w=V(l,w,h)=\left(\dfrac{62}{3}\right)^3=\dfrac{238328}{27}\approx8826.96$ cubic inches. The box of

greatest volume has dimensions 20.667 inches by 20.667 inches by 20.667 inches.

21. (a) Maximize production $P(K,L)=1.01K^{0.25}L^{0.75}$

Subject to the constraint $g(K,L)=175K+125L-125{,}000=0$

We use the Method of Lagrange Multipliers and construct the function

$$F(K, L, \lambda) = 1.01K^{0.25}L^{0.75} + \lambda(175K + 125L - 125{,}000)$$

The system of equations formed by the partial derivatives is

$$F_K(K, L, \lambda) = (1.01)(.25)K^{-.75}L^{.75} + 175\lambda = 0 \quad \Rightarrow \quad \lambda = -\frac{(1.01)(.25)K^{-.75}L^{.75}}{175} \quad (1)$$

$$F_L(K, L, \lambda) = (1.01)(.75)K^{.25}L^{-.25} + 125\lambda = 0 \quad \Rightarrow \quad \lambda = -\frac{(1.01)(.75)K^{.25}L^{-.25}}{125} \quad (2)$$

$$F_\lambda(K, L, \lambda) = 175K + 125L - 125{,}000 = 0 \quad (3)$$

From (1) and (2) we find

$$\frac{\cancel{(1.01)}(.25)K^{-.75}L^{.75}}{175} = \frac{\cancel{(1.01)}(.75)K^{.25}L^{-.25}}{125}$$

$$\frac{1}{\cancel{4}} \cdot \frac{L^{.75}}{K^{.75}} \cdot \frac{1}{175} = \frac{3}{\cancel{4}} \cdot \frac{K^{.25}}{L^{.25}} \cdot \frac{1}{125}$$

$$\frac{L}{175} = \frac{3K}{125}$$

$$L = 4.2K$$

Substitute for L in (3)

$$175K + 125(4.2K) = 125{,}000$$

$$175K + 525K = 125{,}000$$

$$700K = 125{,}000$$

$$K = \frac{1250}{7} = 178.57$$

Back-substituting for K, we get $L = 4.2K = 750$.

To maximize total production, the company should use 178.87 units of capital and 750 units of labor.

(b) The maximum production is $P\left(\dfrac{1250}{7}, 750\right) = (1.01)\left(\dfrac{1250}{7}\right)^{.25}(750)^{.75} = 529.140$.

23. Minimize the amount of material used $A(l, w, h) = 2lw + 2lh + 2wh$

Subject to the constraint $g(l, w, h) = lwh - 175 = 0$

We use the Method of Lagrange Multipliers and construct the function

$$F(l, w, z, \lambda) = 2lw + 2lh + 2wh + \lambda(lwh - 175)$$

The system of equations formed by the partial derivatives is

$$F_l(l,w,h,\lambda)=2w+2h+wh\lambda=0 \quad\Rightarrow\quad \lambda=-\frac{2w+2h}{wh} \quad (1)$$

$$F_w(l,w,h,\lambda)=2l+2h+lh\lambda=0 \quad\Rightarrow\quad \lambda=-\frac{2l+2h}{lh} \quad (2)$$

$$F_h(l,w,h,\lambda)=2l+2w+lw\lambda=0 \quad\Rightarrow\quad \lambda=-\frac{2l+2w}{lw} \quad (3)$$

$$F_\lambda(l,w,h,\lambda)=lwh-175=0 \quad (4)$$

To solve the system of equations we first use (1) and (2)

$$\frac{2w+2h}{wh}=\frac{2l+2h}{lh}$$

$$\frac{w+h}{w}=\frac{l+h}{l}$$

$$lw+lh=lw+wh$$

$$lh=wh$$

$$l=w$$

We use (2) and (3) and find

$$\frac{2l+2h}{lh}=\frac{2l+2w}{lw}$$

$$\frac{l+h}{h}=\frac{l+w}{w}$$

$$lw+wh=lh+wh$$

$$lw=lh$$

$$w=h$$

We substitute for h and l in (4), and get $w^3=175$ or $w=\sqrt[3]{175}\approx5.593$. Since $w=l=h$, we have the dimensions of the container that uses the least material and holds 175 cubic feet are $\sqrt[3]{175}\approx5.593$ feet by $\sqrt[3]{175}\approx5.593$ feet by $\sqrt[3]{175}\approx5.593$ feet.

25. Minimize the cost C of making a box

$$C(l,\text{w},h)=2lw+lw+2lh+2wh=3lw+2lh+2wh$$

Subject to the constraint $V(l,\text{w},h)=lwh-18=0$

We use the Method of Lagrange Multipliers and construct the function

$$F(l,\text{w},z,\lambda)=3lw+2lh+2wh+\lambda(lwh-18)$$

The system of equations formed by the partial derivatives is

$$F_l(l,w,h,\lambda)=3w+2h+wh\lambda=0 \quad\Rightarrow\quad \lambda=-\frac{3w+2h}{wh} \quad (1)$$

$$F_w(l,w,h,\lambda)=3l+2h+lh\lambda=0 \quad\Rightarrow\quad \lambda=-\frac{3l+2h}{lh} \quad (2)$$

$$F_h(l,w,h,\lambda)=2l+2w+lw\lambda=0 \quad\Rightarrow\quad \lambda=-\frac{2l+2w}{lw} \quad (3)$$

$$F_\lambda(l,w,h,\lambda)=lwh-18=0 \quad (4)$$

We solve the system of equations.

From (1) and (2) we find $\dfrac{3w+2h}{wh}=\dfrac{3l+2h}{lh}$

$$\dfrac{3w+2h}{w}=\dfrac{3l+2h}{l}$$

$$3lw+2lh=3lw+2wh$$

$$2lh=2wh$$

$$l=w$$

From (2) and (3) we find $\dfrac{3l+2h}{lh}=\dfrac{2l+2w}{lw}$

$$\dfrac{3l+2h}{h}=\dfrac{2l+2w}{w}$$

$$3lw+2wh=2lh+2wh$$

$$3w=2h \quad \text{or} \quad h=\dfrac{3}{2}w$$

We substitute for l and h in (4), and get $w \cdot w \cdot \left(\dfrac{3w}{2}\right)=18$

$$w^3=12 \quad \text{or} \quad w=\sqrt[3]{12}\approx 2.289$$

So we find $l=w=2.289$ and $h=\dfrac{3}{2}w=\dfrac{3}{2}\sqrt[3]{12}\approx 3.434$.

The cost is minimized if the box has a bottom measuring 2.289 feet by 2.289 feet and a height measuring 3.343 feet.

8.6 The Double Integral

1.
$$\int_0^2 \left(xy^3+x^2\right)dx = \left(\dfrac{x^2y^3}{2}+\dfrac{x^3}{3}\right)\Bigg|_0^2 = \left[\dfrac{4y^3}{2}+\dfrac{8}{3}\right]-[0]=2y^3+\dfrac{8}{3}$$

3.
$$\int_2^4 \left(3x^2y+2x\right)dy = \left(3x^2 \cdot \dfrac{y^2}{2}+2xy\right)\Bigg|_2^4 = \left[3x^2 \cdot \dfrac{16}{2}+2x \cdot 4\right]-\left[3x^2 \cdot \dfrac{4}{2}+2x \cdot 2\right]$$
$$= 24x^2+8x-6x^2-4x=18x^2+4x$$

5.
$$\int_2^3 (x+3y)\,dx = \left(\dfrac{x^2}{2}+3xy\right)\Bigg|_2^3 = \left[\dfrac{9}{2}+9y\right]-\left[\dfrac{4}{2}+6y\right]=\dfrac{5}{2}+3y$$

7.
$$\int_2^4 (4x-6y+7)\,dy = \left(4xy-\dfrac{6y^2}{2}+7y\right)\Bigg|_2^4 = [16x-3 \cdot 16+28]-[8x-3 \cdot 4+14]=8x-22$$

9. $\displaystyle\int_0^1 \frac{x^2}{\sqrt{1+y^2}}\,dx = \frac{1}{\sqrt{1+y^2}}\int_0^1 x^2\,dx = \frac{1}{\sqrt{1+y^2}}\cdot\frac{x^3}{3}\Big|_0^1 = \frac{1}{\sqrt{1+y^2}}\left[\frac{1}{3}-0\right] = \frac{1}{3\sqrt{1+y^2}}$

11. $\displaystyle\int_0^2 e^{x+y}\,dx = e^y\int_0^2 e^x\,dx = e^y\left[e^x\Big|_0^2\right] = e^y\left[e^2-e^0\right] = e^y\left(e^2-1\right)$

13. $\displaystyle\int_0^4 e^{x-4y}\,dx = e^{-4y}\int_0^4 e^x\,dx = e^{-4y}\left[e^x\Big|_0^4\right] = e^{-4y}\left(e^4-1\right)$

15. $\displaystyle\int_0^2 \frac{x}{\sqrt{y+6}}\,dx = \frac{1}{\sqrt{y+6}}\int_0^2 x\,dx = \frac{1}{\sqrt{y+6}}\left[\frac{x^2}{2}\Big|_0^2\right] = \frac{1}{\sqrt{y+6}}(2-0) = \frac{2}{\sqrt{y+6}}$

17. $\displaystyle\int_0^2\left[\int_0^4 y\,dx\right]dy$

Evaluating the inner integral first.

$$\int_0^4 y\,dx = y\int_0^4 dx = y\cdot\left(x\Big|_0^4\right) = y[4-0] = 4y$$

Then

$$\int_0^2\left[\int_0^4 y\,dx\right]dy = \int_0^2 4y\,dy = \frac{4y^2}{2}\Big|_0^2 = 8-0 = 8$$

19. $\displaystyle\int_1^2\left[\int_1^3\left(x^2+y\right)dx\right]dy$

Evaluating the inner integral first.

$$\int_1^3\left(x^2+y\right)dx = \left(\frac{x^3}{3}+yx\right)\Big|_1^3 = \left[(9+3y)-\left(\frac{1}{3}+y\right)\right] = \frac{26}{3}+2y$$

Then

$$\int_1^2\left[\int_1^3\left(x^2+y\right)dx\right]dy = \int_1^2\left(\frac{26}{3}+2y\right)dy = \left(\frac{26}{3}y+\frac{2y^2}{2}\right)\Big|_1^2$$

$$= \left(\frac{52}{3}+4\right)-\left(\frac{26}{3}+1\right) = \frac{26}{3}+3 = \frac{35}{3}$$

21. $\displaystyle\int_0^1\left[\int_1^2\left(x^2+y\right)dx\right]dy$

Evaluating the inner integral first.

$$\int_1^2\left(x^2+y\right)dx = \left(\frac{x^3}{3}+yx\right)\Big|_1^2 = \left(\frac{8}{3}+2y\right)-\left(\frac{1}{3}+y\right) = \frac{7}{3}+y$$

Then

$$\int_0^1\left[\int_1^2(x^2+y)dx\right]dy = \int_0^1\left(\frac{7}{3}+y\right)dy = \left(\frac{7}{3}y+\frac{y^2}{2}\right)\Big|_0^1 = \left(\frac{7}{3}+\frac{1}{2}\right)-(0) = \frac{14+3}{6} = \frac{17}{6}$$

23. $\int_1^2\left[\int_3^4(4x+2y+5)dx\right]dy$

Evaluating the inner integral first.

$$\int_3^4(4x+2y+5)dx = \left(\frac{4x^2}{2}+2yx+5x\right)\Big|_3^4 = (32+8y+20)-(18+6y+15) = 2y+19$$

Then

$$\int_1^2\left[\int_3^4(4x+2y+5)dx\right]dy = \int_1^2(2y+19)\,dy = \left(\frac{2y^2}{2}+19y\right)\Big|_1^2 = (4+38)-(1+19) = 22$$

25. $\int_2^4\left[\int_0^1(6xy^2-2xy+3)dy\right]dx$

Evaluating the inner integral first.

$$\int_0^1(6xy^2-2xy+3)dy = \left(\frac{6xy^3}{3}-\frac{2xy^2}{2}+3y\right)\Big|_0^1 = (2x-x+3)-(0) = x+3$$

Then

$$\int_2^4\left[\int_0^1(6xy^2-2xy+3)dy\right]dx = \int_2^4(x+3)dx = \left(\frac{x^2}{2}+3x\right)\Big|_2^4 = (8+12)-(2+6) = 12$$

27.

$$\iint_R(y+3x^2)\,dx\,dy = \int_1^3\left[\int_0^2(y+3x^2)\,dx\right]dy = \int_1^3\left[\left(yx+\frac{3x^3}{3}\right)\Big|_0^2\right]dy$$

$$= \int_1^3(2y+8)\,dy = \left(\frac{2y^2}{2}+8y\right)\Big|_1^3$$

$$= (9+24)-(1+8) = 24$$

29.

$$\iint_R(x+y)\,dy\,dx = \int_0^2\left[\int_1^4(x+y)dy\right]dx = \int_0^2\left[\left(xy+\frac{y^2}{2}\right)\Big|_1^4\right]dx$$

$$= \int_0^2\left[(4x+8)-\left(x+\frac{1}{2}\right)\right]dx = \int_0^2\left(3x+\frac{15}{2}\right)dx$$

$$= \left(\frac{3x^2}{2} + \frac{15}{2}x \right) \Bigg|_0^2 = (6+15) - (0) = 21$$

31.

$$V = \iint\limits_R f(x,\, y)\, dy\; dx = \int_1^2 \left[\int_3^4 (2x + 3y + 4)\, dy \right] dx = \int_1^2 \left[\left(2xy + \frac{3y^2}{2} + 4y \right) \Bigg|_3^4 \right] dx$$

$$= \int_1^2 \left[(8x + 24 + 16) - \left(6x + \frac{27}{2} + 12 \right) \right] dx$$

$$= \int_1^2 \left(2x + \frac{29}{2} \right) dx = \left(\frac{2x^2}{2} + \frac{29}{2}x \right) \Bigg|_1^2$$

$$= (4 + 29) - \left(1 + \frac{29}{2} \right)$$

$$= 32 - \frac{29}{2} = \frac{64 - 29}{2} = \frac{35}{2} \text{ cubic units}$$

Chapter 8 Review

TRUE-FALSE ITEMS

1. True   **3.** False

FILL-IN-THE-BLANKS

1. surface   **3.** $x = x_0$

REVIEW EXERCISES

1. We use the distance formula with $(x_1,\, y_1,\, z_1) = (1,\, 6,\, -2)$ and $(x_2,\, y_2,\, z_2) = (2,\, 4,\, 0)$.

$$d = \sqrt{(x_2 - x_1)^2 + (y_2 - y_1)^2 + (z_2 - z_1)^2}$$
$$d = \sqrt{(2-1)^2 + (4-6)^2 + (0 - (-2))^2}$$
$$d = \sqrt{1 + 4 + 4} = \sqrt{9} = 3$$

3. We use the distance formula with $(x_1,\, y_1,\, z_1) = (4,\, 6,\, 8)$ and $(x_2,\, y_2,\, z_2) = (6,\, 2,\, 1)$.

$$d = \sqrt{(x_2 - x_1)^2 + (y_2 - y_1)^2 + (z_2 - z_1)^2}$$

$$d = \sqrt{(6-4)^2 + (2-6)^2 + (1-8)^2}$$
$$d = \sqrt{4+16+49} = \sqrt{69}$$

5. We use the distance formula with $(x_1, y_1, z_1) = (-3, 7, -1)$ and $(x_2, y_2, z_2) = (0, 3, -1)$.

$$d = \sqrt{(x_2-x_1)^2 + (y_2-y_1)^2 + (z_2-z_1)^2}$$
$$d = \sqrt{(0-(-3))^2 + (3-7)^2 + (-1-(-1))^2}$$
$$d = \sqrt{9+16+0} = \sqrt{25} = 5$$

7. The radius is the distance between the center and a point on the edge of the sphere.

$$r = \sqrt{(x_1-x_0)^2 + (y_1-y_0)^2 + (z_1-z_0)^2}$$
$$r = \sqrt{(3-2)^2 + (4-2)^2 + (0-2)^2}$$
$$r = \sqrt{1+4+4} = \sqrt{9} = 3$$

9. The standard equation of the sphere with radius $r = 2$ and center at $(x_0, y_0, z_0) = (-6, 3, 1)$ is

$$(x-x_0)^2 + (y-y_0)^2 + (z-z_0)^2 = r^2$$
$$(x-(-6))^2 + (y-3)^2 + (z-1)^2 = 2^2$$
$$(x+6)^2 + (y-3)^2 + (z-1)^2 = 4$$

11.
$$(x-1)^2 + (y+3)^2 + (z+8)^2 = 25$$

Compare the equation to the standard equation of the sphere

$$(x-x_0)^2 + (y-y_0)^2 + (z-z_0)^2 = r^2$$
$$(x-1)^2 + (y-(-3))^2 + (z-(-8))^2 = 5^2$$

The center of the sphere is $(1, -3, -8)$ and its radius is 5.

13. (a) Complete the squares and put the equation into standard form.

$$x^2 + y^2 + z^2 - 2x + 8y - 6z = 10$$
$$(x^2-2x) + (y^2+8y) + (z^2-6z) = 10$$
$$(x^2-2x+1) + (y^2+8y+16) + (z^2-6z+9) = 10+1+16+9$$
$$(x-1)^2 + (y+4)^2 + (z-3)^2 = 36$$

(b) The center of the sphere is $(1, -4, 3)$ and its radius is 6.

15. $f(x, y) = 2x^2 + 6xy - y^3$

(a) $f(1, -3) = 2(1^2) + 6(1)(-3) - (-3)^3 = 2 - 18 + 27 = 11$

(b) $f(4, -2) = 2(4^2) + 6(4)(-2) - (-2)^3 = 32 - 48 + 8 = -8$

17. $f(x, y) = \dfrac{x + 2y}{x - 3y}$

(a) $f(1, -3) = \dfrac{1 + 2(-3)}{1 - 3(-3)} = \dfrac{-5}{10} = -\dfrac{1}{2}$ (b) $f(4, -2) = \dfrac{4 + 2(-2)}{4 - 3(-2)} = 0$

19. $x^2 + 3y + 5$ is a polynomial, and polynomials are defined for all real numbers. So the domain of $z = f(x, y)$ is the entire xy-plane.

21. Since only logarithms of positive numbers are allowed, $y - x^2 - 4 > 0$ or $y > x^2 + 4$. The domain of $z = f(x, y)$ is the set or ordered pairs $\{(x, y) \mid y > x^2 + 4\}$. This describes the set of points (x, y) inside the parabola $y = x^2 + 4$.

23. Since only square roots of nonnegative numbers are allowed in the real number system,
$$x^2 + y^2 + 4x - 5 \geq 0$$
$$\left(x^2 + 4x\right) + y^2 \geq 5$$
$$\left(x^2 + 4x + 4\right) + y^2 \geq 5 + 4$$
$$\left(x + 2\right)^2 + y^2 \geq 9$$
The domain of $z = f(x, y)$ is the set or ordered pairs $\left\{(x, y) \mid (x + 2)^2 + y^2 \geq 9\right\}$. This describes the set of points either on the circle centered at $(-2, 0)$ and having a radius of 3 or outside the circle.

25. $z = f(x, y) = x^2 y + 4x$

$\quad f_x(x, y) = 2xy + 4$ $f_y(x, y) = x^2$

$\quad f_{xx}(x, y) = 2y$ $f_{yy}(x, y) = 0$

$\quad f_{xy}(x, y) = 2x$ $f_{yx}(x, y) = 2x$

27. $z = f(x, y) = y^2 e^x + x \ln y$

$\quad f_x(x, y) = y^2 e^x + \ln y$ $f_y(x, y) = 2y e^x + \dfrac{x}{y}$

$\quad f_{xx}(x, y) = y^2 e^x$ $f_{yy}(x, y) = 2 e^x - \dfrac{x}{y^2}$

$$f_{xy}(x,y) = 2y\,e^x + \frac{1}{y} \qquad\qquad f_{yx}(x,y) = 2y\,e^x + \frac{1}{y}$$

29. $z = f(x,y) = \sqrt{x^2+y^2} = \left(x^2+y^2\right)^{1/2}$

$$f_x(x,y) = \frac{1}{2}\left(x^2+y^2\right)^{-1/2} \cdot 2x = \frac{x}{\left(x^2+y^2\right)^{1/2}} = \frac{x}{\sqrt{x^2+y^2}}$$

$$f_y(x,y) = \frac{1}{2}\left(x^2+y^2\right)^{-1/2} \cdot 2y = \frac{y}{\left(x^2+y^2\right)^{1/2}} = \frac{y}{\sqrt{x^2+y^2}}$$

$$f_{xx}(x,y) = \frac{1\cdot\left(x^2+y^2\right)^{1/2} - x\cdot\frac{1}{2}\left(x^2+y^2\right)^{-1/2}\cdot 2x}{x^2+y^2} = \frac{\left(x^2+y^2\right)-x^2}{\left(x^2+y^2\right)^{3/2}} = \frac{y^2}{\left(x^2+y^2\right)^{3/2}}$$

$$f_{yy}(x,y) = \frac{\left[1\cdot\left(x^2+y^2\right)^{1/2}\right] - \left[y\cdot\frac{1}{2}\left(x^2+y^2\right)^{-1/2}\cdot 2y\right]}{x^2+y^2}$$

$$= \frac{\left[x^2+y^2\right] - \left[y^2\right]}{\left(x^2+y^2\right)^{3/2}} = \frac{x^2}{\left(x^2+y^2\right)^{3/2}}$$

$$f_{xy}(x,y) = -\frac{1}{2}x\left(x^2+y^3\right)^{-3/2}\cdot 2y = -\frac{xy}{\left(x^2+y^3\right)^{3/2}}$$

$$f_{yx}(x,y) = -\frac{1}{2}y\left(x^2+y^3\right)^{-3/2}\cdot 2x = -\frac{xy}{\left(x^2+y^3\right)^{3/2}}$$

31. $z = f(x,y) = e^x \ln(5x+2y)$

$$f_x(x,y) = e^x\cdot\frac{5}{5x+2y} + e^x\ln(5x+2y) = \frac{5\,e^x}{5x+2y} + e^x\ln(5x+2y)$$

$$f_y(x,y) = e^x\cdot\frac{2}{5x+2y} = \frac{2e^x}{5x+2y}$$

$$f_{xx}(x,y) = \frac{5e^x(5x+2y)-25e^x}{(5x+2y)^2} + \frac{5e^x}{5x+2y} + e^x\ln(5x+2y)$$

$$= \frac{10e^x(5x+2y)-25e^x}{(5x+2y)^2} + e^x\ln(5x+2y)$$

$$= \frac{5e^x(10x+4y-5)}{(5x+2y)^2} + e^x \ln(5x+2y)$$

$$f_{yy}(x, y) = \frac{-2e^x \cdot (2)}{(5x+2y)^2} = -\frac{4e^x}{(5x+2y)^2}$$

$$f_{xy}(x,y) = \frac{-5e^x(2)}{(5x+2y)^2} + e^x \cdot \frac{2}{5x+2y} = \frac{-10e^x + 2e^x(5x+2y)}{(5x+2y)^2}$$

$$= \frac{2e^x(10x+2y-5)}{(5x+2y)^2}$$

$$f_{yx}(x, y) = \frac{2e^x(5x+2y)-2e^x(5)}{(5x+2y)^2} = \frac{2e^x(5x+2y-5)}{(5x+2y)^2}$$

33. $f(x, y, z) = 3xe^y + xye^z - 12x^2y$

$$f_x(x, y, z) = 3e^y + ye^z - 24xy$$
$$f_y(x, y, z) = 3xe^y + xe^z - 12x^2$$
$$f_z(x, y, z) = xye^z$$

35. The slope of the line tangent to the intersection of z and the plane $y = 2$ is
$$f_x(x, y) = 3y^2$$
At the point $(1, 2, 12)$ the slope is
$$f_x(1, 2) = 3 \cdot 2^2 = 12$$

37. The slope of the line tangent to the intersection of z and the plane $x = 1$ is
$$f_y(x, y) = x^2 e^{xy}$$
At the point $(1, 0, 1)$ the slope is
$$f_y(1, 0) = 1^2 \cdot e^{1 \cdot 0} = 1$$

39. $z = f(x, y) = xy - 6x - x^2 - y^2$

(a) Find the partial derivatives of z, set each equal to zero, and solve the system of equations.

$$f_x(x, y) = y - 6 - 2x = 0 \qquad\qquad f_y(x, y) = x - 2y = 0$$

$$\begin{cases} y - 6 - 2x = 0 & (1) \\ x - 2y = 0 \quad \Rightarrow \quad x = 2y & (2) \end{cases}$$

Substituting (2) into (1) gives $y - 6 - 2(2y) = 0$
$$y - 4y = 6$$
$$-3y = 6$$
$$y = -2$$

Solving for x in (2) we get $x = 2(-2) = -4$.
The critical point is $(-4, -2)$.

(b) Find the second-order partial derivatives, and find the value of D.

$$f_{xx}(x, y) = -2 \qquad\qquad f_{yy}(x, y) = -2 \qquad\qquad f_{xy}(x, y) = 1$$

$$D = f_{xx}(-4, -2) \cdot f_{yy}(-4, -2) - \left[f_{xy}(-4, -2) \right]^2 = (-2)(-2) - 1^2 = 3 > 0$$

Since $f_{xx}(-4, -2) < 0$ and $D > 0$, the function z has a local maximum at $(-4, -2)$. The value of the local maximum is $z = 12$.

41. $z = f(x, y) = 2x - x^2 + 4y - y^2 + 10$

(a) Find the partial derivatives of z, set each equal to zero, and solve the system of equations.

$$f_x(x, y) = 2 - 2x = 0 \qquad\qquad f_y(x, y) = 4 - 2y = 0$$

$$x = 1 \qquad\qquad\qquad\qquad y = 2$$

The critical point is $(1, 2)$.

(b) Find the second-order partial derivatives, and find the value of D.

$$f_{xx}(x, y) = -2 \qquad\qquad f_{yy}(x, y) = -2 \qquad\qquad f_{xy}(x, y) = 0$$

$$D = f_{xx}(1, 2) \cdot f_{yy}(1, 2) - \left[f_{xy}(1, 2) \right]^2 = (-2)(-2) - 0^2 = 4 > 0$$

Since $f_{xx}(1, 2) < 0$ and $D > 0$, the function z has a local maximum at $(1, 2)$. The value of the local maximum is $z = 15$.

43. $z = f(x, y) = x^2 - 9y + y^2$

(a) Find the partial derivatives of z, set each equal to zero, and solve the system of equations.

$$f_x(x, y) = 2x = 0 \qquad\qquad f_y(x, y) = -9 + 2y = 0$$

$$x = 0 \qquad\qquad\qquad\qquad y = \frac{9}{2}$$

The critical point is $\left(0, \frac{9}{2} \right)$.

(b) Find the second-order partial derivatives, and find the value of D.

$$f_{xx}(x, y) = 2 \qquad f_{yy}(x, y) = 2 \qquad f_{xy}(x, y) = 0$$

$$D = f_{xx}\left(0, \frac{9}{2} \right) \cdot f_{yy}\left(0, \frac{9}{2} \right) - \left[f_{xy}\left(0, \frac{9}{2} \right) \right]^2 = 2 \cdot 2 - 0^2 = 4 > 0$$

Since $f_{xx}\left(0, \frac{9}{2} \right) > 0$ and $D > 0$, the function z has a local minimum at $\left(0, \frac{9}{2} \right)$. The value of the local minimum is $z = -\dfrac{81}{4}$.

45. **STEP 1** Find the maximum of $z = f(x, y) = 5x^2 + 3y^2 + xy$

Subject to the constraint $g(x, y) = 2x - y - 20 = 0$

STEP 2 Construct the function $F(x, y, \lambda) = f(x, y) + \lambda g(x, y)$

$$F(x,\ y,\ \lambda\ ) = 5x^2 + 3y^2 + xy + \lambda(2x - y - 20)$$

STEP 3 Set up the system of equations

$$\begin{aligned} F_x(x,y,\lambda) &= 10x + y + 2\lambda = 0 & (1)\\ F_y(x,y,\lambda) &= 6y + x - \lambda = 0 & (2)\\ F_\lambda(x,y,\lambda) &= 2x - y - 20 = 0 & (3) \end{aligned}$$

STEP 4 Solve the system of equations for x and y.
Add twice (2) to (1) to eliminate λ.

$$(10x + y + 2\lambda) + 2(6y + x - \lambda) = 0$$
$$10x + y + 2\lambda + 12y + 2x - 2\lambda = 0$$
$$12x + 13y = 0 \qquad (1)$$

Subtract six times (3) from (1) to eliminate x.

$$(12x + 13y) - 6(2x - y - 20) = 0$$
$$12x + 13y - 12x + 6y + 120 = 0$$
$$13y + 6y = -120$$
$$19y = -120$$
$$y = \frac{-120}{19}$$

Substituting for y in (3) we get

$$2x - \left(\frac{-120}{19}\right) - 20 = 0 \qquad (3)$$

$$2x = \frac{260}{19} \quad \text{or} \quad x = \frac{130}{19}$$

STEP 5 Evaluate $z = f(x,\ y)$ at $\left(\dfrac{130}{19}, \dfrac{-120}{19}\right)$.

$$z = = 5\left(\frac{130}{19}\right)^2 + 3\left(\frac{-120}{19}\right)^2 + \left(\frac{130}{19}\right)\left(\frac{-120}{19}\right) = \frac{112,100}{19^2} \approx 310.526$$

The maximum value of z subject to the condition is 310.526.

47. **STEP 1** Find the minimum of $z = f(x,\ y) = x^2 + y^2$
     Subject to the constraint $g(x,\ y) = 2x + y - 4 = 0$
STEP 2 Construct the function $F(x,\ y,\ \lambda\ ) = f(x,\ y) + \lambda\ g(x,\ y)$
$$F(x,\ y,\ \lambda\ ) = x^2 + y^2 + \lambda(2x + y - 4)$$

STEP 3 Set up the system of equations

$$\begin{aligned} F_x(x,y,\lambda) &= 2x + 2\lambda = 0 & \Rightarrow & \quad \lambda = -x & (1)\\ F_y(x,y,\lambda) &= 2y + \lambda = 0 & \Rightarrow & \quad \lambda = -2y & (2)\\ F_\lambda(x,y,\lambda) &= 2x + y - 4 = 0 & & & (3) \end{aligned}$$

STEP 4 Solve the system of equations for x and y.
From (1) and (2) we find $x = 2y$.
Substituting for x in (3) gives $2(2y) + y - 4 = 0$

$$5y = 4$$

$$y = \frac{4}{5} \quad \text{and} \quad x = 2y = \frac{8}{5}$$

STEP 5 Evaluate $z = f(x, y)$ at $\left(\frac{8}{5}, \frac{4}{5}\right)$.

$$z = \left(\frac{8}{5}\right)^2 + \left(\frac{4}{5}\right)^2 = \frac{64 + 16}{25} = \frac{80}{25} = \frac{16}{5} = 3.2$$

The minimum value of z subject to the constraint is 3.2.

49.

$$\int_0^2 \left(4x^2 y - 12y\right) dx = \left(\frac{4x^3 y}{3} - 12xy\right)\Bigg|_0^2 = \frac{4 \cdot 2^3 y}{3} - 12 \cdot 2y = \frac{32}{3} y - 24y = -\frac{40}{3} y$$

51.

$$\int_{-1}^3 \left(6x^2 y + 2y\right) dy = \left(\frac{6x^2 y^2}{2} + \frac{2y^2}{2}\right)\Bigg|_{-1}^3 = \left(3x^2 \cdot 3^2 + 3^2\right) - \left(3x^2(-1)^2 + (-1)^2\right)$$

$$= 27x^2 + 9 - 3x^2 - 1 = 24x^2 + 8$$

53.

$$\int_1^2 \left[\int_0^3 \left(6x^2 + 2x\right) dy\right] dx$$

Evaluating the inner integral first,

$$\int_0^3 \left(6x^2 + 2x\right) dy = \left(6x^2 y + 2xy\right)\Big|_0^3 = 18x^2 + 6x$$

Then

$$\int_1^2 \left[\int_0^3 \left(6x^2 + 2x\right) dy\right] dx = \int_1^2 \left(18x^2 + 6x\right) dx = \left(\frac{18x^3}{3} + \frac{6x^2}{2}\right)\Bigg|_1^2$$

$$= \left(6 \cdot 2^3 + 3 \cdot 2^2\right) - (6 + 3) = 48 + 12 - 9 = 51$$

55.

$$\int_0^2 \left[\int_1^8 \left(x^2 + 2xy - y^2\right) dx\right] dy$$

Evaluating the inner integral first,

$$\int_1^8 \left(x^2 + 2xy - y^2\right) dx = \left(\frac{x^3}{3} + \frac{2x^2 y}{2} - xy^2\right)\Bigg|_1^8$$

$$= \left(\frac{8^3}{3} + 64y - 8y^2\right) - \left(\frac{1}{3} + y - y^2\right) = \frac{511}{3} + 63y - 7y^2$$

Then

$$\int_0^2 \left[\int_1^8 \left(x^2 + 2xy - y^2 \right) dx \right] dy = \int_0^2 \left(\frac{511}{3} + 63y - 7y^2 \right) dy = \left(\frac{511}{3} y + \frac{63y^2}{2} - \frac{7y^3}{3} \right) \Big|_0^2$$

$$= \frac{1022}{3} + 126 - \frac{56}{3} = \frac{966}{3} + 126 = 448$$

57.

$$\iint_R f(x, y) \, dy \, dx = \int_{-1}^1 \int_1^3 (2x + 4y) \, dy \, dx = \int_{-1}^1 \left[\left(2xy + \frac{4y^2}{2} \right) \Big|_1^3 \right] dx$$

$$= \int_{-1}^1 \left[(6x + 18) - (2x + 2) \right] dx$$

$$= \int_{-1}^1 (4x + 16) \, dx = \left(\frac{4x^2}{2} + 16x \right) \Big|_{-1}^1$$

$$= (2 + 16) - (2 - 16) = 32$$

59.

$$\iint_R f(x, y) \, dy \, dx = \int_0^3 \int_1^2 (2xy) \, dy \, dx = \int_0^3 \left[\left(\frac{2xy^2}{2} \right) \Big|_1^2 \right] dx = \int_0^3 (4x - x) \, dx$$

$$= \int_0^3 3x \, dx = \frac{3x^2}{2} \Big|_0^3 = \frac{27}{2}$$

61.

$$V = \iint_R f(x, y) \, dy \, dx = \int_1^8 \int_0^6 (2x + 2y + 1) \, dy \, dx = \int_1^8 \left[\left(2xy + \frac{2y^2}{2} + y \right) \Big|_0^6 \right] dx$$

$$= \int_1^8 (12x + 36 + 6) \, dx = \int_1^8 (12x + 42) \, dx$$

$$= \left(\frac{12x^2}{2} + 42x \right) \Big|_1^8$$

$$= 6 \cdot 64 + 42 \cdot 8 - 6 - 42 = 672 \text{ cubic units}$$

63.

(a) $\dfrac{\partial z}{\partial K} = 80 \cdot \dfrac{1}{4} K^{-3/4} L^{3/4} = 20 \left(\dfrac{L}{K} \right)^{3/4}$ $\dfrac{\partial z}{\partial L} = 80 \cdot \dfrac{3}{4} K^{1/4} L^{-1/4} = 60 \left(\dfrac{K}{L} \right)^{1/4}$

(b) When $K = \$800,000$ and $L = 20,000$ labor hours

$$\frac{\partial z}{\partial K} = 20 \left(\frac{20,000}{800,000} \right)^{3/4} = 20 \left(\frac{1}{40} \right)^{3/4} \approx 1.257$$

$$\frac{\partial z}{\partial L} = 60\left(\frac{800{,}000}{20{,}000}\right)^{1/4} = 60(40)^{1/4} \approx 150.892$$

(c) The factory should increase the use of labor. Explanations will vary.

65. $C(x, y) = 1050 + 40x + 45y$

$C_x(x, y) = 40$ $\qquad\qquad\qquad$ $C_y(x, y) = 45$

Explanations will vary.

67. (a) $R(x, y) = px + qy$

$$= (350 - 6x + y)x + (400 + 2x - 8y)y$$
$$= 350x - 6x^2 + xy + 400y + 2xy - 8y^2$$
$$= 350x - 6x^2 + 3xy + 400y - 8y^2$$

(b) $R_x(x, y) = 350 - 12x + 3y$

$R_y(x, y) = 3x + 400 - 16y$

Explanations will vary.

69. (a) $P(x, y) = R(x, y) - C(x, y)$

$$= \left[350x - 6x^2 + 3xy + 400y - 8y^2\right] - \left[1050 + 40x + 45y\right]$$
$$= 310x - 6x^2 + 3xy + 355y - 8y^2 - 1050$$

(b) $P_x(x, y) = 310 - 12x + 3y$

$P_x(50, 30) = 310 - 12(50) + 3(30) = -200$

$P_y(x, y) = 3x + 355 - 16y$

$P_y(50, 30) = 3(50) + 355 - 16(30) = 25$

Explanations will vary.

71. (a) $P(x, y) = R(x, y) - C(x, y)$

First we find R. $R(x, y) = px + qy = (9 - x)x + (21 - 2y)y = 9x - x^2 + 21y - 2y^2$

$$P(x, y) = \left[9x - x^2 + 21y - 2y^2\right] - (x + y + 225)$$
$$= 8x - x^2 + 20y - 2y^2 - 225$$

We find the critical points of function P.

$P_x(x, y) = 8 - 2x = 0$ $\qquad\qquad\qquad$ $P_y(x, y) = 20 - 4y = 0$

$\qquad\qquad\quad x = 4$ $\qquad\qquad\qquad\qquad\qquad\qquad y = 5$

The critical point is (4, 5).

We find the second-order partial derivatives and D to determine the character of the critical point.

$$P_{xx}(x, y) = -2 \qquad\qquad P_{yy}(x, y) = -4 \qquad\qquad P_{xy}(x, y) = 0$$

$$D = P_{xx}(4, 5) \cdot P_{yy}(4, 5) - \left[P_{xy}(4, 5) \right]^2 = (-2)(-4) - 0 = 8$$

Since $P_{xx}(4, 5) < 0$ and $D > 0$, there is a local maximum at $(4, 5)$. The supermarket should sell 4000 units of juice x and 5000 units of juice y to maximize profit.

(b) The maximum profit attainable from the orange juice sales is given by $P(4, 5)$.

$$P(4, 5) = 8(4) - 4^2 + 20(5) - 2(5^2) - 225 = -159$$

Since x and y are in thousands, the maximum profit is a loss of $159,000.

73. We want to find the maximum production $P(K, L) = 10 K^{0.3} L^{0.7}$

$$\text{subject to the condition} \qquad g(K, L) = 50K + 100L - 51,000 = 0$$

We use the method of Lagrange multipliers, and construct the function

$$F(K, L, \lambda) = 10 K^{0.3} L^{0.7} + \lambda(50K + 100L - 51,000)$$

The system of equations formed by the partial derivatives is

$$F_K(K, L, \lambda) = 3K^{-0.7} L^{0.7} + 50\lambda = 0 \qquad \Rightarrow \quad \lambda = -\frac{3K^{-0.7} L^{0.7}}{50} \quad (1)$$

$$F_L(K, L, \lambda) = 7K^{0.3} L^{-0.3} + 100\lambda = 0 \qquad \Rightarrow \quad \lambda = -\frac{7K^{0.3} L^{-0.3}}{100} \quad (2)$$

$$F_\lambda(K, L, \lambda) = 50K + 100L - 51,000 = 0 \qquad\qquad\qquad (3)$$

From (1) and (2) we find

$$\frac{3K^{-0.7} L^{0.7}}{50} = \frac{7K^{0.3} L^{-0.3}}{100}$$

$$\frac{300L^{0.7}}{K^{0.7}} = \frac{350K^{0.3}}{L^{0.3}}$$

$$300L = 350K$$

$$L = \frac{350}{300}K = \frac{7}{6}K$$

Substitute for L in (3)

$$50K + 100\left(\frac{7}{6}K\right) = 51,000$$

$$\frac{1000}{6}K = 51,000$$

$$K = 306$$

Substituting back and solving for L gives $L = \frac{7}{6}K = \frac{7}{6} \cdot 306 = 357$

To maximize productivity $50(306) = $15,300 should be allotted for capital and $100(357) = $35,700 should be allotted for labor.

(b) The maximum number of units that can be produced is 3409 units.
$$P(306, 357) = 10(306^{0.3})(357^{0.7}) = 3408.66$$

CHAPTER 8 PROJECT

1. Total cost C is the sum of holding cost and reorder cost. If we let x denote the lot size of the vacuum cleaners and y denote the lot size of the microwave ovens, we find

$$C(x, y) = \left(\frac{30x}{2} + \frac{15y}{2} \right) + \left(40\left(\frac{500}{x} \right) + 60\left(\frac{800}{y} \right) \right)$$

$$= 15x + 7.5y + \frac{20,000}{x} + \frac{48,000}{y}$$

3. Assuming that there are $\frac{37}{2} = 19$ vacuum cleaners in the store and $\frac{80}{2} = 40$ microwave ovens in the store, you would need
$$19\ (20) + 40\ (10) = 780 \text{ cubic feet of storage.}$$

5. The problem is to now minimize cost $C(x, y) = 15x + 7.5y + \frac{20,000}{x} + \frac{48,000}{y}$

subject to the constraint $\quad g(x, y) = 20x + 10y - 1000 = 0$

To use the method of Lagrange multipliers we need to construct function F.
$$F(x, y, \lambda) = 15x + 7.5y + \frac{20,000}{x} + \frac{48,000}{y} + \lambda(20x + 10y - 1000)$$

The system of equations that are used to determine the minimum C is formed by the partial derivatives of F.

$$F_x(x, y, \lambda) = 15 - \frac{20,000}{x^2} + 20\lambda = 0 \quad \Rightarrow \quad \lambda = \frac{20,000 - 15x^2}{20x^2} \quad (1)$$

$$F_y(x, y, \lambda) = 7.5 - \frac{48,000}{y^2} + 10\lambda = 0 \quad \Rightarrow \quad \lambda = \frac{48,000 - 7.5y^2}{10y^2} \quad (2)$$

$$F_\lambda(x, y, \lambda) = 20x + 10y - 1000 = 0 \quad (3)$$

7. If the demand for vacuum cleaners is 500 units, and they are ordered in lots of 24, then there will be 21 orders placed per year. $\left(\frac{500}{24} = 20.833 \right)$

If the demand for microwave ovens is 800 units, and they are ordered in lots of 52, then there will be 15 orders placed per year (with the last order of size 20).

On the average there will be 12 vacuum cleaners and 26 microwave ovens in the store. The space occupied by these items is 20 (12) + 10(26) = 500 cubic feet.

Appendix A
Graphing Utilities

A.1   The Viewing Rectangle

1. Using the window shown, X-scale = 1, and Y-scale = 2, we get the point (-1, 4).

3. Using the window shown, X-scale = 1, and Y-scale = 1, we get the point (3, 1).

5. Xmin = -6 Ymin = -4   7. Xmin = -6 Ymin = -1
    Xmax = 6 Ymax = 4 Xmax = 6 Ymax = 3
    Xscl = 2 Yscl = 2 Xscl = 2 Yscl = 1

9. Xmin = 3 Ymin = 2
    Xmax = 9 Ymax = 10
    Xscl = 1 Yscl = 2

11. Answers will vary, but an appropriate viewing window would be
    Xmin = -12 Ymin = -4
    Xmax = 6 Ymax = 8
    Xscl = 1 Yscl = 1

13. Answers will vary, but an appropriate viewing window would be
    Xmin = -30 Ymin = -100
    Xmax = 50 Ymax = 50
    Xscl = 10 Yscl = 10

15. Answers will vary, but an appropriate viewing window would be
    Xmin = -10 Ymin = -20
    Xmax = 110 Ymax = 180
    Xscl = 10 Yscl = 20

A.2   Using a Graphing Utility to Graph Equations

1. (a)   (b)   (c)   (d)

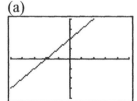 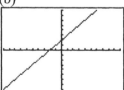 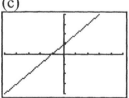 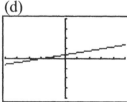

3.   (a)   (b)   (c)   (d)

5.   (a)   (b)   (c)   (d)

7.   (a)   (b)   (c)   (d)

9.   (a)   (b)   (c)   (d)

11. (a)   (b)   (c)   (d)

13. (a)   (b)   (c)   (d)

15. (a)   (b)   (c)   (d)

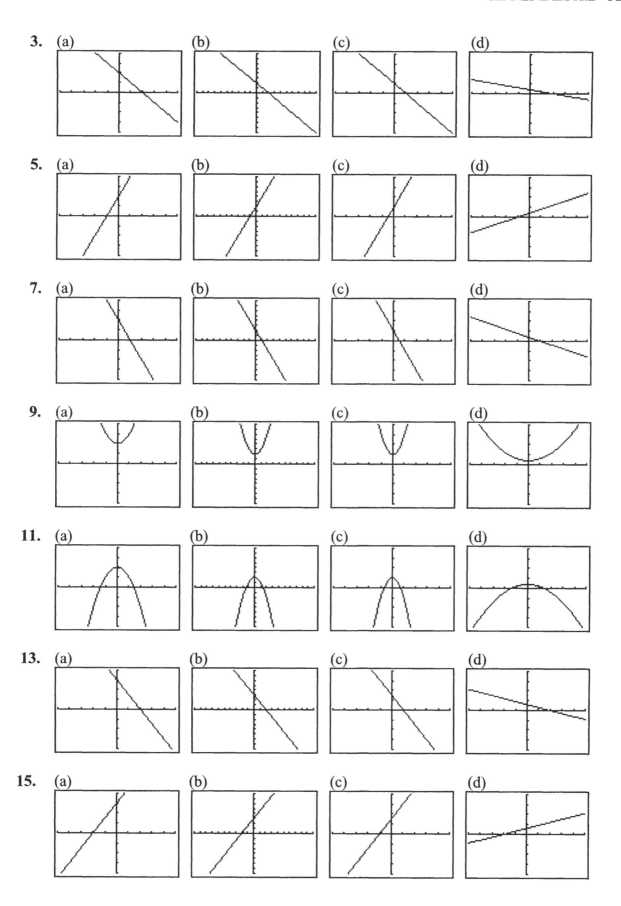

17. $y = x + 2$

X	Y1
-3	-1
-2	0
-1	1
0	2
1	3
2	4
3	5

X=-3

19. $y = -x + 2$

X	Y3
-3	5
-2	4
-1	3
0	2
1	1
2	0
3	-1

X=-3

21. $y = 2x + 2$

X	Y5
-3	-4
-2	-2
-1	0
0	2
1	4
2	6
3	8

X=-3

23. $y = -2x + 2$

X	Y7
-3	8
-2	6
-1	4
0	2
1	0
2	-2
3	-4

X=-3

25. $y = x^2 + 2$

X	Y9
-3	11
-2	6
-1	3
0	2
1	3
2	6
3	11

X=-3

27. $y = -x^2 + 2$

X	Y1
-3	-7
-2	-2
-1	1
0	2
1	1
2	-2
3	-7

X=-3

29. $3x + 2y = 6$

X	Y3
-3	7.5
-2	6
-1	4.5
0	3
1	1.5
2	0
3	-1.5

X=-3

31. $-3x + 2y = 6$

X	Y5
-3	-1.5
-2	0
-1	1.5
0	3
1	4.5
2	6
3	7.5

X=-3

A.3 Square Screens

1. A square screen results if $2(X\text{max} - X\text{min}) = 3(Y\text{max} - Y\text{min})$
In this window we test
$$2(3 - (-3)) \ ? \ 3(2 - (-2))$$
$$2 \cdot 6 \qquad 3 \cdot 4$$
$$12 \ = \ 12$$
The window is square.

3. A square screen results if $2(X\text{max} - X\text{min}) = 3(Y\text{max} - Y\text{min})$
In this window we test
$$2(9 - 0) \ ? \ 3(4 - (-2))$$
$$2 \cdot 9 \qquad 3 \cdot 6$$
$$18 \ = \ 18$$
The window is square.

5. A square screen results if $2(X\text{max} - X\text{min}) = 3(Y\text{max} - Y\text{min})$
   In this window we test

   $$2(6 - (-6)) \ ? \ \ 3(2 - (-2))$$
   $$2 \cdot 12 \qquad 3 \cdot 4$$
   $$24 \ \neq \ 12$$

   The window is not square.

7. A square screen results if $2(X\text{max} - X\text{min}) = 3(Y\text{max} - Y\text{min})$
   In this window we test

   $$2(9 - 0) \ ? \ \ 3(3 - (-3))$$
   $$2 \cdot 9 \qquad 3 \cdot 6$$
   $$18 \ = \ 18$$

   The window is square.

9. Answers may vary.

   If $X\text{min} = -4$ and $X\text{max} = 8$, then $2(X\text{max} - X\text{min}) = 2(8 - (-4)) = 2 \cdot 12 = 24$.

   To make the screen square, the difference between $Y\text{min}$ and $Y\text{max}$ must be $24 \div 3 = 8$.

   To include the point $(4, 8)$, the y-axis must go at least as high as 8. A possible choice is $Y\text{min} = 1$ and $Y\text{max} = 9$.

A.4 Using a Graphing Utility to Locate Intercepts and Check for Symmetry

1.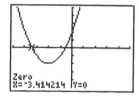

   The smaller x-intercept is $(-3.41, 0)$.

3.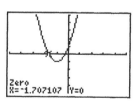

   The smaller x-intercept is $(-1.71, 0)$.

5.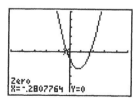

   The smaller x-intercept is $(-0.28, 0)$.

7.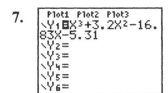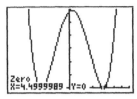

The positive x-intercept is $(3, 0)$.

9.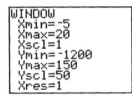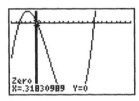

The positive x-intercept is $(4.50, 0)$.

11.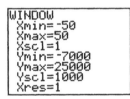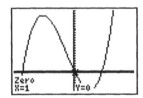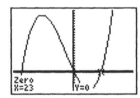

There are two positive x-intercepts; they are $(0.32, 0)$ and $(12.3, 0)$.

13.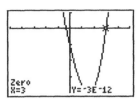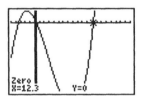

There are two positive x-intercepts; they are $(1, 0)$ and $(23, 0)$.

15. (a) The x-intercepts are $(-1, 0)$ and $(1, 0)$. The y-intercept is $(0, -1)$.
(b) The graph is symmetric with respect to the y-axis.

17. (a) There is no x-intercept and there is no y-intercept.
(b) The graph is symmetric with respect to the origin.

Printed and bound by CPI Group (UK) Ltd, Croydon, CR0 4YY

20/10/2024

14576725-0004